D1326686

Understand Technical Maths

Owen Bishop

Newnes
An imprint of Butterworth-Heinemann Ltd
Linacre House, Jordan Hill, Oxford OX2 8DP

 A member of the Reed Elsevier plc Group

OXFORD LONDON BOSTON
MUNICH NEW DELHI SINGAPORE SYDNEY
TOKYO TORONTO WELLINGTON

First published 1994

British Library Cataloguing in Publication Data
Bishop, O. N.
Understand Technical Maths
I. Title
519

ISBN 0 7506 1955 4

Composition by Genesis Typesetting, Rochester, Kent
Printed and bound in Great Britain by Bath Press, Avon

Contents

Using this book

Maths features widely in many branches of technology. This book aims to help those beginning their studies in technological subjects to understand the maths involved. The book covers the BTEC unit in Mathematics at F level, and also the maths required for more specialized aspects of technology, such as Mathematics for Engineering at National level. The worked examples show applications to many branches of technology, not only in building, construction and engineering but in electronics, food technology, and design.

The book is in four parts. The first part covers essential maths which you will already have done at school. If you ever found this difficult, never quite understood why things are done in a particular way, or were slow at calculating things in your head, spend some time on this part. Then you will be really proficient in the maths essentials, so that you are ready for the maths in the rest of the book.

Part 1 also introduces the elements of algebra and geometry. Each chapter begins with a *Try these first* test. This is to help you gauge whether or not you need to study that chapter. If you find that you can work all the questions easily (answers are given at the back of the book), you probably do not need to work through the chapter. Just glance through it and try a few questions as revision. If you can do only some of the questions, look at the parts of the chapter which deal with the questions you could not do. If the questions in the test make little sense to you, study the chapter in detail. Each topic within a chapter is followed by *Test yourself*, a batch of questions to help you check on your progress. Where appropriate, there are suggestions in the chapter for exploratory work, using a calculator or computer to assist understanding of maths concepts.

In Part 2 we cover maths topics that are likely to be required by all students of technology. You should make yourself familiar with all the chapters in this part. As in Part 1, there are *Try these first* and *Test yourself* exercises to help you monitor your progress. Scattered throughout the text are guides to using calculators. Although it is not possible for the book to deal with every type of calculator on the market, we give outline instructions that should apply to most models. If in doubt, consult the user manual for your calculator. At this stage you should begin to use a scientific calculator. This has functions such as automatic square roots, powers, trig ratios, and logs as well as being programmable to deal with short calculation sequences. If you do not know what these things are at the moment, wait until you come to the particular chapter concerned, where they are fully explained. Later in the book, it will be helpful if you can have the use of a graphic calculator or a computer running graphic software. But never use a calculator or computer blindly. Always make sure you *understand* what the calculator is doing for you. Work a few examples on paper first, step by step. When you know what it is all about, from then on use the calculator to save time, and to avoid the risk of arithmetic errors.

The range of maths topics is extended in Part 3 to include those which are important in many, but not necessarily in all, branches of technology. Here you should select the chapters that you need for your own studies.

Part 4 deals with the elements of statistics, the handling of data. This is an important field in many aspects of technology. If statistics form part of your technology syllabus, study the whole of this part.

Throughout the book the language used is as informal as possible and the descriptions and explanations are rather longer than you will find in most maths books. This is to make things clearer to you and to answer the small points that often cause doubts and worries. We hope that this approach will be of benefit to you and help you more quickly and easily understand technical maths.

Acknowledgement

The author wishes to thank the Casio Computer Co. Ltd for their valuable assistance.

Part 1 – Maths Essentials

Working with numbers quickly and accurately, and an introduction to some of the fundamentals of maths.

1 Plus and minus

Try the test questions below and check your answers at the back of the book. If you score full marks, or nearly, you probably do not need to work on the rest of this chapter. Go on to the next chapter.

Try these first

Work on paper or in your head, not with a calculator.

1 Which number added to 4 makes 10?

2 Which number added to 7 makes 10?

3 Sum these pairs of numbers.

a 8 + 2 **b** 4 + 7 **c** 5 + 9

d 12 + 45 **e** 76 + 87 **f** 432 + 389

4 Sum this string of numbers.

4 + 5 + 6 + 5 + 9 + 2 + 3 + 2 + 7 + 1

5 Evaluate.

a 7 − 3 **b** 9 − 4 **c** 57 − 23

d 62 − 47 **e** 4 − 9 **f** 23 − 67

Finding sums

Working with numbers

There are three ways of working with numbers:

- in your head
- on paper
- with a calculator (or computer).

Calculators are fast and accurate, BUT there are many times when it is better to use your brain. In some parts of this book, even in the more advanced parts, working things out in your head is the *only* practicable way of getting the answer quickly. This is why you should start NOW to develop your skills of calculating by using memory and reasoning. And the keys to this are *memorising* and *practice.*

There are also times when it is more convenient to do a short calculation on paper rather than reach out for the calculator. Even when you are working on a calculator, it is a good idea to check your results roughly in your head, just to be sure that you have pressed the right keys. So, to be successful at maths, you must practice *all* three ways of calculating.

In this book there are:

- things to learn to help you to calculate in your head
- techniques for working on paper
- instructions for using a calculator effectively
- tips on short-cut methods and quick checking

and plenty of exercises for practising all three ways of calculating. At the back of the book there are short type-in computer programs to give you even more practice in fast and accurate working.

When the exercises say 'in your head' do not be tempted to use a calculator instead. If you find it hard to use your brain to work things out, then you really *need* the practice.

When we add two or more numbers together, the result is their sum. Below is a list of the pairs of numbers which add up to 10. Their sum is 10. We will call them **ten-pairs** to distinguish them from other pairs of numbers. Learn them well. When you know them, it makes all kinds of calculations much easier.

Ten-pairs

These six pairs of numbers all add up to 10:

0 and 10
1 and 9
2 and 8
3 and 7
4 and 6
5 and 5

Learn these pairs to make calculations easier.

Test yourself 1.1

Do these in your head, without using a calculator.

1 For each of these numbers, write the number which adds up to 10. Use the ten-pairs list if you need to.

2 4 1 0 7 2 5 10 6 4 3 8 9 1 7

2 Do the same for these numbers, *without* referring to the list of ten-pairs.

6 2 10 4 7 3 7 1 6 9 3 0 2 5 8

3 Find the sums of these pairs of numbers.

a 3 + 7 **b** 6 + 4 **c** 5 + 5 **d** 8 + 2 **e** 7 + 3

4 Find the sums of these pairs of numbers.

a 3 + 8 **b** 5 + 6 **c** 7 + 5 **d** 3 + 6 **e** 2 + 6

Question 4 above demonstrates how the ten-pairs are useful in adding pairs that do *not* add up to 10. In 4a, you *know* that 3 + 7 = 10, but here we have 8 instead of 7. Eight is one *more* than seven, so the sum is one *more* than ten. The sum is 11.

In 4d, we know that 3 + 7 = 10. Six is one *less* than seven so the sum is one *less* than ten. The sum is 9.

Explore these

1 Play this game to practise number pairs. Your partner calls out any number between 0 and 10. You have to reply immediately with the number which makes it up to 10. If you are right, your friend calls another number, and so on. Keep a score on paper of how many times you answer correctly. If you are wrong, or you score 10 correct answers in a row, it is your turn to call numbers for your partner to match. The winner is the one with the most correct answers in 5 minutes.

When you have both become really good at this game, vary it by making the pairs add up to 12 or some other number.

2 If you find it hard to remember the ten-pairs, there is a computer program on page 467 which will give you practice. It should run on any computer which has BASIC, but you (or your teacher) should easily be able to adapt it to your form of BASIC or to another computer language, if necessary.

3 Here is an **addition table** of numbers up to 10. Some of the sums have been filled in. They show, for example, that 2 + 6 = 8 (column 2, row 6) and 7 + 4 = 11 (column 7, row 4). Copy out the table and fill in the rest of the sums.

+	1	2	3	4	5	6	7	8	9	10
1	2	3								
2										
3			6							
4							11			
5										
6		8								
7										
8									17	
9										19
10					15					

Summing many numbers

When there are more than two numbers to add:

6 + 3 + 4 = ?

Group the numbers into ten-pairs:

6 + 3 + 4 = ?
(6 and 4 bracketed)

There is one ten-pair and a 3:

6 + 3 + 4 = 10 + 3 = 13
(6 and 4 bracketed)

The sum is 13.

Here is a longer string of numbers to add:

$4 + 2 + 6 + 5 + 7 + 5 = ?$

Mark off the ten-pairs (or pair them in your head):

$\underbrace{4 + 2 + 6}\ + \underbrace{5 + 7 + 5} = ?$

There are two ten-pairs, also a 2 and a 7:

$$\begin{aligned} 4 + 2 + 6 + 5 + 7 + 5 &= 20 + 2 + 7 \\ &= 20 + 9 = 29 \end{aligned}$$

The sum is 29.

Here is another example:

$6 + 3 + 4 + 8 + 2 + 7 + 1 + 9 + 9 = ?$

This has four ten-pairs:

$6 + 3 + 4 + 8 + 2 + 7 + 1 + 9 + 9 = 40 + 9 = 49$

The sum is 49.

Two-figure numbers

For example, $30 + 11 = ?$ The stages are:

$$\begin{array}{r} 30 \\ +\ \underline{11} \end{array} \qquad \begin{array}{r} 30 \\ +\ \underline{11} \\ 1 \end{array} \qquad \begin{array}{r} 30 \\ +\ \underline{11} \\ 41 \end{array}$$

Set out the numbers with their units and tens in the same columns. Add the units, then the tens. The sum is 41.

The technique has an extra stage if the figures in a column add up to 10 or more. Look at this example:

$37 + 25 = ?$

$$\begin{array}{r} 37 \\ +\ \underline{25} \end{array} \qquad \begin{array}{r} 37 \\ +\ \underline{25} \\ {\scriptstyle 1}\ \ \\ 2 \end{array} \qquad \begin{array}{r} 37 \\ +\ \underline{25} \\ {\scriptstyle 1}\ \ \\ 62 \end{array}$$

The sum of the units is 12. We write the 2 in the units column, but there is no space there for the 1. As the 1 stands for 10, the 1 is *carried* into the tens column, to be added in with the other tens figures.

Here is another example:

$43 + 69 = ?$

$$\begin{array}{r} 43 \\ +\ \underline{69} \end{array} \qquad \begin{array}{r} 43 \\ +\ \underline{69} \\ {\scriptstyle 1}\ \ \\ 112 \end{array}$$

The 1 (standing for 100) carried over from the tens column is written in the hundreds column. There is no need to write it small as there is nothing to add it to.

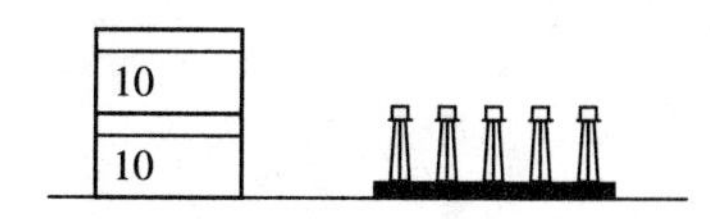

Figure 1.1

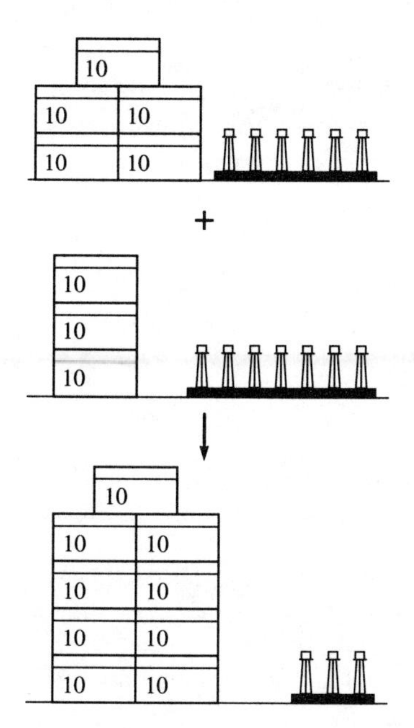

Figure 1.2

Standard packs

Selling items in standard packs makes it easier for suppliers to fill an order quickly. It makes it easier to check stocks. Standard packs help arithmetic too. In arithmetic, the standard pack is a box of 10 items.

Suppose that you have 2 packs of ten transistors and 5 single transistors. How many transistors do you have? Call the single transistors units. Write down the number of packs (tens) followed by the number of units:

25

You have 25 transistors. Writing down two numbers is much quicker than tipping the transistors out of the packs and counting them all from 1 to 25.

In the number 25, the 2 does not mean 2 units. It means 2 packs of ten. In the number 44, the right-hand 4 means 4 units; the left-hand 4 means 4 packs of ten. The value of the figure depends on its position in the number. This is a really compact way of representing numbers as a row of figures. It is much more compact than the old Roman system which, for example, represents 1947 (the date of invention of the transistor) by a long string of symbols: MCMXLVII. No wonder that the Romans found maths difficult.

If you have 5 packs of transistors and 6 units already, and then receive 3 packs and 7 units from the supplier, how many transistors do you have then?

Put the unit transistors on the bench, count out 10 transistors from them and put them in a spare empty pack. You had $6 + 7 = 13$ units, but now have 1 extra pack and 3 units. This gives you $5 + 3 + 1 = 9$ packs altogether. Write down the number of packs followed by the number of units:

93

You have a total of 93 transistors.

Set out as an addition:

$$\begin{array}{r} 56 \\ +\ 37 \\ \hline {\scriptstyle 1} \\ 93 \end{array}$$

Add the units, make up as many packs from them as you can, carry over the number of new packs and add this in with the existing packs.

Some suppliers package goods in superpacks of 100. If you have 2 superpacks of bolts, 5 packs and 7 units, you have:

257 bolts

We can also have superduper packs holding 1000 items. Each type of pack holds ten times as many items as the next smaller size. This is the basis of the **decimal** system of numbers. In a decimal number each figure is worth ten times more by shifting it one place to the left.

There are two carries in this addition of two 4-figure numbers:

$$\begin{array}{r} 2672 \\ +\ 3565 \\ \hline {\scriptstyle 11\ \ } \\ 6237 \end{array}$$

But numbers with so many figures are better added on a calculator.

Test yourself 1.2

Do these on paper or in your head, except when the question asks you to use a calculator.

1 Add these pairs of numbers.

a 4 + 6 **b** 7 + 3 **c** 2 + 9 **d** 4 + 5

e 8 + 2 **f** 9 + 3 **g** 3 + 8 **h** 5 + 7

2 Add these pairs of numbers.

a 4 + 9 **b** 7 + 7 **c** 5 + 8 **d** 8 + 4

e 8 + 9 **f** 10 + 5 **g** 5 + 7 **h** 3 + 2

3 Add these strings of numbers.

a 3 + 4 + 7 + 6 + 3 + 5 + 2 + 7

b 5 + 10 + 8 + 5 + 2 + 3 + 6 + 3 + 4 + 0 + 5

[*Hint*: sometimes its even quicker to find *three* numbers that add up to 10.]

c 4 + 3 + 4 + 7 + 6 + 5 + 2 + 2 + 1 + 6 + 8 + 9 + 2

Have you noticed that . . . ?

3 + 7 + 2 + 6 = 18

and

7 + 3 + 6 + 2 = 18

and

2 + 6 + 3 + 7 = 18

and so on.

In addition, the order of the numbers makes no difference to the total.

4 Add these pairs of numbers.

a 12 + 13 **b** 22 + 33 **c** 57 + 35 **d** 73 + 17 **e** 36 + 25

f 73 + 54 **g** 82 + 87 **h** 45 + 65 **i** 29 + 12 **j** 77 + 77

k 91 + 39 **l** 70 + 31 **m** 34 + 99 **n** 44 + 58 **o** 95 + 86

Check your answers with a calculator.

Like with like

You can only add or subtract two numbers if they both refer to the same kind of thing. You add 4 bolts to 6 bolts, making 10 bolts. You can subtract 3 transistors from 8 transistors, leaving 5 transistors.

But, you cannot add 4 bolts to 9 transistors, or take 4 transistors from 9 bolts. Additions and subtractions of unlike objects do not make sense.

5 Add these strings of three numbers.

[*Hint*: In some of these you need to carry 2.]

a 11 + 32 + 45 **b** 35 + 20 + 24 **c** 28 + 34 + 41

d 52 + 16 + 25 **e** 41 + 33 + 17 **f** 45 + 31 + 52

g 70 + 23 + 52 **h** 35 + 72 + 27 **i** 81 + 29 + 23

j 82 + 59 + 78 **k** 73 + 87 + 62 **l** 87 + 98 + 46

Check your answers with a calculator.

6 Add these pairs of numbers.

a 215 + 478 **b** 503 + 272 **c** 444 + 327

d 378 + 243 **e** 465 + 717 **f** 542 + 458

Check your answers with a calculator.

Explore this

The computer program on page 468 will give you plenty of practice in adding pairs of numbers.

Negative numbers

There is a long straight paved path which goes for miles in both directions, east and west. The paving-blocks are numbered, the numbers increasing toward the right (Figure 1.3a). You are standing on block 0, facing east. Now you are given an instruction:

+ 4

This means 'Walk 4 paces forward'. We will assume that one pace is the width of one block, so this takes you to block 4 (Figure 1.3b).

Now go back to block 0 and follow this instruction:

+ 2 + 5

This means 'Walk 2 paces, then walk 5 paces'. You arrive at block 7 (Figure 1.3c). Walking along the path to the east is **addition**:

2 + 5 = 7

Note that, by convention, we usually do not bother to write the first + symbol; we assume that it is there. Now for a slightly different instruction:

10 – 3

The 'minus' or 'negative' symbol means 'walk backward'. Start on block 0 facing east, walk 10 paces forward, then walk three paces backward. Walking

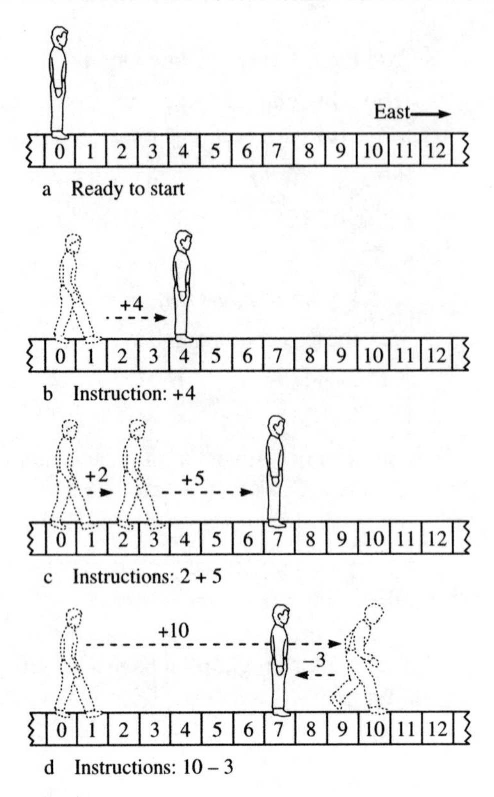

Figure 1.3

backward along the path in a westerly direction is **subtraction** (Figure 1.3d). You finish at block 7, demonstrating that:

$$10 - 3 = 7$$

Taking away 3 from 10 leaves 7.

Going west

So far, you have only wandered along the path to the east of block 0. Let us try something a bit more adventurous:

$$5 - 8$$

'Start at 0 facing east, walk 5 paces forward, walk 8 paces backward'. Figure 1.4 shows that this brings you to a block to the west of block 0.

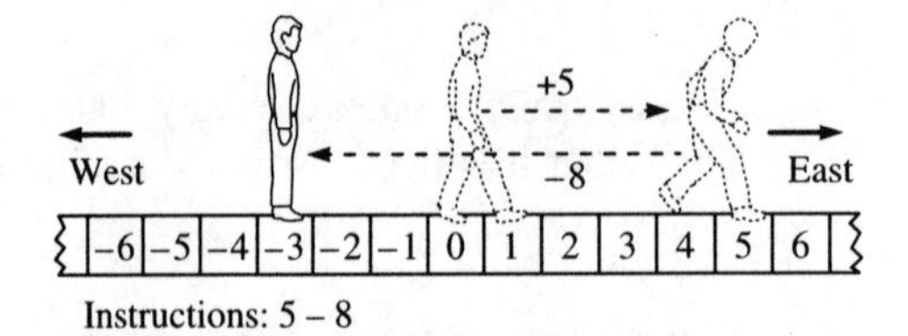

Figure 1.4

Negatives on a calculator

With most simple calculators you can enter the numbers and symbols just as they are printed in this book.

To calculate: 3 + 8 – 5 = ?

Key: [3] [+] [8] [–] [5] [=]

The display shows the result at each stage and the final answer, 6.

You may also be able to *begin* the calculation with a negative:

To calculate: –5 + 6 – 4 = ?

Key: [–] [5] [+] [6] [–] [4] [=]

The display shows the result, –3.

Some advanced calculators have separate keys for subtract (usually –) and for negate. The negate key is used when we want to *begin* a calculation with a negative number.

If the negate kay is marked +/–, it changes the sign of the number displayed. Use it *after* entering the number.

To calculate: –5 + 6 – 4 = ?

Use the negate key to make the 5 negative. Use the subtract key to subtract 4.

Key: [5] [+/–] [+] [6] [–] [4] [=]

On some other calculators, the negate key is marked (–). Use this *before* entering the 5. But use the subtract key to subtract 4.

Key: [(–)] [5] [+] [6] [–] [4] [=]

You could have got to the same block by beginning at block 0, and walking 3 paces backward. This is written as:

–3

Since both sets of instructions bring us to the same place we can say that:

5 – 8 = –3

Subtracting 8 from 5 gives negative 3.

Try this one:

7 – 12 = ?

'Walk 7 forward, walk 12 backward'. You reach block –5.

7 – 12 = –5

Subtracting 12 from 7 gives negative 5. It is just the same as if you had walked 5 paces backward.

Test yourself 1.3

Solve exercises 1 to 3 in your head or use the path in Figure 1.4.

1 Evaluate (find the value of).

a 7 – 3	**b** 12 – 11	**c** 6 – 7	**d** 8 – 8
e 12 – 5	**f** 0 – 4	**g** 4 – 11	**h** 9 – 3
i –3 – 4	**j** 5 – 0	**k** –4 + 6	**l** –7 + 2

2 Evaluate.

a 7 + 5 – 8	**b** 12 – 8 – 7	**c** 4 – 7 + 8
d 3 – 3 – 2	**e** –4 + 5 – 1	**f** –6 – 1 – 4

3 Use ten-pairs (see box) to evaluate.

a 4 + 6	**b** 10 – 3	**c** 2 + 9	**d** 10 – 6
e 10 – 5	**f** 3 – 10	**g** 2 + 8 – 4	**h** 3 + 7 – 5
i 8 – 10	**j** 2 + 6 – 10	**k** 6 + 4	**l** 10 – 9
m 6 + 4 – 5	**n** 6 – 10	**o** 6 – 7 – 3	**p** 8 – 4 – 6

[*Hint*: Yes, you can group negative numbers into ten-pairs, equivalent to –10.]

Rules for ten-pairs

- Pairs add up to 10

Example

6 + 4 = 10

- One of the pair subtracted from 10, leaves the other

Examples

10 – 4 = 6
10 – 8 = 2

- 10 subtracted from one of the pair, leaves the *negative* of the other

Examples

4 – 10 = –6
9 – 10 = –1

4 Evaluate.

a 4 + 7	**b** 11 – 3	**c** 5 – 9	**d** 9 – 5
e –3 + 4	**f** 12 – 7	**g** 1 + 9 – 4	**h** 9 + 9
i 3 – 8	**j** 8 – 3	**k** 5 – 4 – 6	**l** 0 – 4
m –7 – 3	**n** 11 – 5 – 6	**o** 7 – 11	**p** 10 – 3 – 8

Check your results with a calculator.

Have you noticed . . . ?

$$6 - 4 = 2$$
$$4 - 6 = -2$$

$$12 - 7 = 5$$
$$7 - 12 = -5$$

When we reverse the order of the numbers, we reverse the sign of the answer.

Explore these

1 Give yourself more subtraction practice. Write down two or three numbers: write them quickly without thinking about them. Then put a + or – symbol in front of each number, again without thinking about exactly what you write. Work out the result of the set of numbers. Check your result by using a calculator.

2 The computer program on page 468 gives you practice in subtraction and in handling negative numbers.

Two-figure subtractions

The technique is similar to that for addition:

```
  86          86          86
- 32        - 32        - 32
  --          --          --
               4          54
```

Set out the numbers in columns. Subtract the units, then the tens. The difference between 86 and 32 is 54.

Borrowing a standard pack

The store has 6 packs of transistors (ten in each pack) and 3 unit transistors. Someone comes to the store to collect 47 transistors.

```
  63
- 47
  --
```

There is no problem in collecting 4 packs from 6 packs, but how can they take 7 units from 3 units? Negative transistors are impossible.

The solution is to break open one of the packs and add its contents to the units pile, so that there are now 13 units.

```
  5̶6̶ ¹3
-  4 7
   ---
   1 6
```

We *borrow* a pack from the tens pile, leaving 5 packs. The small '1' indicates that there are now 13 unit transistors to take 7 from. The small 5 shows that there are now only 5 packs. Subtracting units: 13 – 7 = 6. Subtracting tens: 5 – 4 = 1. This leaves 1 pack and 6 units, 16 transistors are left in the store.

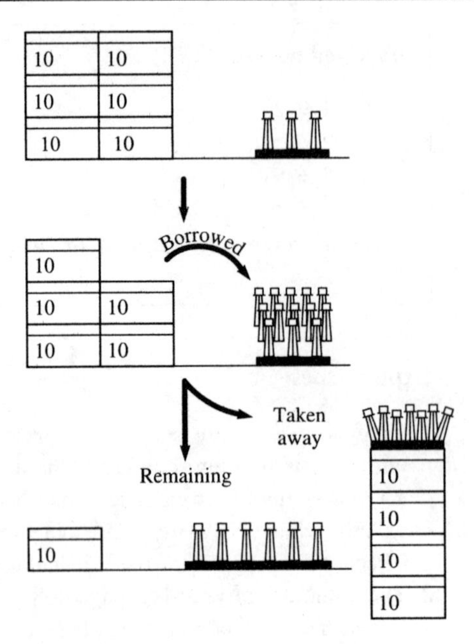

Figure 1.5

In this example, subtracting the units would give a negative result. So we borrow from the tens:

$$\begin{array}{r} 54 \\ -\ \underline{16} \end{array} \qquad \begin{array}{r} 5^{1}4 \\ -\ \underline{1\ 6} \\ 8 \end{array} \qquad \begin{array}{r} {}^{4}\not{5}^{1}4 \\ -\ \underline{1\ 6} \\ 3\ 8 \end{array}$$

Borrow ten, so that we can take 6 from 14, result 8. Cross out the 5 and substitute 4 toshow that 10 has been borrowed. Subtract 1 from 4, result 3. The difference between 54 and 16 is 38.

Another example:

$$\begin{array}{r} 41 \\ -\ \underline{25} \end{array} \qquad \begin{array}{r} 4^{1}1 \\ -\ \underline{2\ 5} \\ 6 \end{array} \qquad \begin{array}{r} {}^{3}\not{4}^{1}1 \\ -\ \underline{2\ 5} \\ 1\ 6 \end{array}$$

The difference between 41 and 25 is 16.

Test yourself 1.4

Work these problems on paper or in your head. Then check the results with a calculator.

1 Evaluate.

a 58 – 34 **b** 99 – 37 **c** 32 – 12 **d** 42 – 30

e 78 – 11 **f** 67 – 47 **g** 58 – 52 **h** 71 – 69

2 Evaluate:

a 58 – 39	**b** 45 – 28	**c** 70 – 46	**d** 38 – 19
e 34 – 23	**f** 10 – 7	**g** 51 – 49	**h** 67 – 34

3 Use a calculator to evaluate:

a 97 + 32 – 47 – 61	**b** 104 – 56 – 72 + 17
c 97 + 83 – 42 + 67 – 130 – 75	**d** 2314 + 4489 – 9861 + 1001

Negative differences

The technique for subtracting two-figure numbers does not work if the result is negative:

45 – 83 = ?

The solution is first to turn this into:

83 – 45 = ?

Find the *difference*, which is the smaller number subtracted from the larger number.

$$\begin{array}{r} 83 \\ -\ \underline{45} \end{array} \qquad \begin{array}{r} {}^{7}\not{8}\,{}^{1}3 \\ -\ \underline{4\ 5} \\ 3\ 8 \end{array}$$

But we have swapped the numbers round, so we must reverse the sign of the answer (see box):

45 – 83 = –38

The rule is:

Swap the numbers, subtract, reverse the sign

Test yourself 1.5

1 Evaluate.

a 84 – 56	**b** 56 – 84	**c** 33 – 55
d 24 – 45	**e** 100 – 57	**f** 57 – 100

2 Evaluate:

a 25 – 34	**b** 68 – 102	**c** –56 + 77
d –43 + 12	**e** 39 – 41	**f** 4 – 98

3 To assess if you have covered this chapter fully, answer the questions in *Try these first* page 3.

Explore this

1 If you start at block 0 on the number path (Figure 1.4) and run along it toward the east, two blocks at a time, you step on blocks:

0, 2, 4, 6, 8, 10, 12, . . .

This is a **sequence**. Which are the next three blocks you come to? What name is given to the numbers in this sequence?

2 Now suppose you run along three blocks at a time. The sequence begins 0, 3, 6, 9, 12, 15, . . . Complete the sequence for the next 8 blocks you land on.

3 Taking the numbers of blocks visited in 2, write down the sums of their figures:

Block	0	3	6	9	12	15 . . .
					1 + 2 =	1 + 5 =
Sum	0	3	6	9	3	6 . . .

Say as much as you can about the sums.

4 Write down the sequence when you leap along 5 at a time. What do you notice about the units figure of the block numbers?

5 Write down the sequence when you leap along 9 at a time.

6 Sum the figures of the 9-steps sequence. What do you notice?

7 Write down the units figure only of the blocks in the 9-steps sequence. What do you notice?

8 Write down the tens figure of the blocks in the 9-steps sequence. What do you notice. Find a rule for adding 9 to any number.

9 Work out a rule for adding 8 to any number.

2 Times and divide

Addition and subtraction (Chapter 1) are two of the four basic operations of arithmetic. Multiplication and division are the other two.

For this chapter, you need to know about addition, subtraction and negative numbers, all of which are dealt with in Chapter 1.

Try these first

Your success with this short test will tell you which parts of this chapter you already know. Do not use a calculator.

1 Multiply the following pairs of numbers *in your head.*

a 2×4 **b** 3×5 **c** 11×9

2 Multiply the following pairs of numbers on paper.

a 34×6 **b** 77×3 **c** 91×32

3 Divide the first number by the second number *in your head.*

a 15/3 **b** 24/2 **c** 121/11

4 Divide the first number by the second number, on paper.

a 92/4 **b** 85/5 **c** 576/12

5 Evaluate.

a -4×7 **b** -6×-7 **c** $-76/4$

d $-324/-12$

6 Evaluate.

a $2 \times 46 - 12$ **b** $9 \times (6 + 2)$ **c** $54/9 - 4$

Another look at addition

Speeding along the number path in *Explore this*, section 2 (page 16), you touch down on blocks:

3, 6, 9, 12, 15, 18, 21, 24, 27, 30, 33, 36, ...

This sequence is produced by starting at 0 and leaping three blocks a given number of *times*. Each time you leap you add three. For example, if you leap 7 times, you reach block 21. Writing this as an addition, we see that:

$$3 + 3 + 3 + 3 + 3 + 3 + 3 = 21$$

This is 3 added to itself 7 times. Adding 3 to itself 7 times is equivalent to multiplying it by 7. We say that 7 *times* 3 is 21. Or 3 *multiplied by* 7 is 21. Or, in symbols:

$$7 \times 3 = 21$$

Below we list the number of times you leap and the number of blocks this takes you to:

Times	1	2	3	4	5	6	7	8	9	10	11	12
Block	3	6	9	12	15	18	21	24	27	30	33	36

This table is the **multiplication table** for 3. From this, you can read off the results of multiplying 3 by various numbers. For example:

$4 \times 3 = 12$
$7 \times 3 = 21$
$12 \times 3 = 36$

Have you noticed . . . ?

In the multiplication tables:

$3 \times 7 = 21$
and $7 \times 3 = 21$

also:

$5 \times 12 = 60$
and $12 \times 5 = 60$

and many other examples.

The *order* of multiplication does not affect the result.

The result of multiplying two or more numbers together is called the **product**. Multiplication tables for other numbers can be made up in the same way. Here is the multiplication table for 7:

Times	1	2	3	4	5	6	7	8	9	10	11	12
Product	7	14	21	28	35	42	49	56	63	70	77	84

The products are calculated by adding 7 each time. Given this table, we can tell that $4 \times 7 = 28$, and $9 \times 7 = 63$, for example.

Explore this

You need to know the multiplication tables of all the numbers from 2 to 12, taking each table as far as 12 times. Work them out for yourself, and write them in your notebook. Then learn them. Multplications *can* be done on a calculator but there are many instances where it is essential to be able to multiply in your head. The computer program on page 468 gives you table-learning practice.

Test yourself 2.1

As far as possible, work these exercises in your head.

1 Find the products of these pairs of numbers.

a 2×4	**b** 6×3	**c** 3×6	**d** 5×5
e 11×7	**f** 4×8	**g** 7×12	**h** 3×9
i 12×12	**j** 8×10	**k** 9×12	**l** 5×7

2 Below, you are given two numbers. For each pair say how much the first number must be multiplied by to give the second number. For example, given 4 and 12, the number required is 3.

a 7, 56	**b** 5, 35	**c** 3, 36	**d** 2, 22
e 12, 60	**f** 11, 121	**g** 4, 36	**h** 6, 54
i 8, 48	**j** 9, 18	**k** 10, 100	**l** 7, 42

Bigger numbers

When one or both of the numbers to be multiplied together are greater than 12, use a calculator, or work out the product on paper, using **long multiplication**.

Example

$7 \times 28 = ?$

This is a two-figure number multiplied by a one-figure number.

$$\begin{array}{r} 2\ 8 \\ \times\quad 7 \\ \hline \\ \end{array} \qquad \begin{array}{r} 2\ 8 \\ \times\quad 7 \\ \hline {}_5\quad \\ 6 \end{array} \qquad \begin{array}{r} 2\ 8 \\ \times\quad 7 \\ \hline {}_5\quad \\ 1\ 9\ 6 \end{array}$$

Multiply the units by 7: $8 \times 7 = 56$. Write down the 6, carry the 5 to the tens column. Now multiply the tens by 7: $2 \times 7 = 14$. But we have carried 5 so: $14 + 5 = 19$. Write this in the tens and hundreds columns. Product is 196.

If the multiplier has two figures, we multiply by its tens figure and then by its units figure. Finally, we add the two products.

Example

$54 \times 32 = ?$

First multiply 54 by 30. Write a 0 in the units column (equivalent to 'times 10'), then multiply 54 by 3:

$$\begin{array}{r} 5\ 4 \\ \times\ 3\ 2 \\ \hline \\ \end{array} \qquad \begin{array}{r} 5\ 4 \\ \times\ 3\ 2 \\ \hline \\ 0 \end{array} \qquad \begin{array}{r} 5\ 4 \\ \times\ 3\ 2 \\ \hline {}_1\qquad \\ 1\ 6\ 2\ 0 \end{array}$$

$54 \times 3 = 162$, followed by the 0, making 1620.

Multiply 54 by 2:

```
   5 4           5 4
 × 3 2           3 2
   1           1
 1 6 2 0       1 6 2 0
   1 0 8     +   1 0 8
               1 7 2 8
```

54 × 2 = 108. Finally, add 108 to 1620.

Product = 54 × 32 = 1728.

Test yourself 2.2

1 Find these products, working on paper.

a	42 × 3	**b**	31 × 9	**c**	82 × 5	**d**	47 × 6
e	28 × 6	**f**	55 × 10	**q**	352 × 4	**h**	83 × 12
i	42 × 16	**j**	88 × 19	**k**	73 × 36	**l**	36 × 73
m	62 × 89	**n**	452 × 23	**o**	122 × 74	**p**	621 × 38

2 Find these products, using a calculator.

a	367 × 34	**b**	754 × 22	**c**	905 × 83	**d**	412 × 731
e	777 × 333	**f**	128 × 128	**g**	1472 × 504	**h**	8525 × 362

Multiplying negatives

First look at the multiplication 5 × 3, as illustrated by the number path Figure 2.1a. You leap 3 blocks 5 times, and arrive at block 15:

5 × 3 = 15

Leaping backwards 3 blocks 5 times takes you to block –15 (Figure 2.1b):

5 × –3 = –15

How do we interpret the instruction:

–5 × 3 = ?

It is not possible to leap a negative number of times, but you could try facing the other way. The negative before the 5 means that the leaping is done facing west (Figure 2.1c). The result is:

–5 × 3 = –15

Facing east and leaping backward, or facing west and leaping forward comes to the same thing. It now remains to work out:

–5 × –3 = ?

You face west and leap 3 paces backward 5 times (Figure 2.1d). Leaping backward while facing west makes you travel east and you finish at block 15:

–5 × –3 = 15

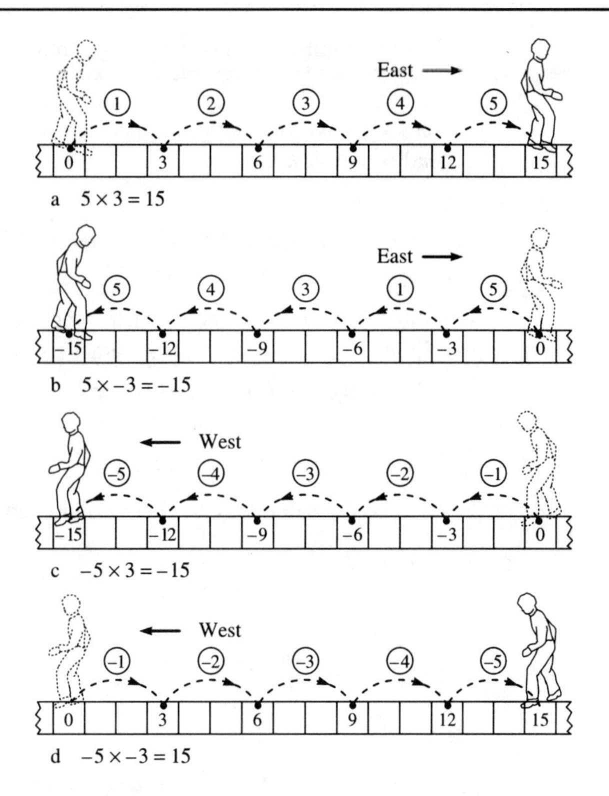

Figure 2.1

Multiplying a negative number by another negative number gives a **positive** product. Once you have understood these leapings in various directions, you need not remember exactly what you did. Just remember the result, which is summarized in the box.

Signs rules for products

Signs of numbers	*Sign of product*
+ times +	+
+ times –	–
– times +	–
– times –	+

Multiply and add

You are asked to evaluate:

$$3 \times (4 + 5) = ?$$

The rule is:

Evaluate all expressions in brackets first

Add the numbers in brackets, *then* multiply 3 by their sum:

$3 \times (4 + 5) = 3 \times 9 = 27$

Here we have the same numbers but the 3 and 4 are bracketed together, instead of the 4 and 5:

$(3 \times 4) + 5 = ?$

Multiply the numbers in brackets, *then* add 5:

$(3 \times 4) + 5 = 12 + 5 = 17$

Usually we do not bother to write brackets around numbers that are to be multiplied together. We omit the brackets and make another rule:

Multiply, *then* add or subtract

You are asked to evaluate:

$3 \times 4 + 5 = ?$

Solve it in the same way as the second example above; result 17.

More examples

$3 \times 7 - 5 = 21 - 5 = 16$
$3 \times (7 - 5) = 3 \times 2 = 6$
$7 + 2 \times 9 = 7 + 18 = 25$
$2 \times 6 + 3 \times 3 = 12 + 9 = 21$
$2 \times (6 + 3) \times 3 = 2 \times 9 \times 3 = 54$
$2 + 6 \times 3 + 3 = 2 + 18 + 3 = 23$

Test yourself 2.3

1 Evaluate these in your head.

a 2×-3 **b** -8×3 **c** 12×-6 **d** -3×-6
e -11×6 **f** -9×-9 **g** 5×7 **h** 8×-7

2 Evaluate these on paper.

a 24×-23 **b** 89×-37 **c** -56×72 **d** -20×-30
e 39×-57 **f** -78×41 **g** -69×-33 **h** -13×-87

Check your answers with a calculator.

3 Evaluate these in your head.

a $5 \times 6 + 2$ **b** $5 + 6 \times 2$ **c** $3 \times 6 - 10$
d $7 + 2 \times 12$ **e** $2 \times 6 + 4 \times 5$ **f** $12 - 3 \times 4$
g $2 \times 7 \times 5$ **h** $3 + 11 \times 11$ **i** $4 \times 5 \times 10$

4 Evaluate these on paper.

a $3 \times 7 + 3 + 12$ **b** $4 \times 5 + 7 \times 9$
c $23 + 46 \times 3 - 57$ **d** $(4 + 12) \times 6 - 59$
e $35 \times (5 + 3) + 6$ **f** $(5 - 7) \times 4 + 8$
g $(4 - 10) \times (12 - 34)$ **h** $23 - 5 \times (6 + 12 - 7)$

Check your answers with a calculator.

Chain calculations

Some calculators evaluate a chain calculation correctly, giving multiplying and dividing priority over adding and subtracting. For example:

$3 \times 4 + 8 \times 2 = ?$

Key this in just as it is written:

[3] [×] [4] [+] [8] [×] [2] [=]

The calculator displays the correct result, 28. After evaluating 3×4 it 'holds' the product 12 until it has evaluated 8×2.

Other calculators work out a result at each stage, and use this for the next stage. Such calculators multiply 3 by 4 but then add 8 to the product *immediately*. This gives 20, which they then multiply by 2 to give 40, an incorrect result. Check to see if your calculator evaluates chains correctly.

Dividing

The idea of dividing is to find out how many of one number are contained in another number. For example, you leap along the number path 4 blocks at a time, and reach block 36. How many leaps have you made? The multiplication table for 4 supplies the answer. Knowing that:

$36 = 9 \times 4$

we also know that thirty six contains nine fours. Leaping 4 at a time we divide 36 blocks into 9 leaps. In other words:

36 *divided by* 4 = 9

One commonly-used symbol for division is ÷, so we write:

$36 \div 4 = 9$

It is also true that:

$36 \div 9 = 4$

You can reach block 36 in a more energetic way by dividing the distance into 4 leaps, 9 blocks at a time.

Another symbol for division is a line, either horizontal or sloping (a 'slash'):

$$36 \div 4 = \frac{36}{4} = 36/4$$

All mean the same thing. We shall often use the slash form in this book.

The result of dividing one number by another is called the **quotient**.

Not all divisions work out exactly. For example: 79/11 = ? We know that $7 \times 11 = 77$. Seven leaps of 11 blocks brings you to block 77; you still have 2 blocks to go. Another leap would take you too far. We call the left-over part the **remainder**. The quotient is 79/11 = 7, the remainder is 2.

For dividing larger numbers by numbers up to 12, we set out the division as shown below. For example:

852/3 = ?

$3 \,\lvert\, \underline{8\ 5\ 2}$	$3 \,\lvert\, \underline{8^2 5\ 2}$	$3 \,\lvert\, \underline{8^2 5\ 2}$	$3 \,\lvert\, \underline{8^2 5^1 2}$
	2	2 8	2 8 4

Divide 3 into 8; 3 goes 2 times because 2 × 3 = 6. Write 2 below the line. Take 6 from 8, leaving 2 to carry to the next stage.

Divide 3 into 25; 3 goes 8 times because 8 × 3 = 24. Write 8 below the line. Take 24 from 25, leaving 1 to carry to the next stage.

Divide 3 into 12; 3 goes 4 times because 4 × 3 = 12. Write 4 below the line. Take 12 from 12, leaving nothing to carry. The division is exact.

Result: 852/3 = 284, exactly. Check by multiplying: 3 × 284 = 852.

Here is another example: 138/6 = ?

$6 \,\lvert\, \underline{1\ 3\ 8}$	$6 \,\lvert\, \underline{1\ 3^1 8}$	$6 \,\lvert\, \underline{1\ 3^1 8}$
	2	2 3

Divide 6 into 1. It does not go because 6 is bigger than 1.

Divide 6 into 13; 6 goes 2 times. Write 2 below the line, carry 1.

Divide 6 into 18; 6 goes 3 times. Write 3 below the line. Nothing to carry.

Result: 138/6 = 23 exactly. Check by multiplying: 23 × 6 = 138.

This example does not divide out exactly: 4533/11 = ?

$11 \,\lvert\, \underline{4\ 5\ 3\ 3}$	$11 \,\lvert\, \underline{4\ 5^1 3\ 3}$	$11 \,\lvert\, \underline{4\ 5^1 3^2 3}$	$11 \,\lvert\, \underline{4\ 5^1 3^2 3}$
	4	4 1	4 1 2

Remainder = 1

At the last stage we are dividing 11 into 23. It goes 2 times, but not exactly: 23 – 22 = 1. Result 4533/11 = 412, remainder 1.

Test yourself 2.4

1 Evaluate these quotients in your head.

a 12/3	**b** 55/5	**c** 72/6	**d** 36/6
e 36/4	**f** 57/8	**g** 132/11	**h** 100/9
i 84/7	**j** 121/2	**k** 29/3	**l** 47/6

2 Evaluate these quotients on paper.

a 561/11	**b** 245/7	**c** 433/6	**d** 1016/8
e 169/13	**f** 200/14	**g** 999/5	**h** 1123/6

Check your answers with a calculator.

Remainders on calculators

There is a problem when you check some of these exercises. Calculators do not know about remainders. They do not know when to stop dividing.

For example, instead of62/7 = 8 remainder 6,
the calculator gives 62/7 = 8.857142857

Why it does this, and what to do about it, is explained in the next chapter.

Dividing negative numbers

Because division is related to multiplication, the signs rules (page 21) apply to division too:

57/3 = 19   –57/3 = –19   57/–3 = –19   –57/–3 = 19

Priorities

The priorities listed on page 22 extend to include division (see box).

Examples

24/(2 + 4) = 24/6 = 4
12 + 56/7 = 12 + 8 = 20
3 × 8 – 45/5 = 24 – 15 = 9
8 × 6/3 = 48/3 = 16

In the last example, the multiplication and division can be done in any order. Usually we work from left to right, but sometimes it is easier to work from right to left:

8 × 6/3 = 8 × 2 = 16

In a chain, the order of multiplying and dividing makes no difference to the result.

Negatives on calculators

If you enter 5 × –3 just as it is written, you may find that the calculator forgets the × and acts only on the –. Result 5 – 3 = 2.

If your calculator has a +/– key, to change the sign of the displayed number, press this key *after* entering a number that is to be negative:

[5] [×] [3] [+/–] [=]

The display shows –15.

Other calculators have a (–) negate key, used when you want to enter a negative sign but are not doing a subtraction. Use this *before* entering the negative number:

[5] [×] [(–)] [3] [=]

The display shows –15.

Test yourself 2.5

1 Evaluate these expressions in your head.

a $144/12$	**b** $30/-6$	**c** $-50/5 + 3$
d $4 \times 5/2$	**e** $40 - 4 \times 9$	**f** $56/7 - 9$
g $63/-9 - 4$	**h** $-99/-11 - 8$	**i** $3 \times 10/-5$

2 Evaluate these expressions on paper.

a $216/9 + 6$	**b** $100 - 45/5$	**c** $(46 + 18)/8$
d $-801/9 + 100$	**e** $81/(56 - 59)$	**f** $(26 + 30)/(2 + 6)$
g $21 \times 35/7$	**h** $-(600 - 48)/12 + 4$	**i** $276/12 + 189/-9$

Check your results with a calculator.

3 Evaluate these expressions, using a calculator.

a $45 \times 63/35$	**b** $(56 + 72)/64$	**c** $-893/47 + 2106/27$

Long division

If you are dividing by 12 or less, use the technique on page 24. Most of the working is done in your head, using your knowledge of the multiplication tables. When dividing by numbers greater than 12, more of the work has to be done on paper. The technique is called **long division**.

Example

Divide 8648 by 23.

Stage 1: Try dividing 23 into 8. This does not go, so try dividing it into 86. 23 × 3 = 69, 23 × 4 = 92. It goes 3 times, but not 4 times. Write 3 above the line. Write 69 below the 86. Subtract, leaving 17.

```
       3
23 | 8648
     69
     --
     17
```

Stage 2: Copy down the next figure from 8648. This is the 4; write it beside the 17, making 174. Divide 23 into 174. 23 × 7 = 161, 23 × 8 = 184. It goes 7 times, but not 8 times. Write 7 above the line. Write 161 below the 174. Subtract, leaving 13.

```
       37
23 | 8648
     69↓
     ---
     174
     161
     ---
      13
```

Stage 3: Copy down the next (and last) figure. Write it beside the 13, making 138. Divide 23 into 138. 23 × 6 = 138. It goes exactly 6 times. Write the 6 above the line. Write 138 below the 138. Subtract, leaving 0.

```
       37
23 | 8648
     69 |
     ---
     174
     161↓
     ----
      138
      138
      ---
        0
```

Result: 8648/23 = 37

This is the end of the division which, in this example, has no remainder.

Test yourself 2.6

Work these exercises on paper, checking your result with a calculator.

1 Evaluate.

a 3068/13 **b** 9979/17 **c** 6975/31 **d** 5512/52

2 Evaluate.

a 895/19 **b** 61 241/86 **c** –29 970/74 **d** 1850/–43

3 If you need more practice at long division, set yourself some exercises, work them on paper and check your result with a calculator.

4 To assess if you have covered this chapter fully, answer the questions in *Try these first*, page 17.

3 Fractions

In the first two chapters we work with whole numbers of things – whole numbers of transistors, whole numbers of bolts, whole numbers of paces along the number path. A fraction of a transistor is not much use, neither can we take part of a pace along the pathway. But we can have half a kilogram of nails, a girder that is two and a quarter metres long, and find many other instances in which fractions have meaning. This chapter explains what fractions are and how to handle them.

You need to know about the four arithmetic operations described in Chapters 1 and 2, particularly division.

Try these first

Your success with this test will show you which parts of this chapter you already know.

1 Express 23/5 as a decimal fraction.

2 What decimal fraction equals 5/6?

3 Express 102/12 as a mixed fraction.

4 Given the fraction 4/5, what equivalent fraction has 25 as the denominator?

5 Evaluate:

a $1.47 + 3.29$

b $2\frac{1}{2} + 3\frac{5}{7}$

c 7.2×3.4

d $3\frac{1}{3} \times 4\frac{7}{8}$

e $9.146 \div 1.7$

f $7\frac{3}{4} \div 3\frac{1}{2}$

6 How many significant figures are there in

a 2300 **b** 1.071 **c** 3.100?

7 Round 7.374 to

a 2 significant figures **b** 2 decimal places.

8 **a** Convert 1.73 m to mm.

b Convert 7.57 mm to m.

What to do with a remainder

You are asked to evaluate 9/4. The result is 2, remainder 1. You divide 9 by 4, giving 2. At this point you stop dividing and leave a remainder of 1. This is illustrated in Figure 3.1a and b, where we divide 9 cakes into 4 groups of

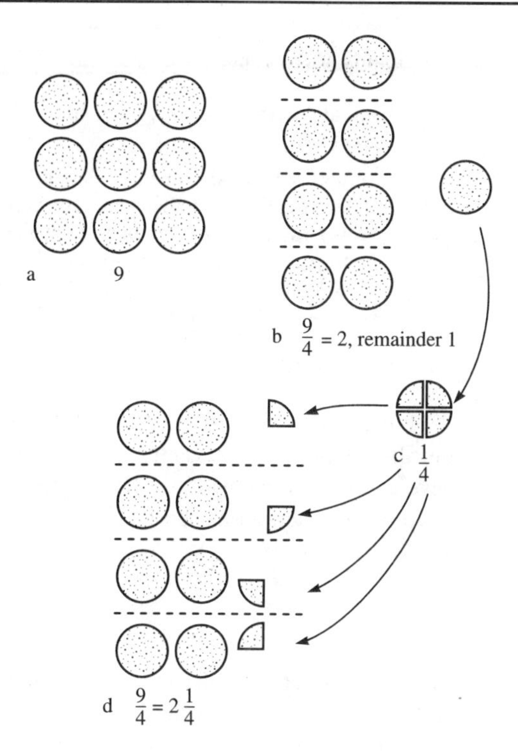

Figure 3.1

2, with 1 cake remaining. But why stop there? The cake can be divided into 4, by cutting it into 4 pieces of equal size (Figure 3.1c). Add one piece to each of the groups, as in Figure 3.1d.

Result: 9 divided by 4 gives 2 and one fourth (or quarter). We write this:

$9/4 = 2\frac{1}{4}$

The **fraction**, $\frac{1}{4}$, is written in the way that shows what has to be done to complete the division. It simply says 'Divide 1 by 4'. A fraction written in this way is a **common fraction**.

Common fractions

A **common fraction** is written as two numbers separated by a horizontal or sloping line. We write the fraction 'two sevenths' like this:

$\frac{2}{7}$ or 2/7

The number to the right of or below the line is called the **denominator**. It tells us what **denomination** (or kind) of fraction it is. In this example, we are dealing with *sevenths*. All fractions that are made up of sevenths have a 7 below the line. In a fraction such as 3/5, the 5 tells us we are working with fifths.

Kinds of fraction

Decimal fractions: tenths, hundredths, thousandths, . . . after a decimal point

Common fractions: numerator/denominator
Vulgar fractions: another name for common fractions
Proper fractions: numerator smaller than denominator
Improper fractions: numerator equal to or bigger than denominator
Mixed fractions: whole number plus common fraction
Equivalent fractions: see text

The number to the left of or above the line is the **numerator**. It tells us what **number** (how many) sevenths, fifths or other fraction we have. With 2/7 we have 2 sevenths, with 3/5 we have 3 fifths.

To make descriptions shorter, we refer in future to the numerator as N and to the denominator as D.

Decimal fractions

There is another way that we can complete a division. Returning to the first example, 9/4, why stop when we have found that there are two fours in nine? Just carry on dividing but, first of all, put a dot (the **decimal point**) on the line to show that this is where the fraction begins.

```
4|9
  2.25
```

The remainder, 1, is taken to be 10 tenths. Dividing 10 tenths by 4 gives 2 tenths (write 2 after the decimal point), remainder 2. The new remainder, 2, is taken to be 20 hundredths. We carry on dividing in this way, writing down the numbers of hundredths, thousandths, ten-thousandths and so on, until the

Shifting the decimal point

Shifting the decimal point one place to the right multiplies a number by 10:

$$1.234 \times 10 = 12.34$$

Shifting it two places multiplies by 100:

$$1.234 \times 100 = 123.4$$

Shifting it three places multiplies by 1000:

$$1.234 \times 1000 = 1234$$

Conversely, shifting the decimal point one place to the left divides a number by 10:

$$1.234/10 = 0.1234$$

and $1.234/100 = 0.01234$

division ends exactly or we decide to stop. The result is a **decimal fraction**. In this example the division ends exactly with 5 hundredths, so the result is 2.25 exactly.

Recurring decimals

The example above divides out exactly. When there are two figures after the decimal point we say there are two decimal places. Some divisions produce more numbers after the decimal point, for example: 79/64 = 1.234375. In other examples, the division *never* works out exactly. For example:

4/3 = 1.3333333333333333 . . .

At every stage after the first, the remainder is 1. Taking this as 10 tenths, 10 hundredths, 10 thousandths, or 10 of whatever stage we are at, the division always gives 3, remainder 1. The threes after the decimal point go on for ever. We say that this is a **recurring** decimal fraction. There is no need to write out an indefinite row of threes; just write *one* 3 and place a dot above it. This indicates that it recurs:

$4/3 = 1.\dot{3}$

$0.\dot{3}$ is the decimal equivalent of one third, 1/3. Dividing by 3 can also produce a row of sixes:

$5/3 = 1.\dot{6}$

$0.\dot{6}$ is the decimal equivalent of two thirds, 2/3. Dividing by any multiple of 3 always produces recurring decimals:

Sometimes more than one figure recurs:

3/111 = 0.027027027027027 . . .

We indicate the recurring part by placing a dot over the first and last figure:

$3/111 = 0.\dot{0}2\dot{7}$

Common and decimal fractions

These are the common and decimal versions of often-used fractions.

Common fraction	*Decimal fraction*
1/2	0.5
1/3	$0.\dot{3}$
1/4	0.25
1/5	0.2
1/6	$0.1\dot{6}$
1/7	$0.\dot{1}4285\dot{7}$
1/8	0.125
1/9	$0.\dot{1}$
1/10	0.1
1/11	$0.\dot{0}\dot{9}$
1/12	$0.08\dot{3}$

Test yourself 3.1

1 Evaluate the quotients below as common fractions and as decimal fractions. Do not use a calculator.

a 7/2	**b** 8/3	**c** 22/7	**d** 14/10
e 6/5	**f** 13/4	**g** 5/9	**h** 1/8
i 17/6	**j** 2/11	**k** 3/32	**l** 25/12

3 Using the information given in the box, what common fractions are equal to these decimal fractions?

a 0.2	**b** 0.5	**c** 0.75	**d** $0.\dot{1}$
e $0.\dot{2}$	**f** $0.\dot{2}85714$	**g** $0.\dot{6}$	**h** $0.\dot{2}\dot{7}$

3 Using the information given in the box, what decimal fractions are equal to these common fractions:

a 2/5	**b** 3/8	**c** 2/9	**d** 3/4
e 7/10	**f** 1/16	**g** 5/12	**h** 2/3

Converting improper fractions

Improper to mixed
Divide N by D as far as possible. Express any remainder as a proper fraction:

$$\frac{139}{8} = 17\text{, remainder } 3 = 17\tfrac{3}{8}$$

Mixed to improper
Multiply the whole number by D, add N. Place the total over N:

$$4\tfrac{3}{5} = \frac{23}{5}$$

Equivalent fractions

A quarter of a cake (Figure 3.2) cut into two equal pieces, gives two eighths:

$$\frac{1}{4} = \frac{2}{8}$$

Now divide each eighth into two, giving four sixteenths:

$$\frac{1}{4} = \frac{2}{8} = \frac{4}{16}$$

Although the quarter cake has been cut into four, there is still just as much cake. 1/4, 2/8, and 4/16 have equal values; they are **equivalent fractions**.

Here is another example: take 1/2 of a cake, and cut each half into three pieces, giving three sixths (Figure 3.3). Cut these again into five pieces each, giving fifteen thirtieths. 1/2, 3/6 and 15/30 are equivalent fractions.

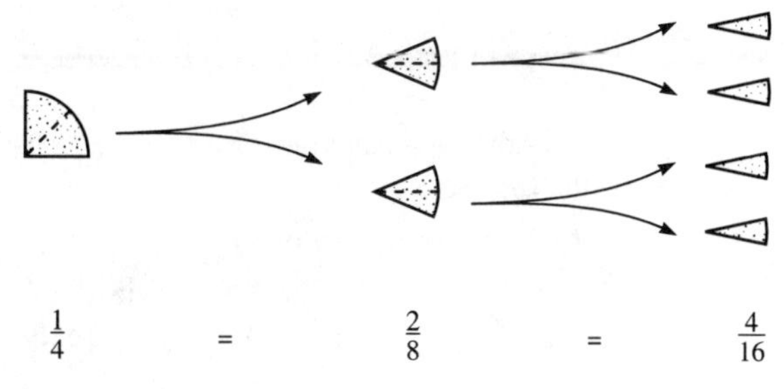

Figure 3.2

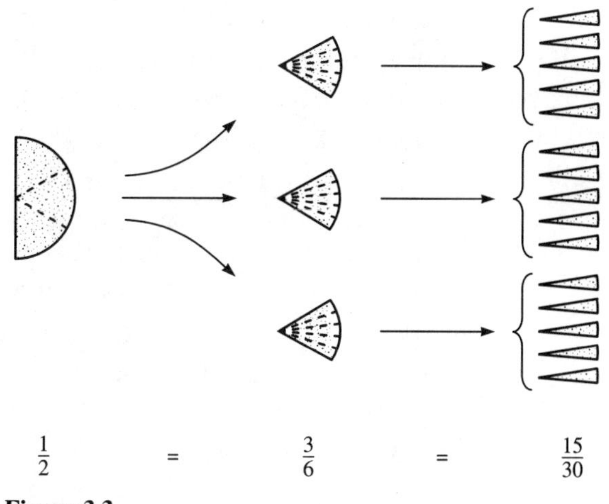

Figure 3.3

The reverse process is sometimes used to simplify a fraction. We divide the N and the D by the same number:

Given $\frac{3}{15}$, divide both by 3 to obtain $\frac{1}{5}$.

Given $\frac{12}{16}$, divide both by 4 to obtain $\frac{3}{4}$.

This is called **cancelling**. Knowing the multiplication tables well makes cancelling much easier.

Forming equivalent fractions

Multiplying N and D by the *same number* gives an equivalent fraction.

To form an equivalent fraction with a given denominator:
Find how many times its D goes into the new D (it must go exactly); multiply both D and N by this amount.

Example
Given 2/3, find the equivalent which has 12 as the new D.

Answer
Dividing 12 by 3, gives 4. Multiply N and D by 4:

$$\frac{2}{3} = \frac{2 \times 4}{3 \times 4} = \frac{8}{12}$$

Test yourself 3.2

Do not use a calculator for these exercises. Work in your head as often as you can.

1 Cancel these fractions.

a 3/6	**b** 2/10	**c** 12/24	**d** 4/6
e 5/15	**f** 14/21	**g** 44/88	**h** 10/35

2 Convert these improper fractions into mixed fractions. Simplify by cancelling, if possible.

a 45/11	**b** 100/12	**c** 54/17	**d** 48/15
e 66/64	**f** 78/40	**g** 57/19	**h** 27/8

3 Convert these mixed fractions into improper fractions.

a $1\frac{3}{4}$	**b** $3\frac{2}{5}$	**c** $4\frac{7}{8}$	**d** $2\frac{3}{11}$
e $4\frac{5}{7}$	**f** $1\frac{19}{20}$	**g** $4\frac{5}{12}$	**h** $23\frac{4}{5}$

4 For each of the fractions, find the equivalent fractions that have the numbers in brackets as denominators.

Example: 1/3 (6, 9, 12) *Answer*: 1/3 = 2/6 = 3/9 = 4/12

a 1/2 (4, 6, 8)	**b** 1/4 (8, 12, 16)
c 1/7 (14, 21, 28, 35)	**d** 1/5 (10, 15, 45, 60)
e 2/3 (6, 9, 12, 15)	**f** 3/4 (8, 12, 16, 20)

5 For each of the fractions, find the equivalent fractions that have the numbers in brackets as numerators.

Example: 1/5 (2, 3, 5, 7) *Answer*: 1/5 = 2/10 = 3/15 = 5/25 = 7/35

a 1/2 (2, 3, 4)	**b** 1/6 (2, 3, 4, 5)
c 1/10 (2, 4, 5, 9)	**d** 1/5 (3, 5, 6, 9)
e 3/8 (6, 9, 30)	**f** 2/7 (4, 6, 10)

Adding and subtracting fractions

Adding numbers with decimal fractions is easy. Just line up the decimal points and add in the usual way, carrying (page 6) if necessary.

Example

3.653 + 2.518 = ?

$$\begin{array}{r} 3.653 \\ +\ \underline{2.518} \\ {\scriptstyle 1\ \ 1} \\ 6.171 \end{array}$$

Subtracting follows the usual routine, with borrowing (page 13) if necessary.

Example

5.736 − 2.254 = ?

$$\begin{array}{r} 5\,.\,{}^{6}\not{7}\,{}^{1}3\ 6 \\ -\ \underline{2\,.\ 2\ 5\ 4} \\ 3\,.\ 4\ 8\ 2 \end{array}$$

Decimal fractions are added one column at a time. We add thousandths to thousandths, hundredths to hundredths, tenths to tenths. We always add the same kinds or (same **denominations**) of fraction together.

Adding common fractions which are of the same denomination is also easy. Here we add ninths:

$$2\tfrac{2}{9} + 3\tfrac{5}{9} = 5\tfrac{7}{9}$$

The 2 and 3 units are added to give 5. The 2-ninths and 5-ninths are added to give 7-ninths. Sometimes the N of the added fraction is larger than or equal to the D. If so, convert this to a proper fraction (page 32). Look at this example, working in fifths:

$$2\tfrac{4}{5} + 1\tfrac{3}{5} = 3 + \tfrac{7}{5} = 3 + 1\tfrac{2}{5} = 4\tfrac{2}{5}$$

The 2 and 1 units are added to give 3. The 4-fifths and 3-fifths come to 7-fifths, an improper fraction. Convert this to one and 2-fifths. Total units are 3 + 1 = 4. Result is four and 2-fifths.

We often have to add common fractions that are *not* of the same kind.

Example

$$\frac{1}{2} + \frac{2}{3} = ?$$

Adding halves to thirds is not possible, just as we cannot add transistors to bolts (page 8). The solution is to find equivalent fractions for 1/2 and 2/3, both equivalent fractions having the *same* new D, instead of 2 and 3. The simplest rule for deciding on the new D is to multiply the original Ds together:

$$2 \times 3 = 6$$

Work in sixths.
Find the equivalent of 1/2, in sixths: 1/2 = 3/6.
Find the equivalent of 2/3, in sixths: 2/3 = 4/6.
Now we can add:

$$\frac{1}{2} + \frac{2}{3} = \frac{3}{6} + \frac{4}{6} = \frac{7}{6} = 1\tfrac{1}{6}$$

At the last stage we convert 7/6 to $1\tfrac{1}{6}$.

Example

$$\frac{3}{5} + \frac{4}{7} = ?$$

The new D is 5 × 7 = 35. Work in thirty-fifths. The equivalent of 3/5 in thirty-fifths is 21/35. The equivalent of 4/7 in thirty-fifths is 20/35. Adding:

$$\frac{3}{5} + \frac{4}{7} = \frac{21}{35} + \frac{20}{35} = \frac{41}{35} = 1\tfrac{6}{35}$$

The same technique is used when subtracting.

$$\frac{4}{5} - \frac{2}{3} = ?$$

Work in fifteenths.

$$\frac{4}{5} = \frac{12}{15} \text{ and } \frac{2}{3} = \frac{10}{15}$$

$$\frac{4}{5} - \frac{2}{3} = \frac{12}{15} - \frac{10}{15} = \frac{2}{15}$$

Subtraction may result in a negative result:

$$\frac{2}{5} - \frac{3}{4} = \frac{4}{20} - \frac{15}{20} = \frac{-11}{20}$$

In the next example, subtracting the whole numbers leaves a positive result, but subtracting the fractions leaves a negative result:

$$5\tfrac{1}{4} - 2\tfrac{1}{3} = 3 + \frac{3}{12} - \frac{4}{12} = 3 - \frac{1}{12}$$

Borrow 1 from the 3 and convert the 1 to an improper fraction:

$$= 2 + \frac{12}{12} - \frac{1}{12} = 2\tfrac{11}{12}$$

Test yourself 3.3

1 Evaluate.

a 1.24 + 3.58 **b** 0.772 + 1.218 **c** 0.3143 + 4.7621

d 5.73 − 1.61 **e** 1.832 − 0.642 **f** 1.34 − 2.65

[*Hint*: See page 13]

2 Evaluate.

a $\frac{1}{5} + \frac{3}{5}$ **b** $\frac{1}{9} + \frac{4}{9}$ **c** $\frac{1}{3} + \frac{2}{3}$ **d** $\frac{1}{2} + \frac{1}{3}$

e $\frac{1}{4} + \frac{1}{5}$ **f** $\frac{3}{7} + \frac{1}{5}$ **g** $\frac{7}{12} + \frac{4}{5}$ **h** $\frac{5}{6} + \frac{3}{4}$

i $\frac{6}{10} - \frac{5}{10}$ **j** $\frac{7}{12} - \frac{2}{5}$ **k** $\frac{5}{11} + \frac{4}{9}$ **l** $\frac{2}{5} - \frac{2}{3}$

3 Evaluate.

a $\frac{1}{6} + \frac{1}{3}$ **b** $2\tfrac{3}{4} + 5\tfrac{2}{5}$ **c** $4\tfrac{5}{6} - 3\tfrac{1}{3}$ **d** $3\tfrac{11}{16} - 1\tfrac{3}{4}$

e $3\tfrac{7}{12} + 4\tfrac{9}{10}$ **f** $2\tfrac{2}{3} - 3\tfrac{5}{6}$ **g** $5\tfrac{5}{16} + 4\tfrac{7}{8}$ **h** $7\tfrac{1}{4} - 6\tfrac{15}{16}$

[*Hint*: See page 13]

Adding rules

Compare this with the box on page 40.

Steps	*Example* $3\frac{3}{5} + 1\frac{7}{8} = ?$
Add whole numbers	$= 4$
Find new D	New D $= 5 \times 8 = 40$
Convert to equivalents	$= 4 + \frac{24}{40} + \frac{35}{40}$
Add equivalents	$= 4 + \frac{59}{40}$
If improper, convert to mixed	$= 4 + 1 + \frac{19}{40}$
Cancel if possible	(nothing to cancel)
Write answer as mixed fraction	$= 5\frac{19}{40}$

Multiplying decimal fractions

This is easier to understand if we convert the multiplier into a whole number. Multiply it by 10 as many times as is needed to get rid of the decimal point (page 30).

Example

$0.25 \times 0.15 = ?$

Convert 0.15 into a whole number; it has two figures after the decimal point. Multiply it by 100 by shifting the decimal point 2 places to the *right*. (page 30) : $0.15 \times 100 = 15$. Here is the multiplication:

$$\begin{array}{r} 0.25 \\ \times \quad 15 \\ \hline 2.50 \\ + \quad 1.25 \\ \hline 3.75 \end{array}$$

$0.25 \times 15 = 3.75$. Since 0.15 was multiplied by 100 to start with, we must divide 3.75 by 100 to obtain the correct result. Shift the decimal point 2 places to the *left*.

$0.25 \times 0.15 = 0.0375$

Confirm the position of the decimal point by applying this rule:

The number of decimal places in the product equals the total of the numbers of decimal places in the numbers multiplied

In the example above, both numbers have 2 decimal places, so the product has four.

Multiplying common fractions

In Figure 3.4 we show how multiplying **whole numbers** together is represented by building up rectangles from unit squares. In every case the number of squares equals the number of columns multiplied by the number of rows.

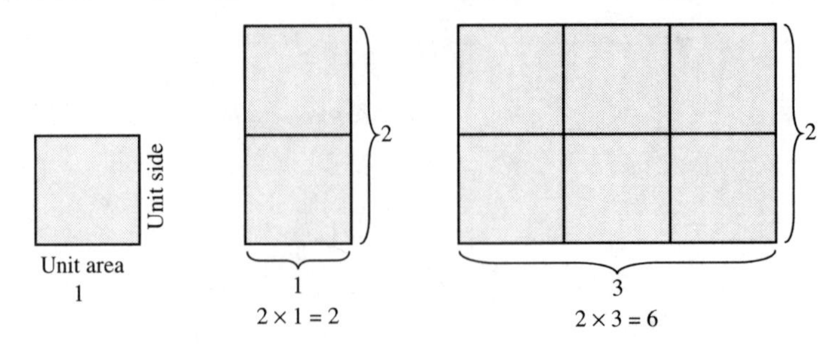

Figure 3.4

What applies to whole numbers also applied to fractions. In Figure 3.5a, the unit square is divided so that the height of the shaded part is 2/3 of the side of the square, and its width is half the side of the square. If you imagine that the square is divided as shown by the dashed lines, each small rectangle is one sixth the area of the square. The shaded part comprises two small rectangles. Putting this as a multiplication:

$$\frac{2}{3} \times \frac{1}{2} = \frac{2}{6}$$

The product 2/6 cancels to 1/3. In Figure 3.5b we have multiplied 1/4 by 1/3. Here the square is divided into twelfths, and the shaded area comprises one twelfth:

$$\frac{1}{4} \times \frac{1}{3} = \frac{1}{12}$$

Figure 3.5c shows the result of multiplying 3/4 by 4/5. The square is divided into twentieths, of which 12 are in the shaded area:

$$\frac{3}{4} \times \frac{4}{5} = \frac{12}{20}$$

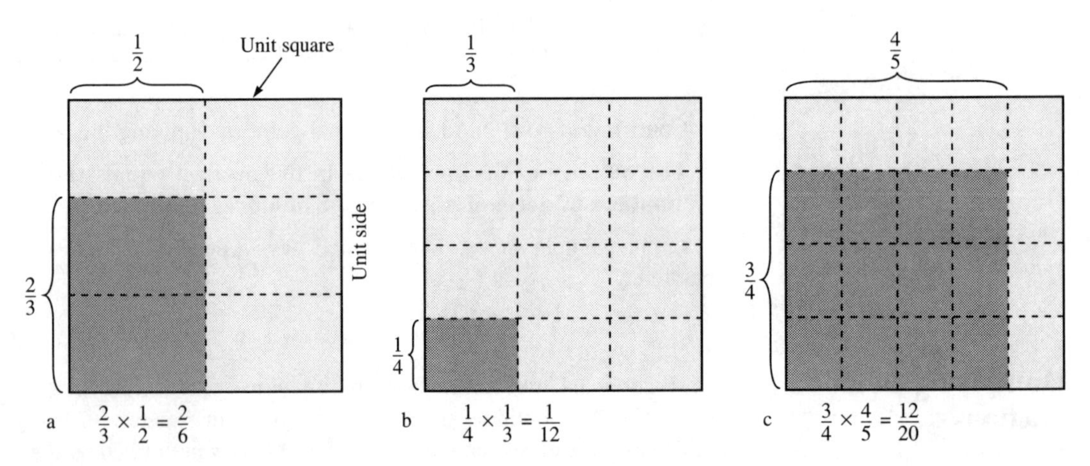

Figure 3.5

In every example, the number of rectangles in the unit square is the product of the Ds. In every example, the number of shaded rectangles is the product of the Ns. This leads to the simple multiplication rule:

$$\textbf{Product} = \frac{\textbf{Product of Ns}}{\textbf{Product of Ds}}$$

This rule also applies to mixed fractions. Figure 3.6 shows the result of:

$2\frac{1}{2} \times 3\frac{1}{4}$

The thing to note is that when we multiply, we multiply *all* of the $2\frac{1}{2}$ by *all* of the $3\frac{1}{4}$. We multiply the whole numbers along with the fractions. With *addition* of fractions, we could add up the whole numbers separately, but this is not allowed when multiplying. This means that we have to convert mixed fractions into improper fractions *before* we start to multiply. The full working of the example of Figure 3.6 is:

$$2\frac{1}{2} \times 3\frac{1}{4} = \frac{5}{2} \times \frac{13}{4} = \frac{65}{8}$$

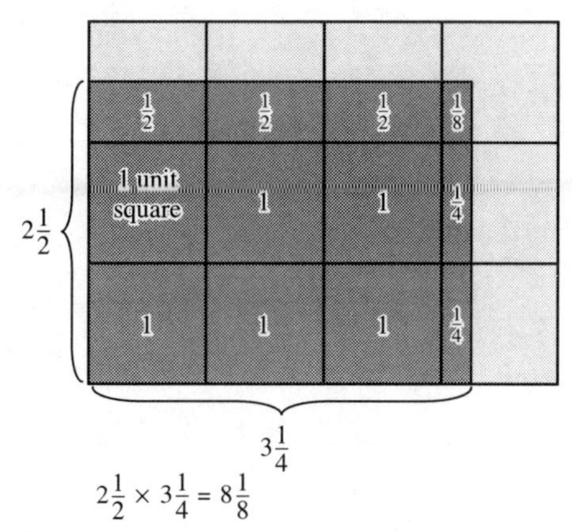

Figure 3.6

Multiplying the Ns: $5 \times 13 = 65$. Multiplying the Ds: $2 \times 4 = 8$, the product is 65/8. To complete the calculation, convert the product back into a mixed fraction:

$2\frac{1}{2} \times 3\frac{1}{4} = 8\frac{1}{8}$

If you count up the unit squares and fractions of unit squares shown in Figure 3.6, you obtain the same result.

The usual sign rules (see box, page 21) apply to multiplication by negative fractions.

Multiplying rules

Compare this with the box on page 37.

Steps	*Example* $3\frac{3}{5} \times 1\frac{7}{8} = ?$
Convert to improper	$= \frac{18}{5} \times \frac{15}{8}$
Set out as multiplication of Ns and Ds	$= \frac{18 \times 15}{5 \times 8}$
Cancel if possible	Cancel 2's; cancel 5's $= \frac{{}^{9}\not{18} \times \not{15}^{3}}{{}_{1}\not{5} \times \not{8}_{4}}$
Multiply	$= \frac{27}{4}$
Convert to mixed	$= 6\frac{3}{4}$

Test yourself 3.4

1 Evaluate in your head.

a 2.4×6 **b** 5.7×0.5 **c** 3.6×1.2
d 4.56×-0.2 **e** -3.2×-0.3 **f** 0.712×0.11

2 Evaluate in your head.

a $\frac{1}{5} \times \frac{2}{3}$ **b** $\frac{3}{7} \times \frac{4}{5}$ **c** $\frac{2}{9} \times \frac{7}{11}$
d $\frac{2}{3} \times \frac{5}{8}$ **e** $\frac{6}{10} \times \frac{5}{9}$ **f** $\frac{5}{12} \times \frac{3}{11}$

Cancelling helps

Cancelling means smaller figures, easier calculations. Cancel whenever you can. Sometimes you can cancel many times.

$$\frac{9}{21} \times \frac{7}{6}$$

Cancel 3's; cancel 7's

$$\frac{{}^{3}\not{9}}{{}_{3}\not{21}} \times \frac{\not{7}^{1}}{\not{6}_{2}}$$

Now cancel 3's again

$$\frac{{}^{1}\not{3}}{{}_{1}\not{3}} \times \frac{1}{2} = \frac{1}{2}$$

3 Evaluate.

a $2\frac{1}{2} \times \dfrac{3}{4}$ **b** $4\frac{3}{5} \times -1\frac{5}{8}$ **c** $3\frac{1}{5} \times \dfrac{5}{16}$

d $-6\frac{3}{4} \times -2\frac{3}{4}$ **e** $1\frac{1}{2} \times 3\frac{3}{5} + \dfrac{7}{10}$ **f** $-\dfrac{5}{6} \times \left(\dfrac{1}{5} - \dfrac{3}{4}\right)$

Reciprocals

When a number is multiplied by its *reciprocal, the product is 1.*

Examples

$2 \times \dfrac{1}{2} = 1$ The reciprocal of 2 is 1/2
The reciprocal of 1/2 is 2

$7 \times \dfrac{1}{7} = 1$ The reciprocal of 7 is 1/7
The reciprocal of 1/7 is 7

$\dfrac{3}{2} \times \dfrac{2}{3} = 1$ The reciprocal of 3/2 is 2/3
The reciprocal of 2/3 is 3/2

Rule: to find the reciprocal of a number, express the number as a common fraction or improper fraction and turn it upside down.

Reciprocal of 3: $3 = \dfrac{3}{1}$ Reciprocal of 3 is $\dfrac{1}{3}$

Reciprocal of $2\frac{1}{2}$: $2\frac{1}{2} = \dfrac{5}{2}$ Reciprocal of $2\frac{1}{2}$ is $\dfrac{2}{5}$

Reciprocals do not have to be common fractions:

Reciprocal of 4: 1/4 = 0.25 Reciprocal of 4 is 0.25
Reciprocal of 0.25 is 4

Reciprocal of 0.27: 1/0.27 = 3.703 Reciprocal of 0.27 is $3.\dot{7}0\dot{3}$
Reciprocal of $3.\dot{7}0\dot{3}$ is 0.27

Dividing decimal fractions

Treat the calculation as an ordinary long division (page 26). Estimate where the decimal point should go by dividing the whole numbers, in your head.

Example

26.474/6.2 = ?

```
      427
62 | 26474
     248
      167
      124
       434
       434
         0
```

26/6 is a little more than 4. The decimal point goes after the 4.
Result: 26.474/6.2 = 4.27.

As a check note that, *if the division goes out exactly, with no remainder*, the number of decimal places in the N equals the total of decimal places in the D and the quotient. Here 3 + 1 + 2. Correct.

There is no need to continue the division once the answer has enough decimal places, or begins to recur. On page 44 we discuss how many decimal places to work to.

If the N or the D both have several zeros after the decimal point, shift the point by the same amount in both N and D. This does not make any difference to the result, because you are in effect multiplying both N and D by 10, 100, 1000 . . . They are equivalent fractions.

Example

0.005 876/0.000 13 = ?

Shift the decimal point 4 places to the right and obtain: 58.76/1.3

```
     452
13 | 5876
     52
      67
      65
       26
       26
        0
```

58/1 is 58, so the answer must be rather less than 50. The decimal point comes after the 5:

0.005 876/0.000 13 = 45.2

It divides out exactly, so check decimal places: 6 = 5 + 1. Correct.

Division is used for converting common fractions into decimal fractions. For example, the common fraction, 23/47, means 'Divide 23 by 47'. If we do this using decimals, we obtain the decimal equivalent. By long division, or using a calculator: 23/47 = 0.489 361 702. The division does not go exactly at the final 2, but 9 decimal places are usually far more than we need.

Dividing common fractions

The rule for division is:

To divide by a given fraction, multiply by its reciprocal

For example:

$$3\tfrac{1}{2} \div \frac{1}{2} = ?$$

The reciprocal of 1/2 is 2 (page 41) so, instead of dividing by 1/2, multiply by 2:

$$3\tfrac{1}{2} \times 2 = \frac{7}{2} \times 2 = 7$$

The rule is explained on page 43.

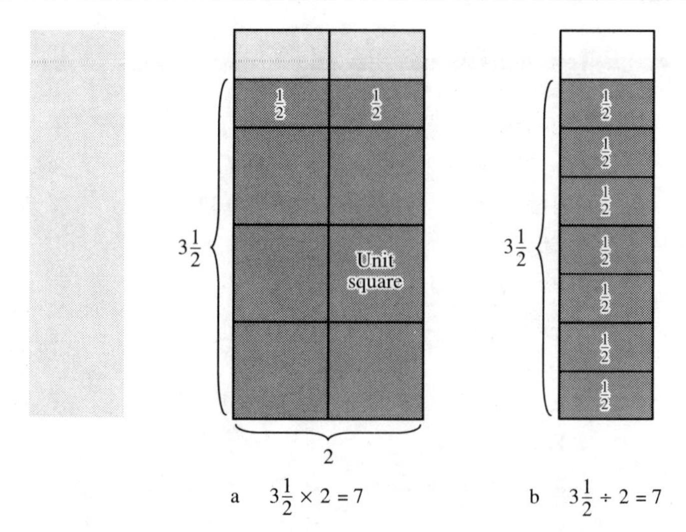

Figure 3.7

A more easily remembered form of this rule is:

To divide by a fraction, turn it upside down and multiply

Here is another example:

$$\frac{3}{7} \div \frac{4}{5} = \frac{3}{7} \times \frac{5}{4} = \frac{3 \times 5}{7 \times 4} = \frac{15}{28}$$

Another example is shown in the box.

Dividing rules

Compare this with the box on page 40.

Steps	**Example** $3\frac{3}{5} \div 1\frac{7}{8} = ?$
Convert to improper	$= \frac{18}{5} \div \frac{15}{8}$
Reciprocal of second number* (change to multiply)	$= \frac{18}{5} \times \frac{8}{15}$
Set out as multiplication of Ns and Ds	$= \frac{18 \times 8}{5 \times 15}$
Cancel if possible	Cancel 3's $= \frac{{}^{6}\cancel{18} \times 8}{5 \times \cancel{15}_{5}}$
Multiply	$= \frac{48}{25}$
Convert to mixed	$= 1\frac{23}{25}$

*Except for this step, the routine is exactly the same as multiplying.

Test yourself 3.5

1 Evaluate.

a 42.12/5.4 **b** 11.04/–2.3 **c** 200.6/3.4

d 5.2/130 **e** 0.45/0.11 **f** 165.9093/2.11

Check your results with a calculator.

2 Evaluate.

a $\frac{3}{4} \div \frac{5}{7}$ **b** $\frac{7}{16} \div \frac{4}{5}$ **c** $\frac{14}{20} \div 2\frac{1}{10}$

d $3\frac{5}{6} \div 7\frac{2}{3}$ **e** $4\frac{5}{8} \div 1\frac{2}{5}$ **f** $9\frac{6}{11} \div 8\frac{3}{4}$

3 Evaluate.

a $3\frac{3}{4} \times 4\frac{2}{5} - 7\frac{1}{4}$ **b** $2\frac{1}{4} + 4\frac{3}{8} \div 2\frac{1}{2}$

c $3\frac{1}{3} \times 4\frac{3}{5} \div 7\frac{2}{3}$ **d** $3\frac{4}{5} \times 2\frac{1}{7} - 7\frac{1}{2} \div 1\frac{1}{2}$

Rounding

Suppose that you have worked out an answer that comes to 254.3. You may decide that it is 'near enough' if the answer is to the nearest whole number. Figure 3.8a shows that the nearest whole number is 254. So 254.3 is **rounded down** to 254.

Another number to be rounded might be 163.7. Figure 3.8b shows that the nearest whole number is 164. So 163.7 is **rounded up** to 164.

If the number is 923.5, it is equally near to 923 and 924. By convention, such a number is rounded up. 923.5 is rounded up to 924 (Figure 3.8c).

We can do the same thing at any position in the number.

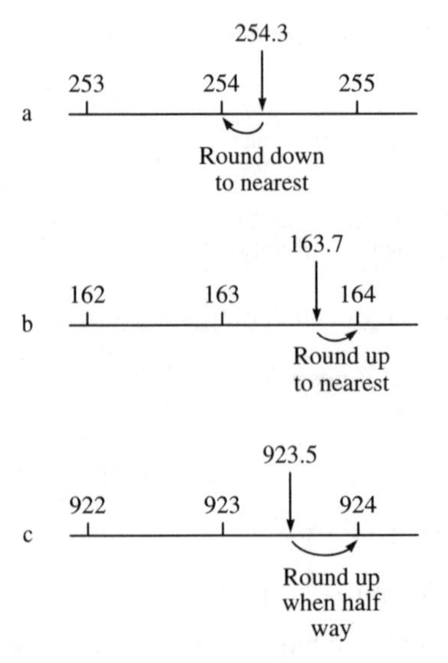

Figure 3.8

Examples

Round 45.636 to 1 decimal place.

Cut off every figure after the 1st decimal place:

45.6 |36

Leaves 45.6

The *first* figure in the cut-off part is 3, so we round down. The 6 *before* the cut-off is not changed. Note that any other figures in the cut-off part are ignored.

45.636 rounded to 1 decimal place is 45.6 (1 dp)

The (1 dp) after the number tells us that this was originally a longer number but has been rounded to 1 decimal place.

Round 647.6382 to 2 decimal places.

647.63 |82

Leaves 647.63

In this example the first cut-off figure is 8, so we round up. The 3 before the cut-off is increased to 4.

647.6382 rounded to 2 dp is 647.64 (2 dp)

Round 12.654 to 1 dp.

12.6 |54

Leaving 12.6

Here the first cut-off figure is 5, so we round up. The 6 before the cut-off is increased to 7.

12.654 rounded to1 dp is 12.7 (1 dp)

If the figure before the cut-off is 9 and it has to be rounded up, rounding up makes this 10. So 1 has to be carried back, as in this example.

Round 67.96 to 1 dp.

67.9 |6

Leaving 67.9

Rounding up the 9 turns it into a zero, with 1 carried back.

67.96 rounded to 1 dp is 68.0 (1 dp)

Note that we include the 0 as part of the rounded number.

There may be even more carrying. For example, 199.996 rounded to 2 dp is 200.00

It is essential to round a number in *one step*. Given 3.748, rounded to 1 dp, we obtain 3.7, which is the correct answer. But if we do it in stages (which we must not) we first round 3.748 to 2 dp, and obtain 3.75. Then, rounding this to 1 dp gives 3.8, an incorrect answer.

Rounding can be done at positions to the left of the decimal point.

Examples

Round 347 to the nearest 10.

34 |7

Leaving 34

This has to be rounded up (giving 35) but we must also put in a zero as a 'place keeper'.

347 rounded to the nearest 10 is 350.

Round 5482 to the nearest thousand.

5 |482

Leaving 5

Round down, inserting three zeros as place keepers.

5482 rounded to the nearest thousand is 5000.

Calculator rounding

When a calculation gives more figures than the display can show, a calculator may or may not round the last figure. Check your calculator by asking it to divide 2 by 3. The full answer is $0.\dot{6}$. If the calculator rounds correctly, the last figure in the display is rounded to 7. If there is no rounding, the last figure is 6.

Test yourself 3.6

1 Round these numbers to the nearest whole number.

a 57.241 **b** 8.3226 **c** 8.635 **d** 65.82

e 5.007 **f** 545.5 **g** 21.998 **h** 49.996

2 Round these numbers to 2 dp.

a 34.761 **b** 7.654 **c** 0.8888 **d** 300.765

e 0.01572 **f** 200.8383 **g** 2.534999 **h** 17.8551

3 Round these numbers to the nearest 100.

a 432 **b** 2354 **c** 12476 **d** 49951

Metres and millimetres

One metre equals 1000 millimetres. If a measurement is given in metres and we want it to be in millimetres, we simply multiply it by 1000. If the number has no decimal point, write three zeros after it:

15 m = 15000 mm

If the number has a decimal point, shift this three places to the right:

23.732 m = 23732 mm

If there are fewer than three decimal places in the original number, we have to add zome zeros to fill up the blank spaces before the decimal point:

5.72 m = 5720 mm

Conversely, millimetres are converted to metres by dividing by 1000. Shift the decimal point three places to the left:

8432 mm = 8.432 m

In the example above it is possible to count along the number to find where to put the decimal point. This example does not have enough figures:

3.5 mm = ?

Count along like this and mark the decimal point:

. 3 5

Fill in the empty places after the point with zeros:

.0035

Then write a zero in front of the point:

0.0035

This is needed because beginning a number with a decimal point can often lead to error. It is very easy not to notice the point if there is nothing in front of it. Result of conversion:

3.5 mm = 0.0035 m

If you have to add or subtract measurements which are in different units, convert them all to the same unit first:

3.5 m + 672 mm + 0.24 m = ?

Convert 672 mm to metres:

3.5 + 0.672 + 0.24 = ?

Write numbers in a column, with decimal points aligned:

```
  3.5
+ 0.672
+ 0.24
  -----
  4.412
```

The sum is 4.412 m.

There are other pairs of units in which one is 1000 times the other, and we use the same technique for converting them:

Unit pairs	*Example*
1 kilometre = 1000 metre	2.3 km = 2300 m
	237 m = 0.237 km
1 kilogram = 1000 gram	74.2 kg = 74200 g
	450 g = 0.45 kg
1 volt = 1000 millivolt	2.53 V = 2530 mV
	128 mV = 0.128 V
1 amp = 1000 milliamp	7.09 A = 7090 mA
	83 mA = 0.083 A
1 megahertz = 1000 kilohertz	89.5 MHz = 89500 kHz
	1546 kHz = 1.546 MHz

Test yourself 3.7

1 Convert from metres to millimetres.

a 35	**b** 57	**c** 54.8	**d** 6.789
e 56.45	**f** 1000	**g** 0.046	**h** 40.008

2 Convert from millimetres to metres.

a 6423	**b** 70024	**c** 563	**d** 871
e 562.9	**f** 83	**g** 9	**h** 0.67

3 Add these distances and express the result in the stated unit.

a 4.5 m + 345 mm + 1045 mm, in m

b 0.562 m + 34 mm + 1.54 m, in mm

c 83 m + 0.034 m + 32 mm + 1 m, in mm

4 Make the following conversions:

a 34.5 V, to mV

b 5 MHz, to kHz

c 632 mA to A

5 To assess if you have covered this chapter fully, answer the questions in *Try these first*, page 28.

4 Letters for numbers

This chapter introduces a branch of maths known as *algebra*. In this we make letters of the alphabet take the place of numbers. Algebra opens the way to many of the powerful maths techniques that we use in technology.

You need to know the four arithmetical operations (Chapters 1 and 2), and fractions (Chapter 3).

Try these first

Your success with this short test will show you which parts of this chapter you already know.

1 Solve $x = y(4 + 2z)$, when $y = 3$ and $z = 7$

2 Solve $a = 4.2(3.3b - 2)$, when $b = 1.7$

3 Solve $p = \frac{7q - 1}{4}$, when $q = 3$

4 Solve $m = 3(2n + 7)$ when $n = -3$

5 Simplify $4(n + m) - n(1 - m) + 5m$

6 Simplify $3r(2s + r) - 4s(3r + 7)$

An unknown quantity

A caterer is asked to provide snacks for a party of tourists. The party has two guides but the number of tourists is not yet known. Because we do not know the number of tourists, we cannot say how many snacks will be needed. But we can write down *something*, even if it is incomplete. Instead of a known number of tourists, such as 12 or 17, we use a letter, such as a.

If there are a tourists, the caterer has to supply a snacks for the tourists and 2 for the guides. We use another letter such as b for the number of snacks required. Putting the sentence above into letters and numbers:

$$b = a + 2$$

This is an **equation**. It shows us two things that are *equal*. The number of snacks needed (b) *equals* the number of tourists (a) plus 2.

The equation covers all possibilities. If there are 10 tourists, we need 12 snacks; if there are 5 tourists, we need 7 snacks; if there are 250 tourists, we need 252 snacks.

Some names

Variable: a quantity which varies, represented by a letter.

Examples

x, a, p.

Term: one or more variables and/or numbers, multiplied or divided by each other (if there is more than one).

Examples

xy, $2g$, $a/2$, 4, q, $5ab$

Expression: usually two or more terms with + or – signs between them.

Examples

$2x + y$
$3 - 4ax$

Equation: Made up of two expressions that have equal value.

Examples

$y = 3 + 5x$
$p = 4a/2 - 6$

Letters can stand also for numbers that are mixed fractions or decimal fractions. A plastic tray weighs 62.37 g empty and is used for weighing out quantities of fertilizer. If r is the reading shown on the dial of the balance, the equation for f, the number of grams of fertilizer, is:

$$f = r - 62.37$$

This equation shows that every time we weigh out a quantity of fertilizer, we have to subtract 62.37 from the dial reading to allow for the empty weight of the tray. If, for example, the dial reads 155.71 g, $r = 155.71$. We calculate:

$$f = r - 62.37 = 155.71 - 62.37 = 93.34$$

There is 93.34 g of fertilizer in the tray.

We might have several different trays with different empty weights. This introduces a third variable t into the equation, where t is the empty weight of a tray:

$$f = r - t$$

Here we have a general-purpose equation for calculating the weight of fertilizer. What it says is:

To find the weight of fertilizer, subtract the empty weight of the tray from the dial reading.

The advantage of using an equation with letters f, r, and t is that it is easier to follow than the lengthy sentence.

Test yourself 4.1

1 If $a = b + 7$, find a when $b = 4$.

2 If $m = n + 8$, find m when $n = 6$.

3 If $x = 10 + y$, find x when $y = 34$.

4 If $j = k - 6$, find j when $k = 7$.

5 If $y = z - 8$, find y when $z = 6$.

6 If $d = 5.62 + e$, find the value of d when $e = 3.03$.

7 If $p - 7 = q$, find the value of q when $p = 10$.

8 If $a = b + c$, find the value of a when $b = 5$ and $c = 7$.

9 If $x = y - z$, find the value of x when $y = 13$ and $z = 4$.

10 A mail order firm charges £1 each for mirror tiles, plus a standard charge of £5 per order for packing and postage. Write an equation to show how to find t, the total cost of an order, given n, the number of tiles ordered. Use the equation to find the cost of an order for:

a 12 tiles **b** 30 tiles

11 A person giving a party needs a chair for each person. He already has 4 chairs but needs to borrow the rest. Write an equation to show how to find b, the number of chairs to be borrowed, given p the number of people at the party (including the host). Use the equation to find how many chairs must be borrowed when the number at the party is:

a 8 people **b** 50 people

Multiplying

Back to the tourists! The snack is to consist of 3 sandwiches. How many sandwiches are needed? The two couriers need 6 sandwiches and, for every one of the a tourists, we need 3 sandwiches. If c is the number of sandwiches:

$$c = 3a + 6$$

The term '$3a$' means '3 multiplied by a'. We could have written this as '$3 \times a$' but it is usual to leave out the multiplying sign when it is part of a single term. Now we can calculate that if, for example, there are 20 tourists, the number of sandwiches is:

$$c = 3a + 6 = 3 \times 20 + 6 = 66$$

The caterer must supply 66 sandwiches.

We could look at this in another way. Sandwiches are needed by a tourists plus 2 guides. The total number of *people* is $(a + 2)$. To find how many sandwiches are needed, multiply this total by 3:

$$c = 3(a + 2)$$

We do not need to write the multiplication sign between the '3' and the '$(a + 2)$'. If there are 20 tourists, as before, the number of sandwiches is:

$$c = 3(a + 2) = 3(20 + 2) = 3 \times 22 = 66$$

The caterer must supply 66 sandwiches.

This is exactly the same result as before. Note that, because of the bracket, we add 20 and 2 *before* multiplying by 3 (see page 21).

The result is the same as before because $3(a + 2)$ has exactly the same value as $3a + 6$. We multiply each of the terms in the brackets by the number in front of the brackets:

$$3 \times a = 3a$$
$$3 \times 2 = 6$$
$$\Rightarrow \quad 3 \times (a + 2) = 3a + 6$$

Arrow symbol

$\Rightarrow$ shows that the next statement follows from the one before. In words it is the same as 'gives' or 'with the result that ' or 'and so' or 'implies that '.
In some books, the symbol $\therefore$ is used instead.

The advantage of using the equation with brackets is that we can easily change the number of sandwiches each person is to have. For example, if we decide to give them 5 sandwiches each the equation becomes:

$$c = 5(a + 2)$$

With 20 tourists:

$$c = 5(a + 2) = 5(20 + 2) = 5 \times 22 = 110$$

The caterer must supply 110 sandwiches.

We can take this one stage further and have a *variable* number of sandwiches. If c is the total *number* of sandwiches, a is the *number* of tourists and d is the *number* of sandwiches for each person (note that all letters stand for *numbers*), then:

$$c = d(a + 2)$$

Brackets

To clear brackets, multiply every term inside the brackets by the term in front of the brackets:

$$a(b + c) = ab + ac$$

Observe the sign rules for multiplication (page 21).

$$-x(3 - 4y) = -3x + 4xy$$

In division, the line counts as a pair of brackets. To clear the line divide every term above the line by the term or terms below the line:

$$\frac{a + 2b + 5}{c} = \frac{a}{c} + \frac{2b}{c} + \frac{5}{c}$$

If there are 15 tourists and every person has 4 sandwiches:

$$c = d(a + 2) = 4(15 + 2) = 4 \times 17 = 68$$

The caterer must supply 68 sandwiches.

We can also work with fractions. A sheet of paper weighs 3.5 g and an envelope weighs 4.1 g. Find the weight (paper plus envelope) of a 3-page letter, and a 4½-page letter. The equation is:

$$w = 3.5p + 4.1$$

w is the number of grams the letter weighs, and p is the number of pages. If $p = 3$:

$$w = 3.5 \times 3 + 4.1 = 10.5 + 4.1 = 14.6$$

A 3-page letter weighs 14.6 g.

If $p = 4.5$:

$$w = 3.5 \times 4.5 + 4.1 = 15.75 + 4.1 = 19.85$$

A 4½-page letter weighs 19.85 g.

Test yourself 4.2

1 If $x = 5y$, find x when $y = 4$.
2 If $a = 2b + 6$, find a when $b = 13$.
3 If $f = 2.5g - 8.1$, find f when $g = 6.3$.
4 If $m = 4(n + 6)$, find m when $n = 5$.
5 If $x = 5(7 - y)$, find x when $y = 6$.
6 If $p = q(8 + r)$, find p when $q = 3$ and $r = 4$.
7 If $h = i(j - 7.25)$, find h when $i = 2.5$ and $j = 9.5$.
8 If $w = v(x + y)$, find w when $v = 2$, $x = 4$ and $y = 1.5$.
9 If $s = t(8 + u)$, find s when $t = 4$ and $u = 0$.
10 A shelf requires a bracket to support it at each end and two additional brackets for each metre of its length. Write an equation to show how to calculate n the number of brackets needed to support x shelves that are each y metres long. Find the number of brackets needed to support:
a 10 shelves 2 m long **b** 5 shelves 6 m long.

Division

A box of 300 nails is divided equally among a number of students. How many do they get each? This depends upon the number of students. If there are 10 students, they get 30 nails each; if there are 50 students they get only 6 nails. The equation is:

$$n = \frac{300}{m}$$

The number of nails, n, is 300 divided by the number of students, m. The equation shows how to find the number of nails, given the number of students. We could also have a variable number of nails, t:

$$n = \frac{t}{m}$$

For example, if there are 450 nails and 9 students:

$$n = \frac{450}{9} = 50$$

Each student gets 50 nails.

What happens if there are 12 students?

$$n = \frac{450}{12} = 37.5$$

Half a nail does not make sense. Each student has 37 nails and 6 nails are left over.

What happens if there are *no* students?

$$n = \frac{450}{0}$$

If there are no students, there are no people to share the nails. This illustrates a general rule that:

Division by zero is impossible

Try dividing a number by 0 on your calculator or computer; the display will show 'E' or some other error message.

Solving equations

Solving an equation is finding the value of one variable, when you are given the values of the other variables. This is just what you have been doing in the exercises above, but we have not used the word 'solve'. From now on we put the questions in a shorter form. Instead of asking:

If $x = y + 2$, find x when $y = 3$.

we will say:

Solve $x = y + 2$, when $y = 3$.

Test yourself 4.3

Solve these equations.

1 $a = \frac{8}{b}$, when $b = 2$.

2 $c = \frac{23}{d}$, when $d = 4$.

3 $m = \frac{n}{7}$, when $n = 84$.

4 $x = \frac{2y}{5}$, when $y = 35$.

5 $s = \frac{5t + 1}{6}$, when $t = 7$.

6 $p = \frac{4q - 3}{r}$, when $q = 6$ and $r = 7$.

7 $j = \frac{3k + 2}{2m - 2}$, when $k = 10$ and $m = 3$.

8 $x = \frac{4y}{3z - 11}$, when $y = 3$ and $z = 4$.

Adding letters

If you have a term such as $4x$, and add a number such as 7 to it, the operation is represented by:

$$4x + 7$$

This expression shows what we want to do, but does not actually do it. We cannot add x's and numbers because (like bolts and transistors, page 8) they are not the same kind of thing. It is only after we have been given a numerical value for x and have turned $4x$ into a *number* that we can do the adding:

$$\text{Given } x = 3$$

$$\Rightarrow \quad 4x + 7 = 4 \times 3 + 7 = 12 + 7 = 19$$

Some expressions may have more than one term with x, for example:

$$5x + 3x$$

Here we are adding x's to x's. This might be the number of sandwiches needed by x tourists, given 5 for lunch and 3 for tea. Both the x's represent the same number of tourists, so we can add the terms:

$$5x + 3x = 8x$$

We need 8 sandwiches for each tourist.

The number in front of the x is called the **coefficient**. To add two terms in x, just add the coefficients. This applies to any variable not just to x, but it must always be the *same* variable in both terms. An expression such as $5x + 3y$, for example, can not be added since x and y are different variables. We can add them only when we are told the values of x and y and can turn the whole expression into numbers.

Conversely, to subtract two terms in x (or any other variable) subtract the coefficients:

$$9x - 4x = 5x$$

If x tourists start the day with 9 sandwiches each and then eat 4, they each have 5 left. The total left is $5x$ sandwiches.

The same rules are applied to expressions with brackets. For example, given the expression:

$$4(x + 7) + 3(x - 3)$$

First *clear the brackets* (page 52):

$$\Rightarrow \quad 4x + 28 + 3x - 9$$

Next, *collect terms*, by which we mean add (or subtract) terms that are of the *same kind*:

$$\Rightarrow \quad 7x + 19$$

The expression now contains two terms that are of different kinds, one in x and one a pure number, so it cannot be simplified any further.

Here is an expression to simplify that has two variables in it:

$$2(a + 2b) + 3(a - 1) + 3b$$

Clear the brackets:

$$\Rightarrow \quad 2a + 4b + 3a - 3 + 3b$$

Note that in clearing the first bracket we multiply the coefficient of b by the number in front of the bracket. Collect terms:

$$\Rightarrow \quad 5a + 7b - 3$$

This is as far as we can go in simplifying the expression.

Adding and subtracting terms

The number, if any, in front of the letter or letters is the **coefficient**.

Examples

In $4x$, the coefficient is 4.
In $7bc$, the coefficient is 7.

To add terms (with the *same* variable), add the coefficients.

Examples

$4a + 5a = 9a$
$y + 3y + 7z = 4y + 7z$

To subtract terms (with the *same* variable), subtract coefficients.

Examples

$8f - 3f = 5f$
$12x - 4y - 10x = 2x - 4y$

Adding or subtracting terms with the same variable is known as **collecting terms**.

If a bracket has a negative sign in front of it, everything inside the bracket is reversed in sign. For example:

$-2(j - k + 4)$

Multiplying the terms in the bracket by –2:

$-2 \times j = -2j$ (– times + gives –, see box page 21)
$-2 \times -k = +2k$ (– times – gives +)
$-2 \times 4 = -8$ (– times + gives –)

The result of clearing the bracket is:

$\underline{-2j + 2k - 8}$

Here is another example to simplify:

$3p - 5(p + q - 12) + 2(p - 3q)$

Clear brackets (the first one has a negative sign in front):

$\Rightarrow \quad 3p - 5p - 5q + 60 + 2p - 6q$

Collect terms:

$\Rightarrow \quad 0p - 11q + 60$

$0p$ equals zero, so we have no term in p:

$\Rightarrow \quad \underline{-11q + 60}$

Now look at this example:

$2x(y + 2) + y(x + 4) - x$

Clear brackets:

$\Rightarrow \quad 2xy + 4x + xy + 4y - x$

Two of the terms are in xy, which is different from x and from y. Terms in xy can be added only to other terms in xy. Collecting terms:

$\Rightarrow \quad \underline{3xy + 3x + 4y}$

Test yourself 4.4

Simplify these expressions.

1 $2m + 3n + 4m$
2 $5p - 2q + 3p + 6q$
3 $7(x + 1) - 4x$
4 $5g + 2(5 + 2g)$
5 $3 + 4j - 3(j + 2)$
6 $4x + 2(x + 3y) - y$
7 $2(2d - 4) + d(1 + e)$
8 $5x - 4(x + 2)$
9 $4(y + 1) - 3(y - 2)$
10 $5(2x + 3y) + 3(4 - 5y)$
11 $2a(2 + 3b) + 5ab$
12 $x + y - 2(x - y)$

Order does not matter for . . .

- *Variables in terms*

Examples

ab is equivalent to ba
xy is equivalent to yx
But we prefer alphabetical order: ab, ac, xy

- *Terms in brackets*

Examples

$(2 + p)$ is equivalent to $(p + 2)$
$(-2j + 5)$ is equivalent to $(5 - 2j)$
But we prefer not to start with a negative symbol: $(5 - 2j)$

- *Terms in expressions*

Example

$4 + 6b - 7c$ is equivalent to $6b - 7c + 4$ and to $-7c + 6b + 4$
$4x^2 - 3x^3 + 2x$ is equivalent to $-3x^3 + 4x^2 + 2x$
But we prefer not to start with a negative symbol.

Solve these equations, simplifying the right-hand side first.

13 $x = 2y + 3(5y - 5)$, when $y = 3$

14 $v = 4(w + 2) - 2(3w - 3)$, when $w = 2$

15 $a = b(c + 1) + c(2b - 4)$, when $b = 3$ and $c = 2$

Negative variables

Although negative tourists and negative sandwiches are unimaginable, there are instances when a variable can have a negative value. For example, the temperature of a greenhouse is 20°C and rises by r°C each hour. The equation shows how to find t, its temperature after h hours:

$$t = 20 + rh$$

For example, if $r = 3$ and $h = 5$, $t = 20 + 3 \times 5 = 35$.

The temperature is 35°C.

On another occasion, the temperature is falling. We have defined r as a *rise* in temperature. To use the same equation for a *falling* temperature, we make r negative. If the temperature falls by 2°C each hour, we say that $r = -2$. If $h = 3$, then:

$$t = 20 + rh = 20 + (-2) \times 3 = 20 - 6 = 14$$

The temperature is 14°C.

Note that we do not change the equation. We just give the variable r a negative value.

When using negative values for variables, it is important to observe the sign rules.

Example

Solve $a = 17 - 3b$, when $b = 4$ and when $b = -4$

when $b = 4$,

$$a = 17 - 3(4) = 17 - 12 = 5$$

when $b = -4$,

$$a = 17 - 3(-4) = 17 - (-12) = 17 + 12 = 29$$

Test yourself 4.5

Solve these equations.

1 $y = 10 + 2x$, when $x = -3$
2 $A = 15 - 3b$, when $b = -4$
3 $f = 5g + 4$, when $g = -2$
4 $p = -2q - 7$, when $q = 4$
5 $x = 2(y + 2) - 4y$, when $y = -1$
6 $X = 4(2Y - 3) + Z(Y - 1)$, when $Y = -2$ and $Z = -3$

Capital letters

Capital letters can also be used in equations, but are more often used in practical formulae. Capital letters obey exactly the same rules of algebra as small letters.

Examples

$V = IR$ (Ohm's Law formula)
$T = t + 273$ (converts ° Celsius to kelvin)

When both capital and small versions of the same letter are used (as in second example), take care not to confuse them, especially when writing out formulae by hand.

Multiplying a variable by itself

To illustrate the point, we simplify this expression:

$$a(5 + a)$$

Following the usual rule, we would write:

$$5a + aa$$

The term 'aa' is taken to mean 'a multiplied by a'. Instead of writing it this way we write:

$$a^2$$

The figure 2 means that there are 2 a's multiplied together. This is spoken as 'a to the power of 2', or more briefly as 'a to the 2'. In Chapter 6, we shall see that this can also be spoken as 'a squared'.

We need not stop at two a's, for:

$$a \times a^2 = a \times a \times a = a^3$$

This is 'a to the power of three', or 'a cubed'.

These powers of a and other variables is a topic that we continue in Chapter 13.

The power of one

a^2 is $a \times a$.

a^1 is just a.

When a is to the power of one, we usually do not write the ^{1}.

Only terms in the same variable or combinations of variables may be added or subtracted (page 55). Similarly, only powers of the same kind can be added or subtracted.

We add terms in a^2 to terms in a^2, but not to terms in a or terms in a^3.

Have you noticed . . .

The square of a variable is always *positive*, whether the variable itself is positive or negative.

If $x = 4$, then $x^2 = x \times x = 4 \times 4 = 16$

If $x = -4$, then $x^2 = -4 \times -4 = 16$ (minus times minus give plus, page 21)

Examples

Simplify $x(x + 2) + 4(x - 2) + 3x^2$

Clear brackets:

$$\Rightarrow \quad x^2 + 2x + 4x - 8 + 3x^2$$

Collect terms, x's with x's, and x^2's with x^2's:

$$\Rightarrow \quad \underline{4x^2 + 6x - 8}$$

If $x = 3$, we can find the value of each term and *then* we are able to add these to find the value of the expression:

$$(4 \times 3 \times 3) + (6 \times 3) - 8 = 36 + 18 - 8 = 46$$

Simplify: $3p^2 - 5p + 9p^3 + 2p - 2p^3 + 5p^2$

Collect terms:

$$\Rightarrow \quad \underline{7p^3 + 8p^2 - 3p}$$

If $p = 2$, the value of the expression is:

$$(7 \times 2 \times 2 \times 2) + (8 \times 2 \times 2) - (3 \times 2) = 56 + 32 - 6 = 82$$

Test yourself 4.6

Simplify these expressions.

1 $2a + 3a^2 - a + 4 + 2a^2$

2 $3f - 4 + f^3 - 4f + 3 - f^2$

3 $2x^2 - 3xy + 5y^3 - 2xy + 6x^2$

4 $a(a - 2) + 4(a + 2)$

5 $a(3a + b) + b(4a - 2b)$

6 $j(4 + 2j) - 4(1 + j)$

7 $r^2(r + 3) - 2(r^2 + 4r) + 9(r + 2) - 17$

8 $2m^2(m + 3) - m(m + 2m^2) - 5m^2$

Solve these equations.

9 $f = g^2 + 3$, when $g = 4$.

10 $i = 4j - j^2$, when $j = 2$.

11 $x = y(y + 1)$, when $y = 5$.

12 $r = S(S^2 - S + 2)$, when $S = 3$.

13 $V = 3w + 2w^2 - 1$, when $w = 1.2$.

14 $b = a^2 + 2a + 4$, when $a = -3$.

15 $g = h(4 + i) - 2i(3 - h)$, when $h = 2$, and $i = 3$.

16 $D = d^2 - 3(d + e^3)$, when $d = 6$ and $e = 4$.

17 To assess if you have covered this chapter fully, answer the questions in *Try this first*, page 49.

Explore this

A microcomputer can help you to practice solving equations. It needs to have BASIC and is used in Direct Mode (that is, running without a program, so that you can type in commands or statements directly). Most simple computers go into Direct Mode automatically when switched on. With others, you may have to call up BASIC first.

Make up an equation of your own, for example:

$$a = 3(2b + 3) - 5b$$

Decide on a value for the variable on the right (b). Solve the equation on paper or in your head, to find the value of a.

Now use the computer to check your result. Type in the equation, exactly as you have written it. BUT you have to insert all the multiplication signs that we usually take for granted. In BASIC, the multiplication sign is *. So the equation above is:

```
a = 3*(2*b + 3) - 5*b
```

Press RETURN or ENTER. Next type in the value of the variable *b*. For example, if *b* equals 5, then type:

$b = 5$

Press RETURN or ENTER. To solve the equation with this value of *b*, type:

PRINT *a*

Press RETURN and ENTER. The computer replies by displaying the value of *a*. On some computers, instead of typing the word PRINT, you can simply type:

?*a*

Then press RETURN or ENTER. Repeat this with different values of *b* and each time you will get the corresponding value of *a*.

You can also use this method with two or even more different variables on the right.

This is a quick way of testing yourself with equations. It is also a useful technique if you have the same equation to solve several times with lots of different values of *b*, or other variables.

5 Lines, angles and shapes

Most technologists spend their time *designing* and *making* things – cars, bridges, dams, aircraft, washbasins, lipsticks, cakes, guitars and tens of thousands of other artifacts. Planning and making depends a lot on being able to make accurate drawings of what is to be made, its parts, and how to fit them together. In this chapter we deal with the basic elements of plans and drawings – lines, angles and shapes.

You need to know about the four rules of arithmetic (Chapter 1 and 2), how to work with fractions (Chapter 3), and a little algebra (Chapter 4).

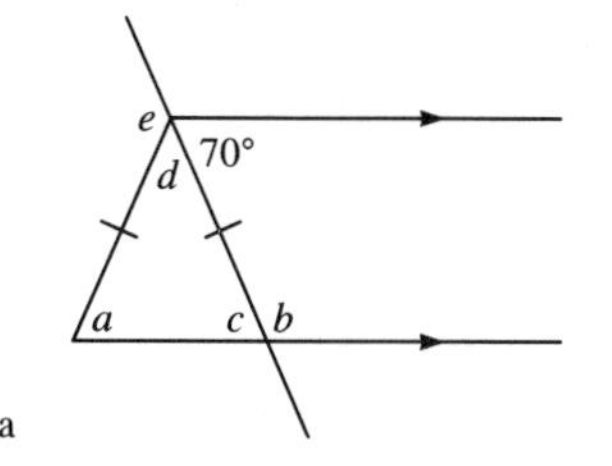

a

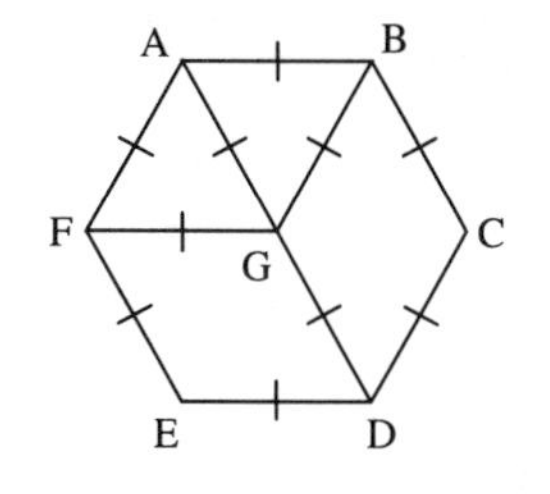

b

Figure 5.1

Try these first

Your success with this short test will show you which parts of this chapter you already know.

1 Calculate the sizes of the angles a to e in Figure 5.1a. You cannot measure them with a protractor because the drawing is only a sketch.

2 Name these shapes in Figure 5.1b.

a AGF **b** ABGF **c** ABCD **d** ABCDEF

Lines and angles

Most diagrams, plans, maps and other technical drawings are made with lines. In many of them, as in all of the diagrams in this chapter, the lines are **straight**. We identify lines by placing capital letters at their ends. In Figure 5.2a, two lines AB and BC meet at B. Where they meet there is an **angle** between them. The size of an angle is measured in **degrees**. If two lines meet end-to-end, so that they appear to be one continuous straight line, the angle between them is 180 degrees (Figure 5.2b). We use the symbol ° for degrees, so the angle is written as 180°.

An angle of half this size is called a **right angle** (Figure 5.2c). A right angle is 90°. In diagrams, a right angle is usually marked in a special way (Figure 5.2d).

An angle of 180° is equivalent to half a turn (turning to face in the opposite direction). A complete turn is twice this, so a complete turn is 360° (Figure 5.2e).

There are names for the in-between angles:

Acute – between 0° and 90° (Figure 5.2a)
Obtuse – between 90° and 180° (Figure 5.2f)
Reflex – between 180° and 360° (Figure 5.2g)

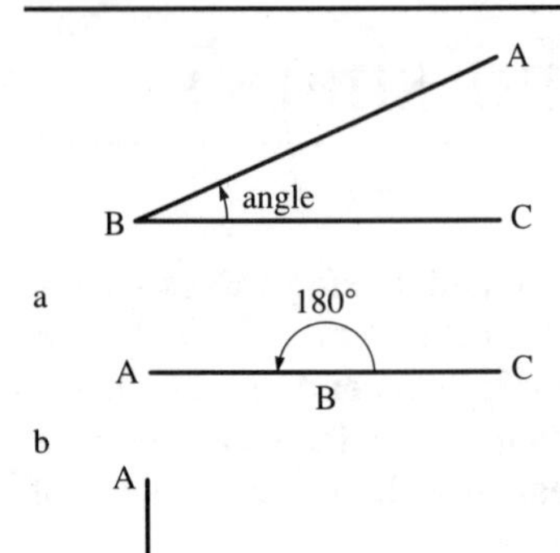

A
90°
B
C
c

A
B
C
d

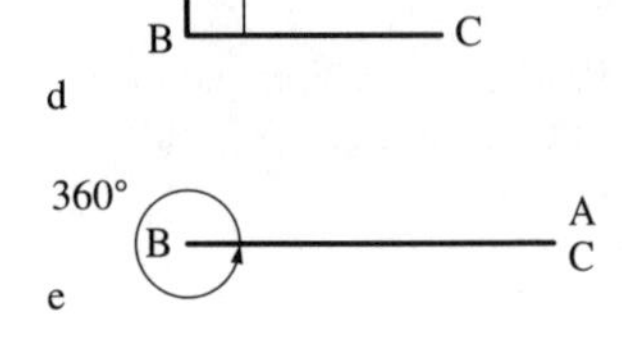

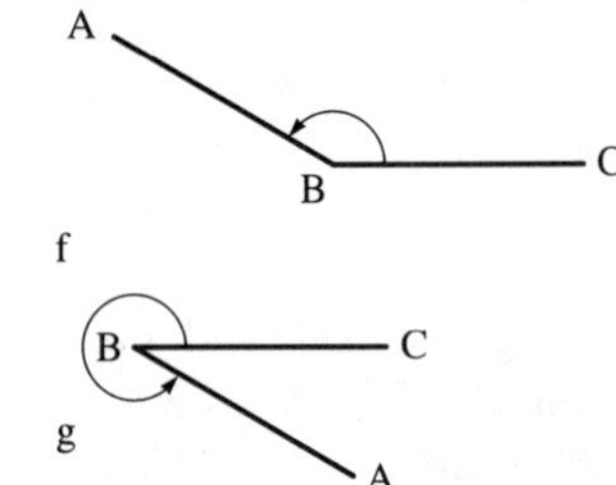

Figure 5.2

In Figure 5.3, a line AB meets another line CD at point B, which is on CD. There are two angles at B. One way of identifying the angles is by writing *small* letters in them. In the figure, the angles are marked x and y. Angles can also be identified by writing the letters of the lines which enclose the angle. In the figure, $\hat{x}$ can also be identified as ABC, and $\hat{y}$ as ABD. Note how we put a ^ over the letter to show that the letter refers to an angle.

Since CD is a straight line, the total angle at B is 180°:

$$\hat{x} + \hat{y} = 180°$$

We say that $\hat{x}$ and $\hat{y}$ are **supplementary** angles.

Similarly, in Figure 5.4, where CBD is a right angle:

$$\hat{x} + \hat{y} = 90°$$

We say that $\hat{x}$ and $\hat{y}$ are **complementary** angles.

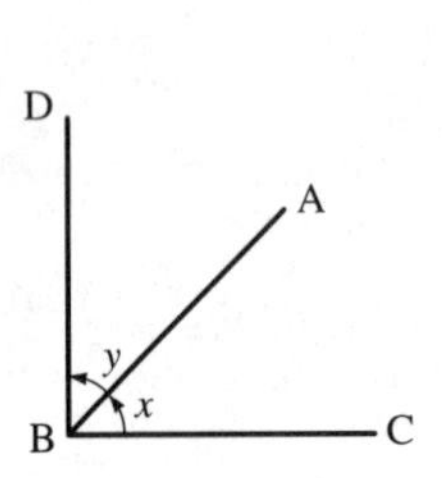

Figure 5.4

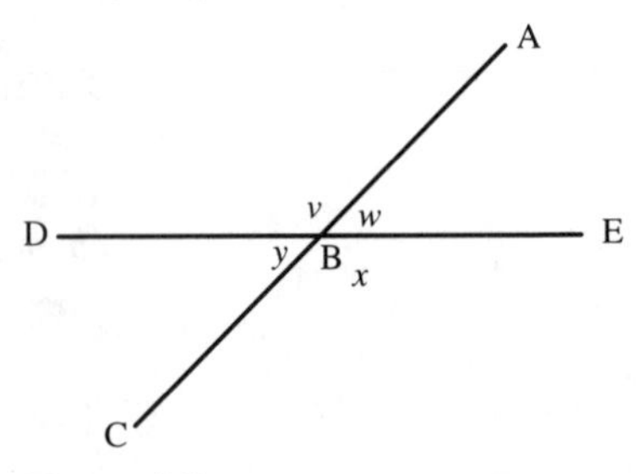

Figure 5.5

In Figure 5.5, where two lines AC and DE cross at B, the four angles add up to 360°.

$$\hat{v} + \hat{w} = 180 \quad \text{(supplementary angles)}$$
$$\text{and so} \quad \hat{v} = 180 - \hat{w}$$
$$\text{similarly} \quad \hat{x} + \hat{w} = 180 \quad \text{(supplementary angles)}$$
$$\text{and so} \quad \hat{x} = 180 - \hat{w}$$

If $\hat{v}$ equals $180 - \hat{w}$, and also $\hat{x} = 180 - \hat{w}$, then:

$$\hat{v} = \hat{x}$$

Lines and angles

Angles are measured in degrees (°)

An **acute** angle is less than 90°.
A **right angle** is 90°.
An **obtuse** angle is between 90° and 180°.
A **half a turn** is 180°
A **reflex** angle is between 180° and 360°.
A **whole turn** is 360°.

An angle and its complement add up to 90°. (Complementary angles)
An angle and its supplement add up to 180°. (Supplementary angles)
Where two lines cross, opposite angles are equal.

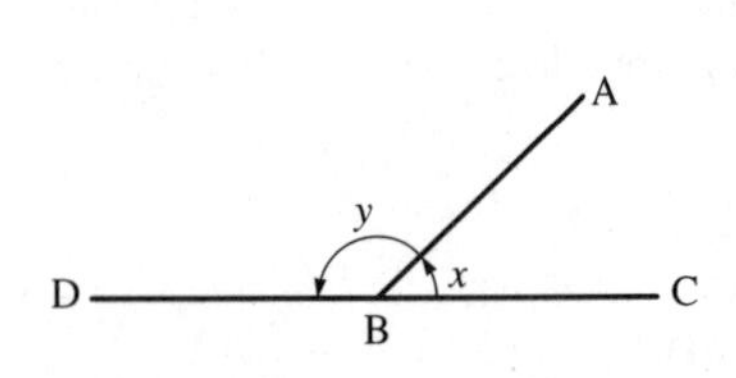

Figure 5.3

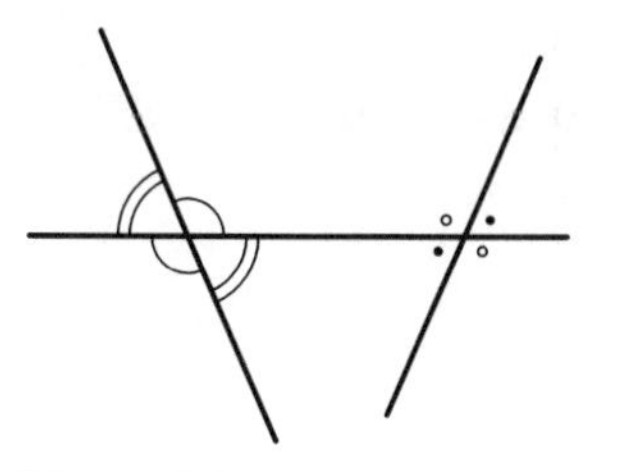

Figure 5.6

These are **opposite** angles. The other pair of opposite angles is:

$$\hat{w} = \hat{y}$$

In diagrams we show that angles are equal by marking them similarly, for example with one, two or three arc lines or with dots or small circles (Figure 5.6).

Parallel lines

Parallel lines run in the same direction but keep a fixed distance apart. They never meet. In a drawing we indicate that lines are parallel by drawing arrowheads on them (Figure 5.7). If there is more than one set of parallel lines in a diagram, we mark the second set with double arrows.

If a line is drawn across parallel lines, we can immediately mark in a pair of angles that are equal (Figure 5.8). these are equal because the parallel lines run in the same direction and they are both set at the same angle to this crossing line. The angles are called **corresponding angles**.

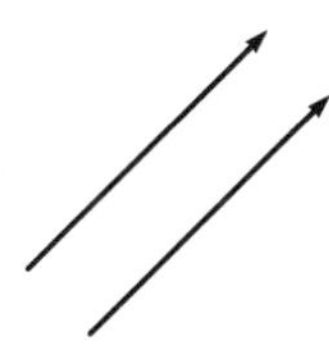

Figure 5.7

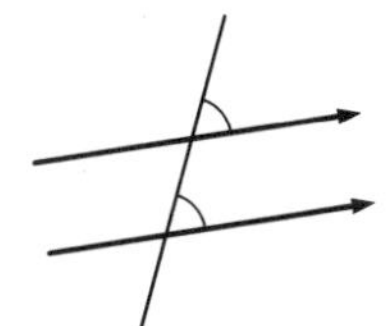

Figure 5.8

We have shown in Figure 5.6 that opposite angles are equal, so there are actually *four* equal angles where a line crosses parallel lines. Figure 5.9 shows all four angles. Of these, the two marked with dots are called **alternate angles**. As well as these, there is another set of four equal angles, marked in Figure 5.10. Again, the two marked with dots are the alternate angles.

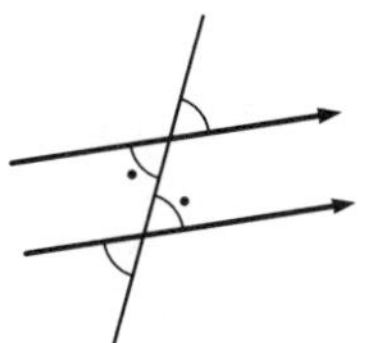

Figure 5.9

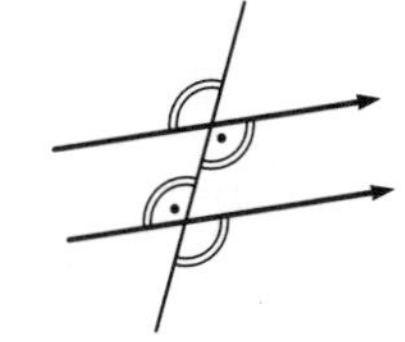

Figure 5.10

Test yourself 5.1

1 Name the angles $\hat{a}$ to $\hat{f}$ in Figure 5.11.

2 Measure the angles $\hat{a}$ to $\hat{e}$ in Figure 5.12, using a protractor.

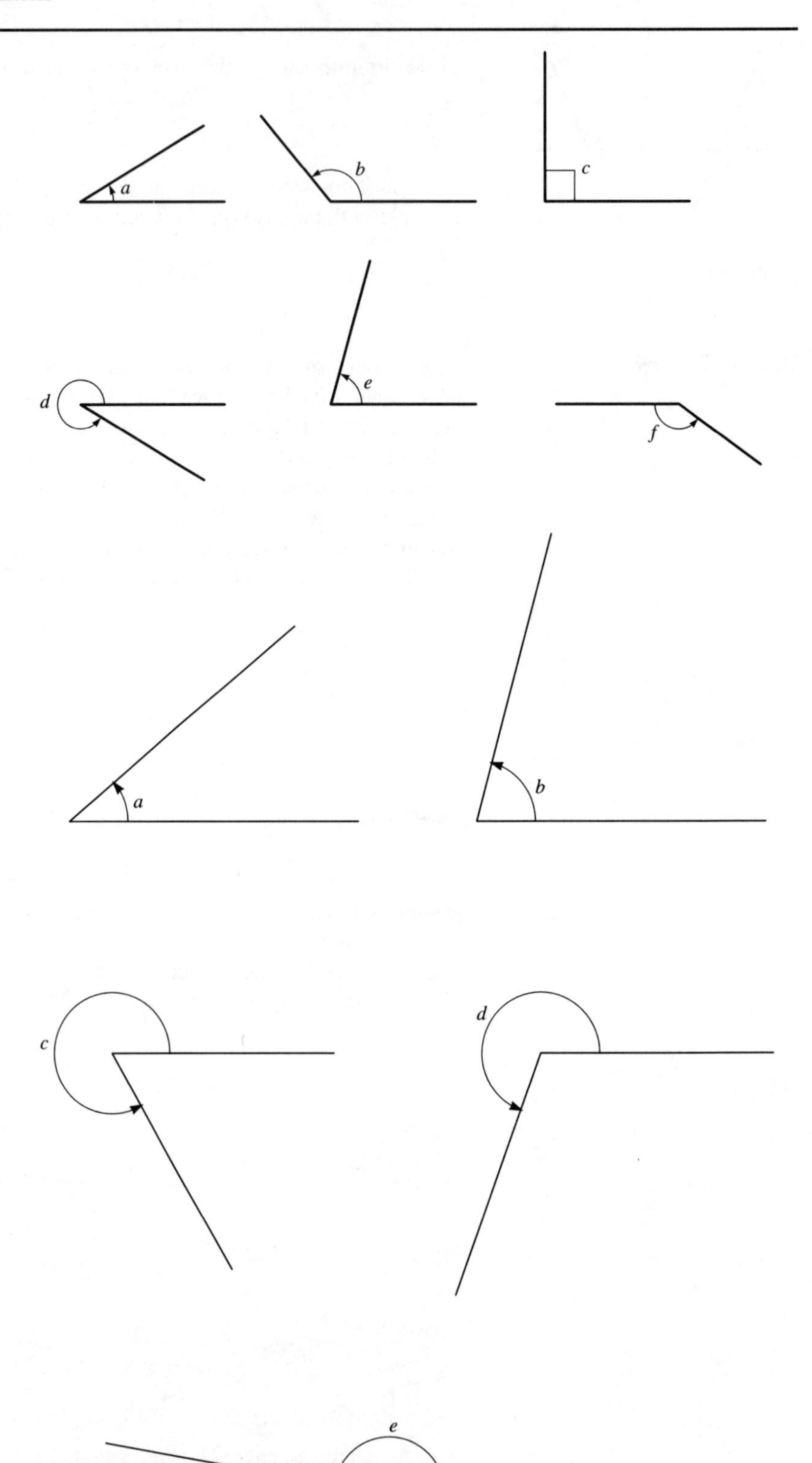

Figure 5.11

Figure 5.12

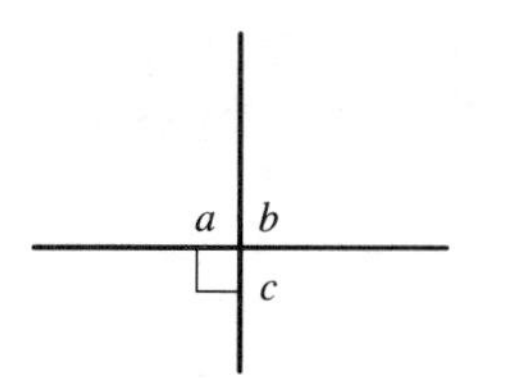

Figure 5.13

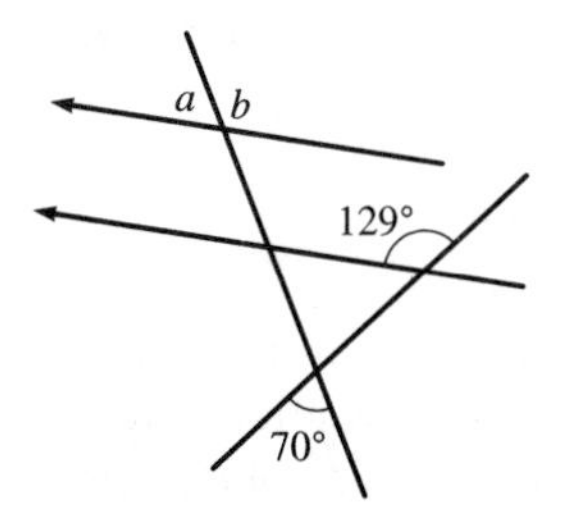

Figure 5.14

3 In Figure 5.13, one of the angles is marked as a right angle. Which of the angles $\hat{a}$ to $\hat{c}$ are also right angles?

4 Copy Figure 5.14 (you need not draw the angles accurately). What do we call the pair of angles $\hat{a}$ and $\hat{b}$? Mark with a all other angles that are equal to $\hat{a}$. Mark with b all angles that are equal to $\hat{b}$. The sizes of two of the angles are marked. Write in the sizes of as many angles as you can.

5 Five roads meet at a roundabout. The angles between the roads are equal. What is the size of this angle?

6 A rotary switch has 6 positions, with equal angles between them. What is the angle between adjacent positions?

7 What angle is complementary to: **a** 73° **b** 45° **c** 1°

8 What angle is supplementary to: **a** 50° **b** 160° **c** 90°

Triangles

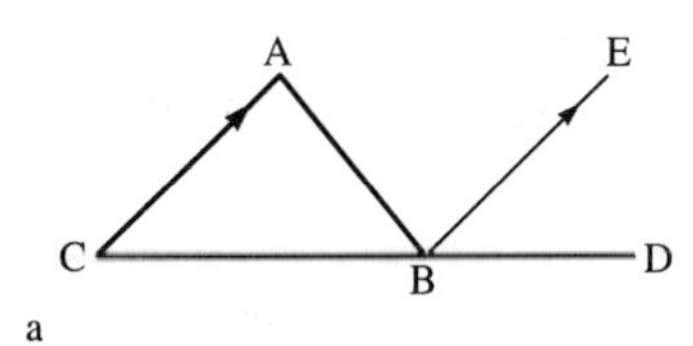

a

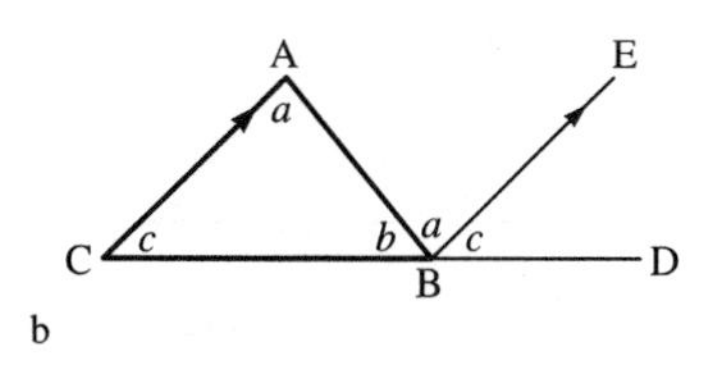

b

Figure 5.15

Triangles are identified by marking each corner (or **vertex**) with a letter. In Figure 5.15a we identify the triangle as △ABC. Now to find out more about this triangle. We continue side CB out as far as D (it does not matter exactly how far it is to D). Next we draw a line BE through B, parallel to CA. Using what we know about angles, we can mark in various angles that are equal. This has been done in Figure 5.15b:

- Angles $\hat{a}$ are equal because they are **alternate** angles. (*Remember*: CA is parallel to BE.)
- Angles $\hat{c}$ are equal because they are **corresponding** angles.

Angle $\hat{b}$ is not equal to any other angle *but* note that: $\hat{a} + \hat{b} + \hat{c}$ together fit into angle CBD, which is 180°.

Looking at the triangle, we see that its three **interior** (= inside) angles are $\hat{a}$, $\hat{b}$ and $\hat{c}$. Because $\hat{a} + \hat{b} + \hat{c} = 180°$, we can state the rule that:

The sum of the interior angles of a triangle is 180°

The exact shape of the triangle makes no difference. This result applies to *all* triangles.

The diagram proves another fact. The angle ABD is outside the triangle, and it is called an **exterior angle**. This angle consists of $\hat{a}$ and $\hat{c}$, which are equal to the two **interior** angles opposite to vertex B. This leads to another rule:

The exterior angle at any vertex of a triangle equals the sum of the two opposite interior angles

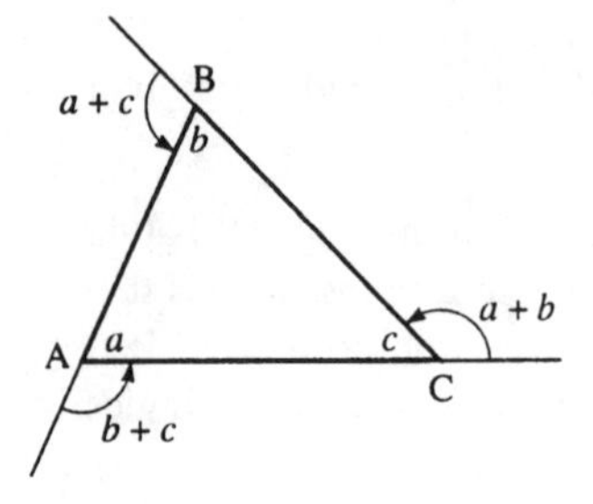

Figure 5.16

Figure 5.16 shows all three exterior angles. According to the rule just given, the exterior angle at vertex A equals $\hat{b} + \hat{c}$. Similarly the exterior angles at B and at C equal $\hat{a} + \hat{c}$ and $\hat{a} + \hat{b}$.
Totalling all three exterior angles we have:

$$\text{Sum of exterior angles} = \hat{b} + \hat{c} + \hat{a} + \hat{c} + \hat{a} + \hat{b}$$

This total includes every angle *twice* so the sum of the exterior angles equals *twice* the sum of the interior angles $\hat{a}$, $\hat{b}$ and $\hat{c}$. The sum of the interior angles is 180°, with the result that:

The sum of the exterior angles of a triangle is 360°

Think of this in another way. Walking along side AC toward C, you turn sharply left when you get to C. Continuing along side CB, you turn sharply left at B. Finally, you turn sharply left at A to face in your original direction. Walking once round the triangle you have made a complete turn of 360°. This also shows that the sum of the exterior angles is 360°.

Kinds of triangle

Triangles are classified according to their shape (see boxes). Note how we have indicated which sides are equal in length by drawing one or two (sometimes more) short strokes across them.

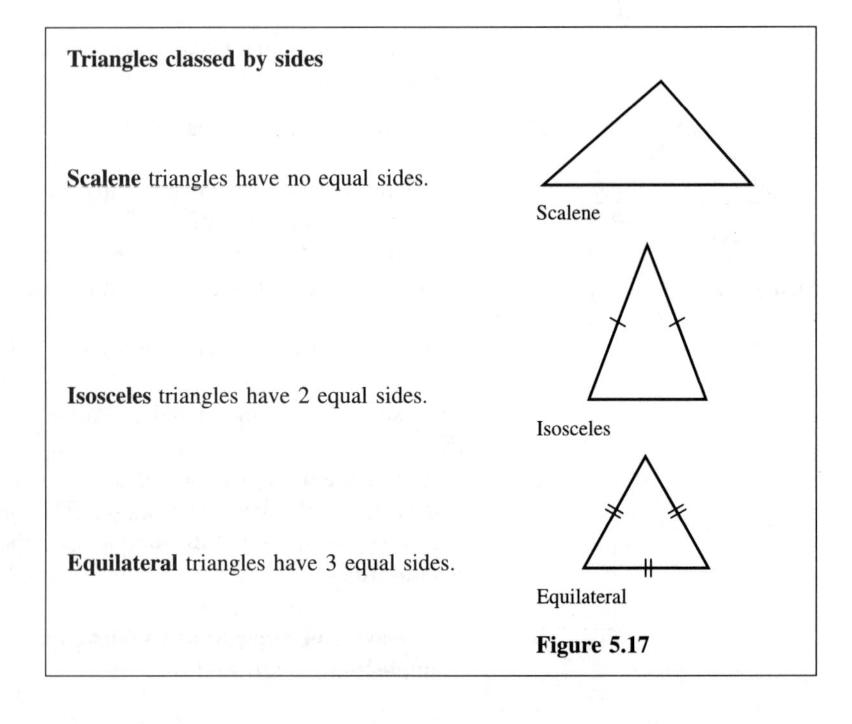

Figure 5.17

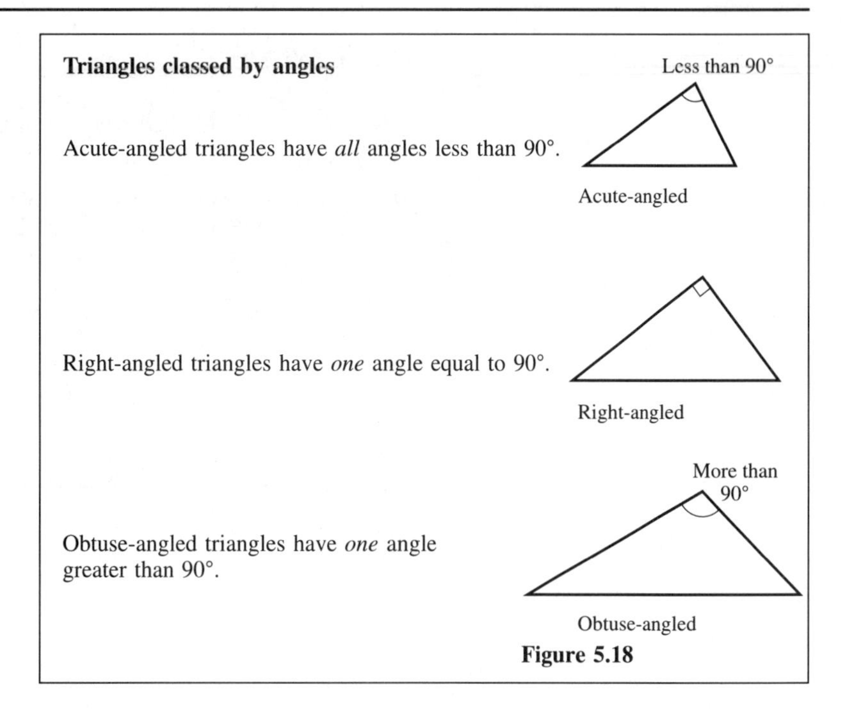

Figure 5.18

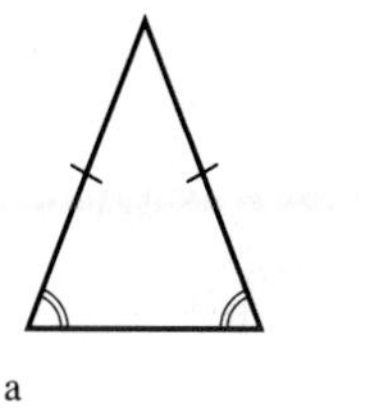
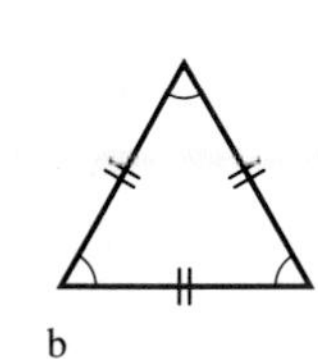

Figure 5.19

An isosceles triangle not only has two equal sides but also two equal angles (Figure 5.19a). An equilateral (equal-sided) triangle has all three angles equal. Dividing their total by 3 we find that:

$$\text{each angle} = \frac{180}{3} = 60°$$

All equilateral triangles have exactly the same shape (Figure 5.19b).

Describing triangles

The vertices of a triangle ABC (shortened to $\triangle$ABC) are A, B, and C. These capital letters, with a ^ sign above them, are used to identify the angle at each vertex: $\hat{A}$, $\hat{B}$, $\hat{C}$.

If there is more than one angle at a vertex, we list the letters of the lines which enclose the angle. In Figure 5.20, the three angles at vertex B are $A\hat{B}D$, $D\hat{B}E$ and $E\hat{B}C$. The vertex is the middle letter, with the ^ above it.

Earlier in this chapter we have identified angles by a *small* letter, such as *a*, *b* or *x*, as in Figure 5.15. This method is more often used when we are specially interested in the *size* of the angle. It may be used to pick out two or more angles that are the same size. For example, the two angles marked *a*, and the two marked *c* in Figure 5.15.

Two more terms used with triangles, **base** and **height**, are explained in Figure 5.20.

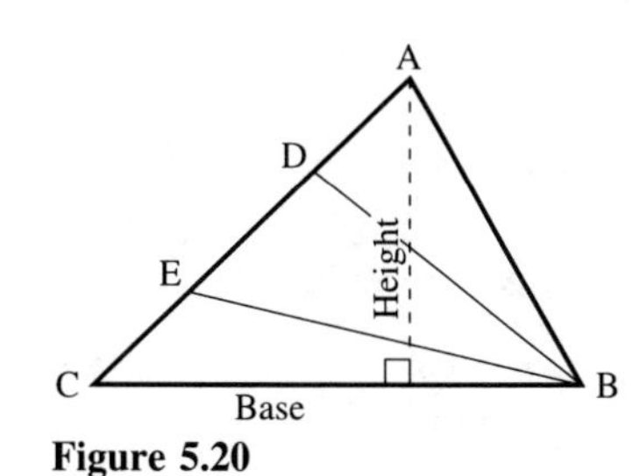

Figure 5.20

Test yourself 5.2

1 How big are the marked angles in Figure 5.21? Note that these are only sketches. You cannot find their size by using a protractor.

2 Figure 5.22 is a street map. The distances BF, FG and BG are equal. The angle BCF is a right angle. List the angles you would turn through in driving from point A to point G, along the route ABCDEFG.

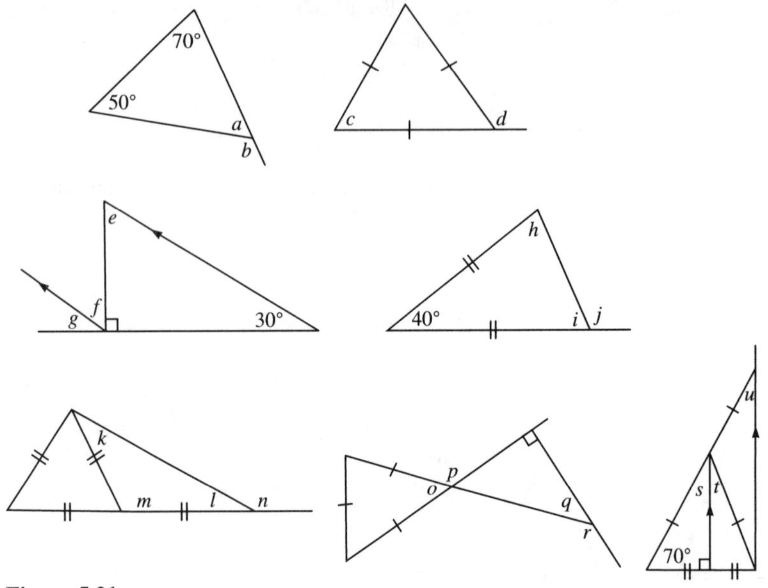

Figure 5.21

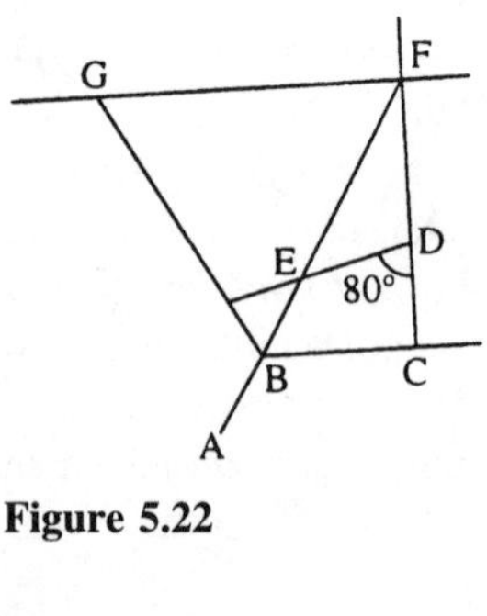

Figure 5.22

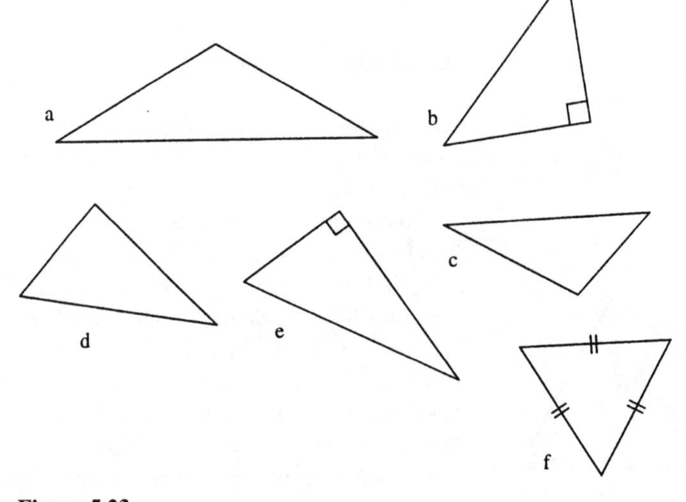

Figure 5.23

3 In Figure 5.23, identify the triangles which belong to these classes:

a scalene **b** isosceles **c** equilateral

d right-angled **e** acute-angled **f** obtuse angled

There may be more than one triangle of some of the kinds. Also some of the triangles could belong to two *or more* classes.

Four-sided shapes

A figure with four sides is called a **quadrilateral**. In Figure 5.24 a point is placed inside a quadrilateral and lines join it to all four vertices. This divides the quadrilateral into four triangles. The angles of each triangle add up to 180°, so the sum of all the angles in the drawing is $4 \times 180 = 720°$. But the four angles around the point (a, b, c, and d) add up to one complete turn, which is 360°. Therefore the angles at the vertices of the quadrilateral must total $720 - 360 = 360°$. This applies to all quadrilaterals, even to the oddly-shaped one in Figure 5.22. The rule is:

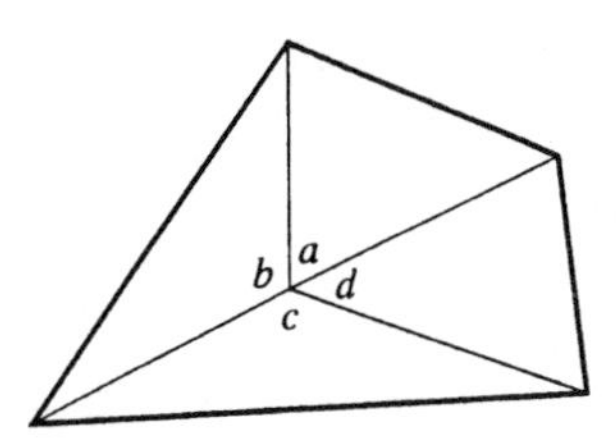

Figure 5.24

The sum of the interior angles of a quadrilateral is 360°

If you walk round a quadrilateral, turning left at each corner, you make one complete turn. For this reason, what we said about the exterior angles of a triangle (Figure 5.16) also applies to the exterior angles of a quadrilateral:

The sum of the exterior angles of a quadrilateral is 360°

But this rule does *not* apply to quadrilaterals like that in Figure 5.25. This is a **convex** quadrilateral because one of its angles is a reflex angle (page 64), and you turn *right* at that vertex.

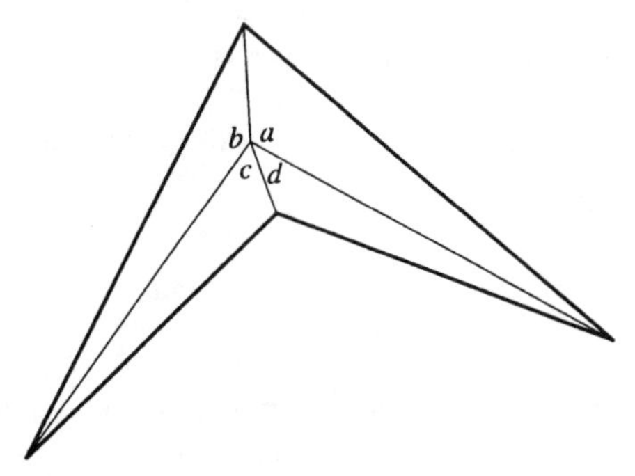

Figure 5.25

Some special quadrilaterals and their properties are illustrated in Figure 5.26. A **parallelogram** (a) is formed between two pairs of parallel lines. The diagram shows that its opposite sides are equal in length and its opposite angles are equal. Contrast this with a **rectangle** (b), which is a parallelogram with all angles equal (to 90°). Contrast it with a **square** (c), which is a

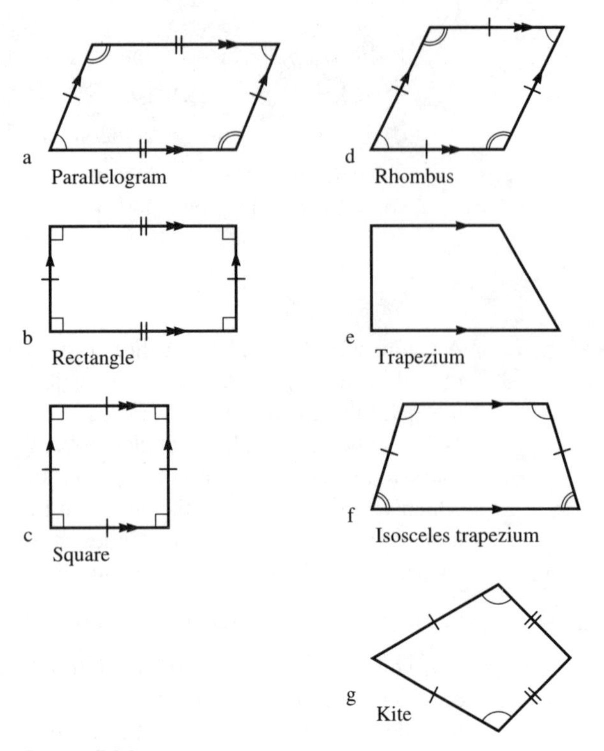

Figure 5.26

parallelogram with all angles *and* all sides equal. A **rhombus** (d) is a parallelogram with all sides equal (like a square) but is not right-angled.

A **trapezium** (e) has only one pair of opposite sides parallel, and these differ in length. Often the non-parallel sides differ in length too. If they are equal, we have an **isosceles trapezium** (f), which is like an isosceles triangle with the apex cut off.

a **kite** (g) has no parallel sides but there are equal sides and angles as shown in the diagram.

Polygons

A polygon is *any* closed (joined-up) figure with three or more straight sides. Triangles and quadrilaterals are polygons with three and four sides, respectively. But the sides must not cross each other. Figure 5.27 is not a polygon, although we could take it to be *two* polygons, a triangle and a quadrilateral, touching together at their apices. **Hexagons** (6-sided) and **octagons** (8-sided) are two types of polygon that we often see. A polygon is described as *regular* if all its *sides and angles* are equal. Figure 5.28 shows a regular hexagon and a regular octagon. Other examples of regular polygons are an equilateral triangle (Figure 5.19b) and a square (Figure 5.26c). Figure 5.28d shows a convex octagon, with one reflex angle. Figure 5.25 is another example of a reflex polygon.

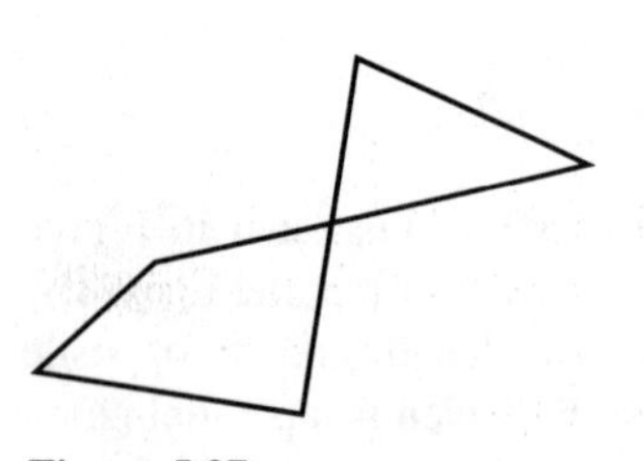

Figure 5.27

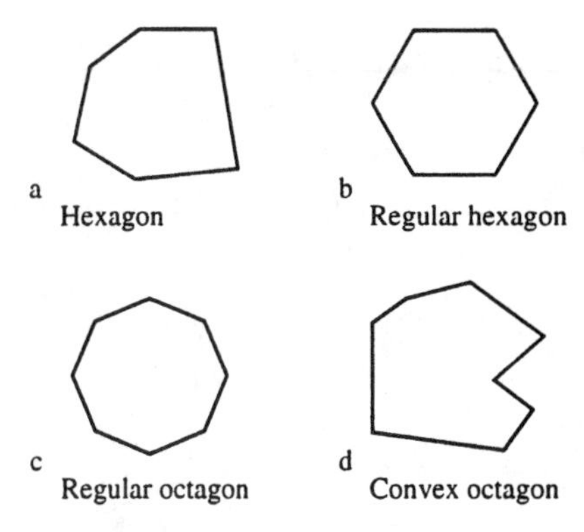

Figure 5.28

Explore this

Fill out a table like this for polygons with up to 10 sides:

Number of sides	*Sum of interior angles*	*Size of one interior angle**
3	180	60
4	360	90
5		
6		
7		
8		
9		
10		

*when the polygon is regular (examples: Figures 5.19b, 5.26c, 5.28b and c). Figure 5.24 is a clue to calculating the angles.

Work out a rule for calculating these angles for a polygon with any number of sides.

Test yourself 5.3

1 How big are the angles $\hat{a}$ to $\hat{m}$ in Figure 5.29? Note that the figures are only sketches and the angles are not drawn precisely. You cannot find their size by using a protractor.

2 What is the sum of the exterior angles of a hexagon?

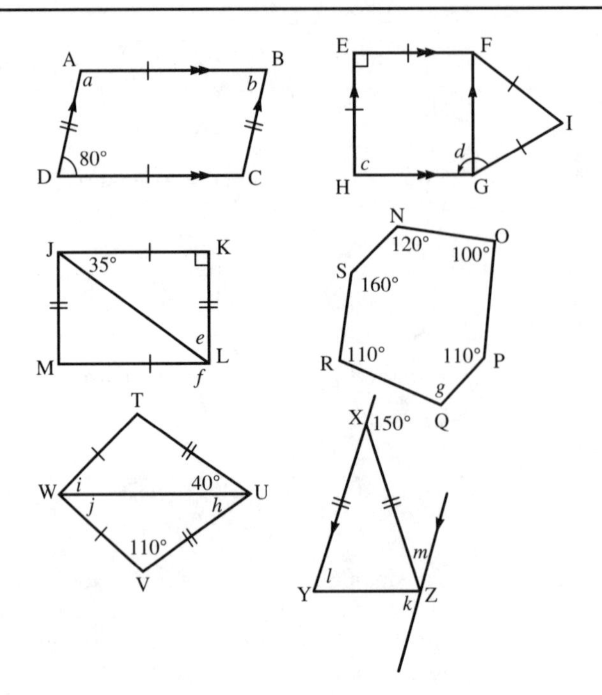

Figure 5.29

3 What is the size of an exterior angle of a regular octagon?

4 Name the polygons in Figure 5.29.

a ABCD	**b** EFGH	**c** GFI
d JKLM	**e** JKL	**f** NOPQRS
g TUVW	**h** TUW	**i** XYZ

5 To assess if you have covered this chapter fully, answer the questions in *Try these first*, page 63.

6 Areas and volumes

The areas of shapes and surfaces often need to be known, perhaps for deciding how many sheets of plastic board are needed for making a model, or how much paint is needed for the roof of a garage. For similar reasons, we also need to know how to work out volumes. This chapter shows how to calculate the areas of all kinds of shapes, and the volumes of simply shaped objects.

You need to know about multiplication (Chapter 2) and shapes (Chapter 5).

Try these first

Your success with this short test will tell you how much of this chapter you already know.

1 A car park is 250 m long and 45 m wide. What is the area of the park?

2 Find the area of the polygon in Figure 6.1a.

3 An instrument case measures 110 mm long, 60 mm wide and 45 mm high. What is its volume?

4 A swimming pool is 25 m wide; its section is shown in Figure 6.1b. What volume of water does it hold when full? How deep is the water at the deepest part when the pool contains half its maximum volume?

5 Convert 612750 mm^2 to m^2.

6 Convert 3.72 m^3 to mm^3.

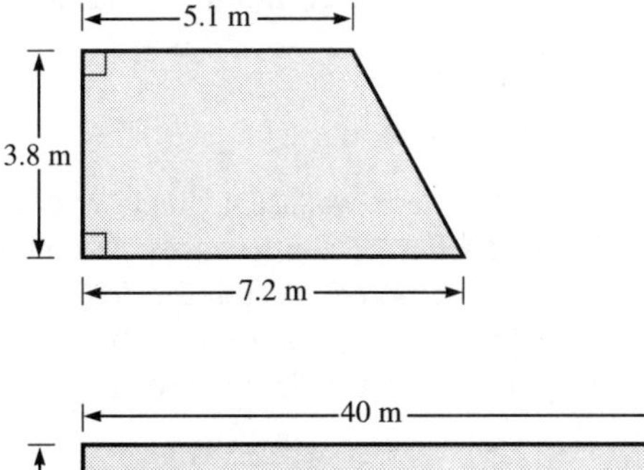

a

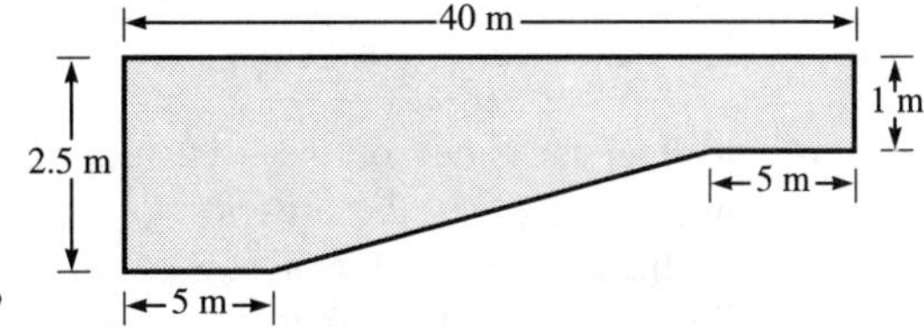

b

Figure 6.1

Paving squares

A patio is surfaced with square paving slabs (Figure 6.2). Each slab is 1 m square and there are 3 rows of 4 slabs. By counting the squares, we can see that there are 12 slabs. The area of the patio is 12 square metres. We shorten the words 'square metres' to 'm^2'.

Instead of actually counting the slabs, a quicker way is to multiply the number of rows by the number of slabs in a row:

$3 \times 4 = 12$

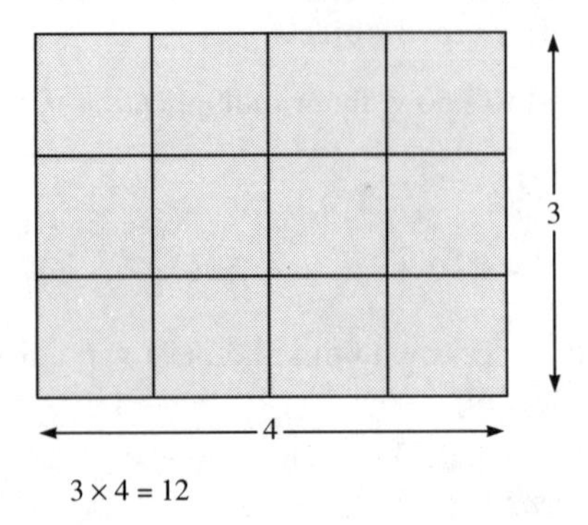

Figure 6.2

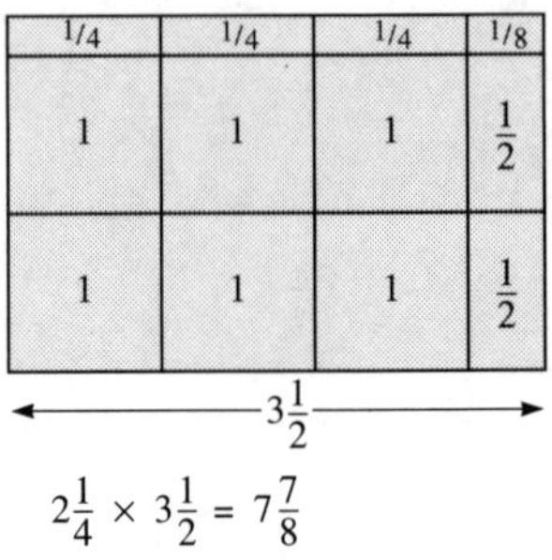

$$2\frac{1}{4} \times 3\frac{1}{2} = 7\frac{7}{8}$$

Figure 6.3

The same applies if the row contains fractions of slabs and one of the rows is made of slabs that are a fraction of a metre wide. In Figure 6.3 there are $3\frac{1}{2}$ slabs in each row. There are 2 rows of full-width slabs, and the top row is only $\frac{1}{4}$ of a slab wide. Counting up the whole slabs and fractions of slabs gives a total of $7\frac{7}{8}$ slabs, an area of $7\frac{7}{8}\,m^2$. This is the same area as obtained by multiplying:

$2\frac{1}{4} \times 3\frac{1}{2} = 7\frac{7}{8}$

Summing up, we state this rule:

The area of a rectangle is found by multiplying length by width

In symbols:

area = $\ell \times w$

Note that here we use a 'handwriting' letter 'el' rather than a typed one because letter 'l' can so easily be confused with figure '1'.

It is not necessary for the length and width to be expressed as mixed fractions; decimal fractions can be used too. Given a rectangle which is 7.57 m long and 3.81 m wide, the area is:

$7.57 \times 3.81 = 28.8417\,m^2$

When the sides of the rectangle are measured in metres, the result of multiplying gives the area in square metres. If the sides are measured in millimetres, the result is in square millimetres (mm^2). If they are measured in kilometres, the result is in square kilometres (km^2). What you *cannot* do is

Squares

A square with sides 2 cm long has an area of $2 \times 2 = 4\,\text{cm}^2$.

4 is 2 *multiplied by itself*
or 2 *squared*

The multiplication tables tell you the squares of numbers from 2 to 12. Here they are:

Number	*Number squared*
2	4
3	9
4	16
5	25
6	36
7	49
8	64
9	81
10	100
11	121
12	144

Learn to recognize these squared numbers so that, whenever you see 49 (for example), you *immediately* remember that it is 7 squared.

multiply lengths and widths that are in different units. For example, a piece of roadside verge is 2.6 km long and 40 cm wide: what is its area? Before multiplying length by width, both must be expressed in the same units. We could choose to work in metres. 2.6 km = 2600 m. 40 cm = 0.4 m. Now multiply:

$$2600 \times 0.4 = 1040$$

The area of the verge is $1040\,\text{m}^2$.

Area of a triangle

For the rectangle ABCD of Figure 6.4, we use the rule we have just stated:

area = length × width

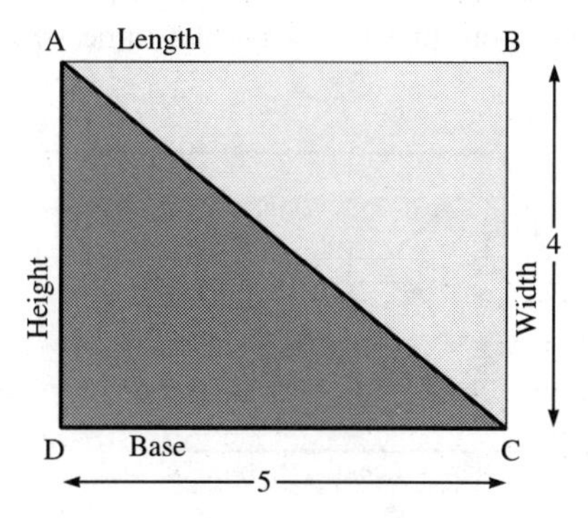

Figure 6.4

Line AC divides the rectangle into two identical parts, so that the area of the triangle ACD is half that of the rectangle. The *base* of the triangle equals the *length* of the rectangle. The *height* of the triangle equals the *width* of the rectangle.

Area of triangle = half area of rectangle = $\frac{1}{2} \times$ length $\times$ width
= $\frac{1}{2} \times$ base $\times$ height

This leads to another rule:

The area of a triangle equals half of the base multiplied by the height

In Figure 6.4, the area of the rectangle is:

length $\times$ width = $5 \times 4 = 20$

The area of the triangle ACD is:

$\frac{1}{2} \times$ base $\times$ height = $\frac{1}{2} \times 5 \times 4 = 10$

area ACD = 10

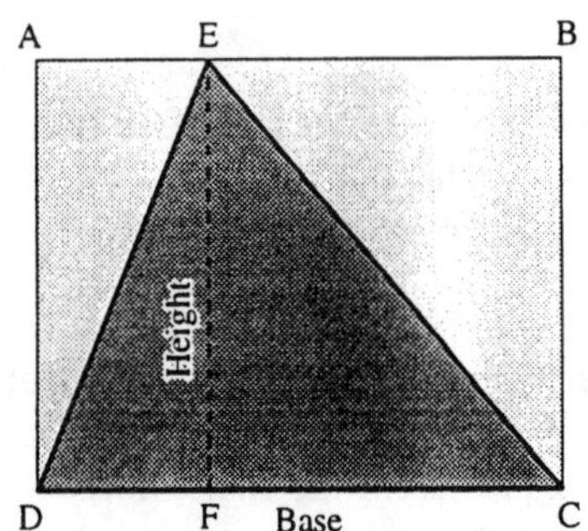

Figure 6.5

This result applies to the right-angled triangle of Figure 6.4, but does it apply to other kinds of triangle? In Figure 6.5, the apex of the triangle is at E, instead of at A. ECD is a scalene triangle. Drop a line vertically down from the apex E to meet the base at F. This divides the triangle into two parts, both of which are right-angled triangles.

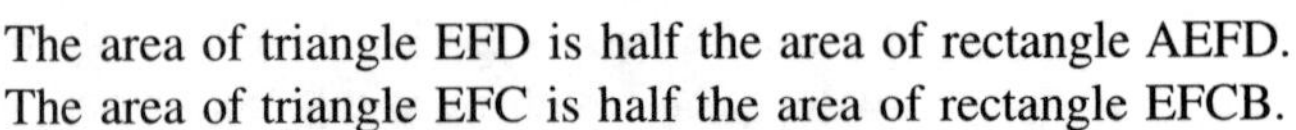

The area of triangle EFD is half the area of rectangle AEFD.
The area of triangle EFC is half the area of rectangle EFCB.
The area of the whole triangle ECD is half the area of the whole rectangle ABCD so, as above, the area of the triangle ECD is

$\frac{1}{2} \times$ base $\times$ height.

The same reasoning can be applied to other types of triangle.

Other polygons

The rules for the areas of rectangles and triangles can be applied to many other polygons. Given a polygon, we are usually able to divide it into rectangles and triangles. We calculate the area of each part then sum the areas to find the area of the whole. For example, the parallelogram in Figure 6.6 is divided into two triangles, both of which have the same size, shape, and area.

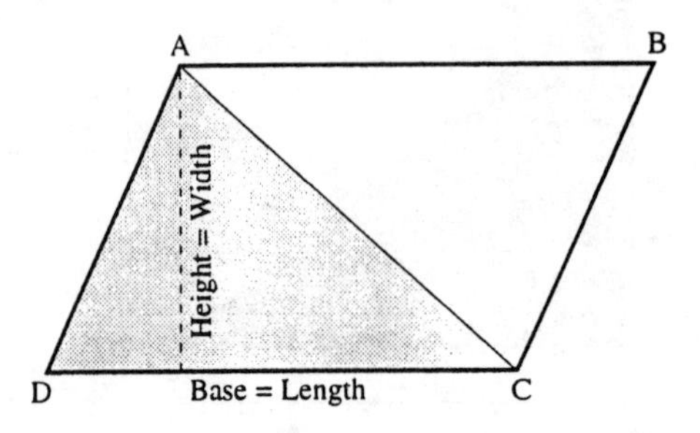

Figure 6.6

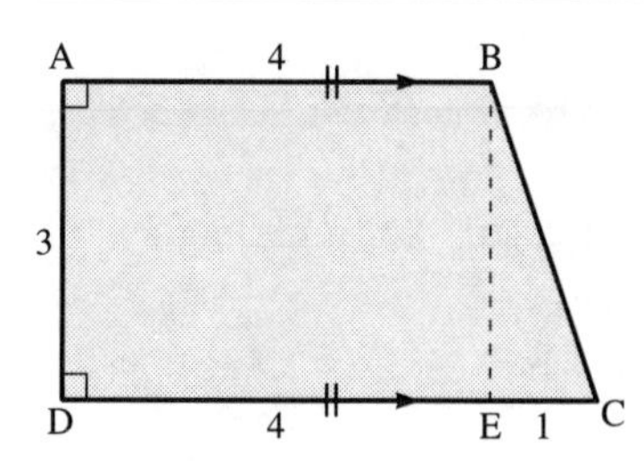

Figure 6.7

$$\text{area ACD} = \tfrac{1}{2} \times \text{base} \times \text{height}$$
$$= \tfrac{1}{2} \times \text{length of parallelogram} \times \text{width}$$

The parallelogram consists of two such triangles:

$$\text{area ABCD} = 2 \times \text{area ACD} = \text{length} \times \text{width}$$

The area of a parallelogram equals length multiplied by width

This rule is almost the same as the rule for rectangles, *but* the width of a parallelogram is the vertical distance between its two sides, *not* the distance measured along one of its sloping ends.

We divide the trapezium in Figure 6.7 into a rectangle ABED and a triangle BCE. Since AB is parallel to DC, the height BE of triangle BCE is 3:

$$\text{area ABED} = 4 \times 3 = 12$$
$$\text{area BCE} = \tfrac{1}{2} \times 1 \times 3 = 1.5$$

Summing the areas:

$$\underline{\text{area ABCD} = 12 + 1.5 = 13.5}$$

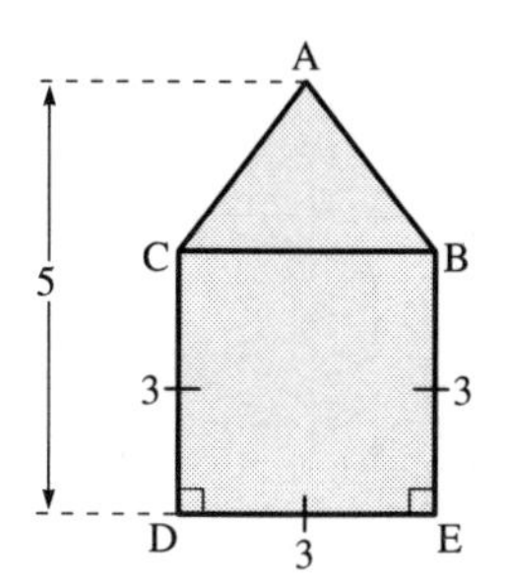

Figure 6.8

In Figure 6.8, the polygon consists of a square and an isosceles triangle. The base of the triangle is 3; its height is 2.

$$\text{area ABC} = \tfrac{1}{2} \times 3 \times 2 = 3$$
$$\text{area CBED} = 3 \times 3 = 9$$

Summing the areas:

$$\underline{\text{area ABEDC} = 3 + 9 = 12}$$

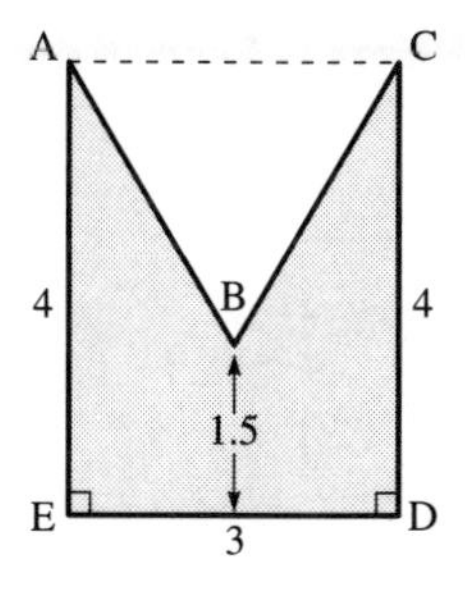

Figure 6.9

It may be simpler to think of an area as the *difference* between two areas. In Figure 6.9, the area of the reflex polygon ABCDE is the difference between the area of rectangle ACDE and triangle ABC (height = 2.5):

$$\text{area ACDE} = 3 \times 4 = 12$$
$$\text{area ABC} = \tfrac{1}{2} \times 3 \times 2.5 = 3.75$$

The differences of the areas:

$$\underline{\text{area ABCDE} = 12 - 3.75 = 8.25}$$

Test yourself 6.1

In questions 1 to 4 give the exact answer, then round it to a suitable number of decimal places or significant figures.

1 A sheet of A3 drawing paper is 293 mm by 430 mm. What is its area?

2 A garden plot is 20 m by 60 m. It is divided into two by a diagonal line on one side of which it is cultivated as a triangular lawn. What is the area of the lawn?

3 A roll of fax paper is 30 m long and 208 mm wide. What is its area? {*Hint*: For conversions, see page 46]

4 Fabric to make a belt for a dress is cut as a strip 820 mm long and 60 mm wide. What is the area of the fabric?

5 A can holds enough paint to cover 25 m^2. If a fence is 1.4 m high, what length of fence can be painted?

6 Find the areas of the shapes in Figure 6.10a to e.

7 Find the areas of the three zones A, B, and C, outlined by squares in Figure 6.10f.

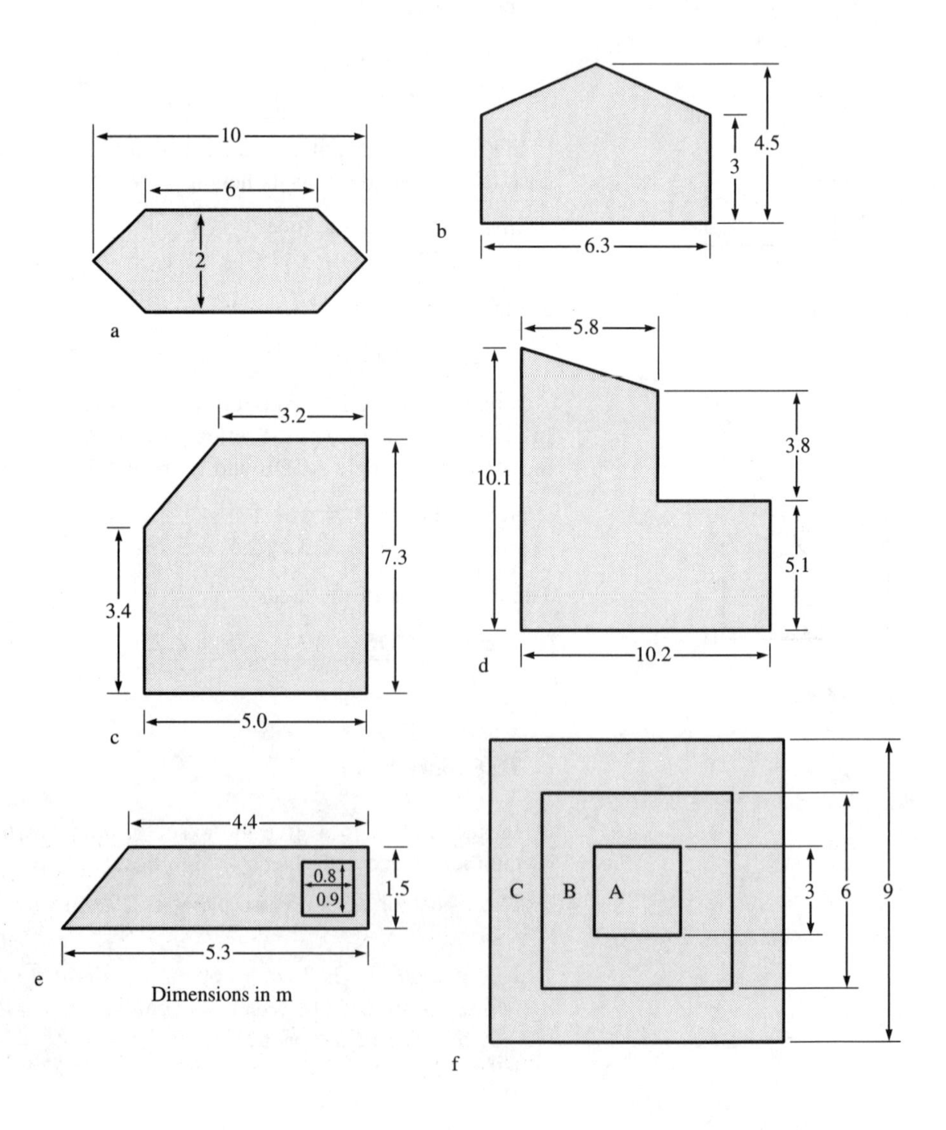

Figure 6.10

Irregular shapes

Some shapes are complicated or have irregularly curved edges. They are not easily broken down into rectangles and triangles. An approximate method of estimating area is to divide the area into small squares and count how many squares it contains. In Figure 6.11 an irregular area (a leaf) is traced on to squared paper. The paper is divided into 2 mm squares. The area of each square is 4 mm^2.

Around the edges of the area we look at each small square and:

- if half of it or more is inside the area, we count it
- if less than half of it is in the area, we ignore it.

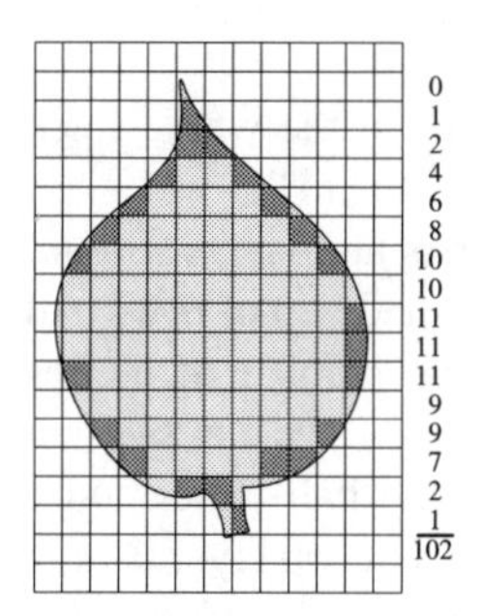

Figure 6.11

In Figure 6.11 the edge squares to be counted have been shaded in. Count along each row of squares, writing the number of squares on the right. Then sum the row totals to give the grand total.

Area = number of squares × area of 1 square
= 102 × 4 = 408

Area = 408 mm^2

Explore this

Collect 20 typical leaves from the same kind of tree or bush. Place them on squared paper and draw a pencil line around each leaf, as closely as you can. Use the square-counting technique to find the area of each leaf. Make a table of these measurements:

Leaf no.	*Length (mm)*	*Width at widest part* (mm)	*Area* (mm^2)
1			
2			
3			
.			
.			
20			

Volume

A number of concrete blocks are stacked in a pile (Figure 6.12). Each block is a 1 m cube. In each layer of the pile there are 3 rows of 4 blocks. By counting the blocks in the top layer, we can see that there are 12 blocks. Instead of actually counting the slabs, a quicker way is to multiply the number of rows by the number of blocks in a row:

3 × 4 = 12

We cannot see all the blocks in the other layers but we do not need to. We can count the number of layers and multiply by this number:

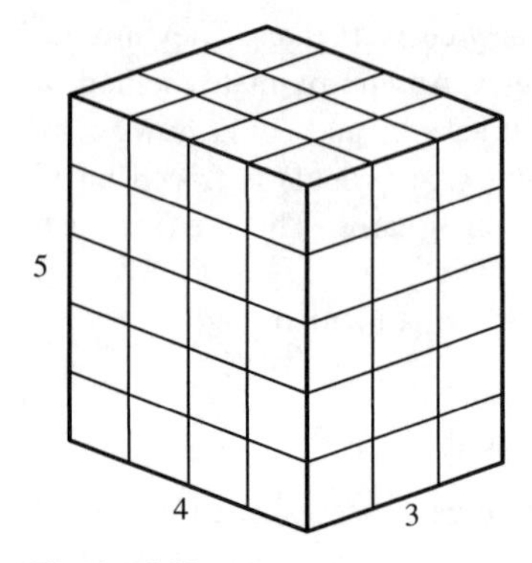

Figure 6.12

Total number of blocks = Blocks in top layer × Number of layers

There are 5 layers, so the total number of blocks is:

$\underline{3 \times 4 \times 5 = 60 \text{ blocks}}$

To calculate the number of blocks we multiply the number of rows, by the number in a row, by the number of layers.

The volume of 1 block is 1 cubic metre, so the volume of the whole stack is 60 cubic metres. We shorten the words 'cubic metres' to 'm^3'.

The same reasoning applies to fractions of rows, blocks and layers. Figure 6.13a shows a concrete dam, which can be thought of as made up of rows and layers of 1 m cubes of concrete. The dam is not a whole number of metres long, wide or high. If we lift off the top layer (Figure 6.13b), we can see that each of the bottom two layers consists of 3 whole cubes, 4 half-cubes and 1 quarter cube. This makes $5\frac{1}{4}$ cubes in each these 2 layers. The top layer is only half a metre high, so it has 3 half-cubes, 4 quarter-cubes and 1 eighth-cube. This makes $2\frac{5}{8}$ cubes.

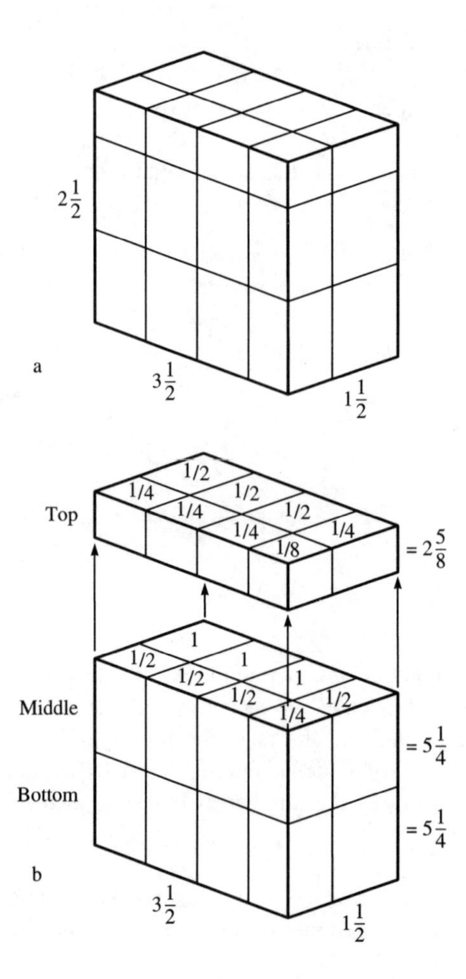

Figure 6.13

Add up the layers:

Top	$2\frac{5}{8}$
Middle	$5\frac{1}{4}$
Bottom	$5\frac{1}{4}$
	$13\frac{1}{8}$

The total volume is $13\frac{1}{8}\,m^3$.

This is the same volume as obtained by multiplying:

$3\frac{1}{2} \times 2\frac{1}{2} \times 1\frac{1}{2} = 13\frac{1}{8}$

Summing up, state this rule:

The volume of a rectangular prism is found by multiplying length by width by height

In symbols:

$\text{volume} = \ell \times w \times h$

In some examples it is more appropriate to use the terms 'depth', or 'thickness' instead of 'height' but the calculation is the same.

Prisms

A prism is a solid body which shows the same shape when it is cut across at right angles, anywhere along its length. Its ends also have that same shape. Regular prisms are described by the shape of their cross-section. Figure 6.14 shows some examples.

Prisms may also have irregular cross-sections, as in Figure 6.15.

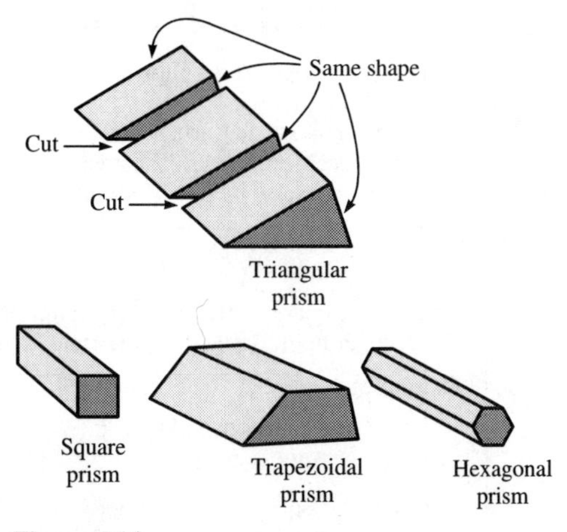

Figure 6.14

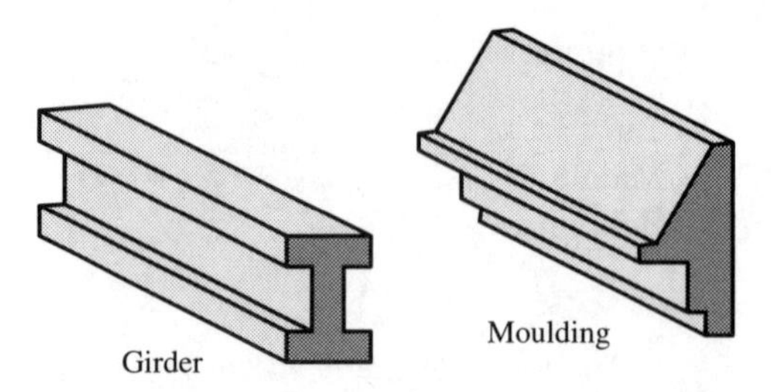

Figure 6.15

Cubes

A cube with sides 2 cm long has a volume of $2 \times 2 \times 2 = 8\,\text{cm}^3$.

8 is *2 multiplied by itself and multiplied by itself again* or *2 cubed*

Here are the cubes of a few small numbers:

Number	*Number squared*
2	8
3	27
4	64
5	125

Learn to recognize these cubed numbers so that, whenever you see 64 (for example), you *immediately* remember that it is 4 cubed.

The size of a rectangular prism can also be expressed in decimal fractions. For example a miniature capacitor measures 5.8 mm long, 3.2 mm wide and 2.1 mm high. Calculate its volume:

$$5.8 \times 3.2 \times 2.1 = 38.976$$

Round the result to the nearest whole number.

$$\underline{\text{Volume} = 39\,\text{mm}^3}$$

Converting areas and volumes

Imagine an area measuring 1 m × 1 m. The sides are divided into 1000 segments, each 1 mm long. The area is ruled into1 mm squares. There are 1000 rows of squares, each with 1000 squares in it. Altogether the number of 1 mm squares is:

$$1000 \times 1000 = 1\,000\,000$$

To convert square metres into square millimetres, multiply by 1 000 000. Shift the decimal point 6 places to the right, adding zeros if necessary.

Examples

$$1\,\text{m}^2 = 1\,000\,000\,\text{mm}^2$$
$$2.4\,\text{m}^2 = 2\,400\,000\,\text{mm}^2$$

To convert square millimetres to square metres, divide by 1 000 000. Shift the decimal point 6 places to the left, inserting zeros if necessary.

Examples

$1\,000\,000\,\text{mm}^2 = 1\,\text{m}^2$
$546\,000\,\text{mm}^2 = 0.546\,\text{m}^2$
$367\,\text{mm}^2 = 0.000367\,\text{m}^2$

Imagine a 1 m cube, cut into 1 mm cubes. There are 1000 layers, each containing 1000 rows of 1000 tiny cubes: Altogether the number of 1 mm cubes is:

$1000 \times 1000 \times 1000 = 1\,000\,000\,000$

To convert cubic metres into cubic millimetres, multiply by 1 000 000 000. Shift the decimal point 9 places to the right, adding zeros if necessary.

Examples

$1\,\text{m}^3 = 1\,000\,000\,000\,\text{mm}^3$
$3.6\,\text{m}^3 = 3\,600\,000\,000\,\text{mm}^3$

To convert cubic millimetres into cubic metres, divide by 1 000 000 000. Shift the decimal point 9 places to the left, inserting zeros if necessary.

Examples

$1\,000\,000\,000\,\text{mm}^3 = 1\,\text{m}^3$
$482\,000\,000\,\text{mm}^3 = 0.482\,\text{m}^3$
$512\,\text{mm}^3 = 0.000\,000\,512\,\text{m}^3$

Summary
m^2 to mm^2: 6 places to right
mm^2 to m^2: 6 places to left
m^3 to mm^3: 9 places to right
mm^3 to m^3: 9 places to left

Checking Count both ways. Check your result by counting in the opposite direction, to make sure you get back to the original number.

Test yourself 6.2

1 Plastic for model-making is sold in sheets 300 mm long, 100 mm wide and 1.6 mm thick. What is the volume of 1 sheet?

2 Three cardboard boxes have the following dimensions (in mm):

Box	*Length*	*Width*	*Height*
A	300	200	100
B	280	240	120
C	350	180	90

Which would hold the greatest volume of polystyrene beads? Convert its volume to m^3.

3 An aquarium is 450 mm long, 350 mm wide and 400 mm deep. What volume of water (in m^3) does it hold when it is half-full?

4 A wooden beam is 6.2 m long, 95 mm wide and 250 mm deep. What is its volume? Round this volume to 2 dp.

5 Iron posts are made by pouring molten iron into a rectangular mould 2 m long and 120 mm wide. If $0.0216\,\text{m}^3$ of molten iron are poured into the mould, how deep will it be?

Volume of a prism

The volume of the rectangular prism in Figure 6.16 is:

$$\text{volume} = \ell \times w \times h$$

It makes no difference to the result if we bracket the width and height together:

$$\text{volume} = \ell \times (w \times h)$$

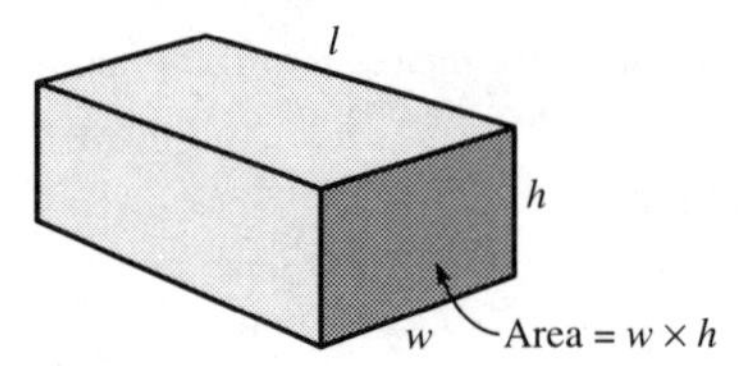

Figure 6.16

But the expression in brackets is the area of one end of the prism, so we can re-write the equation:

$$\text{volume} = \ell \times \text{area of end}$$

This gives the clue to finding the volume of other kinds of prism. An example of a right-angled triangular prism is a wooden wedge (Figure 6.17). The area of one end is:

$$\text{area} = \tfrac{1}{2} \times \text{base} \times \text{height} = 0.5 \times 100 \times 40 = 2000$$

The volume is:

$$\text{volume} = \text{length} \times \text{area} = 90 \times 2000 = 180\,000$$

<u>The volume is 180 000 mm^3</u>

Because the edges of the prism run parallel from one end to the other the shape of the end is exactly the same as that of any *cross-section* (see box,

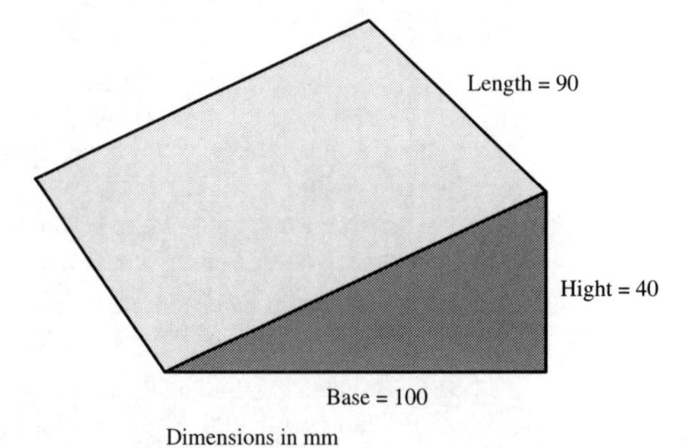

Figure 6.17

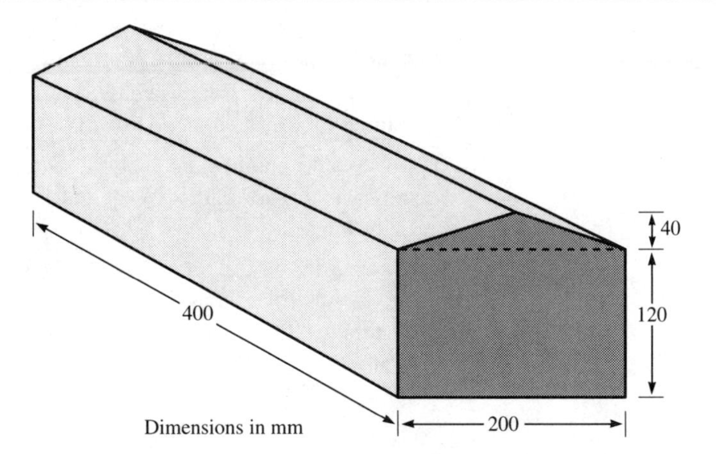

Figure 6.18

page 83). Instead of using the word 'end', which is a little imprecise, we can use the word 'cross-section':

Volume of a prism = area of cross-section × length

The prism of Figure 6.18 is a coping stone. Its volume is found by thinking of the cross-section as a triangle jointed to a rectangle. The area of the rectangle is 24 000 mm^2. The area of the triangle is $0.5 \times 40 \times 200 =$ 4000 mm^2. Total cross-section area is 28 000 m^2, and:

$$\text{Volume} = 400 \times 28\,000 = 11\,200\,000\ \text{mm}^3$$

Test yourself 6.3

1 A girder has a cross-section as shown in Figure 6.19a. What is the area of cross-section? The girder is 5 m long: what is its volume in m^3?

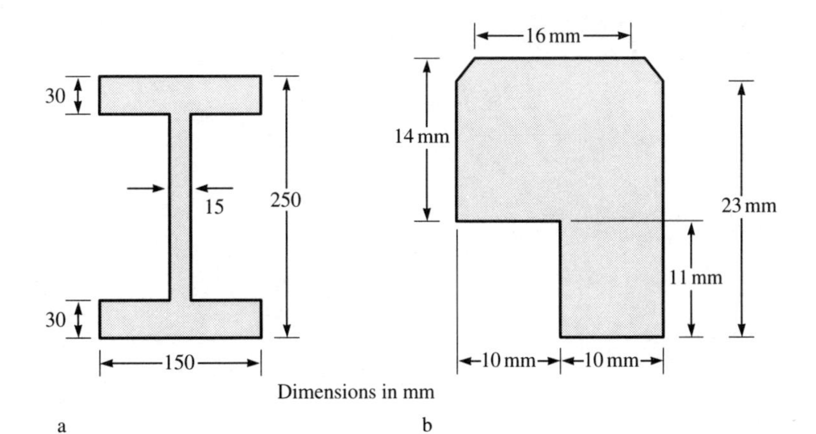

Figure 6.19

2 A swimming pool is 30 m long and 16 m wide. It is 1 m deep at the shallow end and slopes evenly down to 2.5 m at the deep end. What volume of water is needed to fill it?

3 A wooden moulding for making picture frames has the cross-section shown in Figure 6.19b. What is the volume of a strip of moulding 1.8 m long?

4 Rectangular plastic tubing measures 7.9 mm by 6.4 mm. The walls of the tubing are 1.6 mm thick. What is the volume of plastic in a piece of tubing 38 mm long? Round the result to the nearest whole number.

5 To assess if you have covered this chapter fully, answer the questions in *Try these first* on page 75.

7 Graphs

When an equation has two variables in it, a graph shows how changes in the value of one variable cause changes in the value of the other. In this way, graphs are a valuable aid to understanding maths relationships. Graphs are also used for displaying certain kinds of data, to make them easier to follow. In this chapter we explore the techniques of plotting graphs and look at ways in which graphs are useful in maths. Using graphs for displaying data is dealt with in Chapter 32.

You need to know the four arithmetic operations (Chapter 1 and 2), and a little algebra (Chapter 4).

Try these first

Your success with this short test will show you which parts of this chapter you already know.

1 Plot the graph of the equation, $y = 2.5x - 3$, for $x = 0$ to $x = 5$. Use the graph to find the value of y when $x = 3.6$.

2 Figure 7.1 is the graph of an equation.

a Try to discover what the equation is.

Use the graph to find:

b the value of y when $x = 4.5$

c the value of x when $y = 8$.

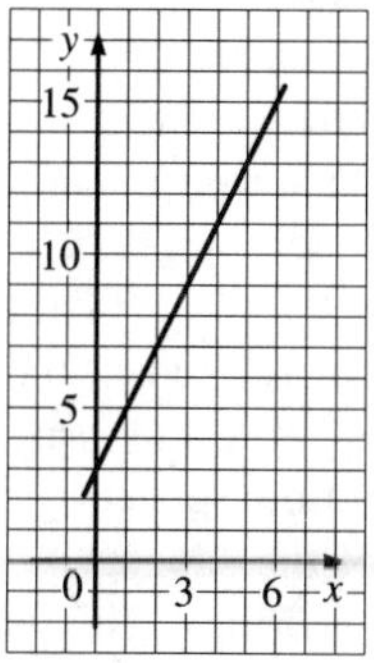

Figure 7.1

Equations and variables

Take this equation as an example:

$$y = 2x + 3$$

A graph shows us how y varies when we vary the value of x. First we need to find values of y for a number of different values of x. In Chapter 4 we saw how to solve an equation of this type, given one or more values of x. For example, if $x = 2$, then $y = 7$. If $x = 4$, then $y = 11$. We could find y for dozens of different values of x. Let us do that for values of x ranging from 1 to 6, in steps of 1, and make a table of the results:

x	1	2	3	4	5	6
y	5	7	9	11	13	15

This set of numbers is made into a graph. Each pair of numbers in this table refers to a **point** on the graph. There are six points, and the pairs of numbers are their **coordinates**. We refer to the points by quoting their coordinates in brackets. A complete list of the points is:

(1,5) (2,7) (3,9) (4,11) (5,13) (6,15)

The first number in the brackets is always the value of x, the **x-coordinate**. The second number is the corresponding value of y, the **y-coordinate**.

One thing to notice about x and y (or any other pair of variables that we may use, such as a and b or p and q) is that we *choose* different values for *one* of the variables (in this case x) and then *calculate* the corresponding values of the *other* variable (y). The value of y depends on the value we decide to give to x. We say that y is the **dependent variable**. By contrast, x is the **independent variable**.

Planning the layout

Although graphs may be plotted on plain paper, using a ruler to measure the distances, it is much more convenient to use paper with printed squares. Graph paper has 1 mm or 2 mm squares, often with slightly thicker lines to indicate 1 cm or 2 cm squares. For the graphs in this chapter, it is easier to use ordinary arithmetic paper, printed with 5 mm or even slightly larger squares.

A good way to learn to draw graphs is to take a sheet of 5 mm squared paper and follow through the steps below, as we describe them. Use a sharp pencil (preferably grade H) or a fine (0.1 mm or 0.2 mm) fibre-tip pen (*not* a ball-point or felt-tip pen).

The first step in plotting a graph is to decide on a point of reference, the **origin**. This is the point with coordinates (0,0). If all values of x and y are positive, as in this example, we locate the origin near the bottom left corner of the paper (Figure 7.2), at the bottom left corner of one of the squares.

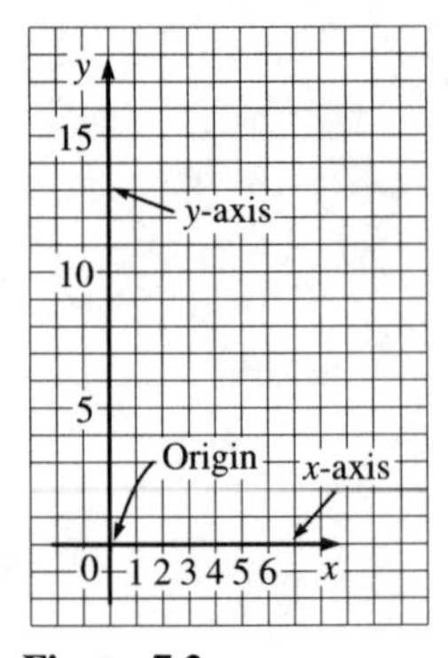

Figure 7.2

Having established the origin, we next need to find our bearings. This is done by drawing two lines through the origin, at right angles to each other. Usually they are drawn horizontally across the paper and vertically up it. These lines are called **axes**. The horizontal line is nearly always used for plotting values of the independent variable, x, so this line is often called the **x-axis**. The vertical line is used for plotting values of the dependent variable, y, so is known as the **y-axis**. Draw over two of the grid lines, so that the axes are accurately at right angles. The axes are labelled to show which refers to x and which to y.

Note that the axes end in arrow-heads. This indicates that, although we are going to plot points for values of x up to 6 only, we could easily plot more points for values greater than 6 (and also for values less than 0, see later). There are no limits to the values of x or y in this equation. The axes could be extended indefinitely.

Scales

The axes in Figure 7.2 are graduated in values of x and y. In this example, assuming that we are drawing on 5 mm squared paper, we use whole squares as the unit of measurement for both x and y.

Sometimes we need to use different scales on each axis, but it is better to use the same scale if possible.

Plotting the points

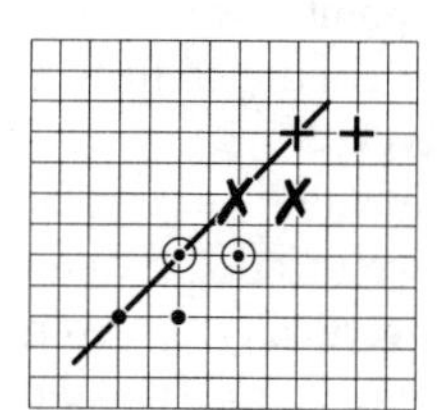

Figure 7.3

The first point has the coordinates (1,5). These numbers instruct us to start at the origin, move 1 unit in the direction of the x-axis, move 5 units in the direction of the y-axis, then plot the point. Start at the origin and count 1 square to the right, along the x-axis, then count 5 squares up. Plot the point. There are several ways of plotting points (Figure 7.3). A fine dot is the best, especially if the dots are to be joined to make a continuous line. But drawing the line may hide the dots. If you need to know where the dots are after a line has been drawn through them, ring the dots with small circles before drawing the line. Crosses can also be used for plotting points but, in practice, it is difficult to make the lines cross at precisely the right point.

Figure 7.4 shows all six points plotted. At this stage we have to decide whether or not to join the dots with a line. When finding values of y for the table, we used whole-number values of x, just to make the calculations easier. But we *could* have used other values, such as $x = 2.5$ (giving $y = 8$), and any other in-between values. These points would be on the same line as the points already plotted. We join a continuous line through all the points.

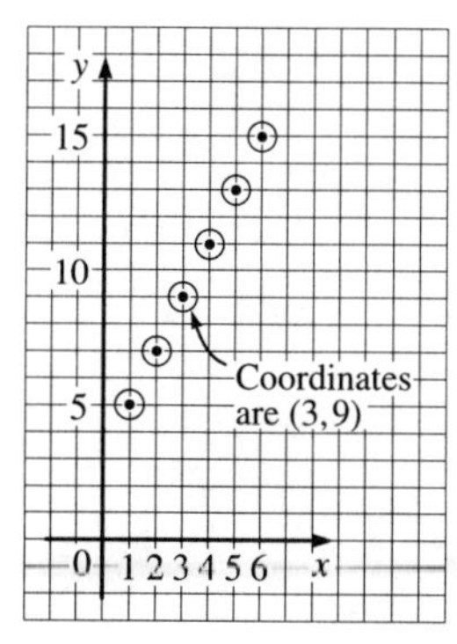

Figure 7.4

There are instances where a variable can take only whole-number values. An example is the sandwich equation, page 49. In this, a represents the number of tourists, and we cannot have 2.5 tourists. In such a case we must leave the graph as a set of unconnected points. This restriction does not apply to the present example. It is clear from looking at Figure 7.4 that the line is straight. To confirm this lay a ruler along the row of points. Draw a straight line running through all the points (Figure 7.5). Other equations may give graphs with curved lines. Note that the line is extended a little way below the point (1,3) and above the point (6,15). This shows that the line really extends to smaller and bigger values of x and y, but we have plotted only a section of the whole graph.

Complete the graph by labelling it; for example, by writing the equation on it.

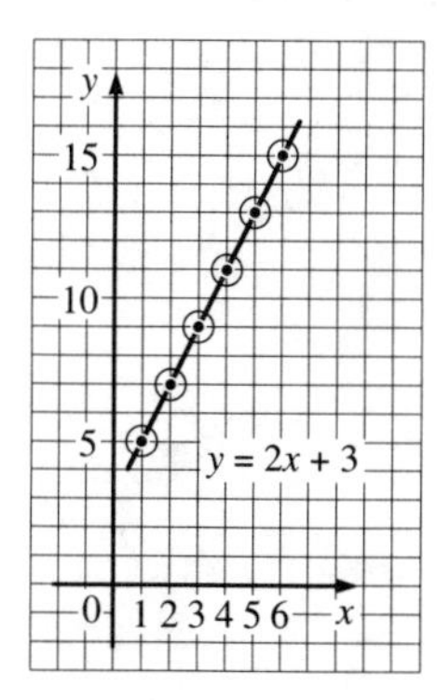

Figure 7.5

Graph-plotting summary

- Equation in the form $y = \ldots$
- Make a coordinates table: given values of x, calculate corresponding y's
- Find best position on the grid for the origin
- Draw and label the axes
- Decide on suitable scales (preferably the same)
- Mark scales on the axes
- Plot points
- Join points if valid
- Label graph with equation or other title.

Remember . . .
x is across
y is high

Interpreting the graph

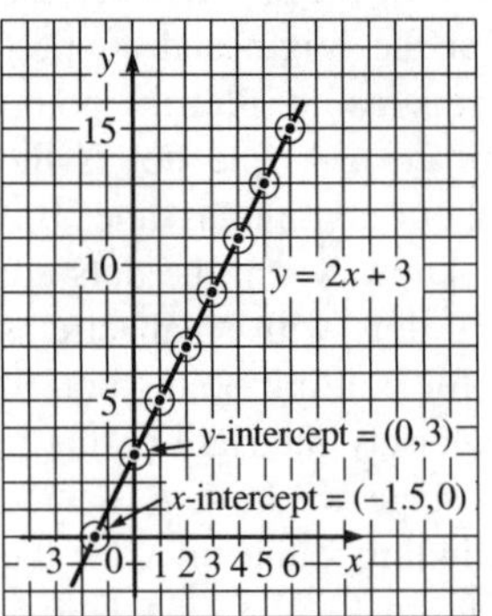

Figure 7.6

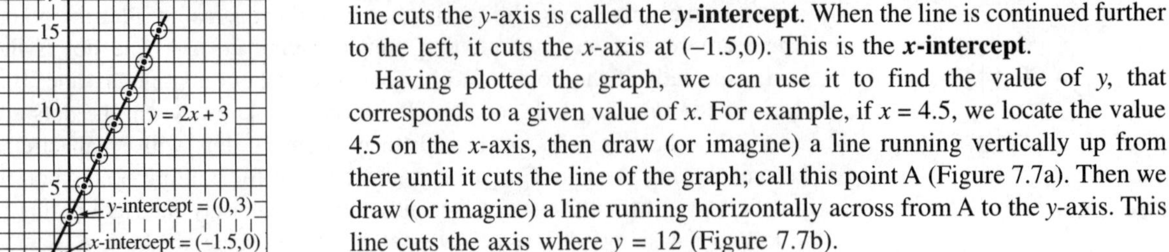

The line slopes up towards the right, showing that when x increases, y increases too. If we continue the line downward to the left (Figure 7.6), it cuts the y-axis at (0,3). This is the point for which $x = 0$. The point at which the line cuts the y-axis is called the **y-intercept**. When the line is continued further to the left, it cuts the x-axis at (–1.5,0). This is the ***x*-intercept**.

Having plotted the graph, we can use it to find the value of y, that corresponds to a given value of x. For example, if $x = 4.5$, we locate the value 4.5 on the x-axis, then draw (or imagine) a line running vertically up from there until it cuts the line of the graph; call this point A (Figure 7.7a). Then we draw (or imagine) a line running horizontally across from A to the y-axis. This line cuts the axis where $y = 12$ (Figure 7.7b).

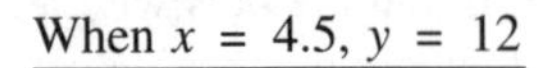

When $x = 4.5$, $y = 12$

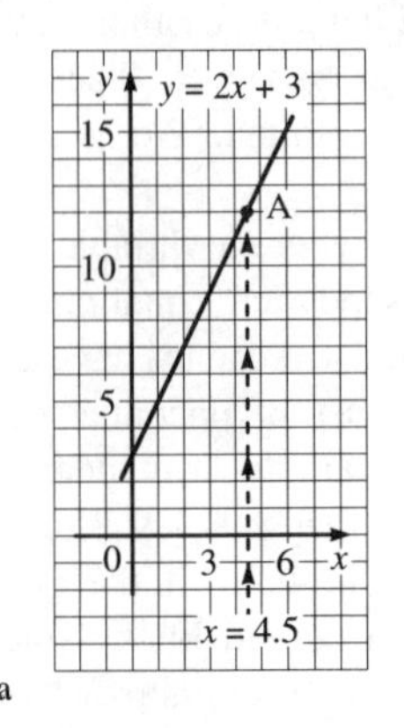

a

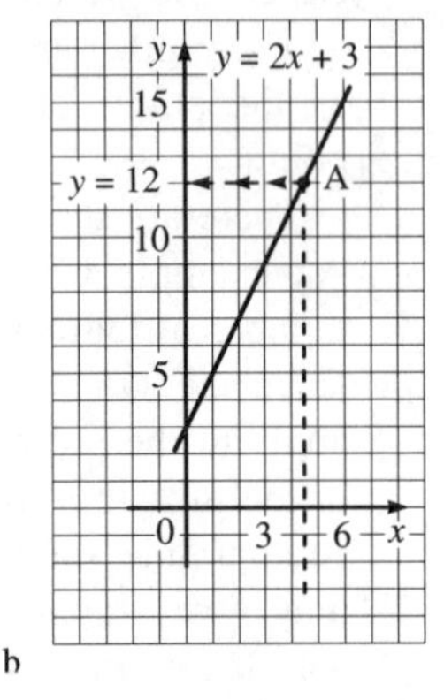

b

Figure 7.7

This is easily checked by substituting $x = 4.5$ in the equation:

$$y = 2x + 3 = 2 \times 4.5 + 3 = 9 + 3 = 12$$

Using the graph to find values of y is quicker than calculating it from the equation. It is specially useful if you have a lot of values to read from the same graph.

Working in the reverse direction, we can find what value of x corresponds to a given value of y. For example, reading the graph as in Figure 7.8, we find that:

When $y = 7.5$, $x = 2.25$

Figure 7.8

More equations

Here is another equation to plot:

$$y = 3x - 7$$

Plot this from $x = 0$ to $x = 5$. First, calculate the values of y:

x	0	1	2	3	4	5
y	–7	–4	–1	2	5	8

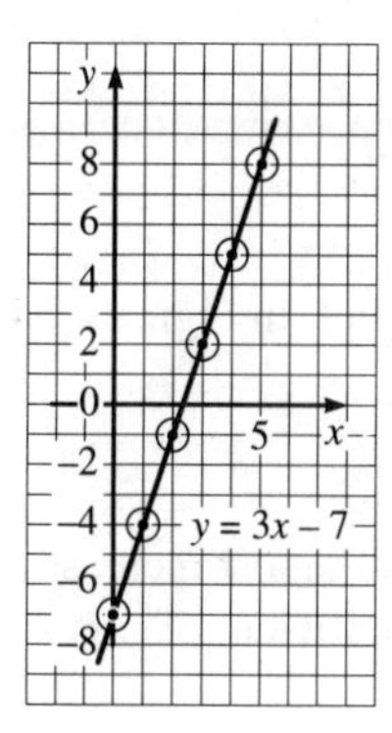

Figure 7.9

Here is an example in which the origin is *not* at the bottom left corner of the paper. There are negative values of y, so the origin is placed about half-way up from the bottom of the paper (Figure 7.9). But all values of x are positive, so the origin is near the left edge of the paper. The axes are drawn through the origin. As with Figure 7.5, the line slopes up towards the right.

The next equation is plotted for both negative and positive values of x, from $x = -4$ to $x = 4$:

$$y = 1.5x + 2$$

x	–4	–3	–2	–1	0	1	2	3
y	–4	–2.5	–1	0.5	2	3.5	5	6.5

Both x and y have negative and positive values, so the origin is located near the centre of the paper (Figure 7.10). The scale is still one square for 1 unit. When we need to plot values of y which end in '.5', we put the point half-way along the square.

Figure 7.10 is drawn partly for negative *values* of x. By contrast Figure 7.11 is the graph of an equation in which x has a negative *coefficient* (page 55). The equation is:

$$y = 10 - 2x$$

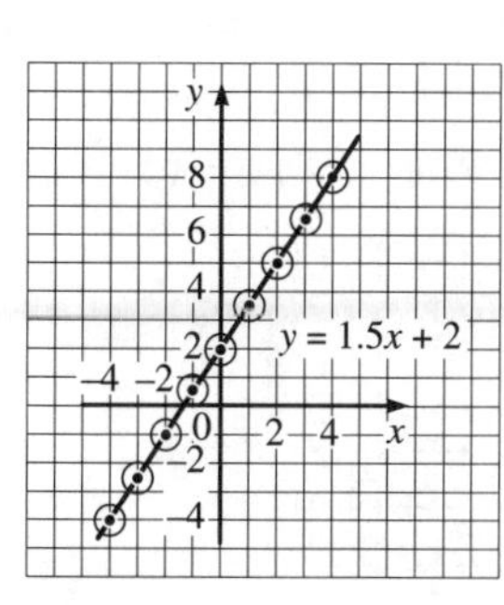

Figure 7.10

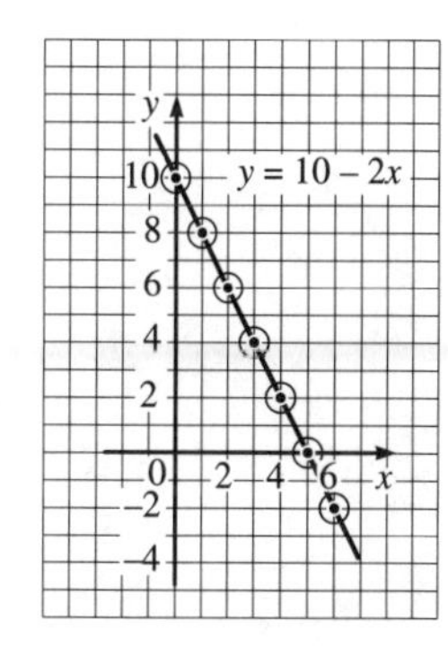

Figure 7.11

The coefficient of x is –2. Here is the table of coordinates for values of x from 0 to 6:

x	0	1	2	3	4	5	6
y	10	8	6	4	2	0	–2

This graph differs from the others; the line slopes *down* to the right. As x increases, y decreases. This is the result of having a negative coefficient of x in the equation. It makes y smaller at each step.

Coefficient and steepness

Look again at the tables of x and y on pages 89–93. In all of these, we made x step one unit at a time:

0, 1, 2, 3, 4, 5, . . .

Or, beginning with a negative value:

$-4, -3, -2, -1, 0, 1, \ldots$

But the corresponding values of y step on by a different amount. In Figure 7.5, for example, y steps on 2 units at a time:

$5, 7, 9, 11, \ldots$

This is because the coefficient of x is 2 in the equation. Each time x increases by 1, y increases by 2. In the equation of Figure 7.9 the coefficient of x is 3, and this makes y step on 3 units at a time:

$-7, -4, -1, 1, \ldots$

By looking at the equation and noting the coefficient of x, we can tell how much y steps on. This is related to the **steepness** of the line. The bigger the steps of y, the steeper the line.

The steepest line is that of the equation:

Special graphs

Learn to recognize these graphs and similar ones.

y = constant (no x in the equation)

Same value of y for all values of x; graph is a horizontal line.

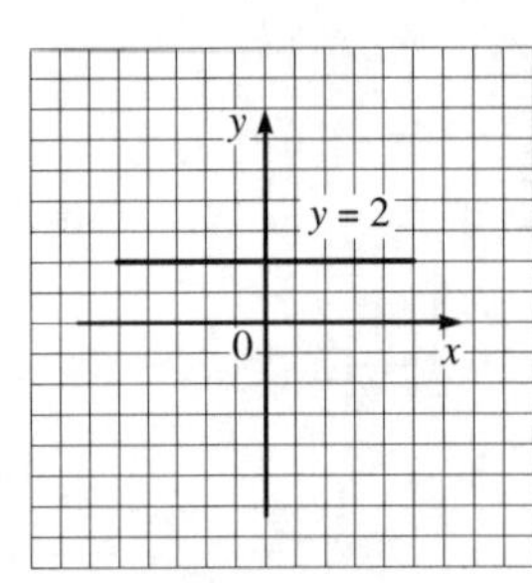

Figure 7.12

x = constant (no y in the equation)

Value of y does not depend on x; graph is a vertical line.

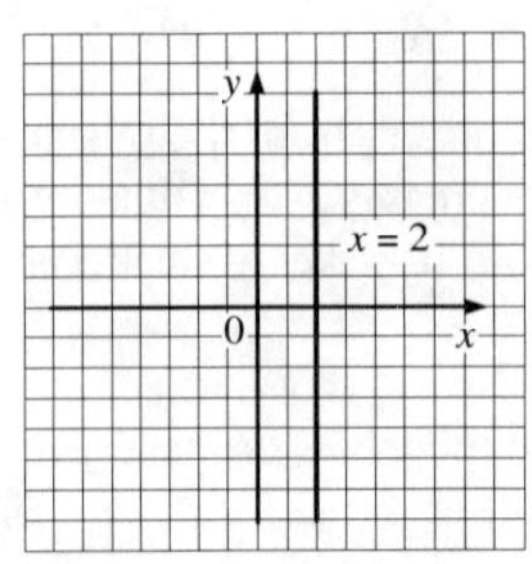

Figure 7.13

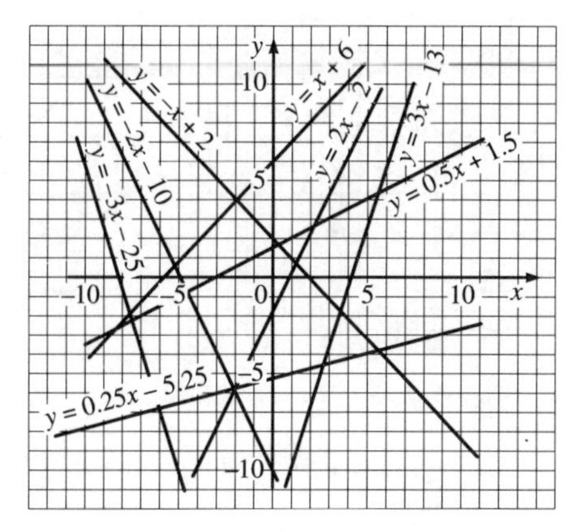

Figure 7.14

This has the biggest coefficient of x, and y steps on 3 units each time. The line with the least slope is that in Figure 7.10, in which the coefficient of x is 1.5 and y increases by 1.5 at each step.

In Figure 7.11, the line slopes *downward* to the right. It has a *negative* slope. In its equation, x has a *negative* coefficient.

A line slopes up if the coefficient of x is positive; it slopes down if it is negative

By simply looking at the coefficient of x, we can tell how steep the line is and if it slopes up or down. There is no need to calculate values of y to find out these simple facts.

To make this point clear, Figure 7.14 shows an assortment of lines with different coefficients of x. Compare the slopes with the coefficients.

Cutting the y-axis

All the lines we have plotted (except Figure 7.13) cut the y-axis when $x = 0$. When $x = 0$ the value of the term in x becomes zero. The value of y becomes that of the constant number. For example, in Figure 7.6, when $x = 0$, then $2x = 0$ and $y = 3$. The line cuts the y-axis at $y = 3$. Similarly, in Figure 7.9, we have $y = 3x - 7$. The line cuts the y-axis at $y = -7$.

By looking at the value of the constant number in the equation, we can tell where the line cuts the y-axis.

More special graphs

Learn to recognize these graphs and similar ones.

$y = x$

Line slopes 45° up (if scales are equal), and passes through origin.

$y = -x$

Line slopes 45° down (if scales are equal), and passes through origin.

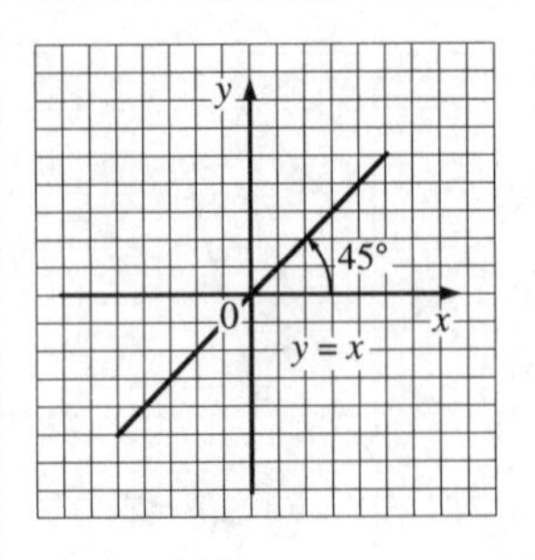

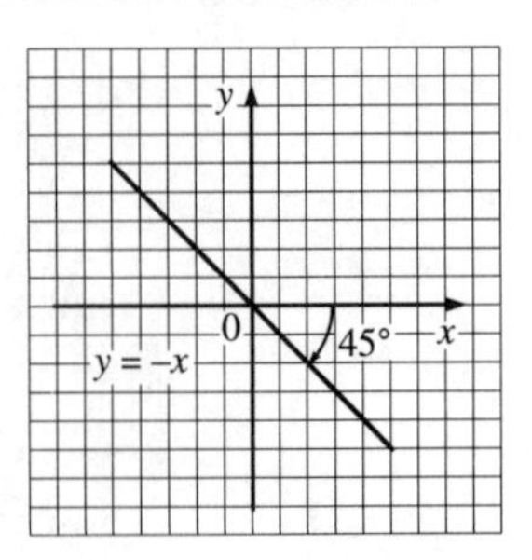

Figure 7.15

These graphs are shifted up or down by including a constant in the equation. For example,

$$y = x + 5$$

Line slopes 45° up and y-intercept is 5.

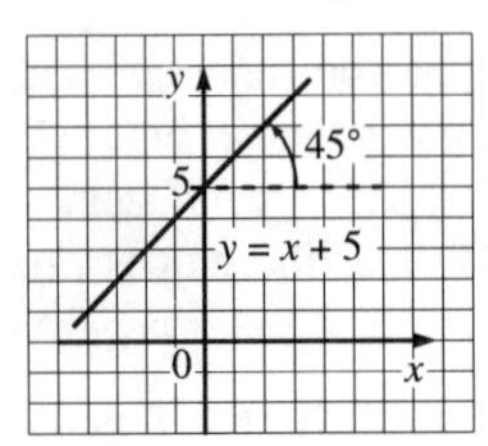

Figure 7.16

Have you noticed?

All graphs in this chapter are straight lines. All have the same type of equation:

$$y = mx + c$$

where m is the coefficient of x and c is a constant number.

All equations of this form give a straight line graph. Learn to recognize this form of equation. Then you will know when an equation gives a straight line.

Test yourself 7.1

Plot the graphs of these equations, for the stated values of x.

1 $y = 2x + 1$, for $x = 0$ to $x = 10$

2 $y = 3x - 4$, for $x = 2$ to $x = 8$

3 $y = 1.5x + 7$ for $x = -2$ to $x = 7$

4 $y = 12 - 2x$ for $x = 0$ to $x = 10$

5 From the graph of question 1, read the value of y when $x = 2.5$, and the value of x when $y = 14$.

6 From the graph of question 4, read the value of y when $x = 8.5$ and the value of x when $y = 5$.

7 Study this list of equations, then answer the questions *without* plotting their graphs:

a $y = 5x + 6$ **b** $y = 4 - 3x$ **c** $y = 7x - 4$

d $y = x + 10$ **e** $y = 5x - 2$ **f** $y = 7 + 2x$

Which graph has the steepest slope? Which has the least steep slope? Which graph slopes down to the right? Which two graphs have the same slope? What can you say about their lines? For each graph, say where it cuts the y-axis.

Quick plotting

If the equation shows that the graph is a straight line, we need calculate only the first and last points, then rule a straight line to join them.

Different scales

The slope of the line and the overall shape of a graph depend on the scales along the two axes. The graphs of Figures 7.1 to 7.16 all have the same scale on both axes. This makes it easier to plot the points and to read off values.

Figure 7.17a is the graph of:

$$y = 5x + 2$$

The coefficient of x is 5, making the line slope steeply upward. With equal scales on both axes, the graph is tall and narrow. Such a shape does not fit well into the paper. It is difficult to read values of x precisely for given values of y.

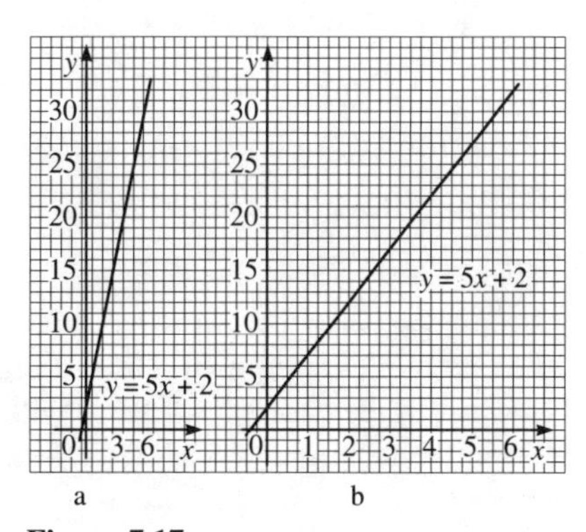

Figure 7.17

Usually it is best to choose scales so that the graph spreads fully across and fully up the available space. Figure 7.17b shows the same equation plotted with different scales on each axis. The y-axis has the same scale as in Figure 7.17a, one square equalling one unit. The scale of the x-axis is expanded by making *four* squares equal to one unit.

Compare the steepness of the line in Figure 7.17a with that in Figure 7.17b.

Changing scales changes steepness

The only time that changing scales does not change steepness is when we change both scales by the same amount, for example double both or treble both.

Test yourself 7.2

Plot the graphs of these equations on paper printed in 5 mm squares. You need two A4 sheets for these 8 graphs. Divide each sheet into quarters, so that each quarter measures about 100 mm × 140 mm, or 20 squares wide by 28 squares high. Each graph must fill the space as much as possible, leaving room at one or two edges, as required for labelling the axes. Some graphs do not fit in easily. For example you may find that 1 square to 1 unit makes the graph too narrow, but 2 squares to 1 unit makes it too wide for the space available. It may help to turn the paper round to make the space 28 squares wide and 20 squares high. Do not mark the scales in fractions of a square, for example $1\frac{1}{3}$ squares to a unit. In such a case, it is better to reduce the scale to 1 square to a unit and not quite fill the space. Always include the origin.

[*Hint*: The graphs do not *have* to be plotted in steps of 1 along the x-axis.]

1 $y = 1.5x - 7$, from $x = 0$ to $x = 8$

2 $y = 3x + 2$, from $x = 0$ to $x = 3$

3 $y = 0.5x + 2$, from $x = 0$ to $x = 10$

4 $y = 25x + 10$, from $x = -3$ to $x = 1$

5 $y = -3x - 5$, from $x = 2$ to $x = 7$

6 $y = 0.1x + 4.2$, from $x = 0$ to $x = 30$

7 $y = 8x - 100$, from $x = -20$ to $x = 60$

8 $y = 1 - 6x$, from $x = -2$ to $x = 14$

9 A tank contains 15 m^3 of water. Water is pumped out of the tank at the rate of 1.2 m^3 an hour. Draw a graph to show how the quantity of water in the tank (y) decreases with time (x). How much water is left in the tank after 4.5 h? After what times does the tank become empty?

10 Weed-killer is made up by dissolving 30 g of solid in a litre of water. The solid is weighed in a plastic scoop which weighs 20 g when empty. Draw a graph to show the total weight (y) of solid and scoop, used to make different volumes (x) of weed-killer. Let the volumes range from 1 litre to 10 litres.

The scoop plus solid weighs 230 g: how much weed-killer can be made from this?

11 An iron bar is put in an oven, and its temperature is rising at a steady rate. Its temperature is measured twice after it has been put in the oven. After 5 minutes its temperature is 50 °C; after 15 minutes it is 110°C. Plot a graph to show the temperature of the bar (y) at different times (x). What was the temperature of the bar when it was put in the oven? How long will it take for the temperature to reach 200 °C?

12 To assess if you have covered this chapter fully, answer the exercises in *Try these first*, page 89.

8 Using essential maths

There is no new maths in this chapter. Instead we present a number of problems in various branches of technology and show how to solve them, using the essential maths of Part 1. We also show how to estimate results approximately and how to check your calculations quickly. The references at the end of each solution tell you where to look for fuller explanations.

At the end of the chapter there are versions of these problems for you to solve on your own.

Marking out

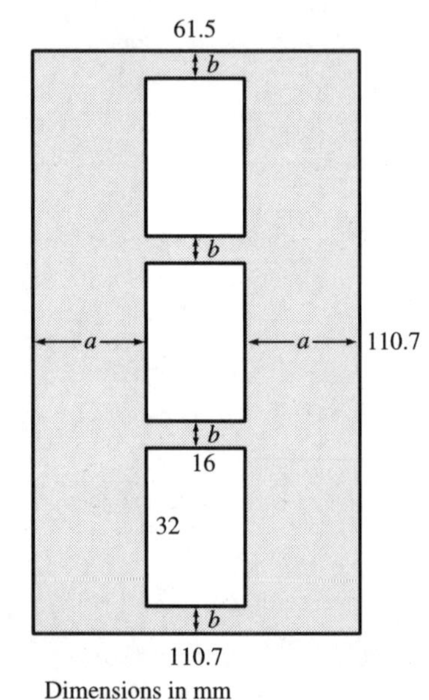

Figure 8.1

Problem 1

Three rectangular apertures are to be made in a panel as shown in Figure 8.1. The distances marked a are to be equal. The distances marked b are to be equal.

a Find a in mm to 1 dp.

b Find b in mm to 1 dp.

c Find the area of panel remaining, in mm^2, to 2 sf.

How many decimal places?

A person weighs a crystal of a salt on a balance. The reading is 23 mg, the balance being accurate to the nearest milligram. The person dissolves the salt in water and then pours equal quantities of the solution into each of 12 beakers. How much salt is in each beaker?

To calculate this, divide 23 by 12. On a 10-digit calculator, the result is:

$$\frac{23}{12} = 1.916\,666\,667$$

Note that the calculator automatically rounds (page 44) the recurring 6 into a 7 at the last decimal place. But 9 decimal places are obviously far too many. How many decimal places is it sensible for the answer to have?

To decide this, we have to look at the idea of **significant figures**.

In maths, a number such a 1.23 means exactly that, no more and no less. But in technology we often use numbers to express measurements. We specify the precision of a number by saying how many **significant figures** (sf) it contains.

Here is a routine for deciding how many significant figures a number contains:

Question	*Response*	*Examples*
1 Number begins and ends with non-zeros?	Count all figures	2.67 has 3 sf 43.07 has 4 sf
2 Number begins with zeros?	Ignore beginning zeros	0.034 has 2 sf 0.00002 has 1 sf
3 Number ends with zeros *after* the decimal point?	Include all ending zeros in the count*	25.100 has 5 sf 5.000 has 4 sf
4 Number ends with zeros *before* the decimal point?	Ignore ending zeros	45000 has 2 sf 6920 has 3 sf

*The reasoning behind this is that, since we have taken the trouble to write all those zeros, they must mean something.

A useful rule is:

A result can have no more sf than the least precise measurement

In the example, the weight is 23 g, which has 2 sf. There can be no more than 2 sf in the result. Rounding 1.916666667 to 2 sf (one before the decimal point and one after it) gives 1.9.

The amount of salt in each beaker is 1.9 mg (2 sf).

Other examples in this chapter make the idea of significant figures clear.

Checking results

While you are working on a problem, develop the habit of checking at every stage that the figures *make sense.*

For example, you will probably work out the quotient of part **a** of Problem 1 on a calculator. But it is easy to make a mistake in keying, especially when putting in the decimal point. If you key 6.15 instead of 61.5 you get a negative result, which is obviously nonsense. But you might not notice the negative sign on the display. Always look carefully to check that a result really is positive when it is expected to be.

If you key 615, you get 299.5, which is longer than the panel. This is more nonsense. Always look out for answers that are impossibly big or small.

It is good technique to have a reasonable idea of what the answer should be. A rough estimate tells you what answer to expect. Round the numbers to 1 sf, as in this estimate of part **a**:

$$\frac{61.5 - 16}{2} \approx \frac{60 - 20}{2} = \frac{40}{2} = 20$$

This is so easy that you do it in your head. It gives the same result as the actual result rounded to 1 sf.

To confirm the results of a division more accurately, multiply the result by the denominator to see if you get back to the original numerator:

$22.75 \times 2 = 45.5$ OK

In part **b**:

$$\frac{110.7 - 3 \times 32}{4} \approx \frac{110 - 3 \times 30}{4} = \frac{110 - 90}{4} = \frac{20}{4} = 5$$

This is rather higher than the true answer, mainly because we rounded (3×32) to 90, when it is really 96. Try rounding it to 100. Then the estimate is:

$$\frac{110 - 100}{4} = \frac{10}{4} = 2.5$$

The true answer lies between these two estimates, so it appears that the calculation is correct.

Solution

a $a = \dfrac{61.5 - 16}{2} = \dfrac{45.5}{2} = 22.75 = 22.8 \text{ (1 dp)}$

$\underline{a = 22.8 \text{ mm (1 dp)}}$

b $b = \dfrac{110.7 - 3 \times 32}{4} = \dfrac{14.7}{4} = 3.675 = 3.7 \text{ (1 dp)}$

$\underline{b = 3.7 \text{ mm (1 dp)}}$

c Area of panel = $110.7 \times 61.5 = 6808.05$

Area of apertures = $32 \times 16 \times 3 = 1536$

$\Rightarrow$ area of panel remaining = $6808.05 - 1536 = 5272.05 = 5300$ (2 sf)

Area of panel remaining = 5300 m^2 (2 sf)

Comments

Note the difference between rounding to a number of decimal places and rounding to a number of significant figures.

Reference pages: 37, 40, 44, 76.

Adjusting quantities

Problem 2

A cake recipe lists the following quantities of ingredients:

1 tablespoon margarine
4 tablespoons water
3 eggs
½ cup sugar
1½ cup flour

There are only 2 eggs available. Adjust the quantities of ingredients to make a smaller cake.

Solution

There are 2 eggs instead of 3, so all quantities have to be multiplied by $\frac{2}{3}$:

Margarine	$1 \times \frac{2}{3} = \frac{2}{3}$ tablespoon
Water	$4 \times \frac{2}{3} = \frac{8}{3} = 2\frac{2}{3}$ tablespoons
Eggs	$3 \times \frac{2}{3} = \frac{6}{3} = 2$ eggs
Sugar	$\frac{1}{2} \times \frac{2}{3} = \dfrac{1 \times 2}{2 \times 3} = \frac{2}{6} = \frac{1}{3}$ cup
Flour	$1\frac{1}{2} \times \frac{2}{3} = \frac{3}{2} \times \frac{2}{3} = 1$ cup

Comments

Quantities are specified in tablespoons and cups. Working in common fractions rather than in decimals gives results that are easier to measure out. For water, we convert an improper fraction to a mixed fraction. Calculating the number of eggs required is a good way of confirming that we are multiplying by the correct fraction. We cancel 2's when working out the sugar. We cancel 2's and 3's when working out the flour. With the flour, we are multiplying a number by its reciprocal, so the product must be 1.

Reference pages: 32, 37, 40, 41.

Costing

Problem 3

A certain type of bolt is manufactured in a number of metric thread sizes and sold in bags:

Thread size	*Quantity per bag*	*Price per bag*
M6	50	£4.20
M8	25	£4.30
M10	10	£2.50
M12	10	£3.40

A store supervisor prepares an order:

Thread size	*Number of bags*
M6	7
M8	12
M10	4
M12	3

a How many bolts are ordered altogether?

b What is the total cost of the order?

c Which thread size costs the least per bolt and what is their unit price to the nearest penny?

Solution

a Number ordered $= 50 \times 7 + 25 \times 12 + 10(4 + 3)$
$= 50 \times 7 + 25 \times 12 + 10 \times 7$
$= 350 + 300 + 70$
$= 720$

The number of bolts ordered is 720

b

Thread size	*Number of bags*	*Price per bag (£)*	*Cost (£)*
M6	7	4.20	$4.20 \times 7 =$ 29.40
M8	12	4.30	$4.30 \times 12 =$ 51.60
M10	4	2.50	$2.50 \times 4 =$ 10.00
M12	3	3.40	$3.40 \times 3 =$ 10.20
			101.20

The total cost is £101.20

c

Thread size	*Price per bag (£)*	*Quantity per bag*	*Cost per bolt (£)*
M6	4.20	50	4.20/50 = 0.084
M8	4.30	25	4.30/25 = 0.172
M10	2.50	10	2.50/10 = 0.250
M12	3.40	10	3.40/10 = 0.340

Converting to pence: $0.084 \times 100 = 8.4$.
Rounding 8.4 to nearest whole number gives 8.

M6 bolts are the cheapest, at 8 p each

Checking Problem 3

A very rough check is: the order is for 26 bags, and the average price per bag is about £4. Total cost should be about $26 \times 4 =$ £104. Calculated answer is close to this.

When checking the addition of costs, round to 1 sf: $30 + 50 + 10 + 10 = 100$. Calculated answer is close to this.

Check divisions by multiplying result by denominator: $0.084 \times 50 = 4.2$, so calculation is correct. Similarly for the other divisions.

Comments

1 Using brackets reduces the number of multiplications from 4 to 3. Clear brackets first, then multiply, then add products.

2 First find products (bags × price), then add products.

3 All quotients divide out exactly, with 2 or 3 dp, so results are given to 3 dp. Multiply by 100 by shifting decimal point 2 places to the right.

Reference pages: 19, 21, 44, 52.

Economical cutting

Problem 4

A sheet of hardboard measuring 2.44 m × 1.22 m is to be cut into boards each measuring 480 mm × 180 mm. To simplify sawing, all boards are oriented the same way on the sheet, so that the sheet may be cut into strips which are then cut into boards.

a Ignoring wastage from the saw-cuts, which way of orienting the boards gives the most rectangles?

b How many boards are obtained?

c If a sheet costs £3.80, what is the cost of each board, to the nearest penny?

Solution

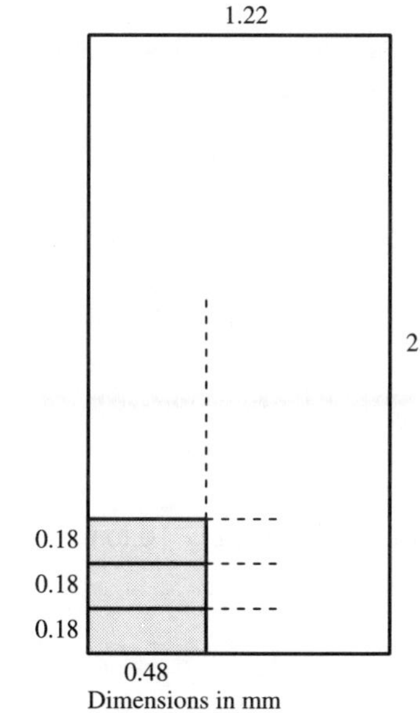

Figure 8.2

a In metres, the boards measure 0.48 m × 0.18 m. If arranged crossways, as in Figure 8.2, the number of rows of boards is:

$$\frac{2.44}{0.18} = 13.\dot{5}$$

The number of columns of boards is:

$$\frac{1.22}{0.48} = 2.54$$

The decimal places represent wasted material and we have 13 rows of 2 boards, a total of 13 × 2 = 26 boards.

If the boards are oriented lengthways, the number of rows is:

$$\frac{2.44}{0.48} = 5.08$$

The number of columns is:

$$\frac{1.22}{0.18} = 6.\dot{7}$$

The number of boards is 5 × 6 = 30. This is more than before, so this the most economical arrangement.

b

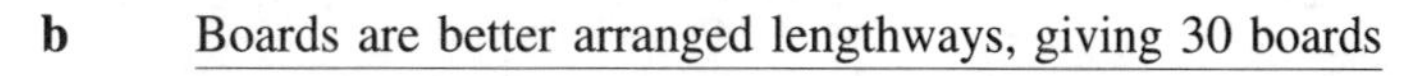
Boards are better arranged lengthways, giving 30 boards

c The cost of one board, in pounds, is:

$$\frac{3.80}{30} = 0.12\dot{6}$$

The fraction ends with a recurring 6. Convert the fraction from pounds to pence:

$$0.12\dot{6} \times 100 = 12.\dot{6}$$

Rounding to the nearest penny gives 13.

Boards cost 13 p each

Checking Problem 4

Rough check by rounding to 1 sf:

$$\text{Number of rows} = \frac{2.44}{0.18} = \frac{2}{0.2} = 10$$

Similarly for the other divisions. Check the divisions exactly by multiplying result by denominator:

$$0.18 \times 13.\dot{5} = 2.44$$

Calculation correct. Similarly for the other divisions. Note the recurring decimal; to check exactly, you must key in several (2 or more) 5's. If you key 13.5 (not recurring) the result is 2.43 which indicates an incorrect result.

Comments

This is a good example of a problem in which remainders from division must be ignored.

Reference pages: 26, 31, 76.

Volume and weight

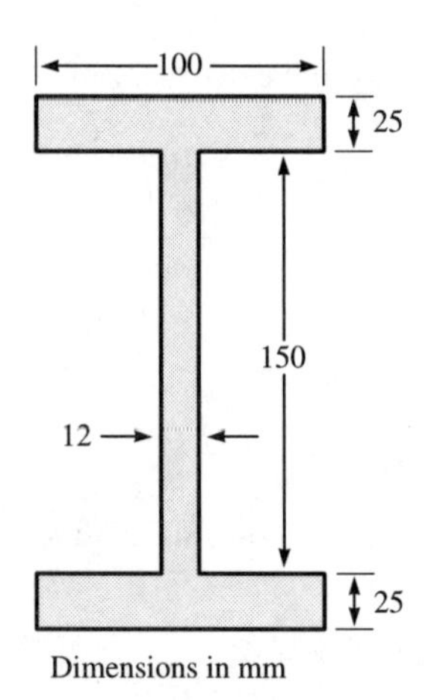

Figure 8.3

Problem 5

Steel girders 4 m long have the cross-section shown in Figure 8.3. 1 m^3 of steel weighs 7800 kg.

a What is the volume of 1 girder? (3 sf)

b What is the weight of 1 girder in kg? (3 sf)

c How many girders could be loaded on to a truck rated to carry 10 tonnes? (1 tonne = 1000 kg)

Solution

a Area of cross-section $= 2 \times 100 \times 25 + 150 \times 12$
$= 5000 + 1800 = 6800\,\text{mm}^2$

Convert 6800 mm^2 to m^2

$$\frac{6800}{1\,000\,000} = 0.0068$$

Volume = area × length $= 0.0068 \times 4 = 0.0272\,\text{m}^3$.

The volume of 1 girder is 0.0272 m^3.

b Weight = Volume × Weight of 1 m^3 $= 0.0272 \times 7800 = 212.16$ kg. Rounding to 3 sf gives 212 kg.

The weight of 1 girder is 212 kg.

c Convert 10 tonnes to kilograms:

$10 \times 1000 = 10\,000\,\text{kg}$

Divide this by 212 to see how many whole girders can be loaded:

Long division:

$$\begin{array}{r} 47 \\ 212 \overline{)10000} \\ \underline{848} \\ 1520 \\ \underline{1484} \\ 36 \end{array}$$

Quotient of 47 remainder 36. The remainder (36) represents a fraction of a girder, but only whole girders can be loaded.

47 girders can be loaded on the truck.

Checking Problem 5

Use a calculator for the long division. A rough check by multiplying result by denominator after rounding to 1 sf:

$47 \times 212 \approx 50 \times 200 = 10\,000$ OK

Exact check without rounding, but add the remainder:

$47 \times 212 + 36 = 9964 + 36 = 10\,000$ OK

Comments

This is an exercise in managing decimal points. It is easier to work in mm to find the cross-section, then to convert to m for the rest of the calculation.

Reference pages: 30, 79, 83, 84.

Areas

Problem 6

A field (Figure 8.4) is divided into two by a fence; the lower area is exactly one third of the area of the whole field. What is the distance d to 3 sf?

Solution

Area of field $= 25 \times 16 = 400$

If the lower area is one third of the whole, then:

$$\text{Lower area} = \frac{400}{3} = 133.\dot{3}$$

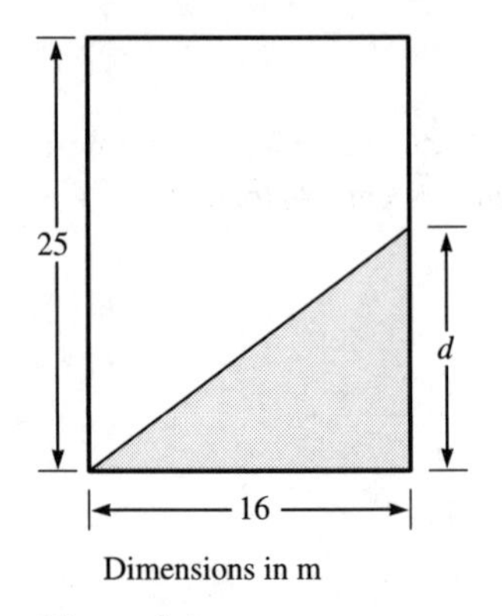

Figure 8.4

But we can also calculate the lower area as the area of a triangle, base 16, height d.

$$\text{Lower area} = \tfrac{1}{2} \times \text{base} \times \text{height} = \tfrac{16}{2} \times d = 8 \times d$$

These two results apply to the same area, so:

$$8 \times d = 133.\dot{3}$$

$$\Rightarrow \quad d = \frac{133.\dot{3}}{8} = 16.\dot{6} = 16.7 \text{ (3 sf)}$$

The length d is 16.7 m (3 sf).

Comment

We have to solve this equation:

$$8 \times d = 133.\dot{3}$$

To find d, we need to know how many 8's make $133.\dot{3}$. We find this by dividing $133.\dot{3}$ by 8.

Reference pages: 40, 54, 70.

Making a kite

Problem 7

A kite frame is made from two sticks of bamboo: AB = 400 mm, CD = 300 mm, glued together to form a cross. The crossing point E is 100 mm from A, and half-way along CD. A sheet of tissue is glued to the cross, as in Figure 8.5. What is the area of the tissue? Does it make any difference to the area if E is at a different distance from A?

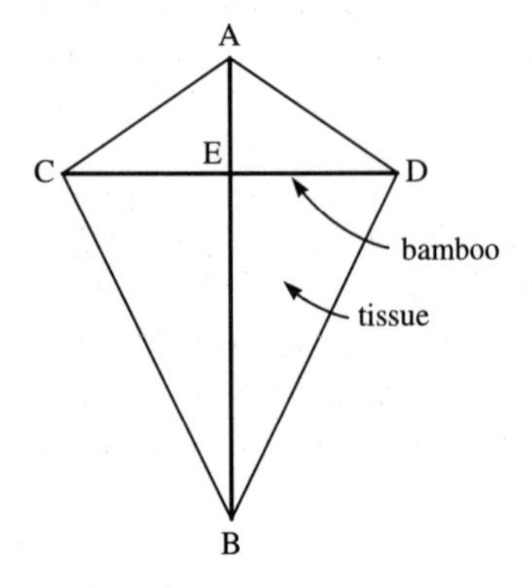

Figure 8.5

Solution

Think of the kite as two triangles, ADB and ACB. The base of $\triangle$ABD is AB, which is 400 mm. The height of $\triangle$ABD is DE, which is half of CD, so it is 150 mm.

$$\begin{aligned}\text{Area } \triangle\text{ADB} &= \tfrac{1}{2} \times \text{base} \times \text{height}\\ &= \tfrac{1}{2} \times 400 \times 150\\ &= 30\,000 \text{ mm}^2\end{aligned}$$

$\triangle$ACB has the same dimensions and so has the same area. The total area is that of the two triangles.

The area of tissue is $60\,000 \text{ mm}^2$.

If E is moved along AB, the base of each triangle is always AB and its height is always half of CD. So the area remains unchanged.

Moving CD along AB makes no difference to the area.

Comment

The position of E on AB did not come into the calculation, so makes no difference to the result. This leads to a general rule for kites, that:

Area = ½ × length × width

Reference pages: 72, 77.

Pulleys

Problem 8

A pulley system is tested by hanging different loads on it and measuring the effort required to lift the load. The results of the tests are:

Load (N)	10	20	30	40	50
Effort (N)	2.5	3.5	4.5	5.5	6.5

Plot a graph of these results, and deduce the equation relating effort to load. What effort is needed to lift the pulley system when:

a no load is attached to it

b the load is 35 N?

Solution

The effort depends on the load, so effort is the dependent variable and is plotted along the y-axis. Load is the independent variable and is plotted along the x-axis. All values are positive, so the origin is at the bottom left. Suitable scales are 1 square to 2 N for load, and 2 squares for 1 N for effort.

When the points are plotted, it is seen that this is a straight line graph. Points may be joined because it is possible to apply loads between the values given in the table. Figure 8.6 shows the finished graph.

The equation has the standard form for a straight-line graph. The line cuts the y-axis at +1.5, so the constant number in the equation is +1.5. The values of effort (y) increase by 1 for every increase of 10 in the values of load (x).

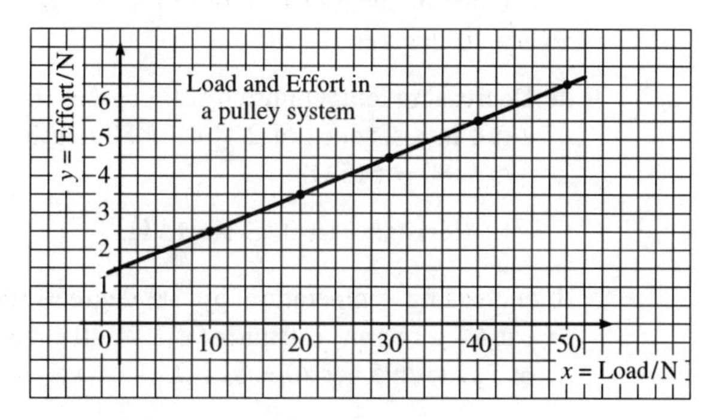

Figure 8.6

This is equivalent to y increasing by 0.1 for every increase of 1 in the value of x. The coefficient of x is +0.1. We can now write the equation:

$$y = 0.1x + 1.5$$

When load (x) is zero, effort (y) is 1.5.

Effort with no load attached is 1.5 N.

When load is 35, effort is 5.

Effort with 35 N load is 5 N.

Comment

The effort with zero load is that required to lift the pulley blocks (assuming no friction). In this example, we were not told the effort with zero load but were able to find it by extending the line to cut the y-axis.

Reference pages: 90, 92, 93, 95.

Resistance and temperature

Problem 9

The resistance of a piece of iron wire is measured at two temperatures. At 20 °C, it is 695 Ω. At 80 °C, it is 935 Ω. Assuming that the graph relating resistance to temperature is a straight line (which is true over a limited range), what is its resistance at

a 0 °C

b 50 °C?

Answer this problem *without drawing the graph.*

Solution

The graph, if drawn, would be a straight line, sloping up to the right. Increasing temperature by 60 degrees (from 20 °C to 80 °C) increases resistance by 240 Ω (from 695 Ω to 935 Ω). Resistance increases by 40 Ω for every 10 degrees rise.

a Decreasing temperature by 20 degrees (two tens) from 20 °C to 0 °C causes a fall in resistance of $2 \times 40 = 80$ degrees. Resistance falls from 695 Ω to $695 - 80 = 615$ Ω.

Resistance at 0 °C is 615 Ω.

b Increasing temperature by 30 degrees (three tens) from 20 °C to 50 °C causes a rise in resistance of $3 \times 40 = 120$ degrees. Resistance rises from 695 Ω to $695 + 120 = 815$ Ω.

Resistance at 50 °C is 815 Ω.

Checking Problem 9

If we increment the temperature from 50 °C to 80 °C (another rise of 30 degrees), the resistance becomes 815 + 120 = 935. This agrees with the data already given for 80 °C.

Comment

If we need to know that the graph of an equation is a straight line, we need to be told the location of only 2 points. We then find the change in the dependent variable (resistance) for unit change of the independent variable (temperature). In these examples all temperatures are rated in tens of degrees, so it is simpler if we find the change in resistance for a ten-degree change in temperature. Starting from one of the *known* points, we find the resistance at other temperatures.

References page: 93.

Writing equations

Problem 10

A box is made from card, with dimensions ℓ, w, and h as shown in Figure 8.7. It is open at the top. It is held together by adhesive tape. 1 m^2 of card weighs 500 g. In the questions which follow, ignore the thickness of the card and the weight of the tape. Assume that the tape runs the full length of each side and overlaps at the corners of the box.

Write equations for finding:

a the volume v of the box

b the area a of card used

c the weight w of the box

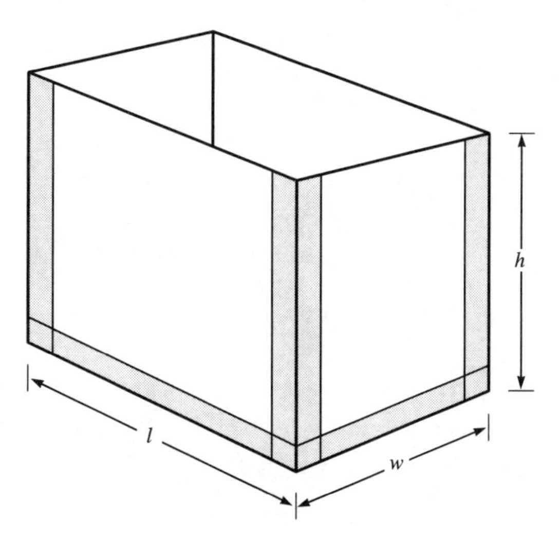

Figure 8.7

d the length t of tape used

e Find the volume and weight of a box for which $\ell = 0.390\,\text{m}$, $w = 0.220\,\text{m}$, and $h = 0.170\,\text{m}$.

f Find the area of card and length of tape needed to make a box for which $\ell = 0.400\,\text{m}$, $w = 0.200\,\text{cm}$ and $v = 0.008\,\text{m}^3$.
Express all results to 3 significant figures.

Solution

a The volume of the box is that of a rectangular prism.

Area of cross-section = width × height = wh
Volume = length × cross-section = ℓwh

$$\underline{v = \ell wh}$$

b Area of one side = ℓh ⇒ Area of two sides = $2\ell h$
Area of one end = wh ⇒ Area of two ends = $2wh$
Area of bottom = ℓw

$$\underline{\text{Total area} = a = 2\ell h + 2wh + \ell w}$$

c Weight = area of card × weight of $1\,\text{m}^2$

$$\underline{w = 500a}$$

d There are 4 vertical joins, length h: Total = $4h$
2 joins, sides-bottom, length ℓ: Total = 2ℓ
2 joins, ends-bottom, length w Total = $2w$

$$\underline{\text{Total length of tape} = t = 4h + 2\ell + 2w}$$

e $v = 0.39 \times 0.22 \times 0.17 = 0.014\,586$

Round to 3 sf.

<u>The volume is $0.0146\,\text{m}^3$ (3 sf).</u>

The area is:

$$a = 2 \times 0.39 \times 0.17 + 2 \times 0.22 \times 0.17 + 0.39 \times 0.22$$
$$= 0.1326 + 0.0748 + 0.0858 = 0.2932$$

The area is $0.2932\,\text{m}^2$.

$$w = 500a = 500 \times 0.2932 = 146.6$$

Round to 3 sf.
<u>The weight is 147 g.</u>

f To find the length of tape we need to know h, which is not given. But we are told that $v = 0.008$. Use the equation for volume:

$$0.008 = 0.4 \times 0.2 \times h$$
$$= 0.08h$$

To find what number multiplied by 0.08 gives 0.008, we divide:

$$h = \frac{0.008}{0.08}$$

Shift the decimal point of numerator and denominator 3 places to the left to eliminate the decimal fractions and make the division easier:

$$h = \frac{8}{80} = 0.1$$

The height is 0.1 m. Now we can calculate t:

$$t = 4 \times 0.1 + 2 \times 0.4 + 2 \times 0.2$$
$$= 0.4 + 0.8 + 0.4 = 1.6$$

Results are correct to 3 sf.

The length of tape is 1.60 m.

The area of card is:

$$a = 2 \times 0.4 \times 0.1 + 2 \times 0.2 \times 0.1 + 0.4 \times 0.2$$
$$= 0.08 + 0.04 + 0.08 = 0.2$$

The area of card is $0.200\,m^2$ (3 sf).

Checking Problem 10

Part **e**: Estimate result after rounding to 1 sf:

$$v = 0.39 \times 0.22 \times 0.17 \approx 0.4 \times 0.2 \times 0.2 = 0.016$$

This is close to the calculated result. Similarly for other multiplications in this part.

Comments

This is an exercise in working with decimal fractions, and learning to keep the decimal point in the right place.

Reference pages: 54, 76, 83.

On your own

Here are versions of the same problems but with different numbers. Try to solve these on your own.

Problem 1

The panel is the same size, but there are now 5 apertures, 20 mm long and 15 mm wide.

Problem 2

Work out the recipe for a larger cake, using 5 eggs.

Problem 3

There has been an increase in prices, and M10 bolts are now packed in bags of 15:

Thread size	*Quantity per bag*	*Price per bag*
M6	50	4.40
M8	25	4.60
M10	15	4.10
M12	10	3.90

The supervisor places this order:

Thread size	*Number of bags*
M6	8
M8	10
M10	7
M12	2

Problem 4

A smaller sheet of hardboard is supplied, 1.83 m × 1.22 m. The boards are to be 370 mm × 190 mm.

Problem 5

Girders are 3.8 m long. The width is increased to 120 mm.

Problem 6

The length of the field is now 27 m. The triangular area is to be one quarter of the whole.

Problem 7

AB = 420 mm, CD = 350 mm.

Problem 8

Using a different pulley system, the results are:

Load (N)	10	30	50	70	90
Effort (N)	4.5	8.5	12.5	16.5	20.5

Problem 9

The wire is copper; its resistance is 535 Ω at 40 °C and 655 Ω at 100 °C. What is its resistance at 0 °C and at 75 °C?

Problem 10

1 m^2 of card weighs 420 g. For part **e** the dimensions of the box are $\ell = 0.410$ m, $w = 0.120$ m, and $h = 0.190$ m. For part **f**, $\ell = 0.300$ m, $w = 0.200$ m, and $v = 0.030\,m^3$.

Part 2 – Basic Maths

The maths in this part is needed in almost all branches of technology. To study this section, you need to know everything that is in Part 1.

9 Factors

Possibly you may have covered this topic already, in other maths courses. Try the test questions below and check your answers at the back of the book. If you score full marks, or nearly, work a few questions from each of the *Test yourself* exercises. Concentrate on questions which seem difficult to you. This is useful revision, after which you will be ready to go on to the next chapter.

Try these first

1 Without using a calculator or paper, and without dividing by 3, say which of these numbers can be divided exactly by 3:

354, 491, 732, 8642, 5328, 73 281, 95 232

2 Say which of these are prime numbers:

9, 7, 35, 17, 51, 2, 77, 47, 23

3 Factorise 90.

4 Find the highest common factor (HCF) of 70 and 385.

5 Find the lowest common multiple (LCM) of 84 and 45.

Products in reverse

One way of thinking about factors is illustrated in the box.

A catering problem

In how many ways can a string of 12 sausages be divided *equally* between host and guests at a barbecue? Sausages are to be served *whole*.

If no-one came to the barbecue, so that there was only the host, the host could have all 12 sausages (Figure 9.1a). If there was one guest, the string could be cut into 2 equal halves, giving 6 sausages each (Figure 9.1b). If two more people came, the two strings of 6 could be divided again into halves and each person would have 3 sausages (Figure 9.1c). Now start again with the string of 12 sausages and let two guests come, making three people to share the sausages equally. This can be done by cutting the string into fours (Figure 9.1d). These could also be cut into two equal halves, sharing the sausages among 6 people (Figure 9.1e). Finally, the host could cut the string into separate sausages, making enough for 12 people, though with only 1 sausage each. These are all the ways in which the sausages can be shared equally.

Figure 9.1 shows that, depending on the number of people at the barbecue, the number of sausages a person may have will be one of:

12, 6, 4, 3, 2, or 1

These numbers are known as the **factors** of 12. These are the only possible ways the sausages may be shared, if we keep to the rules of equal shares and whole sausages. There is no way in which a guest can receive, say, 5 sausages without having more or fewer than the other people at the barbecue.

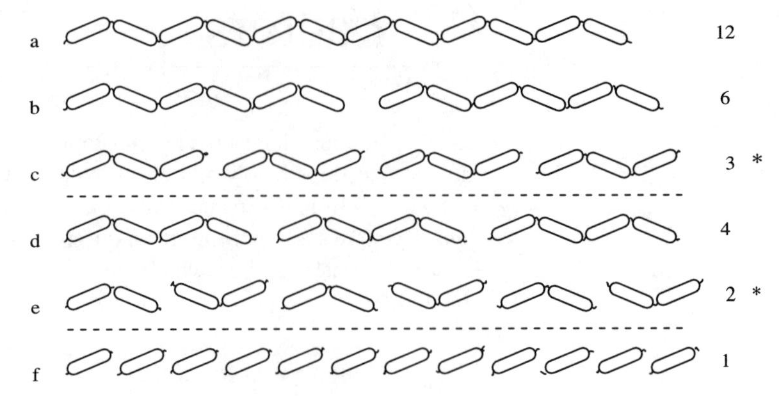

Figure 9.1

Here is another way of looking at the same idea. When we multiply a string of numbers together, the result is called the **product**:

$2 \times 2 \times 3 = 12$

The product of 2, 2, and 3 is 12. Now to pose the reverse problem. Which numbers multiplied together give a product of 12? The numbers we are looking for are called the **factors** of 12.

Multiplication of numbers such as the above is quick and easy on a calculator, or you may even be able to do it in your head. But breaking a number down into factors (we call this **factorizing**) is more of a problem. There is no quick and easy way to do it on a calculator. Factorizing is something we often have to do in almost all branches of maths. This is why this chapter comes first in this section of the book.

To be able to factorize, you need to know:

- the factor rules
- the prime numbers up to 97
- the multiplication tables up to 12×12.

We will look at these requirements individually.

Factor rules

One of these rules (see box) says that:

If 5 is a factor, the number ends in 0 or 5

Here are some examples of numbers which have 5 as a factor:

5 10 65 125 140 3625

The advantage of the rule is that given a long number, such as 6 743 289 043 325, we can say immediately that 5 is a factor of it, without having toactually divide it by 5. Conversely, we can say that 4 052 738 922 does *not* have 5 as a factor.

Factor rules

If 2 is a factor, the number is even and ends in 0, 2, 4, 6, or 8.
If 3 is a factor, the digits add up to 3, 6, or 9.
If 4 is a factor, the last two digits divide by 4.
If 5 is a factor, the number ends in 0 or 5.
If 8 is a factor, the last three digits divide by 8.
If 9 is a factor, the digits add up to 9.
If 10 is a factor, the last digit is 0.

The rule for factor 3 says that:

If 3 is a factor, the digits add up to 3, 6 or 9

Take 45 as an example. The digits 4 and 5 add up to 9, so 3 is a factor of 45 ($45 = 3 \times 15$). Similarly we can tell that 3 is a factor of 51, because $5 + 1 = 6$. If the sum of digits comes to more than 9 we add up the digits in the sum. We repeat this until we obtain 3, 6, or 9, or some other single digit.

Examples

327	Add digits:	$3 + 2 + 7 = 12$
	Add digits again:	$1 + 2 = 3$

Finishing with 3 tells us that 3 is a factor of 327.

5732	Add digits:	$5 + 7 + 3 + 2 = 17$
	Add digits again:	$1 + 7 = 8$

This is not 3, 6 or 9 so 3 is not a factor of 5732.

Prime numbers

A prime number is a positive integer (whole number) which has only 1 and itself as factors. For example, the factors of 7 are 1 and 7:

$$1 \times 7 = 7$$

Limited options

Once again the host is having a barbecue, but has only 7 sausages. There are only two options:

- invite nobody and eat all 7 sausages
- invite 6 other guests, so that each person has 1 sausage.

The number 7 is one of many that offers only *two* options. There is no way of sharing it except to cut it into 7 individual sausages. We call such a number a **prime number**.

If you have only 1 sausage, there is only *one* option – to eat it yourself! So 1 is not a prime number.

In Figure 9.1, the numbers marked with a * are two more examples of prime numbers.

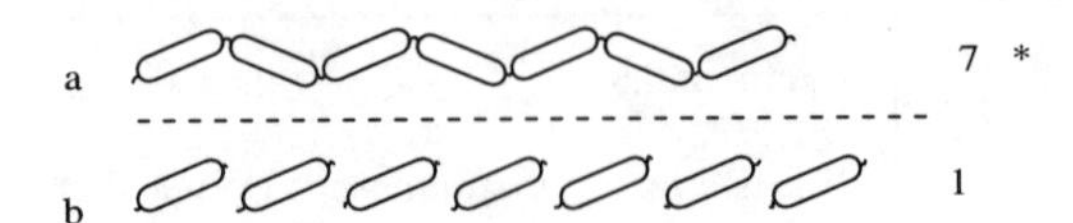

Figure 9.2

There is no other pair of integers which give 7 when multiplied together. By contrast, 6 can be obtained in two ways:

$$1 \times 6 = 6$$
$$2 \times 3 = 6$$

The second pair does not contain 1 or the number itself so 6 is *not* a prime number. The box lists all the prime numbers up to 97. You should try to remember this list; it will prevent you from wasting time trying to factorize a number that is prime.

Prime numbers

The first 20 prime numbers are:

2, 3, 5, 7
11, 13, 17, 19
23, 29
31, 37
41, 43, 47
53, 59
61, 67
71, 73, 79
83, 87
91, 97

All prime numbers are integers and are positive. Except for 2, all prime numbers are odd. Except for 5, no prime number ends in 5.

Easy factorizing

The multiplication tables list many pairs of factors. If you know the tables well, you can factorize most of the smaller numbers immediately. It may help you to think of the number as a string of sausages being shared equally at a barbecue.

Examples

Factorize 55.

We know from the table for 5 that:

$$11 \times 5 = 55$$

This gives two of the factors, the other are the standard ones, 1 and the number itself (55). Factors 5 and 11 are both primes so it is not possible to split them into smaller factors.

The factors of 55 are 1, 5, 11 and 55.

Factorize 24.

From the table for 2:

$12 \times 2 = 24$

From the table for 3:

$8 \times 3 = 24$

From the table for 4:

$6 \times 4 = 24$

Here we have three different pairs of factors. Three of them (6, 8, 12) are not primes so they can be factorized again. But they give only 2, 3 and 4, which we have already had. Including 1 and 24 in the list:

The factors of 24 are 1, 2, 3, 4, 6, 8, 12 and 24.

Test yourself 9.1

1 Without looking at the box, say which of these numbers are prime.
57, 3, 56, 23, 7, 45, 88, 59

2 Which of these numbers has 3 as a factor?
6, 27, 98, 123, 1110, 455, 735, 313

3 Which of these numbers has 4 as a factor?
24, 624, 534, 118, 68, 112, 704, 7010

4 Which of these numbers has 5 as a factor?
15, 100, 51, 65, 4325, 5532, 25, 316

5 Which of these numbers has 8 as a factor?
56, 74, 556, 7186, 60016, 736, 326, 72

6 Which of these numbers has 9 as a factor?
69, 108, 5301, 64251, 399, 102, 5418, 6303

7 Factorize these numbers.

a 33 **b** 50 **c** 63 **d** 84
e 120 **f** 16 **g** 45 **h** 36

Prime factors

In the examples above we have simply listed all the different ways in which a number can be divided exactly. Sometimes we need to split a number into its prime factors. We can do this by listing the factors and crossing out those which are not primes.

Example

Find the prime factors of 24.

List the factors of 24 (see above):

1, 2, 3, 4, 6, 8, 12, 24

Omitting the non-primes leaves only:

2 and 3

Remember that 1 is not a prime. But 2 × 3 does not equal 24. If we are splitting 24 into primes, then either 2 or 3 or both must occur several times. We can find this out by repeated division:

Step	*Division*	*Record the factor*
Divide 24 by 2	24/2 = 12	2
Divide 12 by 2	12/2 = 6	2
Divide 6 by 2	6/2 = 3	2
3 is a prime		

Listing the prime factors as many times as they ocur:

The prime factors of 24 are 2, 2, 2, and 3.

We can also say that:

$2 \times 2 \times 2 \times 3 = 24.$

The number is expressed as the product of its prime factors.

Finding prime factors

1 Divide the number by 2.
2 If it divides exactly, record '2', then repeat **1** with the quotient.
3 If it does not divide exactly, return to **1**, but use the next highest prime.
4 Repeat **1** to **3** until the quotient is prime.

The box summarizes this routine for finding prime factors. At each stage the factor rules help us to decide if the quotient divides out exactly or not. Here is an example of the routine being used for a larger number:

Step	*Division*	*Record the factor*
Divide 126 by 2	126/2 = 63 exactly	2
Divide 63 by 2	63/2 does not go exactly	–
Divide 63 by 3	63/3 = 21 exactly	3
Divide 21 by 3	21/3 = 7 exactly	3
Divide 7 by 3	7/3 does not go exactly	–
Divide 7 by 5	7/5 does not go exactly	–
The next prime is 7		

The prime factors of 126 are 2, 3, 3, and 7.

Check the result, using a calculator: 2 × 3 × 3 × 7 = 126. Result confirmed.

Having found the prime factors we can, if we need to, multiply them together in pairs, threes or more to find the other (non-prime) factors:

Pairs	$2 \times 3 = 6$
	$2 \times 7 = 14$
	$3 \times 3 = 9$
	$3 \times 7 = 21$
Threes	$2 \times 3 \times 3 = 18$
	$2 \times 3 \times 7 = 42$
	$3 \times 3 \times 7 = 63$
All four	$2 \times 3 \times 3 \times 7 = 126$

The other factors are 1 and the prime factors already found. List the factors in numerical order.

The factors of 126 are 1, 2, 3, 6, 7, 9, 14, 18, 21, 42, 63, and 126.

Test yourself 9.2

1 List the prime factors of each of these numbers.

a 18	**b** 45	**c** 161	**d** 147
e 84	**f** 975	**g** 1265	**h** 3509

2 List all the factors of each of these numbers.

a 8	**b** 18	**c** 36	**d** 165

Highest common factor

Given two integers, the highest common factor (or HCF) is the largest number that is a factor of both of them. In terms of barbecues, given two barbecues with different numbers of sausages to be cooked, what is the largest portion of sausages that can be served at *both* barbecues?

Example

What is the HCF of 12 and 30?

We aim to find the highest number that is a factor of 12 and of 30. In other words, the highest factor that the two numbers have *in common*. We start by finding the prime factors of both numbers (using the technique in the box on page 124). Then we list them in two rows:

Prime factors of 12 are:	2	2	3	
Prime factors of 30 are:	2		3	5

Note how we have arranged the factors in columns. 12 contains factor 2 twice, so we write two 2's side by side, in the first two columns. But 30 contains factor 2 only once, so we write 2 in the first column, but not in the second. The threes (common to both numbers) fill the next column. There is one 5 in the last column, because 5 is a factor of 30 but not of 12.

We are looking for **common** factors, so circle the columns in which there are two entries:

(2)	2	(3)	
(2)		(3)	5

The numbers have the factors 2 and 3 in common. We are looking for the *largest* number which is a factor of both. So the HCF is the product of these:

$\text{HCF} = 2 \times 3 = 6$

In terms of barbecues, 6 is the longest string that can be cut from a string of 12 AND from a string of 30 sausages (Figure 9.3).

It is also possible to find the HCF of 3 (or more) numbers.

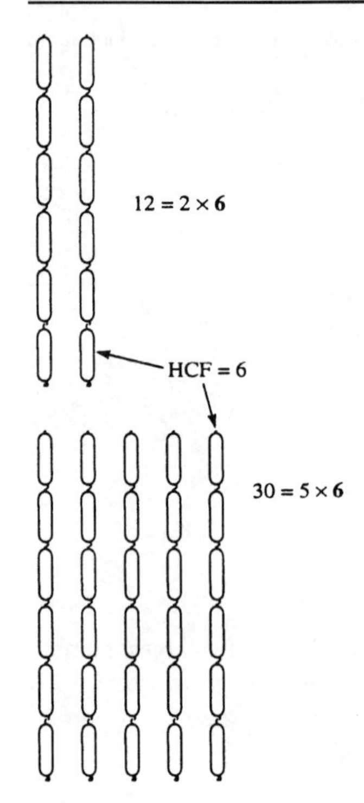

Figure 9.3

Finding the HCF

1 Find the prime factors of each number.
2 List them in columns, matching numbers in the same columns.
3 The *full* columns indicate common factors.
4 Multiply the common factors to find the HCF.

If there are no full columns, there is no HCF.

Examples

Find the HCF of 30, 75 and 45.

Prime factors of 30 are:	2	3		5	
Prime factors of 75 are:		3		5	5
Prime factors of 45 are:		3	3	5	

The full columns show which factors are common to all three numbers: 3 and 5.

$\underline{\text{HCF} = 3 \times 5 = 15}$

Example when one number is a prime:

Find the HCF of 5 and 105.

Prime factor of 5 is itself:		5	
Prime factors of 105 are:	3	5	7

The only common factor is the first number itself.

$\underline{\text{HCF} = 5}$

However, if two numbers are *both* primes there is no HCF.

HCF and cancelling

Cancelling fractions helps to keep the numbers small. This makes it more practicable to work quickly in your head, and to avoid arithmetical errors. Cancelling (page 33) means dividing both numbers by the factors they both contain. In other words, we divide them both by their highest common factor.

Examples

Cancel $\frac{12}{30}$

The HCF of 12 and 30 is 6 (see example, page 125). Cancelling by 6:

$$\frac{12}{30} = \frac{2}{5}$$

Cancel $\frac{18}{45}$

The HCF of 18 and 45 is 9. Cancel by 9:

$$\frac{18}{45} = \frac{2}{5}$$

Lowest common multiple

Given two or more numbers, their lowest common multiple (LCM) is the smallest number into which these given numbers divide exactly. In other words, the LCM is the smallest number that has all of the given numbers as factors.

Example

What is the LCM of 12 and 30? Compare the working of this example with that of the HCF example on page 125.

As before, start by finding the prime factors of both numbers. Then we list them in two rows:

Prime factors of 12 are:	2	2	3	
Prime factors of 30 are:	2		3	5

If the LCM is to have 12 as a factor, then 2, 2 and 3 must be factors of the LCM. Listing its factors so far:

2 2 3 + others?

> **Finding the LCM**
>
> 1 Find the prime factors of each number.
> 2 List them in columns, matching numbers in the same columns.
> 3 *Every* column indicates one factor of the LCM, even if not full.
> 4 Multiply the factors to find the LCM.
>
> If number have no factors in common, the LCM is the product of the numbers.

The LCM must also have 30 as a factor, so it needs 2, 3 and 5 as factors. But we have already listed 2 and 3, for these are common factors and do not need to be listed again. Only 5 needs to be added to the list. The factors of the LCM are:

2 2 3 5

Multiplying the factors to obtain the LCM:

$2 \times 2 \times 3 \times 5 = 60$

The LCM of 12 and 30 is 60.

Check: 60/12 = 5 and 60/30 = 2. A string of 60 sausages is the shortest that can be cut into 12's OR into 30's (Figure 9.4).

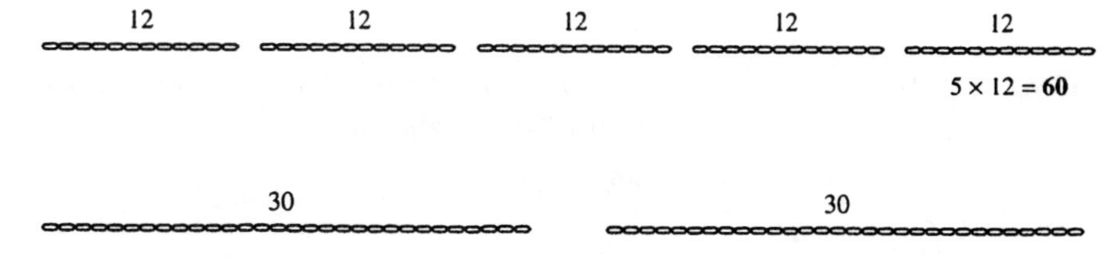

Figure 9.4

Once you have understood the reasoning behind finding the LCM, there is no need to run through it every time. Just look at the columns of factors and write down a factor for every column. The column does *not* have to be full.

Examples

Find the LCM of 28 and 130.

Prime factors of 28 are:	2	2		7	
Prime factors of 130 are:	2		5		13

Multiplying factors:

$2 \times 2 \times 5 \times 7 \times 13 = 1820$

The LCM of 28 and 130 is 1820

An example with no common factors:

Find the LCM of 9 and 14.

Prime factors of 9 are:		3	3	
Prime factors of 14 are:	2			7

These numbers have no factors in common. Multiplying the factors:

$2 \times 3 \times 3 \times 7 = 126$

The LCM of 9 and 14 is 126

Since there are no common factors we have to multiply *all* the factors of 9 and *all* the factors of 14. We are really multiplying 9 by 14, and the LCM is their product.

Test yourself 9.3

1 Find the HCF of:

a 12 and 10 **b** 99 and 66 **c** 56 and 36
d 42 and 66 **e** 102 and 17 **f** 26 and 299
g 23 and 41 **h** 42, 140, and 210 **i** 1155, 3003 and 66

2 Cancel these fractions, using HCFs.

a 14/98 **b** 42/105 **c** 30/78

3 Find the LCM of:

a 12 and 10 **b** 99 and 66 **c** 56 and 36
d 23 and 48 **e** 70 and 15 **f** 174 and 82
g 360 and 21 **h** 20, 42 and 39 **i** 100, 42 and 70

LCMs and fractions

LCMs are useful when adding or subtracting common fractions. To see how, consider these examples.

Examples

Evaluate $\frac{3}{10} + \frac{4}{15}$

The fractions can not be added directly as they have different Ds (denominators, see page 29). They have to be given the *same* D. The rule on page 35 is to multiply the Ds:

$$10 \times 15 = 150$$

Work in 150-ths. But there is an easier way to find a suitable D without using such a large number as 150. The D must be a number that has 10 as a factor and has 15 as a factor. In other words:

Use the LCM as the new D

We call this denominator the **lowest common denominator** (LCD). Using the technique described in the previous section, we find that the LCD of 10 and 15 is 30. Work in thirtieths (Figure 9.5).

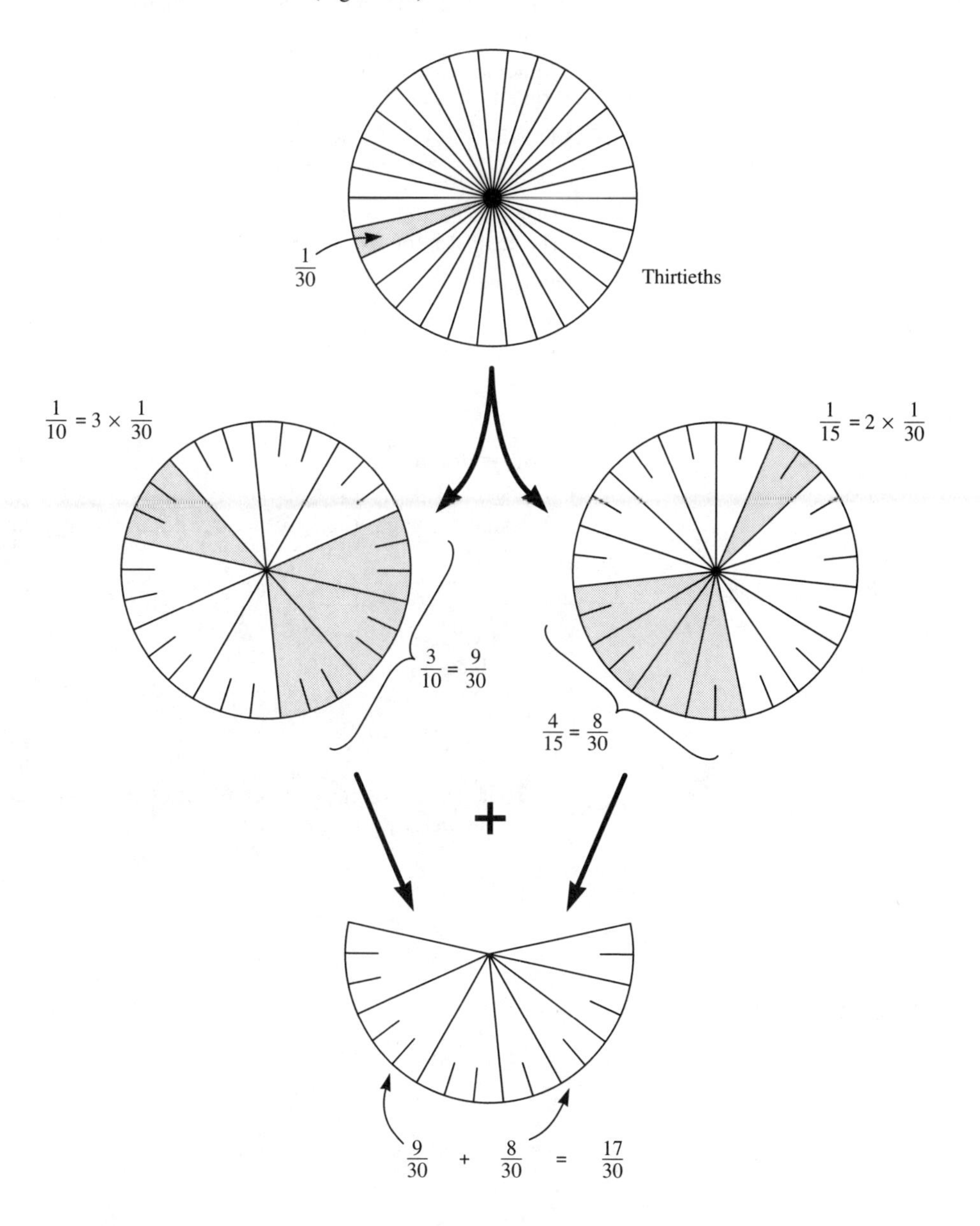

Figure 9.5

Find the equivalent of 3/10 (page 33), in thirtieths: 3/10 = 9/30
Find the equivalent of 4/15, in thirtieths: 4/15 = 8/30

Add the equivalents:

$$\frac{3}{10} + \frac{4}{15} = \frac{9}{30} + \frac{8}{30} = \frac{17}{30}$$

We obtain the same result as by working in 150-ths, but using much smaller numbers, so making the calculation much easier and with less risk of making mistakes. Most examples you will be able to work in your head.

Evaluate $\frac{33}{40} - \frac{11}{16}$

According to the rule on page 35, we work in 640-ths, but this is a very large number. Find the LCD instead:

Prime factors of 40 are:	2	2	2		5
Prime factors of 16 are:	2	2	2	2	

LCD is $2 \times 2 \times 2 \times 2 \times 5 = 80$.

Work in 80-ths.

The equivalent of 33/40 in 80-ths: 33/40 = 66/80
The equivalent of 11/16 in 80-ths: 11/16 = 55/80

Subtracting:

$$\frac{33}{40} - \frac{11}{16} = \frac{66}{80} - \frac{55}{80} = \frac{11}{80}$$

Test yourself 9.4

1 Evaluate, using the LCD, and without using a calculator.

a $\frac{2}{3} + \frac{5}{6}$ **b** $\frac{5}{6} - \frac{3}{4}$ **c** $\frac{3}{10} + \frac{5}{6}$

2 Evaluate, using the LCD.

a $\frac{5}{14} + \frac{13}{35}$ **b** $\frac{7}{9} - \frac{2}{15}$ **c** $\frac{5}{6} - \frac{4}{15} + \frac{3}{10}$

3 To assess if you have covered this chapter fully, answer the questions in *Try these first*, page 119.

10 Ratios

One way of comparing two quantities is to state the *ratio* between them. This chapter describes different kinds of ratios and how to work with them.

You need to know only the Maths Essentials (Part 1) for this chapter.

Try these first

Your success with this short test will tell you which part of this chapter you already know.

1 A box contains 12 red pens and 9 blue pens. What is the ratio of red compared with blue? Simplify the ratio, if possible.

2 A map is drawn on the scale 4 cm to 1 km. Express this as a ratio and as a fraction. On the map, the runway of a small airfield is 37 mm long. What is its actual length?

3 A rectangular chart is drawn 180 mm wide and 200 mm long. A reduced-size photocopy is made that is 150 mm long. What is its width?

4 The angles of a triangle are in the ratio 5:6:7. Find the angles.

5 The volume of oxygen dissolved in a given volume of water decreases by 32% when the water is warmed from 0 °C to 30 °C. If the volume dissolved is 38 mm^3 at 0 °C, what volume is dissolved at 30 °C?

6 The weight of a monthly magazine is directly proportional to the number of pages in it. A 52-page magazine weighs 109.2 g. What is the weight of a 36-page magazine?

7 The current through a resistor is inversely proportional to its resistance (assuming a fixed voltage across it). If the current through a 47 Ω resistor is 2.5 A, what is the current when the resistor is changed to 68 Ω? (2 sf)

Ratios

Figure 10.1 shows examples of ratios. In Figure 10.1a, there are 6 people wearing spectacles *compared with* 4 people not wearing spectacles. To express the idea 'compared with', we use the symbol ':'. So we say The number of spectacle-wearers compared with non-wearers is 6:4. Or we say:

The **ratio** of spectacle-wearers to non-wearers is 6:4.

The minute hand of a clock (Figure 10.1b) is usually long compared with the hour hand. In this clock the ratio of the minute hand to the hour hand is

Figure 10.1

80:50. Although the ratio is derived from measurements in millimetres, we do not use units when quoting a ratio.

Ratios are always quoted in pure *numbers*, without units

In Figure 10.1c we are using a ratio to define the composition of concrete. Take 5 shovelfuls of cement, 10 shovelfuls of sand, and 15 shovelfuls of gravel. The ratio of cement to sand to gravel is 5:10:15. This is an example of a ratio between 3 quantities. This mixture yields a certain quantity of concrete. If we need twice as much concrete, we take 10 shovelfuls of cement, 20 shovelfuls of sand and 30 shovelfuls of gravel. The ratio is 10:20:30. This ratio is the *same* as 5:10:15, for it produces concrete of identical composition. For three times the amount of concrete, the ratio is 15:30:45, which is the same ratio again, 5:10:15.

If we multiply all numbers in the ratio by the same amount, the ratio remains the same

We can also divide the numbers by the same amount. For example, to get half the quantity of concrete we take 2½ shovelfuls of cement, 5 shovelfuls of sand and 7½ shovelful of gravel. The ratio is 2½:5:7½. The best number to divide by is the HCF of 5, 10, 15, which is 5. Dividing by 5 simplifies the ratio to 1:2:3. This ratio is used to describe a popular grade of ready-mixed concrete.

An interesting ratio is shown in Figure 10.1d. This is the ratio of the length to the width of a piece of A4 paper, both dimensions in millimetres. The ratio of length to width is 297:210, giving A4 its characteristic shape. Dividing *both* numbers by 210, we obtain the ratio 1.414:1. A4 paper is 1.414 times as long as it is wide. Now imagine the paper folded in half, along the dashed line. The size of one half is 210 mm long by 148.5 mm wide. The ratio of length to width is 210:148.5. Divide *both* numbers by 148.5, and we obtain 1.414:1. The half sheet too is 1.414 times as long as it is wide. It has exactly the *same shape* as A4 paper. Folding A4 in half gives a sheet with half the area but the same shape. This is the A5 size. We can continue in this way, folding A5 to get A6, and A6 to get A7, but all sizes have exactly the same shape.

Figure 10.1e illustrates a special ratio known as **aspect ratio**. In general, it is the ratio of length to width. Here it is used to describe the wing shape of an aeroplane. It is the ratio of the average wing width (the chord) to the wing span. In Figure 10.1e, the span is given in metres but the chord is given in centimetres. Before we work out the ratio, the two dimensions must be in the same units. Convert 8 m to 800 cm. The aspect ratio is 80:800. Dividing by 80 (the HCF of 80 and 800) simplifies the ratio to 1:10. A high aspect ratio (wide span, narrow wings) is typical of wings with good lifting properties, and is often seen in gliders and high-altitude aeroplanes.

Ratios must be derived from quantities measured in the same units

The use of a ratio to describe shape is extended to 3 dimensions in Figure 10.1f. The ratio of length to width to height is 8:6:4. Note that we have not said what the units are. They might be metres, centimetres, feet, or yards. With different units the boxes are different sizes, but all have the same shape. In other words, the ratio does not define the *size* of the box, but it exactly defines its *shape*.

Ratios are just *numbers*; they do not have units

The fact that the United Kingdom has an area of 245 thousand square kilometres, does not mean much to most people. If we say that the areas of

UK, France and Spain are in the ratio, 245:547:505 it becomes easier to make comparisons. We can see at once that the area of UK is about half that of the other two countries, as shown in Figure 10.1g. This diagram also states the populations in millions, the ratio being 56:54:37. It is clear that the population ratio is not the same as the area ratio and that, on average, UK is far more crowded than Spain.

Assuming the note on the left in Figure 10.1h is the A below middle C, its frequency is 220 Hz. The frequency of the note on the right is 440 Hz. Their ratio is 220:440. Dividing by 220 simplifies the ratio to 1:2. A doubling of frequency gives the musical interval known as an **octave**. The same ratio applies to any other two notes that are an octave apart. Other musical intervals such as thirds and fifths are also defined by frequency ratios.

Ratios are simplified by dividing by the HCF of the numbers

Example

Simplify 9:12.

The HCF of 9 and 12 is 3. Dividing by 3 simplifies the ratio to 3:4.

Very often the smaller or smallest number is a factor of the others, in which event this number simplifies to 1. There are several examples of this above.

Test yourself 10.1

1 The passengers on a bus consist of 12 adults and 3 children. What is the ratio of adults to children? Simplify this ratio.

2 The dimensions of a poster are 500 mm wide and 750 mm long. What is the ratio of width to length? Simplify this ratio.

3 Simplify these ratios.

a 3:15	**b** 100:10	**c** 51:17	**d** 12:8
e 42:7	**f** 3:21:6	**g** 15:50:5	**h** 4:14:10

4 A calculator measures 70 cm × 30 cm × 8 mm. What is the ratio of its dimensions?

Using ratios

Given a ratio and one of the quantities, we can calculate the other quantity or quantities involved.

Examples

A poster is designed with its width and height in the ratio 4:7. Its width is 520 mm. What is its height?

Solve the problem step-by-step, keeping the ratio unchanged. Begin by stating the given ratio.

If the width is 4, the height is 7

If the width is 1, the height is $\frac{7}{4}$ (divide both the width and height by 4)

If the width is 520, the height is $\frac{7}{4} \times 520$ (multiply both the width and the height by 520)

To find the new *height*, we first divide both by the old *width*. Then we multiply both by the new *width*.

Finally evaluate the height: $\frac{7}{4} \times 520 = 7 \times 130 = 910$

The height is 910 mm.

With a mechanical jack, the effort required to lift a given load is in the ratio 3:35. What effort is needed to lift a load of 15 kN?

Begin by stating the given ratio:

An effort of 3 lifts a load of 35

An effort of $\frac{3}{35}$ lifts a load of 1 (divide both effort and load by 35)

An effort of $\frac{3}{35} \times 15$ lifts a load of 15 (multiply both effort and load by 15)

To find the new *effort* we first divide by the old *load*. Then we multiply by the new *load*.

Finally, we evaluate the effort: $\frac{3}{35} \times 15 = \frac{45}{35} = \frac{9}{7} = 1.29$

The effort is 1.29 kN (3 sf).

Scale models

A scale model is usually smaller than the original but has the same shape. For this reason, ratios are important to model-makers. Figure 10.2 shows a house, and our task is to work out the dimensions of a model of that house.

One of the first steps in model-making is to decide on a **scale**. This is the ratio of the dimensions of parts of the model to the dimensions of the corresponding parts of the original. Assume the model house is being made as scenery for a 00 scale model railway. This is rather a strange scale because it is based on the ratio:

4 mm on the model represents 1 foot on the original

Here we have a mix of metric and Imperial units. 1 foot is equivalent to 304.8 mm, so we could express the 00 scale like this:

The 00 scale is 4:304.8

Simplifying, by dividing by 4:

The 00 scale is 1:76

The exact ratio is 1:76.2, but we have rounded it to 2 sf, which is precise enough for most model-makers.

Another way of writing a scale is to express it as a **scale factor**. This is the value of the quotient:

$$\frac{\text{Length on model}}{\text{Length on original}}$$

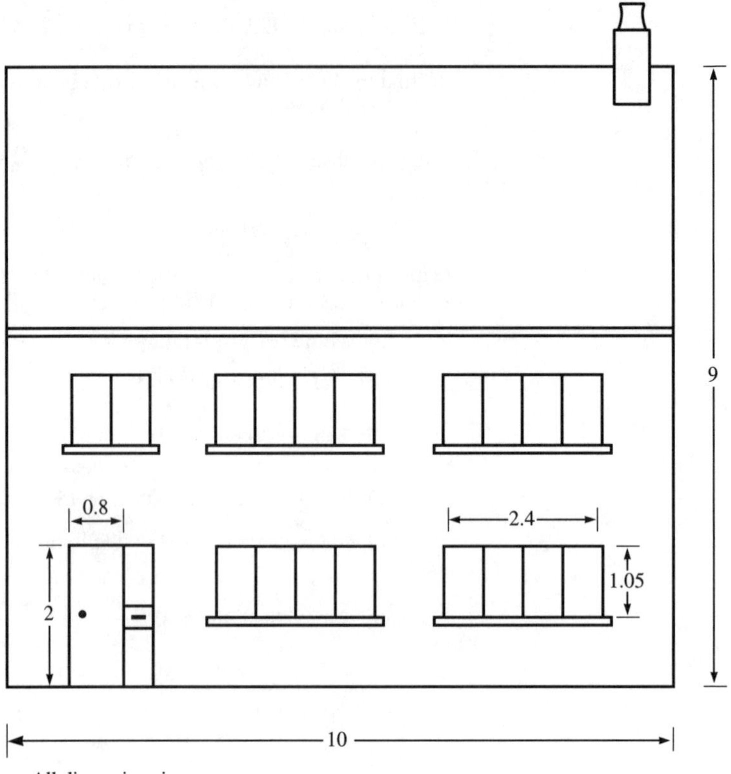

Figure 10.2

If 1 mm represents 76 mm, every dimension on the model is $\frac{1}{76}$ of the corresponding dimension on the original.

The 00 scale factor is $\frac{1}{76}$.

A general rule is that if the ratio is $1:n$, the scale factor is $1/n$. To convert dimensions of the house to dimensions of the model, we can use the method described in the previous section. The same result is obtained more directly if we multiply by the scale factor.

Example

The windows are 2.4 m wide. Since the model is measured in mm, convert 2.4 m to mm:

$$2.4 \times 1000 = 2400$$

Multiply by the scale factor:

$$2400 \times \frac{1}{76} = \frac{2400}{76} = 31.6 \text{ (3 sf)}$$

The windows on the model are 31.6 mm wide.

Converting from model to original is the inverse operation. Instead of multiplying by the scale factor, we multiply by its reciprocal, which is 76:

The chimney is 6.6 mm wide on the model:

$6.6 \times 76 = 500$ (2 sf)

Convert to metres:

$$\frac{500}{1000} = 0.5$$

The chimney of the house is 0.5 m wide.

A working drawing that is actual size has a ratio of 1:1 and the scale factor is 1. The pieces of a pattern for making clothing are an instance where a 1:1 ratio is used. Other models and plans may be larger than life size. For example, the art-work for pcbs is often done at twice the finished size, a ratio of 2:1, scale factor 2.

Test yourself 10.2

1 The ratio of the width of a standard TV screen to its height is 4:3. If a TV set has a screen 250 mm wide, what is its height?

2 Brass consists of copper alloyed with zinc in the ratio 7:3. If a given quantity of brass contains 24 g of zinc, how much copper does it contain?

3 On the 00 scale, what are the dimensions of the door and the height of the model of the house in Figure 10.2? (3 sf)

4 On a 00 scale model of the house the window panels beside the door measure 5.3 mm × 12.5 mm. What are the dimensions of these windows on the original house in mm?

5 The HO model scale is 3.5 mm to 1 foot. Express this as a ratio and as a scale factor. Calculate the height of the door and the width of the house of an HO scale model based on the house of Figure 10.2. (3 sf)

6 A popular scale for model warriors in adventure gaming is 25 mm high. Assuming that this represents a warrior 1.8 m tall, what is the scale of these figures, as a ratio and as a scale factor?

7 The Landranger series of maps by the Ordnance Survey are on the scale of 2 cm to 1 km. Express this as a ratio and as a scale factor. A section of power-line measured on the map is 58 mm long. What is its actual length? A reservoir is 875 m long. What is its length on the map?

8 In the worm gear assembly of Figure 10.3, the sprocket wheel has 15 teeth. It takes 15 turns of shaft A to turn shaft B once. Express this as a ratio. If shaft A turns 90 times, how many times does shaft B turn?

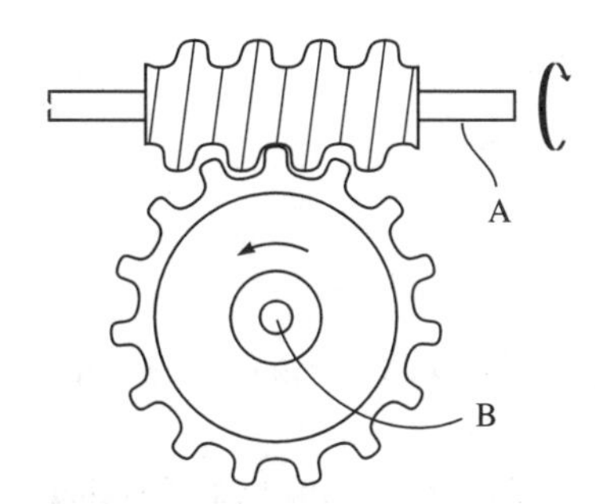

Figure 10.3

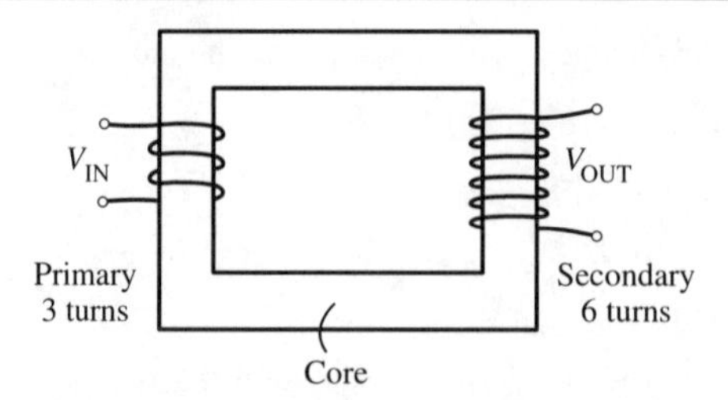

Figure 10.4

9 Figure 10.4 shows a transformer with 3 turns in the primary coil and 6 turns in the secondary coil. The *turns ratio* is 2:1. The ratio of output voltage to input voltage is the same as the turns ratio. If the input voltage is 120 V AC, what is the output voltage (ignoring losses)?

10 The Tornado combat aircraft has a length of 16.7 m. A flying model is to have a length of 500 mm. What is the scale factor to 2 sf?

Proportional parts

We may be asked to divide a quantity into **parts** in a given ratio.

Example

Two garden beds have areas of 7 m^2 and 10 m^2. There is 2.5 kg of fertilizer to be spread on the beds in proportion to their areas. How much is to be applied to each bed, to 3 sf?

We will work in m^2 and in grams of fertilizer. The total fertilizer is 2500 g. Find the total area on which it is to be spread:

Total area is $7 + 10 = 17\,m^2$

Now we proceed as in the examples on page 134:

An area of 17 receives 2500 (begin with totals)

An area of 1 receives $\frac{2500}{17}$ (divide both by 17)

An area of 7 receives $\frac{2500}{17} \times 7$ (multiply both by 7)

Evaluate the fertilizer: $\frac{2500}{17} \times 7 = \frac{17500}{17} = 1029$

The 7 m^2 bed receives 1030 g of fertilizer (3 sf).

Calculate the amount for the 10 m^2 bed in the same way:

$$\frac{2500}{17} \times 10 = \frac{25000}{17} = 1471$$

The 10 m^2 bed receives 1470 g of fertilizer (3 sf).

Check the calculation by adding the two amounts: 1030 + 1470 = 2500, which is the same as the total amount of fertilizer available.

In the example above the ratio was that between the areas of the beds. In other problems we are given the ratio directly, as below.

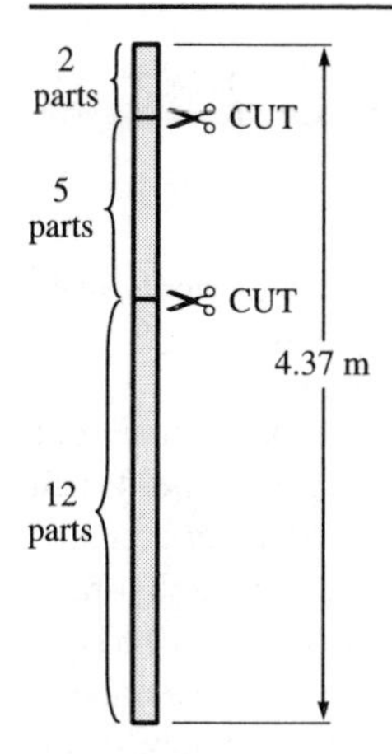

Figure 10.5

Example

A ribbon 4.37 m long is cut into 3 pieces in the ratio 2:5:12. How long is each piece?

Think of the three pieces as made up of 'parts'. We do not know how long a part is yet. One ribbon is 2 parts long, one is 5 parts long, and one is 12 parts long (Figure 10.5).

Begin with totals. The total length of ribbon is 4.37 m. The total number of parts is 19.

The length of 19 parts is 4.37 m

The length of 1 part is $\frac{4.37}{19}$ (divide both by 19)

The length of 2 parts is $\frac{4.37}{19} \times 2$ (multiply both by 2)

Evaluate the length of the shortest piece: $\frac{4.37}{19} \times 2 = 0.46$

Similarly, for the 5-part piece:

The length of 5 parts is $\frac{4.37}{19} \times 5 = 1.15$

And for the longest piece:

The length of 12 parts is $\frac{4.37}{19} \times 12 = 2.76$

Check: $0.46 + 1.15 + 2.76 = 4.37$

The lengths of the pieces of ribbon are 0.46 m, 1.15 m, and 2.76 m.

Test yourself 10.3

1 The ratio of the number of rainy days to dry days in the April of a certain year was 2:3. How many days were rainy and how many were dry?

2 2 kg of ice cream is divided between two parties of children in the same ratio as the numbers of children in each party. Party A has 4 children and Party B has 6 children. How much ice cream does each party receive?

3 A loop of wire 3.5 m long is to be formed into a triangle, with sides in the ratio 3:4.5:5. How long is each side? (2 dp)

4 A publisher decides that the content of a magazine should be news, reviews, readers' letters and advertisements in the ratio 6:4:3:3. The magazine has 64 pages. How many pages are allocated to each? (1 dp)

5 Solder is made from tin and lead in the ratio 3:2. Given 5 kg of tin and a supply of lead, what weight of solder can be made? (1 dp)

Percentages

A percentage is a way of expressing a quantity as the number of parts in 100 parts.

Example

There are 56 g of carbohydrate in every 100 g of coffee whitener. In other words, there are 56 parts of carbohydrate **per hundred** or **percent**. We express this as 56%.

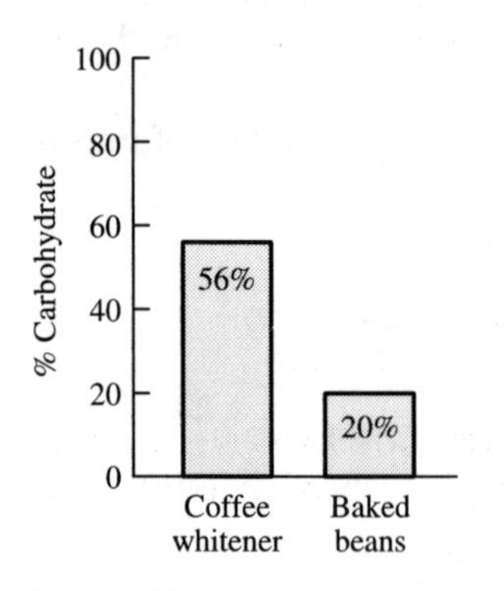

Figure 10.6

Expressing quantities as percentages makes it easier to compare them. For example, 100 g of a brand of baked beans contain 20 g of carbohydrate. When we are told that the beans contain 20% carbohydrate and the whitener contains 56% we can see at once that the whitener is the richer source of carbohydrate. This type of information is readily displayed as a diagram (Figure 10.6).

The conversion of **quantity to percentage** is easy in the examples above because we are told what quantity is found in 100 g of the food. Suppose we are told that there are 6 defective lamps in a batch of 75. What is the percentage of defective lamps? To solve this, use the technique of page 134:

$$6 \text{ defective in } 75$$

$$\Rightarrow \quad \frac{6}{75} \text{ defective in } 1$$

$$\Rightarrow \quad \frac{6}{75} \times 100 \text{ defective in } 100$$

Evaluate the defective number: $\frac{6}{75} \times 100 = 8$

8% of the lamps are defective.

Another aspect of this problem is to ask how many lamps should be purchased to be reasonably sure of obtaining 100 good ones? 6 defective lamps in 75 means 69 good lamps in 75:

$$69 \text{ good lamps in } 75$$

$$\Rightarrow \quad 1 \text{ good lamp in } \frac{75}{69}$$

$$\Rightarrow \quad 100 \text{ good lamps in } \frac{75}{69} \times 100$$

Evaluate the lamps: $\frac{75}{69} \times 100 = 108.7$

Buy 109 lamps to obtain 100 good lamps.

The next example shows how to convert **percentage into quantity**. It refers to the coffee whitener described above. How much carbohydrate is there in a spoonful of whitener, assuming a spoon holds 4 g?

100 g whitener contains 56 g carbohydrate

1 g whitener contains $\frac{56}{100}$ g carbohydrate

4 g whitener contains $\frac{56}{100} \times 4$ g carbohydrate

Evaluate the weight of carbohydrate: $\frac{56}{100} \times 4\,\text{g} = 2.24\,\text{g}$

A spoonful of whitener contains 2.2 g of carbohydrate (2 sf).

Another type of **percentage-to-quantity** conversion is shown by the next example. What weight of baked beans are required to provide 125 g of carbohydrate?

100 (g of) beans provides 20 carbohydrate

$\frac{100}{20}$ beans provides 1 carbohydrate

$\frac{100}{20} \times 125$ beans provides 125 carbohydrate

Evaluate beans: $\frac{100}{20} \times 125 = 625$

625 g of beans are required to provide 125 g of carbohydrate.

Percentage change

A new drilling machine is safer to operate but the time to drill one hole is increased by 18%. If drilling previously took 35 s, how long does it take with the new drill? In this example we have an increase of 18%; the new time is 118% of the old time.

35 is 100%

$\Rightarrow \quad \frac{35}{100}$ is 1%

$\Rightarrow \quad \frac{35}{100} \times 118$ is 118%

Evaluate: $\frac{35}{100} \times 118 = 41.3$

New drilling time is 41 s (2 sf).

When a loaf of bread is baked, the volume of the dough increases from 1.2 ℓ to 2.16 ℓ. What is the percentage increase in the volume? The increase is 2.16 – 1.2 = 0.96. This is to be expressed as a percentage of 1.2.

$$0.96 \text{ parts in } 1.2$$

$$\Rightarrow \quad \frac{0.96}{1.2} \text{ parts in } 1$$

$$\Rightarrow \quad \frac{0.96}{1.2} \times 100 \text{ parts in } 100$$

Evaluate percentage: $\frac{0.96}{1.2} \times 100 = 80$

The volume increases by 80%.

Percentages as fractions

In a batch of seeds, 25% are diseased. What fraction of the seeds is diseased? Imagine a batch of 100 seeds; on average, we expect 25 of these to be diseased. This is the same as saying that 25 *hundredths* of the batch are diseased. The fraction of diseased seeds is:

$$\frac{25}{100} = 0.25$$

25% is equivalent to 0.25 of the batch.

In the same way, any other percentage is converted to a fraction by dividing it by 100.

The inverse operation, multiplying by 100, converts a fraction to a percentage.

Example

A meteorologist reports that 3/10 of the sky is covered by cloud. Express this as a percentage:

$$\frac{3}{10} \times 100 = 30$$

The fraction 3/10 is equivalent to 30%.

Test yourself 10.4

1 A cook mixes 2 tablespoons of sesame oil with 10 tablespoons of sunflower oil to improve the flavour. What is the percentage of sesame oil in the mixture?

2 A box of bolts contains 34 of size M2, 82 of size M4 and 84 of size M6. Express these quantities as percentages of the total number of bolts in the box.

3 Chrome steel contains 2.5% chromium. How much chromium is needed to make 350 kg of chrome steel?

4 A brand of vinegar contains 6% of acetic acid by volume. How much acetic acid is present in 250 ml of the vinegar?

5 An electronic intruder detector has a range of 12 m. When the battery is flat, the range is decreased by 15%. What is its range then?

6 A piece of new fabric 4.2 m long is washed for the first time. It shrinks, and its new length is 4.1 m. By what percentage has its length decreased? (2 dp)

7 The actual capacitance of an electrolytic capacitor may be as much as 80% more than the value printed on the can. It may also be as much as 20% less. What are the greatest and least possible values of a capacitor which has 22 μF printed on the can?

8 Express these percentages as fractions.

a 50% **b** 75% **c** 33⅓% **d** 2% **e** 15%

9 Express these fractions as percentages.

a ⅕ **b** 0.27 **c** $\frac{1}{12}$ **d** 0.98 **e** 0.3721

Proportional quantities

A ratio makes a **comparison** between two or more quantities. Comparisons are done when things are alike. With ratios, we compare like things: lengths with lengths, weights with weights, or volumes with volumes. The same applies with proportional parts and percentages.

Unlike quantities cannot have a ratio but they may still have some kind of numerical relationship. They may be **proportional** to each other.

As an example, take the length of a piece of wire and its resistance. The longer the wire, the greater the resistance. The graph of Figure 10.7 show a typical relationship. It is a straight line, passing through the origin. Doubling the length of wire doubles the resistance. Trebling the length trebles the

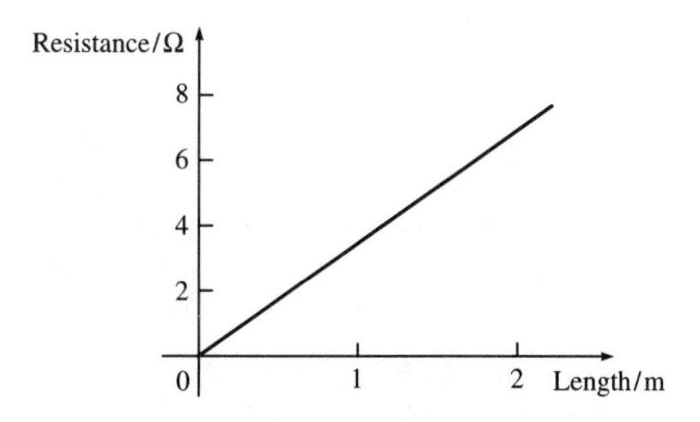

Figure 10.7

resistance. Halving the length halves the resistance. With zero length there is zero resistance. Summarizing these features:

Resistance is *directly proportional* to length

Note that if the line representing two quantities is straight but does *not* pass through the origin, the quantities are related but are *not* proportional. An example is the relationship between temperatures on the Celsius and Fahrenheit scales (Figure 10.8). Doubling the Celsius temperature does *not* double the Fahrenheit temperature.

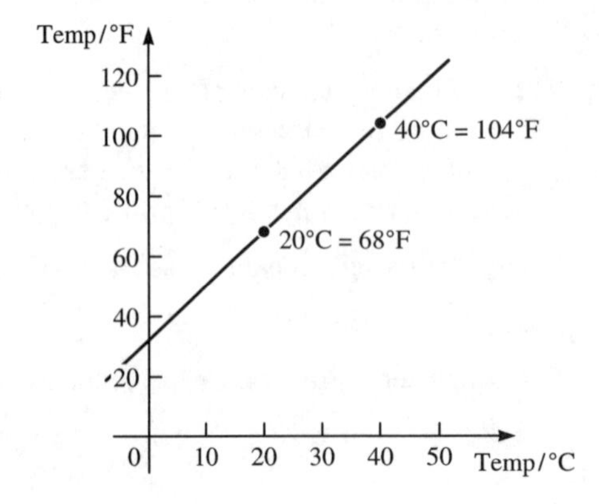

Figure 10.8

Problems in which two quantities are directly proportional are solved using the same technique as for ratios.

Example

A wire 2 m long has a resistance of 5 Ω. What is the resistance of a piece of wire which is 7.2 m long?

length 2 has resistance 5

length 1 has resistance $\frac{5}{2}$

length 7.2 has resistance $\frac{5}{2} \times 7.2$

Evaluate: $\frac{5}{2} \times 7.2 = 18$

A wire 7.2 m long has a resistance of 18 Ω.

Inverse proportion

A quantity of gas is contained in a cylinder with a movable piston (Figure 10.9). If we push the piston down, the volume of the gas decreases and the pressure increases. Doubling the pressure halves the volume. Trebling the pressure reduces the volume to one-third. The volume is **inversely proportional** to the pressure. This table shows the results of such a trial:

Pressure (kPa)	25	50	75	100	125	150	175	200
Volume (cm^3)	400	200	133	100	80	67	57	50

Note that pressure multiplied by volume is a **constant** (Boyle's Law).

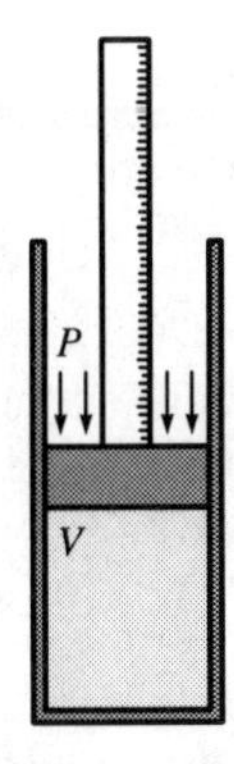

Figure 10.9

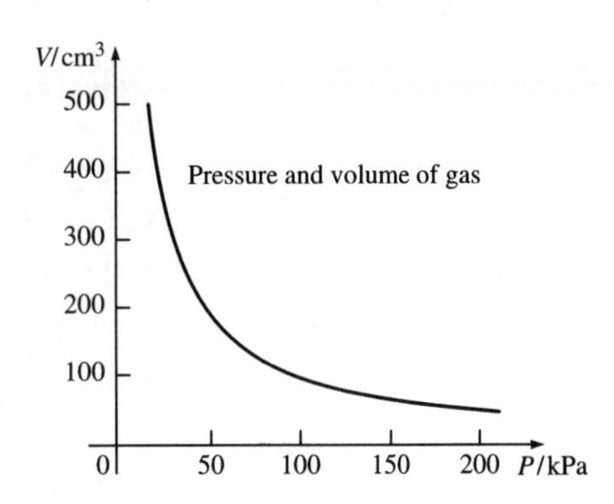

Figure 10.10

Instead of a straight line, the graph of volume against pressure is a curve, as in Figure 10.10. At all points on the curve, $P \times V$ has the same value. This curve is known as a **hyperbola**, page 250) and is found in all examples of inverse proportion.

When we have inverse proportion we multiply one number by a certain amount and divide the other number *by the same amount.*

Examples

The volume of an enclosed quantity of gas is 1.2 mm^3 when its pressure is 200 kPa. What is its volume when the pressure is increased to 300 kPa?

Volume is 1.2 when pressure is 200

Volume is 1.2×200 when pressure is 1 (Multiply by 200 and divide by 200)

Volume is $\dfrac{1.2 \times 200}{300}$ when pressure is 300 (Divide by 300 and multiply by 300)

Evaluate: $\dfrac{1.2 \times 200}{300} = 0.8$

At 300 kPa the volume is 0.8 m^3.

Check: the pressure is increased, so the volume must have decreased.

A builder has enough creosote for 40 m of fencing, 1.4 m high. If the fence is made 1.7 m high, what length of fencing can be creosoted?

Quick check: doubling the height means halving the length of fence, so height and length are inversely proportional.

Length is 40 when height is 1.4

Length is 40×1.4 when height is 1 (Multiply by 1.4 and divide by 1.4)

Length is $\dfrac{40 \times 1.4}{1.7}$ when height is 1.7 (Divide by 1.7 and multiply by 1.7)

Evaluate: $\dfrac{40 \times 1.4}{1.7} = 32.9$ (3 sf)

32.9 m of fence can be creosoted when it is 1.7 m high.

Check: The height is increased, so the length is decreased.

Test yourself 10.5

1 Four eggs of a certain grade weigh 288 g. How much do 17 eggs weigh?

2 If a small weight is hung from a spring, the increase in length of the spring is directly proportional to the weight. The spring stretches by 60 mm when the weight is 50 g. By how much does it stretch when the weight is 80 g? What weight makes the spring stretch by 132 mm?

3 A person machines 24 parts in 52 minutes. How many parts does that person machine in 39 minutes?

4 The belt can be slipped around any of the three drums on shaft B (Figure 10.11). The speed of rotation of B is inversely proportional to the diameter of the drum. With the belt on the middle drum, the shaft rotates 240 times in a minute. What is its rate of rotation when the drum belt is round

a the smaller drum

b the larger drum?

5 The wavelength of a radio station is inversely proportional to the frequency. If the frequency is 2 MHz, the wavelength is 150 m. What is the wavelength when the frequency is 6 MHz? What frequency has a wavelength of 100 m?

6 To assess your progress in this chapter, work the exercises in *Try these first*, page 131.

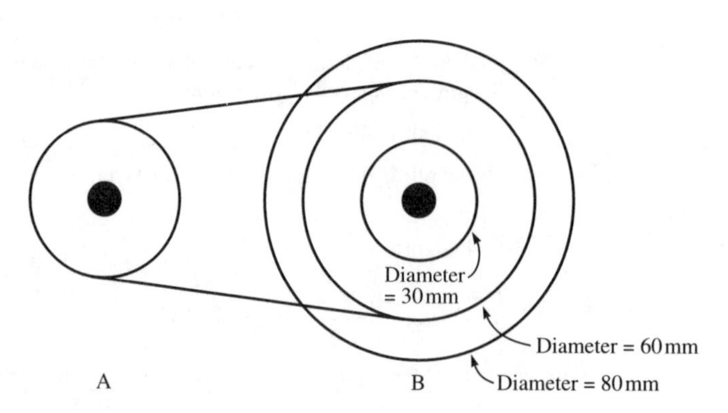

Figure 10.11

Explore this

Use spreadsheet software (Chapter 28) to set up a spreadsheet for direct proportion. The format is:

```
   A....... B....... C....... D.......
 1 Direct proportion
 2
 3       AS                IS TO
 4       SO        1       IS TO
 5      AND                IS TO
```

The formulae are:

```
D4=D3/B3
D5=D3*B5/B3
```

Problems in direct proportion have the form: Given x, which is directly proportional to y, find what y becomes when x is given a new value. The example from page 144 is 'A wire 2 m (= x) long has a resistance of 5 Ω (= y). What is the resistance (= new y) of a piece of wire which is 7.2 m (= new x) long?

Run the spreadsheet. Enter x in cell B3 and y in cell D3. Enter the new x in B5. Update the spreadsheet. The new y appears in cell D5.

This is the display you obtain with the resistance data:

```
Direct proportion

      AS       2.00     IS TO     5.00
      SO        1       IS TO     2.50
     AND       7.20     IS TO    18.00
```

Use the spreadsheet to check the working of other examples from this book, and in any other problems of direct proportion.

Here is the format for a spreadsheet for inverse proportion:

```
   A....... B....... C....... D.......
 1 Inverse proportion
 2
 3       AS                IS TO
 4       SO        1       IS TO
 5      AND                IS TO
```

The formulae are:

```
D4=D3*B3
D5=D3*B3/B5
```

Use this for problems such as the pressure-volume example on page 144 and for any other problems in inverse proportion.

11 More factors

Factorizing is the key to simplifying algebraic expressions. This chapter explains how to recognize different types of expression and how to factorize them.

You need to know the Maths Essentials (Part 1), especially algebra (Chapter 4). You also need to know about factors (Chapter 9).

Try these first

Your success with this short test will show you which parts of this chapter you already know.

1 Find the product $(2p - 3)(3a + 3)$

2 Find the product $(x + 3)(x - 5)$

3 Simplify $\dfrac{n^2 - 5n - 14}{n^2 - 10n + 21}$

4 Factorize $2a^3 - 6a^2 + 4a$

5 $36 - 4y^2$

6 Factorize $x^2 - 8x + 15$

Products

A product is the result of multiplying two or more terms together. For example:

$$2x \times 3y = 6xy$$

Multiply the coefficients, then write the letters, preferably in alphabetical order. If the same letter occurs in both terms, use a small ^{2} to indicate that the letter is squared:

$$5x \times 2xy = 10x^2y$$

These rules are extended to multiplying several terms enclosed in brackets:

$$\begin{aligned} 2x(3x + 4) &= 2x \times 3x + 2x \times 4 \\ &= 6x^2 + 8x \end{aligned}$$

Each of the terms in the bracket is multiplied by the term outside the bracket; the result shows these products added together. Here is an example in which one of the terms is negative:

$$6n(4 - 2n + 3n^2) = 24n - 12n^2 + 18n^3$$

A negative sign *outside* the bracket means that all terms in the bracket change sign (see sign rules, page 21).

Example

$$-2a(5 + 3a - 6a^2) = -10a - 6a^2 + 12a^3$$

Always take care when clearing brackets, as getting the signs wrong is a very common mistake.

Now for a slightly more complicated example:

$$(x + 2)\ (x + 7)$$

Here we have *two* bracketed expressions to be multiplied together. Note that we do not usually write the multiplication sign between the two expressions.

You may find this multiplication easier to understand if you think of this product as the area of the rectangle shown in Figure 11.1. Having decided on a value for x, we make the rectangle 2 units wider and 7 units longer.

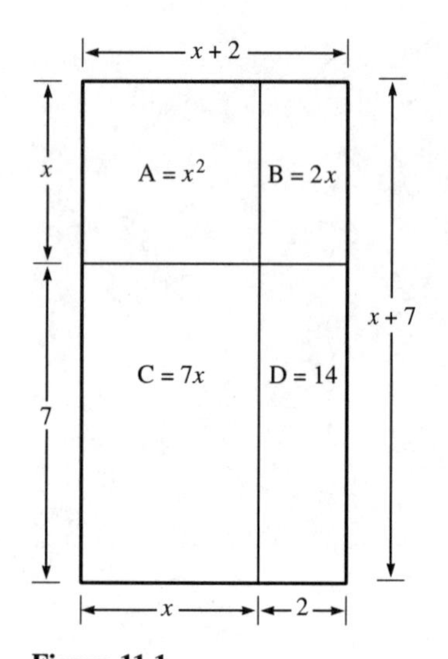

Figure 11.1

The area of the rectangle is made up of four parts:

A $x \times x = x^2$

B $2 \times x = 2x$ } Add these $2x + 7x = 9x$

C $x \times 7 = 7x$

D $2 \times 7 = 14$

Summing these parts shows that:

$$(x + 2)\ (x + 3) = x^2 + 9x + 14$$

Each term in the first expression is multiplied by each term in the second expression. There are four products, of which two can be added since they are both terms in x. This is how we select the pairs to be multiplied:

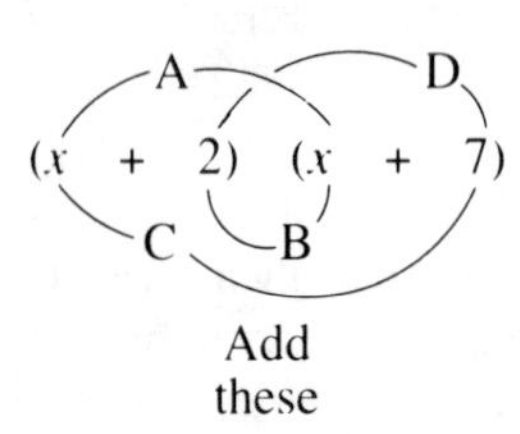

Another way of tacking the calculation, which you may find easier, is to set out a *long multiplication*:

$$\begin{array}{r} x + 2 \\ \times\ \underline{x + 7} \\ x^2 + 2x \end{array} \qquad \begin{array}{r} x + 2 \\ \times\ \underline{x + 7} \\ x^2 + 2x \\ \underline{7x + 14} \end{array} \qquad \begin{array}{r} x + 2 \\ \times\ \underline{x + 7} \\ x^2 + 2x \\ \underline{7x + 14} \\ x^2 + 9x + 14 \end{array}$$

Set it out with x's underneath each other and numbers underneath each other. First multiply $(x + 2)$ by the x belonging to $(x + 7)$. Next multiply $(x + 2)$ by the 7. Finally add the two products.

Test yourself 11.1

1 Find these products.

a $3x \times 7y$ **b** $5a \times 2b$

c $2a \times 3a$ **d** $x \times 2x \times 3y$

e $4n \times n^2$ **f** $5 \times x^2 \times 4y^2$

2 Find these products (clear brackets).

a $3(2 + 3x)$ **b** $4(5 - 2a + b)$

c $2n(3 + n)$ **d** $4p(3r - 2p + 4)$

e $-2j^2(5 - 3j)$ **f** $-t(-2s + 3t^2)$

3 Find these products.

a $(x + 1)(x + 2)$ **b** $(a + 3)(a + 3)$

c $(t + 4)(t + 3)$ **d** $(n + 2)(n + 5)$

Terms with negatives

If all terms in the expressions are positive, all terms in the product are positive. Now for an example with a negative term:

$$(x + 5)(x - 2) = x^2 + 3x - 10$$

Here is another:

$$(x + 2)(x - 5) = x^2 - 3x - 10$$

Both products end with a negative term. The sign of the middle term may be positive or negative. In the first example, it is the sum of $+5x$ and $-2x$, and so is positive. In the second example, it is the sum of $-5x$ and $+2x$, so it is negative.

Product signs

Both positive gives all positive:	$(x + 2)(x + 3) = x^2 + 5x + 6$
One negative gives last negative:	$(x - 2)(x + 3) = x^2 + x - 6$ $(x + 2)(x - 3) = x^2 - x - 6$
Two negatives gives middle negative, last positive	$(x - 2)(x - 3) = x^2 - 5x + 6$

One special case of this type of product is when the numbers in both brackets are the same:

$$(x + 3)(x - 3) = x^2 - 9$$

The middle term has disappeared because it is the sum of $+3x$ and $-3x$. The product is referred to as a **square minus a square** because both x^2 and 9 are square numbers (page 77).

Here we have two negative terms:

$(x - 4)(x - 7) = x^2 - 11x + 28$

In examples such as this, the middle term in the product is always negative. The last term is always positive (a minus times a minus).

Test yourself 11.2

1 Find the products.

a $(a + 5)(a + 7)$ **b** $(n + 4)(n + 4)$

c $(x + 1.5)(x + 2.4)$ **d** $(p + 2)(p - 3)$

e $(n - 4)(n + 2)$ **f** $(q + 6)(q - 2)$

g $(r - 3)(r + 7)$ **h** $(d + 10)(d - 1.5)$

2 Find the products.

a $(k - 3)(k - 4)$ **b** $(n - 2)(n - 1)$

c $(h + 2)(h - 2)$ **d** $(a - 7)(a - 7)$

e $(b - 7)(b + 7)$ **f** $(s - 1)(s - 3)$

g $(x + 1)(x - 1)$ **h** $(r - 0.5)(r - 1.4)$

3 Find the products.

a $2(x + 2)(x + 6)$ **b** $3(a + 1)(a + 7)$

c $n(n + 1)(n + 3)$ **d** $4(k + 3)(k - 2)$

e $5(g + 3)(g - 3)$ **f** $2x(x - 3)(x - 8)$

g $a(b + 1)(b + 4)$ **h** $(x + 1)(x + 2)(x + 3)$

4 Find the products.

a $(a + 1)(b + 1)$ **b** $(x + 2)(y - 4)$

c $(n - 2)(m - 7)$ **d** $(a + b)(a + 2b)$

e $(c + d)(c - d)$ **f** $(2a + 3)(a - 2)$

g $(3x + 2)(4x - 1)$ **h** $3n(2n - 1)(n - 1)$

Common factors

Having seen the results of multiplying different kinds of expression, the next stage is the reverse process. We break down an expression into its factors. When you are given an expression to factorize, look first for common factors. These are factors that occur **in all terms of the expression**.

Examples

Factorize $3x + 6$

The technique is the same as that for HCF, on page 125.

Prime factors of $3x$ are: 3 x
Prime factors of 6 are: 2 3

The full columns show the common factors. Here the only common factor is 3, so this is the HCF. Write 3 outside the bracket:

$3x + 6 = 3(\qquad)$

We have taken out the 3 from each set of factors. For each term, write the remaining factors in the bracket:

$$3x + 6 = 3(x + 2)$$

Check: multiply out the factors: $3(x + 2) = 3x + 6$. This returns us to the original expression. Always check factorizing by multiplying out.

Factorize $4x + 12x^2$

Prime factors of $4x$ are	2	2		x	
Prime factors $12x^2$ are	2	2	3	x	x

The common factors are 2, 2 and x, so the HCF is $4x$. Write this outside the bracket and the remaining terms inside the bracket:

$$4x + 12x^2 = 4x(1 + 3x)$$

For the first term, there are *no* remaining terms, so we write '1' in the bracket. For the second term, the remaining terms are 3 and x, so we write their product, $3x$, in the bracket.

Check: $4x(1 + 3x) = 4x + 12x^2$. Correct.

Not every expression has common factors, but looking for common factors is usually the best first step in factorizing. After a little practice there is no need to set out the table of prime factors. Work this stage in your head. But remember to check your result by multiplying out.

Test yourself 11.3

1 Factorize, if possible.

a $2 + 4a$ **b** $6 + 3n$

c $a - 2a^2$ **d** $5x + 3$

3 $4a^2 + 6a$ **f** $2xy - 4y + 10y^2$

2 Factorize, if possible.

a $6ab - 3ac$ **b** $7n^2 - 4m$

c $2x^3 + 3x^2y - 4xy$ **d** $9a^2b + 6ab^2 + 3ab$

e $2t(s + 1) + 2(2s - t)$
[*Hint*: Clear brackets and collect terms first.]

f $2p(3p + 1) + pq + p(q - 2)$

Square minus a square

At first glance, this expression does not appear to have any factors:

$$x^2 - 4$$

It certainly has no common factors, but it is recognizable as a **square minus a square**. On page 151 we saw that this kind of expression is the product of:

(x + a number) (x – the same number)

In this example, the number is 2, that is to say, the number which gives 4 when squared.

The factors of $(x^2 - 4)$ are $(x + 2)$ and $(x - 2)$.

$$x^2 - 4 = (x + 2)(x - 2)$$

The table of squares on page 77 is an aid to recognizing expressions of this kind. Remember that $1 = 1^2$, so:

$$x^2 - 1 = (x + 1)(x - 1)$$

Looking for a square minus a square is usually the second stage of factorizing.

Quadratic expressions

These are expressions with terms in x^2 and x, and a number. Below we list typical examples and describe how to factorize them.

Examples

Factorize $x^2 + 6x + 5$

In this example, as in all simple examples, the first expression is x^2, so we begin by writing the factors in a pair of brackets:

$$(x \quad)(x \quad)$$

The numbers still to go in the brackets must be a pair which, when multiplied together, give the number term (5) of the expression. The only two factors of 5 are 5 and 1; write these in the brackets:

$$(x \quad 1)(x \quad 5)$$

All we need now are the signs. All signs in the expression are positive, so the signs in both brackets must be positive:

$$(x + 1)(x + 5)$$

Multiplying x by 5, and x by 1 gives $5x$ and x. Their sum is $6x$, which is the middle term of the expression. This confirms that the expression does factorize.

$$\underline{x^2 + 6x + 5 = (x + 1)(x + 5)}$$

Check by multiplying the factors: $(x + 1)(x + 5) = x^2 + 6x + 5$. Correct.

Factorize $x^2 + 5x + 6$

The first term is x^2 so we begin as usual, by writing the brackets and inserting x's:

$$(x \quad)(x \quad)$$

The factors of 6 are: 6 and 1; or 2 and 3. We have to decide which pair to choose. If we select 6 and 1, the coefficient of the middle term is the *sum* of 6 and 1, which is 7. If we select 2 and 3, the coefficient of the middle term is the *sum* of 2 and 3, which is 5. The coefficient in this example is 5, so factors are 2 and 3:

$$(x \quad 2)(x \quad 3)$$

All signs are positive:

$$x^2 + 5x + 6 = (x + 2)(x + 3)$$

Checking by multiplying confirms the result.

Factorize $x^2 - 7x + 10$

The factors of 10 are: 10 and 1; or 2 and 5. But note that the middle term is negative, and the last term is positive. This clue tells us that both brackets contain negative terms (page 152). The factors must be: -10 and -1; or -2 and -5. If we select -10 and -1, the coefficient of the middle term is –11. If we select –2 and –5, the coefficient is –7. The second selection gives the required result, so:

$$\underline{x^2 - 7x + 10 = (x - 2)(x - 5)}$$

As always, check the result by multiplying. In any of these examples, if you are not sure which pair of factors to select, try one of the pairs and see if it works. In this example, we might have tried –10 and –1, using the long multiplication method:

$$\begin{array}{r} x - 10 \\ \times \underline{x - 1} \\ x^2 - 10x \quad\quad \\ \underline{-x + 10} \\ x^2 - 11x + 10 \end{array}$$

This does *not* give the original expression, so –1 and –10 are *not* the required factors. Try again with –2 and –5, which give the result we required.

Factorize $x^2 + 2x - 15$

Factors of 15 are: 1 and 15; or 3 and 5. In this example, the clue is that the last term is negative. This means that one of the factors is positive and one is negative. This makes 4 possibilities. List the possibilities and find what the coefficient of the middle term will be:

Pairs	*Coefficient*
1 and –15	–14
–1 and 15	14
3 and –5	–2
–3 and 5	2

The last set of factors is the one required:

$$\underline{x^2 + 2x - 15 = (x - 3)(x + 5)}$$

The rule is:

The larger factor has the same sign as the middle term

Example

Factorize $x^2 - 4x - 12$

Factors of 12 are: 1 and 12; or 2 and 6; or 3 and 4. There are 3 pairs to choose from. Also, since the last term is negative, the factors are of opposite sign. This makes 6 possible pairs to choose from but, since the middle term is negative, the larger factors must also be negative. This eliminates 3 of the 6 pairs. The remaining 3 are:

Pairs	*Coefficient*
1 and –12	–11
2 and –6	–4
2 and –4	–2

The factors are 2 and –6:

$$\underline{x^2 - 4x - 12 = (x + 2)(x - 6)}$$

With practice, you can easily hit on the correct pair without needing to make a table. If in doubt, test several possibilities by long multiplication until you find the right one.

Factorizing routine

Try these techniques in this order:

1 *Common factors*: find HCF and put outside bracket; remaining terms inside bracket

2 *Square minus a square*: $a^2 - b^2 = (a + b)(a - b)$

3 *Quadratic expressions*:

All terms positive	$\rightarrow$	$(x + ?)(x + ?)$
Last term positive and middle term negative	$\rightarrow$	$(x - ?)(x - ?)$
Last term negative	$\rightarrow$	$(x + ?)(x - ?)$

Larger term has same sign as middle term

Expressions may factorize in more than one way. Some expressions do not factorize at all; always check by multiplying out.

Test yourself 11.4

1 Factorize.

a $x^2 + 3x + 2$ **b** $x^2 + 8x + 7$

c $a^2 + 7a + 12$ **d** $n^2 - 5n + 6$

e $r^2 - 2r + 1$ **f** $b^2 - 11b + 24$

g $q^2 - 16$ **h** $x^2 + 13x + 36$

2 Factorize.

a $k^2 + k - 2$ **b** $n^2 + 3n - 10$

c $a^2 - a - 2$ **d** $q^2 + 9q + 20$

e $b^2 - 11b - 12$ **f** $j^2 - 49$

g $t^2 + 7t - 18$ **h** $p^2 - 6p + 9$

3 Factorize if possible.

a $2y + y^2 - 3$ **b** $36 - f^2$

c $r^2 + 4r + 5$ **d** $2a^2 + 6a - 20$

e $d^3 - 10d^2 - 11d$ **f** $3x^2 - 27$

g $32 - 4s + s^2$ **h** $2p^2q + 2pq - 4q$

i $n^2 + 7n + 7$ **j** $4n^2 - 4n - 24$

Quotients

Factorizing can sometimes be used to simplify a quotient. If the N and D have a common factor, this is cancelled, leaving a simpler quotient.

Examples

Simplify $\dfrac{4x^2y}{2xy^2}$

The factors of $4x^2y$ are:	2	2	x	x	y	
The factors of $2xy^2$ are:	2		x		y	y

The HCF is $2 \times x \times y = 2xy$. Cancel $2xy$ from N and D:

$$\text{Simplified quotient} = \frac{2x}{y}$$

Simplify $\dfrac{2x^2 + 3x}{7x^2 - 4x}$

x is a factor of N and D.

$$\frac{2x^2 + 3x}{7x^2 - 4x} = \frac{x(2x + 3)}{x(7x - 4)}$$

Cancel the x's:

$$\text{Simplified quotient} = \frac{2x + 3}{7x - 4}$$

Simplify $\dfrac{x^2 + x - 2}{x^2 + 3x - 4}$

Factorizing N and D:

$$\frac{x^2 + x - 2}{x^2 + 3x - 4} = \frac{(x + 2)\ (x - 1)}{(x + 4)\ (x - 1)}$$

Cancelling the $(x - 1)$:

$$\text{Simplified quotient} = \frac{x + 2}{x + 4}$$

Test yourself 11.5

1 Simplify.

a $\dfrac{3ab}{6a}$ **b** $\dfrac{4x^2}{2xy}$

c $\dfrac{2x - 6}{4x + 2}$ **d** $\dfrac{x^2 - 2x}{x^3 - xy}$

e $\dfrac{ab}{a(b + 2)}$ **f** $\dfrac{(p + q)(2p - q)}{(2p - q)(4p + 3q)}$

g $\dfrac{y^2 - 4y}{(y + 2)(y - 4)}$ **h** $\dfrac{s^2 - s - 6}{s^2 - 4s + 3}$

i $\dfrac{t^2 - 9}{t^2 + 6t + 9}$ **j** $\dfrac{n^2 - 7n + 10}{n^2 + 3n - 10}$

2 To assess if you have covered this chapter fully, answer the questions in *Try these first*, page 149.

12 Handling equations and formulae

Equations feature in almost every branch of maths. This chapter explains the rules for handling them.

You need to know the Maths Essentials (Part 1) and factors (Chapter 9).

Try these first

Your success with this short test will tell you which parts of this chapter you already know.

1 Solve these equations.

a $4x + 5 = 13$ **b** $5n - 7 = 2(n + 1)$

c $\dfrac{a + 3}{2} - 2 = \dfrac{2a + 2}{5}$

2 For each of these formulae, change the subject to the variable given in brackets.

a $p = \dfrac{3}{4}q + 1$ (q) **b** $4R = 3(2S + T)$ (S)

3 Solve the inequality $4x + 7 \leqslant 15$ and state the largest integer value that x can have.

4 Solve the inequality $13 + 3n > 3 - 2n$ and state the least integer value than n can have.

Equations must balance

An equation is a way of saying that two expressions are equal. A weighing balance shows us when two weights are equal. If it is balanced with equal weights and we add the *same* weight to *both* sides, the balance stays level.

The same applies to equations. Start with an equation:

$$x + 3 = 5$$

Add 4 to *both* sides:

$$x + 3 + 4 = 5 + 4$$
$$\Rightarrow \quad x + 7 = 9$$

The equation remains balanced (Figure 12.1). Both sides remain equal. The equation was true to start with; it is still true now.

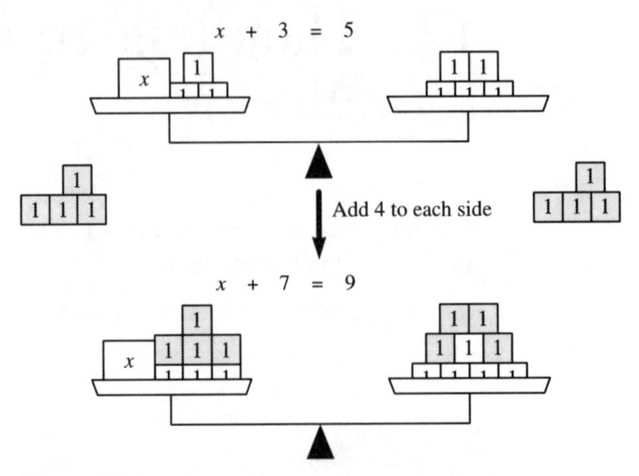

Figure 12.1

This also works with subtraction. Start with another equation:

$$x + 4 = 7$$

Subtract 2 from *both* sides:

$$x + 4 - 2 = 7 - 2$$

$$\Rightarrow \quad x + 2 = 5$$

The equation remains balanced (Figure 12.2).

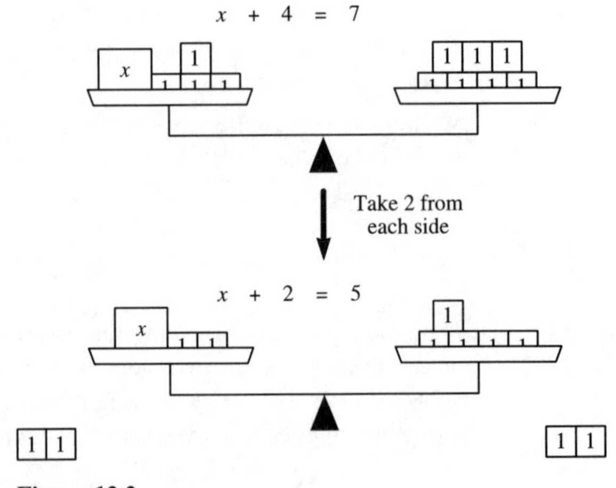

Figure 12.2

These facts lead us to a way of finding the value for x or some other unknown variable. Suppose we are given this equation and asked to find the value of x:

$$x + 2 = 7$$

The idea is to get x on its own on one side of the equation. The easiest way is to remove the 2. Subtract 2 from *both* sides:

$$x + 2 - 2 = 7 - 2$$

$$\Rightarrow \quad \underline{x = 5}$$

The next equation needs the opposite treatment::

$$a - 3 = 6$$

The way to remove the 3 from the left side is to *add* 3 to that side. We must also add 3 to the right side:

$$a - 3 + 3 = 6 + 3$$

$$\Rightarrow \quad \underline{a = 9}$$

Equations are useful for solving problems. For example, 2 people enter a room, and there are then 7 people in the room. How many were in the room before? The unknown quantity is the number of people originally in the room. Represent this by the variable n. There were n to start with, then 2 more entered. The number in the room is $n + 2$. But this number is known to be 7, so we have an equation:

$$n + 2 = 7$$

Solve this to find n. Subtract 2 from both sides:

$$\Rightarrow \quad n = 5$$

<u>There were 5 people in the room.</u>

Test yourself 12.1

1 Solve these equations (find the value of x, a, or other variable) by either adding or subtracting the *same* amount to or from both sides.

a $x + 4 = 11$ **b** $a + 1 = 5$

c $p - 2 = 4$ **d** $y - 12 = 4$

e $x + 3 = 1$ **f** $n - 4 = 0$

g $k + 1 = -1$ **h** $d + 6 = 7$

2 A box contains an unknown number of screws. Five screws are removed from the box. The screws remaining in the box are counted and it is found that there are 12. Write an equation to illustrate this information. Solve the equation to find how many screws were originally in the box.

3 A voltage of a power supply is increased by 6 V, and the reading on the meter is 15 V. What was the reading *before* the increase? Write the equation and solve it.

Rearrangements

It sometimes helps to be able to exchange the left side and the right side of an equation:

$$7 = a + 3$$

Exchange:

$$\Rightarrow \quad a + 3 = 7$$

Having moved a to the left, we solve the equation as above. Subtract 3 from both sides:

$$\Rightarrow \quad a + 3 - 3 = 7 - 3$$
$$\Rightarrow \quad \underline{a = 4}$$

Exchanging sides is equivalent to swapping the contents of the left and right pans of a balance. The balance swings level, as before. Similarly, swapping the sides of an equation makes no difference to their solution.

We can also rearrange the terms in an equation, keeping them on the same sides:

$$5 + n = 6$$
$$\Rightarrow \quad n + 5 = 6$$
$$\Rightarrow \quad \underline{n = 1}$$

This is equivalent to just moving the objects around on the balance pans. It makes no difference to the total weight in each pan.

Unknown quantities

An equation remains balanced when the *same* quantity is added to or subtracted from *both* sides, even if we do not know what that quantity is.

Example

Solve $2x + 3 = x + 7$

This has x's on the left and right, but we want x on the left only. Subtract x from *both* sides:

$$\Rightarrow \quad 2x + 3 - x = x + 7 - x$$

Simplify both sides by collecting terms (page 56):

$$\Rightarrow \quad x + 3 = 7$$

Subtract 3 from both sides:

$$\Rightarrow \quad x + 3 - 3 = 7 - 3$$
$$\Rightarrow \quad \underline{x = 4}$$

When solving equations, the usual aim is to:

Get the variable on its own on the left side of the equation

Test yourself 12.2

1 Solve these equations.

a $n + 8 = 11$ **b** $x - 8 = 0$

c $8 + b = 12$ **d** $5 + t = 5$

e $s + 5 = 4$ **f** $9 = a - 3$

g $2x - 7 = 4 + x$ **h** $7 = 9 + p$

i $4z + 2 = 3z + 5$ **j** $g + 7 = 2 + 2g$

2 A cake takes 6 minutes to cook in a microwave oven. If cooking finished at 12.32, when did cooking begin? Write the equation and solve it.

3 A student heads a poster with instant lettering. After lettering the word ALPHABETICAL, there are 7 of letter A left on the sheet. How many were there before? Write the equation and solve it.

Dividing and multiplying

An equation remains balanced if we *divide* both sides by the *same* amount. This is equivalent on the balance to halving, quartering, or dividing by 10 (or any other number) the amounts in *both* pans. The balance remains swinging level.

Examples

Given $3n + 6 = 15$

Dividing everything on both sides by 3:

$$\Rightarrow \quad \frac{3n}{3} + \frac{6}{3} = \frac{15}{3}$$

$$\Rightarrow \quad n + 2 = 5$$

Subtract 2 from both sides:

$$\underline{n = 3}$$

Dividing everything by the same number *may* produce awkward fractions, so usually it is best to simplify the equation first by using the adding or subtracting rules.

Given $7n - 4 = 8 + 3n$

Subtract $3n$ from both sides:

$\Rightarrow \quad 4n - 4 = 8$

Add 4 to both sides: } Simplify,

$\Rightarrow \quad 4n = 12$

Divide both sides by 4: *then* divide.

$$\underline{n = 3}$$

Similarly, we can *multiply* both sides by the *same* amount.

Given $x/4 + 3 = 5$

Multiply everything on both sides by 4:

$$\Rightarrow \quad x + 12 = 20$$

Subtract 12 from both sides:

$$\Rightarrow \quad \underline{x = 8}$$

Some equations have brackets. It is usually best to clear these as a first step.

$$3(n + 4) - 2 = 25$$

Clear brackets:

$$\begin{aligned} \Rightarrow \quad 3n + 12 - 2 &= 25 \\ 3n + 10 &= 25 \\ \Rightarrow \quad 3n &= 15 \\ \Rightarrow \quad \underline{n} &\underline{= 5} \end{aligned}$$

After clearing the brackets, we simplify the equation, then divide at the final stage.

Now that we know about dividing and multiplying equations, we can solve more difficult problems. For example, it takes 3 g of adhesive to assemble one lampshade. A total of 40 g of adhesive was used, including 7 g which was wasted in trials. How many lampshades were assembled? If n lampshades are assembled, they need $3n$ grams of adhesive. Add to this the wasted amount:

$$3n + 7 = 40$$

Solve this equation:

$$\begin{aligned} \Rightarrow \quad 3n &= 33 \\ \Rightarrow \quad n &= 11 \end{aligned}$$

<u>11 lampshades were assembled.</u>

Test yourself 12.3

1 Solve these equations.

a $2a + 5 = 11$ **b** $4 + 3c = 10$

c $3n - 5 = 7$ **d** $8 + 5a = 33$

e $5x - 2 = 13$ **f** $2y + 7 = 15$

g $5 + 2n = 1$ **h** $4t + 7 = 2t + 13$

i $5y - 3 = 2y$ **j** $7z + 6 = 16 + 2z$

k $5 + 4a = 17 - 2a$ **l** $3p - 3 = 2p - 3$

2 A potter makes 5 identical plates, using 1300 g of clay. 50 g of clay is wasted, remaining in the container or on the wheel. What weight of clay is required for 1 plate? Write the equation and solve it.

3 A recording tape lasts for 15 minutes. Allowing a total of 1.5 minutes of blank tape at the beginning and end, 6 musical items of equal length are recorded. How long is each item? Write the equation and solve it.

The division rule

Now we can explain the rule for dividing by a common fraction (page 42).

Example

Divide 5/6 by 2/3.
Suppose that the result of the division is x.
The equation is:

$$x = \frac{5}{6} \div \frac{2}{3}$$

Multiply both sides by 2/3. On the right, if 5/6 is already divided by 2/3 and is now multiplied by 2/3, this makes it 5/6 again, so:

$$\Rightarrow \qquad x \times \frac{2}{3} = \frac{5}{6}$$

Next we remove the 2/3 from the left, in two stages. First multiply both sides by 3:

$$\Rightarrow \qquad x \times 2 = \frac{5}{6} \times 3$$

Then divide both sides by 2:

$$\Rightarrow \qquad x = \frac{5}{6} \times \frac{3}{2}$$

This shows that, to find x, we *multiply* by 3/2, instead of dividing by 2/3. The same reasoning applies to dividing by any other common fraction.

Changing signs

It may be useful sometimes to multiply both sides of an equation by –1. In other words, *change the signs* of all terms.

Examples

Given $4 - 3a = -2$

Change signs:

$$\Rightarrow \qquad -4 + 3a = 2$$

Add 4 to both sides:

$$\Rightarrow \qquad 3a = 6$$

Divide both sides by 3:

$$\Rightarrow \qquad \underline{a = 2}$$

Given $3(p + 6) = 4(p + 3) - 1$

Remove brackets as the usual first step.

$$\Rightarrow \qquad 3p + 18 = 4p + 12 - 1$$

$$\Rightarrow \qquad 3p + 18 = 4p + 11$$

Subtract $4p$

$$\Rightarrow \qquad -p + 18 = 11$$

Subtract 18

$$\Rightarrow \qquad -p = -7$$

Change the signs:

$$\Rightarrow \qquad \underline{p = 7}$$

Reciprocals

Another thing we can do is to take the reciprocal of both sides of an equation.

Example

$$\frac{12}{a} = 4$$

Take the reciprocal of both sides:

$$\Rightarrow \quad \frac{a}{12} = \frac{1}{4}$$

Multiply by 12:

$$\Rightarrow \quad a = \frac{12}{4}$$

$$\Rightarrow \quad \underline{a = 3}$$

But reciprocals must be taken for the side as a *whole*, not for the individual terms.

$$\frac{10}{x} + 3 = 5$$

Note: The reciprocal of the left side is *not* $\frac{x}{10} + \frac{1}{3}$.

It is best to remove the 3 from the left side first:

$$\Rightarrow \quad \frac{10}{x} = 2$$

Now take the reciprocals:

$$\Rightarrow \quad \frac{x}{10} = \frac{1}{2}$$

$$\Rightarrow \quad x = \frac{10}{2}$$

$$\Rightarrow \quad \underline{x = 5}$$

Fractions

Fractions often make equations complicated. Remove them by multiplying both sides of the equation by the LCM of the denominators, the **lowest common denominator** (LCD, page 127).

Examples

$$\frac{2x + 5}{3} = \frac{4x + 5}{5}$$

The LCM of 3 and 5 is 15; multiply *both* sides by 15:

$$\Rightarrow \quad 15 \times \frac{2x + 5}{3} = 15 \times \frac{4x + 5}{5}$$

Cancelling gives: $5(2x + 5) = 3(4x + 5)$

Clear brackets:

$$\Rightarrow \quad 10x + 25 = 12x + 15$$

$$\Rightarrow \quad 10x = 12x - 10$$

$$\Rightarrow \quad -2x = -10$$

Change signs:

$$\Rightarrow \quad 2x = 10$$

$$\Rightarrow \quad \underline{x = 5}$$

$$\frac{4y - 3}{4} + \frac{1}{3} = \frac{2y + 1}{12} + 2$$

The LCM of 4, 3 and 12 is 12; multiply both sides by 12, *including* the ⅓ on the left and the 2 on the right:

$$\Rightarrow \quad 12 \times \frac{4y-3}{4} + \frac{12}{3} = 12 \times \frac{2y+1}{12} + 12 \times 2$$

Cancelling gives: $3(4y-3) + 4 = 2y + 1 + 24$

$$\Rightarrow \quad 12y - 9 + 4 = 2y + 25$$

$$\Rightarrow \quad 12y - 5 = 2y + 25$$
$$\Rightarrow \quad 12y = 2y + 30$$
$$\Rightarrow \quad 10y = 30$$
$$\Rightarrow \quad \underline{y = 3}$$

$$\frac{3}{y+2} = \frac{6}{3y+1}$$

The LCD is $(y + 2)(3y + 1)$; multiplying both sides by this LCD:

$$\Rightarrow \quad (y+2)(3y+1) \times \frac{3}{y+2} = (y+2)(3y+1) \times \frac{6}{3y+1}$$

One of the factors cancels out on each side:

$$\Rightarrow \quad 3(3y+1) = 6(y+2)$$
$$\Rightarrow \quad 9y + 3 = 6y + 12$$
$$\Rightarrow \quad 9y = 6y + 9$$
$$\Rightarrow \quad 3y = 9$$
$$\Rightarrow \quad \underline{y = 3}$$

Test yourself 12.4

Solve these equations.

1 $4a - 2 = 4 + 3a$

2 $2(5 + 2x) = 5x + 3$

3 $\frac{6}{h} = 3$

4 $\frac{f}{5} + 2 = 4$

5 $\frac{r}{7} + 7 = r - 11$

6 $\frac{3t}{2} + 14 = 2$

7 $5(b + 2) - 6 = 7(b - 2)$

8 $6(r + 5) - 2 = 14r + 44$

9 $\frac{z}{3} - 2 = -3$

10 $\frac{-2y}{6} + 4 = 0$

11 $\frac{3n-9}{8} = \frac{n+11}{12}$

12 $\frac{3-7q}{10} - \frac{2}{5} = \frac{q+7}{2}$

13 $\frac{8}{z-3} = \frac{12}{z-2}$

14 $\frac{4+y}{4} + \frac{y-4}{2} = \frac{4y+2}{8}$

Handling an equation

Always keep it balanced; do the same thing to *both* sides.

Options are:

- Add the same amount to both sides
- Subtract the same amount from both sides
- Divide both sides by the same amount
- Multiply both sides by the same amount*
- Take the reciprocal of both sides
- Change signs of all terms on both sides
- Exchange sides.

Some other options are given on page 188.

Rearranging terms and clearing brackets on one or both sides helps to simplify the equation.

* Multiply by the LCM of the denominators to get rid of fractions.

Formulae

A formula is a type of equation. Usually a formula has two or more variables. We know the values of all the variables except one. We use the known values to find the value of the unknown variable. An example is the formula for calculating the potential difference (or voltage) across a resistor when a given current is flowing through it. The formula is based on Ohm's Law:

$$V = IR$$

The current I is in amps, and the resistance R is in ohms. We use the formula to find the potential difference V, in volts. For example, if $I = 2\,\text{A}$, and $R = 100\,\Omega$, then:

$$V = 2 \times 100 = 200$$

The voltage across the resistor is 200 V.

But suppose we are told the resistance and the voltage and asked to find the current. We need to rearrange the formula so as to bring I on the left, with V and R on the right. A formula is an equation wo we must keep it balanced. Use the rules summarized in the box above.

Start with the given formula. R is already on the right.

$$V = IR$$

Divide both sides by I to remove I from the right:

$$\Rightarrow \quad \frac{V}{I} = R$$

Divide both sides by V to remove V from the left:

$$\Rightarrow \quad \frac{1}{I} = \frac{R}{V}$$

Take the reciprocal of both sides to turn $1/I$ into I:

$$\Rightarrow \qquad I = \frac{V}{R}$$

This is another useful version of the formula. Given $V = 5\,\text{V}$ and $R = 200\,\Omega$, we calculate that:

$$I = \frac{5}{200} = 0.025$$

The current is 0.025 A, or 25 mA.

Temperature conversions

The formula for converting a temperature in degrees Celsius into its equivalent in degrees Fahrenheit is:

$$F = 32 + \frac{9C}{5}$$

Where F and C are the readings taken on the two scales. For example, if the temperature is 25 °C, the Fahrenheit equivalent is:

$$F = 32 + \frac{9 \times 25}{5} = 32 + 45 = 77$$

25 °C is equivalent to 77 °F.

In the formula above, F is said to be the **subject**. If we make this into a formula for calculating Celsius from Fahrenheit we are **changing the subject** of the formula from F to C.

Exchange sides, to get C on the left:

$$\Rightarrow \qquad 32 + \frac{9C}{5} = F$$

Subtract 32, to remove the 32 from the left:

$$\Rightarrow \qquad \frac{9C}{5} = F - 32$$

Multiply by 5, noting that the *whole* of the right side must be multiplied:

$$\Rightarrow \qquad 9C = 5(F - 32)$$

Divide by 9, noting that the *whole* of the right side must be divided:

$$\Rightarrow \qquad C = \frac{5}{9}(F - 32)$$

The subject has been changed to C. If the temperature is 104 °F, the Celsius temperature is:

$$C = \frac{5}{9}(104 - 32) = \frac{5 \times 72}{9} = 40$$

104 °F is equivalent to 40 °C.

Test yourself 12.5

1 Change the subject of each formula to the variable given in brackets.

a $v = u + at \quad (u)$ **b** $W = s - t \quad (s)$

c $a = x + y + z \quad (x)$ **d** $y = 2 + 3x \quad (x)$

e $2a = \frac{3}{b} \quad (b)$ **f** $A = \frac{2B + C}{D} \quad (B)$

g $y = x(w + z) \quad (w)$ **h** $J = \frac{1}{4}(K + 1) \quad (K)$

2 Write a formula for the area A m^2 of a rectangle which is L m long and W m wide. Use the formula to find A when $L = 2.3$ m and $W = 0.8$ m. Change the subject of the formula to W. Use the new formula to find W when $A = 56$ m^2 and $l = 8$ m.

3 An estate has 6 houses and each house has x lighting fittings which need 1 lamp each, and y fittings which need 3 lamps each. Write a formula to find z, the total number of lamps needed on the estate. How many lamps are needed when $x = 8$ and $y = 3$? Change the subject of the equation to y. If 132 lamps are needed, and each house has 10 fittings for 1 lamp, how many fittings in each house hold 3 lamps?

Other kinds of equality

The symbol ≈ is used when two expressions are nearly but not exactly equal. We might use this when we are roughly checking a calculation.

Example

A rectangular plate measures 21.8 mm × 39.2 mm. Its area is found after rounding the dimensions to 1 sf:

$$21.8 \times 39.2 \approx 20 \times 40 = 800$$

$$\underline{\text{Area} \approx 800\,\text{mm}^2}$$

Another use for this symbol occurs when we decide to ignore an expression in an equation because its value is very small compared with others in the equation. Ignoring this small value simplifies the working. For example the effort E needed to raise a load L by a pulley system is:

$$E = \frac{L + w}{5}$$

where w is a constant additional load due to the weight of the pulley block, friction and other forces. If w is very much smaller than L, we can ignore it. The formula becomes:

$$E \approx \frac{L}{5}$$

Consider this equation:

$$x + 2 = 5$$

This is true provided that $x = 3$. Solving the equation tells us the condition or conditions under which it *is* true

Now consider this equation:

$$x + x = 2x$$

This is true for *all* values of x. The two sides are not only equal in value, but they are mathematically **identical**. We are stating a mathematical fact. Such a statement is called an **identity**. We use the symbol ≡:

$$x + x \equiv 2x$$

Examples of identities are given in Chapter 19.

Inequalities

The symbol ≠ means 'is not equal to'.

Example

If $x = 4$, then $3x - 2 \neq 6$

Symbols that we more often use are:

$>$ is greater than
$<$ is less than
$\geqslant$ is equal to or greater than
$\leqslant$ is equal to or less than

Example

If $x = 4$ and $y = 7$, then we can say that:

y is greater than x	$y > x$
x is less than y	$x < y$
x is greater than 2	$x > 2$
y is less than 20	$y < 20$
and many more.	

To remember which symbol is which, note that the bigger number or variable is at the 'wide' end of the symbol, and the smaller number of variable is at the 'point'.

A statement such as:

$$b \geqslant 5$$

means that b may have the value 5, or any value greater than that. But it is not less than 5. Similarly, the statement:

$$c < 7$$

means that c has any value (possibly a negative value) up to but *not* including 7. We also find statements such as:

$$4 \leqslant z < 13$$

This means that z lies in the range from 4 up to, but not including, 13.

The number line

In Chapter 1 we described the number path. Imagine that the paving-blocks are replaced by a continuous strip of concrete. The path is pegged off into units so that we can gauge exactly where we are. Imagine also that, instead of taking steps along the path, we move along it on a skate-board. Now we are not restricted to integers. Instead of having to step from block 8 to block 9, for example, we can glide all the way from peg 8 to peg 9, stopping at *any point* along the way (Figure 12.3). The path is now the number line, on which all possible real numbers (including the integers) are represented.

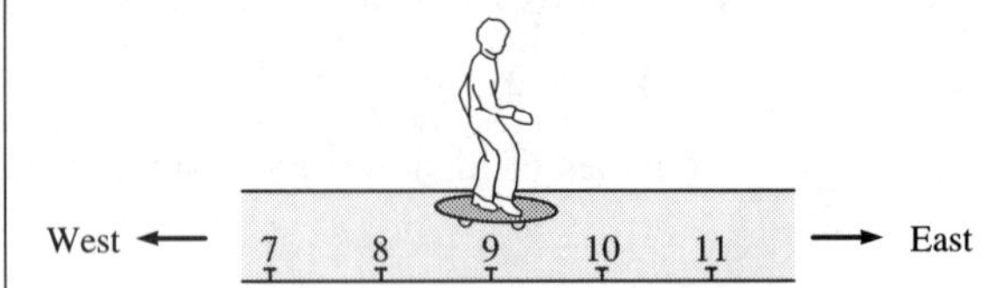

Figure 12.3

The number line is useful for showing the range of inequalities. Given $b \geqslant 5$, for example, the thickened part of the number line in Figure 12.4 shows what values b can have. The solid dot at 5 indicates that b can equal 5, as stated in the inequality.

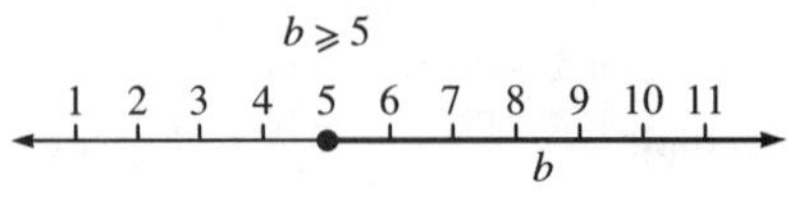

$b \geqslant 5$

1 2 3 4 5 6 7 8 9 10 11

b

Figure 12.4

Given $c < 6$ the thickened line in Figure 12.5 extends indefinitely in the negative direction. The open circle at 6 indicates that c can have any value up to but *not including* 6.

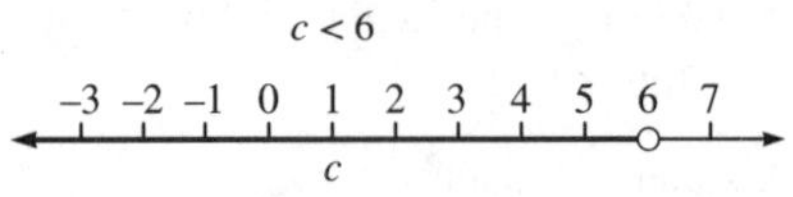

$c < 6$

−3 −2 −1 0 1 2 3 4 5 6 7

c

Figure 12.5

On the line in Figure 12.6 the dots at each integer value show that the variable is an integer.

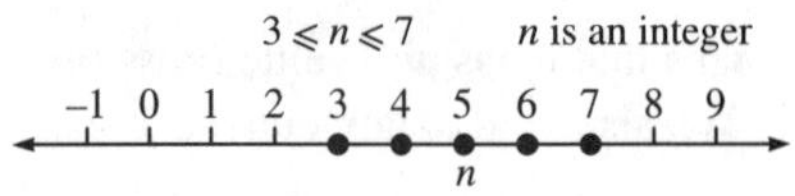

$3 \leqslant n \leqslant 7$ n is an integer

−1 0 1 2 3 4 5 6 7 8 9

n

Figure 12.6

With equalities, we may have to specify what values the variable may take within its range. Take the first example:

$b \geqslant 5$

If b is the amount of rice kept in a store-cupboard in kg, this statement shows that there is always a minimum of 5 kg in stock, but the amount can be more than this. With rice, the amount does not have to be a whole number of kilograms. There might be 6.75 kg, for example, or 9.2 kg. There might be *any* value up to a large amount, depending on the size of the cupboard.

If b is the number of eggs kept in the cupboard, b must be an integer. There might be 6 or 7 eggs, for example, but never 9.2 eggs. If the number must be an integer, this must be stated after the inequality:

$b \geqslant 5$, where b is an integer

Solving inequalities

Solving an inequality is the same as solving an equation. Always do the *same* things to *both* sides of the inequality. The box on page 168 lists the options of things to do. The only difference with inequalities is that changing signs is not allowed.

Examples

Solve the inequality $-4 + 3x > 8$

Add 4 to both sides:

$\Rightarrow \quad 3x > 12$

Divide both sides by 3:

$\Rightarrow \quad \underline{x > 4}$

Solve the inequality $2x + 10 \leqslant 6$

Subtract 10 from both sides:

$\Rightarrow \quad 2x \leqslant -4$

Divide both sides by 2:

$\underline{x \leqslant -2}$

Test yourself 12.6

1 Students are each issued with 4 sheets of paper for a project, and are supplied with more sheets if they need them. Write an inequality for the number of sheets n supplied to each student. Draw a number line and show this range on it.

2 A room heater switches on when the temperature of a room falls to 15 °C and switches off when the temperature reaches 24 °C. Express this as an inequality, with t as the variable. Draw a number line and show this range on it.

3 Write an inequality to show the possible volume v of water in a 15-gallon tank. Draw a number line and show this range on it.

4 When a number is rounded to 1 dp, its value is 12.8. Write an inequality to define all possible numbers n that round to this amount. Draw a number line to show this range.

5 Up to 12 cars can be parked outside an office. Express this as an inequality, and draw the number line to show this range.

6 Solve these inequalities.

a $2 + a \geqslant 3$ **b** $7 > 2j - 5$

c $4 + 3n \leqslant -2$ **d** $9 < 2x + 3$

7 For each of the inequalities in question 6, give the smallest and largest integer value that the variable can have [*Hint*: Some may have no limit.]

8 To assess if you have covered this chapter fully, answer the questions in *Try these first*, page 159.

13 Powers

In Chapter 4 we described squared and cubed quantities, such as x^2 and a^3. Here we take the subject of powers further and find that using them is a great help in all kinds of calculations. Powers are the basis of logarithms, which are described at the end of this chapter.

You need to know Essential Maths (Section 1), especially squares and cubes (page 59).

Try these first

Your success with this short test will tell you which parts of this chapter you already know.

1 Simplify.

a $a^2 \times a^3$ **b** $\dfrac{x^5}{x^3y}$

c $(n^2)^7$ **d** $d^2 \times d^5 \times d^{-6}$

2 Given $p = 121$, evaluate $(p^{1/2})^3$.

3 Express 456.7 in standard (scientific) form.

4 Evaluate 5.9×10^3 multiplied by 2.2×10^2.

5 Evaluate the sum of 4.52×10^4 and 9.032×10^3.

6 Use logs to find x in the equation $5^x = 3$.

7 Use logs to find $4.5^{1.2}$.

Multiplying powers

The product $a \times a$, often referred to as 'a squared', is written as a^2. The ^{2} is called the **index** (plural, **indices**) and it tells us how many a's are multiplied together. Similarly, 'a cubed' is written as a^3. We can have many more than 3 a's, for example:

$$a \times a \times a \times a \times a \times a = a^6$$

Using an index is a quick and compact way of writing the number. Now suppose we are asked to multiply a^6 by a^2. The long way to do this is to write out a string of six a's, followed by a pair of a's:

$$a^6 \times a^2 = a \times a \times a \times a \times a \times a \quad \times \quad a \times a$$

Counting the a's shows that there are now eight a's multiplied together:

$$\underline{a^6 \times a^2 = a^8}$$

Writing out all the a's and counting them is a waste of time. All we need to do is add 6 and 2. You can do this in your head. But writing them out just this once has demonstrated the rule:

To multiply powers, add the indices

Examples

$x^2 \times x^3 = x^{2+3} = x^5$
$n^{12} \times n^7 = n^{12+7} = n^{19}$
$5^3 \times 5^4 = 5^{3+4} = 5^7$
$a^x \times a^{3x} = a^{x+3x} = a^{4x}$

The powers must be powers of the same variable or number. If powers of *different* variables are multiplied, we *do not* add the indices, we just write one after the other, in the usual way:

$a^6 \times b^2 = a^6b^2$

Power of a power

An example of this is $(a^2)^3$, or 'the cube of a-squared'. The term in the bracket is a^2, and the ^{3} following the bracket shows that we have to cube the term that is in the bracket:

$(a^2)^3 = a^2 \times a^2 \times a^2$

Three pairs of a's makes six a's, so

$\underline{(a^2)^3 = a^6}$

The rule is:

To find the power of a power, multiply the indices

Examples

$(n^4)^3 = n^{4\times3} = n^{12}$
$(n^3)^4 = n^{3\times4} = n^{12}$
$(7^2)^5 = 7^{2\times5} = 7^{10}$
$(a^b)^c = a^{b\times c} = a^{bc}$

The first two examples show that it makes no difference which index is inside the bracket and which is outside.

Test yourself 13.1

1 Multiply.

a $p^2 \times p^3$ **b** $t^4 \times t^7$
c $9^6 \times 9^3$ **d** $r^3 \times r^3$
e $x^2 \times x^6 \times x^4$ **f** $q^8 \times p^2q^2$
g $s^t \times s^{5t}$ **h** $c^x \times c^y$

2 Simplify.

a $(n^5)^2$ **b** $(a^7)^2$

c $(h^2)^{10}$ **d** $(k^4)^4$

e $(8^2)^4$ **f** $(x^a)^3$

Dividing powers

As might be expected, dividing one power by another is the inverse of multiplying.

Example

Divide a^7 by a^3.

Set this out as two strings of a's:

$$\frac{a^7}{a^3} = \frac{a \times a \times a \times a \times a \times a \times a}{a \times a \times a}$$

Three a's cancel out in the N and the D, leaving only four a's in the N:

$$\frac{a^7}{a^3} = a^{7-3} = a^4$$

There were 7 a's in the N, but we cancelled (took away) three, leaving four. The rule is:

To divide powers, subtract the indices

Examples

$$\frac{q^6}{q^2} = q^{6-2} = q^4$$

$$\frac{3x^29}{12x^3} = \frac{3}{12} \times x^{9-3} = \frac{x^6}{4}$$

$$\frac{a^4b^3}{a^3b} = a^{4-3}b^{3-1} = ab^2$$

In the second example we also cancel the coefficients as far as possible. In the last example, note that a^1 is the same as a, and b is the same as b^1 (page 60).

The rule still applies if there are more a's in the D than in the N.

Example

Divide a^2 by a^5.

$$\frac{a^2}{a^5} = \frac{a \times a}{a \times a \times a \times a \times a}$$

Two a's cancel out in the N and D, leaving only three a's *in the denominator*:

$$\frac{a^2}{a^5} = \frac{1}{a^3}$$

If, instead of cancelling, we apply the subtraction rule to the original quotient, we find:

$$\frac{a^2}{a^5} = a^{2-5} = a^{-3}$$

Both results come from the same division; they have the same value. We conclude that:

$$a^{-3} \equiv \frac{1}{a^3}$$

In words:

A negative power is the same as the reciprocal (page 41) of the positive power

Examples

$$n^{-4} = \frac{1}{n^4}$$

$$a^{-7} = \frac{1}{a^7}$$

$$3^{-2} = \frac{1}{3^2}$$

Zero power

Any number to the power 0 equals 1.

To prove this, evaluate the quotient $\frac{a}{a}$

Cancel a in both N and D: $\frac{\cancel{a}^{1}}{\cancel{a}_{1}} = 1$

But, by using indices:

$$\frac{a}{a} = a^1 \times a^{-1} = a^{1-1} = a^0$$

Combining both equations:

$$\underline{a^0 = 1}$$

The value of a makes no difference to the proof. $a^0 = 1$, $3^0 = 1$, $24^0 = 1$, $432^0 = 1$, . . . $(\text{any number})^0 = 1$

Test yourself 13.2

1 Divide.

a $\dfrac{n^8}{n^3}$ **b** $\dfrac{15d^6}{5d^2}$

c $\dfrac{q^{12}}{q^3}$ **d** $\dfrac{20x^8}{4x^7}$

e $\dfrac{(t^2)^3}{t^4}$ **f** $\dfrac{r^4 \times r^3}{r^2 \times r}$

2 Write as a reciprocal.

a g^{-7} **b** a^{-5}

c y^{-1} **d** n^{-9}

3 Write as a variable with a negative index.

a $\dfrac{1}{a^2}$ **b** $\dfrac{1}{b^7}$

c $\dfrac{1}{p^{14}}$ **d** $\dfrac{1}{q}$

4 Simplify, writing results with a negative index if appropriate.

a $\dfrac{w^2}{w^5}$ **b** $\dfrac{g^5}{g^9}$

c $\dfrac{(f^4)^3}{f^{14}}$ **d** $\dfrac{(d^2)^3}{(d^4)^5}$

e $a^5 \times a^{-3}$ **f** $\dfrac{s^3 \times s^3}{s^2 \times s^3}$

g $v^5 \times v^3 \times v^{-7}$ **h** $q^{-2} \times 2q^4 \times 3q^{-3}$

Fractional powers

The idea of a fractional power, such as $a^{0.7}$, does not seem to make sense. How can we multiply a by itself 0.7 times? To answer this question, let us see what we can do with another fractional power, $a^{0.5}$. This is also written $a^{1/2}$. If $a^{1/2}$ exists, then the **power of a power** rule (page 176) applies. Try squaring $a^{1/2}$:

$$(a^{1/2})^2 = a^{1/2 \times 2} = a^1 = a$$

$a^{1/2}$ is a quantity which, when **squared**, gives a. We say that $a^{1/2}$ is the **square root** of a.

$$a^{1/2} = \sqrt{a}$$

Finding the square root of a number is the inverse of squaring. The table on page 77 lists numbers and their squares. Use this table in reverse to find square

Square and other roots

A square root is a number or quantity which, when squared, equals a given number or quantity.

The square root of 25 is 5 because, when 5 is squared, we obtain 25. The symbol for square root is $\sqrt{}$

Examples

$\sqrt{25} = 5$ because $5^2 = 5 \times 5 = 25$
$\sqrt{4} = 2$ because $2^2 = 2 \times 2 = 4$
$\sqrt{100} = 10$ because $10^2 = 10 \times 10 = 100$
$\sqrt{2} = 1.414$ (3 dp) because $1.414^2 = 1.414 \times 1.414 = 2$

The symbol for the square root of x is $\sqrt{x}$.

A cube root is a number of quantity which, when cubed, equals a given number or quantity. The symbol for cube root is $\sqrt[3]{}$

Examples

$\sqrt[3]{8} = 2$ because $2^3 = 2 \times 2 \times 2 = 8$
$\sqrt[3]{216} = 6$ because $6^3 = 6 \times 6 \times 6 = 216$

Fourth, fifth, sixth and other roots are defined in a similar way.

Example

$\sqrt[6]{729} = 3$ because $3^6 = 3 \times 3 \times 3 \times 3 \times 3 \times 3 = 729$

roots of the numbers listed in the right hand column. For example, the square root of 49 is 7.

Similarly $a^{1/3}$ is the quantity which, when **cubed**, gives a. It is the **cube root** of a.

$$a^{1/3} = \sqrt[3]{a}$$

Roots on a calculator

Most scientific calculators have a special key for square roots, and also for cube roots.

Examples

Find the square root of 81.

Key strokes	*Display shows*
[8] [1]	81
[$\sqrt{}$]	9

$\underline{\sqrt{81} = 9}$

Find the cube root of 125.

On many calculators the 'Shift' or 'Second function' key must be pressed before the cube root key.

Key-strokes	*Display shows*
[1] [2] [5]	125
[S] [$\sqrt[3]{}$]	5

$\underline{\sqrt[3]{125} = 5}$

The table of cubes on page 84 tells us a few cube roots when used in reverse. For example, the cube root of 64 is 4.

So fractional powers *do* make sense and can be multiplied and divided according to the usual rules.

Example

$$a^{1/3} \times a^{1/3} = a^{1/3 + 1/3} = a^{2/3}$$

There are other ways of arriving at the same result. One is to square the cube root of a, which is what we have just done above:

$$(a^{1/3})^2 = a^{1/3 \times 2} = a^{2/3}$$

Another way is to take the cube root of a-squared:

$$\sqrt[3]{a^2} = a^{2 \times 1/3} = a^{2/3}$$

All give the same result, whatever the value of a:

$$a^{2/3} \equiv (\sqrt[3]{a})^2 \equiv \sqrt[3]{a^2}$$

In a fractional index, the N represents squaring, cubing, . . .; the D represents square-rooting, cube-rooting . . .

Examples

$$p^{4/3} = (\sqrt[3]{p})^4$$
$$16^{3/2} = (\sqrt{16})^3 = 4^3 = 64$$

Summary of the index rules

Rule	*In symbols*	*Example*
Multiplication	$a^n \times a^m = a^{n+m}$	$a^3 \times a^4 = a^7$
Division	$\dfrac{a^m}{a^n} = a^{m-n}$	$\dfrac{a^7}{a^2} = a^5$
Power of power	$(a^m)^n = a^{mn}$	$(a^3)^2 = a^5$
Reciprocal	$a^{-n} = \dfrac{1}{a^n}$	$7^{-2} = \dfrac{1}{7^2} = \dfrac{1}{49}$
Zero power	$a^0 = 1$	$4^0 = 1$
Fractional indices	$a^{1/2} = \sqrt{a}$	$9^{1/2} = \sqrt{9} = 3$
	$a^{1/3} = \sqrt[3]{a}$	$8^{1/3} = \sqrt[3]{8} = 2$
	$a^{n/m} = (\sqrt[m]{a})^n$	$8^{2/3} = (\sqrt[3]{8})^2 = 2^2 = 4$
	or $\sqrt[m]{a^n}$	$\sqrt[3]{8^2} = \sqrt[3]{64} = 4$

Test yourself 13.3

1 Write in the form which uses the root sign.

a $x^{1/4}$ **b** $n^{3/4}$

c $a^{3/2}$ **d** $b^{1.4}$ [*Hint*: Convert the index to an improper fraction]

e $f^{-1/2}$ **f** $t^{-2/3}$

2 Write in the form which uses fractional indices.

a $\sqrt[5]{n}$ **b** $\sqrt[3]{a^2}$

c $(\sqrt[7]{t})^3$ **d** $(\sqrt[9]{s})^4$

e $(\sqrt[5]{x})^2$ **f** $(\sqrt[3]{k})^{-2}$

3 Evaluate (see tables of squares and cubes on pages 77 and 84, or use a calculator).

a $121^{1/2}$ **b** $125^{1/3}$

c $4^{3/2}$ **d** $27^{2/3}$

Powers on a calculator

1 The easiest way to square a number is to use the × key.

Example

Find the square of 13.

Key-strokes	*Display shows*
[1] [3]	13
[×] [=]	169

$13^2 = 169$

2 Some calculators can be made to multiply repeatedly by the same amount. Press the [×] key twice.

Example

Find the first 6 powers of 3.

Key-strokes	*Display shows*	*Equals*
[3] [×] [×]	3	3
[=]	9	3^2
[=]	27	3^3
[=]	81	3^4
[=]	243	3^5
[=]	729	3^6

Each key-stroke displays the next higher power of 3.

The first 6 powers of 3 are 3, 9, 27, 81, 243, and 729.

3 Higher powers and fractional powers are calculated by using the [x^y] key, available on scientific calculators.

Examples

Find 5^{10}

Key-strokes	*Display shows*
[5] [x^y]	5
[1] [0]	10
[=]	9765625

$5^{10} = 9765625$

Find $7^{1.4}$

Key-strokes	*Display shows*
[7] [x^y]	7
[1] [.] [4]	1.4
[=]	15.24534497

$7^{1.4} = 15.24534497$

If your calculator does not have exactly the same facilities as shown above, consult its handbook.

Powers of 10

Here we list powers of 10 in ascending order:

10^1 is 10
10^2 is $10 \times 10 = 100$
10^3 is $10 \times 10 \times 10 = 1000$
10^4 is $10 \times 10 \times 10 \times 10 = 10000$

There is no need to go further, for a rule is clear:

10 to the power of any whole positive number *n* is 1 followed by *n* zeros

The same thing can be done with negative powers:

10^{-1} is 1/10, which is 0.1 in decimal form
10^{-2} is 1/100, which is 0.01 in decimal form
10^{-3} is 1/1000, which is 0.001 in decimal form
10^{-4} is 1/10000, which is 0.0001 in decimal form

10 to the power of any whole negative number *n* has after the decimal point *n*–1 zeros followed by 1

Standard form

Standard form or **scientific form** is a way of writing numbers, using powers of 10. A number in standard form consists of a number with **one** significant figure before the decimal point, multiplied by a power of ten (if necessary) to bring it to its true value. The index is always a whole number but may be positive or negative.

Examples

Express 372.56 in standard form.

This must be written with the decimal place between the 3 and the 7:

$3.7256 \times 10^{?}$

The decimal point has been shifted 2 places to the left. This means that we have divided 372.56 by 100. To bring it back to its true value, multiply by 100, which is 10^2:

$\underline{372.56 = 3.7256 \times 10^2}$

Express 0.00038 in standard form.

Standard form is:

$3.8 \times 10^{?}$

The decimal point has been shifted 4 places to the right, multiplying 0.00038 by 10 000. To restore its true value, we have to divide by 10 000, which is 10^4. Dividing by 10^4 is the same as multiplying by its reciprocal, 10^{-4}:

$\underline{0.00038 = 3.8 \times 10^{-4}}$

Numbers in standard form are multiplied and divided according to the rules for powers.

Examples

Evaluate 2.3×10^2 multiplied by 4.7×10^3.

Multiply the numbers in the usual way; multiply the powers of tens by adding their indices:

$$\begin{aligned} 2.3 \times 10^2 \times 4.7 \times 10^3 &= (2.3 \times 4.7) \times 10^{2+3} \\ &= 10.81 \times 10^5 \end{aligned}$$

But the number now has *two* figures before the decimal point. To convert back to standard form, divide the number by 10, and multiply 10^5 by 10:

$\underline{\text{Result} = 1.081 \times 10^6}$

Evaluate $\dfrac{4.96 \times 10^6}{1.6 \times 10^2}$

Divide the numbers in the usual way; divide the powers by subtracting indices:

$$\begin{aligned} \frac{4.96 \times 10^6}{1.6 \times 10^2} &= \frac{4.96}{1.6} \times 10^{6-2} \\ &= 3.1 \times 10^4 \end{aligned}$$

There is still one figure before the decimal point, so no adjustment is needed.

$\underline{\text{Result} = 3.1 \times 10^4}$

When numbers in standard form are to be added or subtracted, and they have the same power of 10, add the numbers *only*, then correct the power if necessary:

Standard form on a calculator

Scientific calculators are able to display numbers in ordinary form or in standard form. Often they switch automatically to standard form when a number has more figures than the display can show. In standard form, the power of 10 is displayed (usually in smaller figures) on the right.

Example

8.61×10^7 is displayed as

8.61 ^{07}

Remember that this display does *not* mean 8.61^{07}. The small figures indicate that the number is multiplied by powers of *ten*.

Numbers are entered in standard form by using the 'exponent' key (**exponent** is another name for **index**), which is usually labelled E, EXP, or EE (enter exponent).

Example

Find 2.56×10^2 multiplied by 7.42×10^3

Key-strokes	*Display*	*Notes*
[2] [.] [5] [6]	2.56	
[EXP]	2.56 ^{00}	
[2]	2.56 ^{02}	First number entered
[×]	256	Displayed in ordinary form
[7] [.] [4] [2]	7.42	
[EXP]	7.42 ^{00}	
[3]	7.42 ^{03}	Second number entered
[=]	1899520	Displayed in ordinary form

Convert the result to standard form if necessary.

$$\underline{2.56 \times 10^2 \times 7.42 \times 10^3 = 1.89952 \times 10^6}$$

It may be possible to put your calculator into 'Scientific Mode', in which case it displays the result in scientific form automatically. Consult the handbook for details.

Examples

$$\begin{aligned} 2.8 \times 10^3 + 4.1 \times 10^3 &= (2.8 + 4.1) \times 10^3 \\ &= 6.9 \times 10^3 \end{aligned}$$

$$\begin{aligned} 4.8 \times 10^4 + 7.3 \times 10^4 &= (4.8 + 7.3) \times 10^4 \\ &= 12.1 \times 10^4 \\ &= 1.21 \times 10^5 \end{aligned}$$

If two numbers to be added or subtracted do not have the same power of 10, it is better to convert them to ordinary form first:

Example

$6.2 \times 10^3 + 4.9 \times 10^2 + 1.1 \times 10^4$

Convert and list with decimal points aligned:

$$
\begin{aligned}
6.2 \times 10^3 &= 6200 \\
4.9 \times 10^2 &= 490 \\
1.1 \times 10^4 &= \underline{11000} \\
\text{Sum} &= 17690
\end{aligned}
$$

Convert to standard form:

$\underline{\text{Sum} = 1.769 \times 10^4}$

Test yourself 13.4

1 Write out these powers of 10 in full.

a 10^3 **b** 10^6 **c** 10^1

d 10^{-3} **e** 10^{-4} **f** 10^{-6}

2 Write these numbers as powers of 10.

a 100 **b** 10 000 000 **c** 10 000

d 0.1 **e** 0.000 1 **f** 0.000 001

3 Express these standard-form numbers in ordinary form.

a 7.23×10^2 **b** 5.32×10^5 **c** 7.2×10^{-3}

d 6.803×10^{-1} **e** $8.000\,03 \times 10^3$ **f** 7.1×10^{-8}

4 Evaluate these expressions without using a calculator.

a $4.5 \times 10^2 \times 2.7 \times 10^4$

b $9.9 \times 10^5 \times 1.1 \times 10^2$

c $\dfrac{3.285 \times 10^4}{4.5 \times 10^2}$

d $\dfrac{1.904 \times 10^2}{5.6 \times 10^5}$

e $4.2 \times 10^2 + 3.6 \times 10^2$

f $7.61 \times 10^4 + 3.721 \times 10^3 - 2.201 \times 10^4$

Logarithms

Logarithms, or **logs** as they are usually called, express a number as a power of another number.

Example

Express 100 as a power of 10.

We know that 100 is 10^2:

$\log_{10}100 = 2$

The small 10, written as a subscript to the symbol 'log' tells us that we are working in powers of 10. We say that 10 is the **base** of the log. In practice, 10 is so often used as a base that the symbol 'log', without any subscript, is taken to mean 'log to the base 10'. We also use the symbol 'lg' for '$\log_{10}$'.

Logs can be taken to any base, for example:

$\log_2 16 = 4$

This is because 16 is 2^4. But the only bases that are commonly used are 10 and 2.7183. The number 2.7183 is the value of a special constant referred to as the **exponential constant**, symbol 'e'. It is an important constant in maths, which is why it is given its own special symbol. If you continue maths to a more advanced level than this book, you will learn how the value of e is arrived at, and why it is so important. We have quoted its value to 4 dp, but e is one of those numbers that has an indefinitely large number of decimal places. Logs to the base e are known as **natural logarithms**, symbol '$\log_e$' or simply 'ln'.

A log is a power, so logs obey the same rules as indices. For example, to multiply a^2 by a^3, we simply add their indices to obtain a^5. The same applies to logs.

Example

Multiply 2 by 3

Instead of multiplying these numbers, we add their logs:

$$\begin{aligned}\log(2 \times 3) &= \log 2 + \log 3 \\ &= 0.3010 + 0.4771 \text{ (by calculator)} \\ &= 0.7781\end{aligned}$$

The final stage is to find a number which has 0.7781 as its log. We call this number the **antilog**, symbol '$\log^{-1}$'.

$\log^{-1} 0.7781 = 6$ (by calculator)

Result: $2 \times 3 = 6$

This is a rather round-about way to perform a simple multiplication, but we are using this example just to demonstrate one of the properties of logs. For multiplying numbers with many more significant figures, *adding* logs is easier than long multiplication. But using a calculator is easier still.

Rules for logs

Rule	*In symbols*	*Example*
Multiplication	$\log(m \times n) = \log m + \log n$	$\log(3 \times 5) = \log 3 + \log 5$
Division	$\log(m/n)$	$\log(3/5) = \log 3 - \log 5$
Powers	$\log m^n = n \log m$	$\log 3^5 = 5 \log 3$
Reciprocal	$\log(1/m) = -\log m$	$\log(1/3) = -\log 3$
Zero power	$\log 1 = 0$	
Roots	$\log(\sqrt{m}) = \dfrac{\log m}{2}$	$\log(\sqrt{5}) = \dfrac{\log 5}{2}$

Logs on a calculator

For logs to base 10, use the [log] key:

Key-strokes	*Display*
[6]	6
[log]	0.778 151 25

For logs to base e, use the [ln] key:

Key-strokes	*Display*
[6]	6
[ln]	1.791 759 469

For antilogs to base 10, use the [10^x] key. Usually the 'shift' or 'second function' key is pressed first.

Key-strokes	*Display*
[1] [.] [2] [3]	1.23
[S] [10^x]	16.982 436 52

For antilogs to base e, use the [e^x] key. Usually the 'shift' or 'second function' key is pressed first.

Key-strokes	*Display*
[1] [.] [2] [3]	1.23
[S] [e^x]	3.421 229 536

A useful application of logs is to solve equations of this kind: Given $3^x = 7$, find the value of x.

Take logs of *both* sides of the equation:

$$\log(3^x) = \log 7$$

Using the third rule from the box:

$$\Rightarrow \qquad x \log 3 = \log 7$$

$$x = \frac{\log 7}{\log 3}$$

Use a calculator to find the logs:

$$x = \frac{0.8451}{0.4771} = 1.771$$

$$\underline{x = 1.771 \text{ (3 dp)}}$$

In this calculation we took logs of both sides of the equation. This keeps the equation balanced. Taking logs of both sides is another operation to be added to the list on page 168. Incidentally, this equation can be solved by using logs to base e instead of logs to base 10. The logs are different but the result is the same. One fact about logs is that it is not possible to take the log of zero or any negative number.

Test yourself 13.5

1 Evaluate without using a calculator.

a lg 100 **b** lg 1000 **c** lg 1 000 000

2 Use a calculator to evaluate these to 4 dp.

a lg 4 **b** lg 72 **c** ln 72

3 Find x, using logs obtained from a calculator.

a $5^x = 12$ **b** $2.7^x = 89.1$

4 To assess if you have covered this chapter fully, answer the questions in *Try these first*, page 175.

14 Triangles and their ratios

As we saw in Chapter 5, straight-sided shapes can be broken down into two or more triangles. For this reason, understanding the properties of triangles is essential for dealing with straight-sided shapes of all kinds, and for handling problems involving angles. These include surveying, designing buildings and machinery, ceramic and textile design, electronic circuit-boards and many other aspects of technology.

You need to know the Maths Essentials (Part 1) especially the chapter on lines, angles and shapes (Chapter 5). You also need to understand ratios and proportions (Chapter 10).

Try these first

Your success with this short test will tell you which parts of this chapter you already know.

1 In triangle ABC, AC = 70 mm, $\hat{A} = 61°$ and $\hat{C} = 52°$. In triangle PQR, PQ = 70 mm, $\hat{P} = 61°$ and $\hat{Q} = 67°$. In triangle XYZ, XZ = 100 mm, $\hat{Y} = 67°$ and $\hat{Z} = 52°$. Draw sketches of these triangles. Which triangles are congruent and which are similar?

2 Use a calculator to find:

a sin 67° **b** $\tan^{-1}$ 24.72 **c** cos 50°

d $\cos^{-1}$ 0.456 **e** tan 28° **f** $\sin^{-1}$ 0.32

3 A ball rolls 3.4 m down a 14° slope. How much height does it lose?

4 In a triangle ABC, $\hat{B} = 90°$, $\hat{C} = 62°$, and AC = 4.6 m. Find the lengths of AB and BC.

Copying triangles

You are asked to construct a triangle exactly like △ABC, in Figure 14.1 The copy must be exactly the same **size** as the original, and exactly the same **shape**. There are three ways of doing this, depending on what you know about the triangle:

Knowing three sides

Measure the three sides of the triangle. In the original drawing for this book, the sides are: AB = 60 mm, BC = 100 mm, CA = 90 mm. The steps for drawing the copy are:

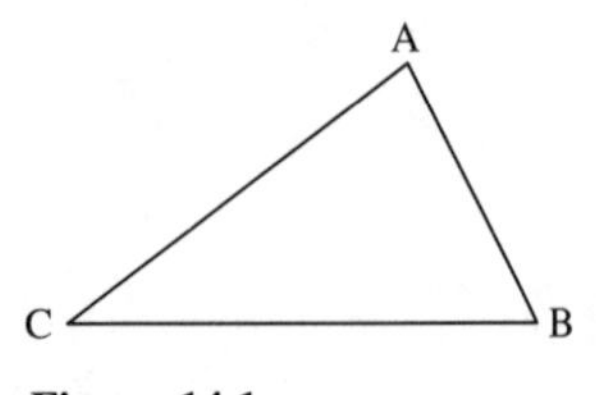

Figure 14.1

a Draw a straight line, BC, 100 mm long.
b Open a pair of compasses to 90 mm, place the point of the compasses on B and draw a part of a circle (an *arc*) where shown in Figure 14.2a. All points on this arc are 90 mm from C.

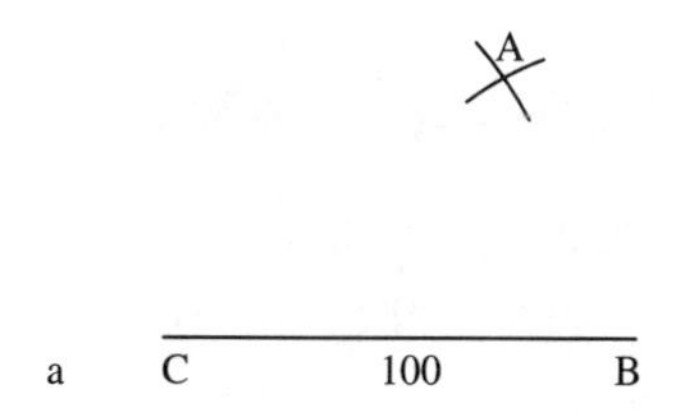

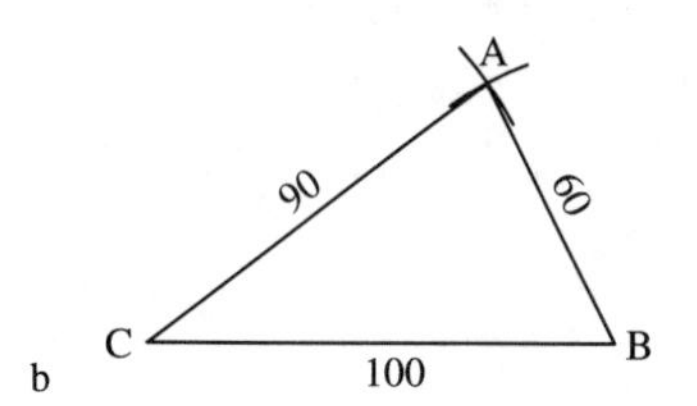

Figure 14.2

c Alter the compasses to 60 mm, place the point of the compasses on B and draw an arc which cuts the first arc. All points on this arc are 60 mm from B.
d Point A is where the arcs cross. This point is 90 mm from C *and* 60 mm from B.
e Draw straight lines to join CA and AB (Figure 14.2b).

Knowing two sides and the angle between them

The angle between the two known sides is called the **included angle**. In the original drawing, the two sides are BC = 100 mm, CA = 90 mm. $\hat{C}$ = 36°. These are the steps for copying the triangle, using the three items of information.

a Draw a straight line, BC, 100 mm long.
b From C, draw a straight line CA, 90 mm long, at an angle of 36° to BC (Figure 14.3a).
c Join AB to complete the triangle (Figure 14.3b).

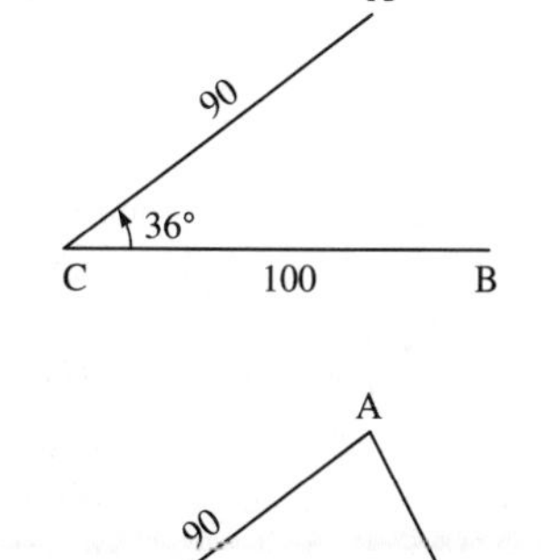

Figure 14.3

Knowing one side and two angles

Measure one side: BC = 100 mm. Measure two angles: $\hat{B}$ = 63°, $\hat{C}$ = 36°. We need to know the angles at the ends of the line that is measured. But the angles of a triangle add up to 180° (page 67) so, if we know *any* two angles, we can always calculate the third angle. The steps for copying the triangle are:

a Draw a straight line, BC, 100 mm long.
b At end C, draw a line at 36° to BC (Figure 14.4a).
c At end B, draw a line at 63° to BC.
d Point A is where the lines cross.
e Join AB and CA to complete the triangle (Figure 14.4b).

Copying a triangle in these ways give other triangles which are exactly the same *size* and have exactly the same *shape* as the original triangle. The original and the copies are called **congruent** triangles.

Congruent triangles do not have to be the same way up. In Figure 14.5, triangle B is rotated with respect to triangle A, but they are still congruent. Triangle C is a mirror image to triangle A but, because their sides are the same length and their angles the same size, triangles A and C are congruent.

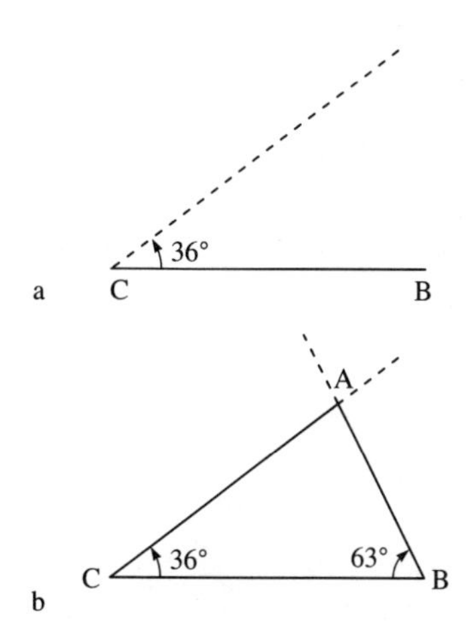

Figure 14.4

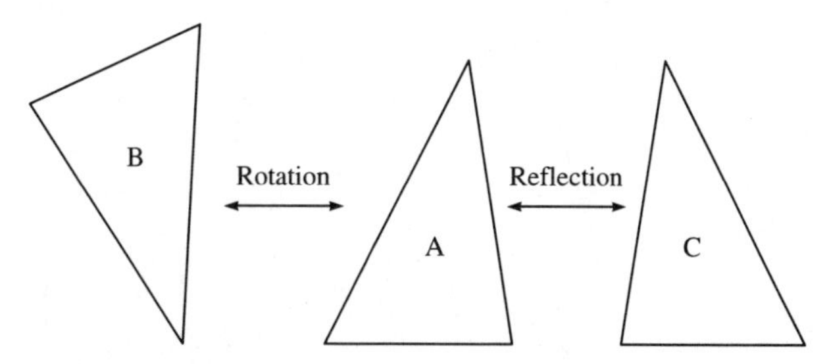

Figure 14.5

Test yourself 14.1

With a ruler and compasses, use the three techniques illustrated by Figures 14.2–14.4 to construct angles, given the following information.

1 a	AB = 50 mm,	BC = 100 mm,	CA = 85 mm.
b	AB = 75 mm,	BC = 25 mm,	CA = 65 mm.
c	AB = 65 mm,	BC = 30 mm,	CA = 80 mm.
2 a	BC = 100 mm,	CA = 75 mm,	$\hat{C} = 50°$.
b	AB = 55 mm,	CA = 85 mm,	$\hat{A} = 80°$.
c	BC = 90 mm,	AB = 35 mm,	$\hat{B} = 110°$.
3 a	BC = 95 mm,	$\hat{C} = 30°$,	$\hat{B} = 85°$.
b	CA = 25 mm,	$\hat{C} = 70°$,	$\hat{A} = 85°$.
c	AB = 45 mm,	$\hat{A} = 60°$,	$\hat{C} = 70°$.

Similar triangles

Whatever technique you use to draw a congruent triangle, you always need to know the length of at least one side. If you are just told the three angles, there is no way of knowing how big the triangle is to be. Figure 14.6 shows six triangles, all with the same three angles, 36°, 63° and 81°. The triangles differ in *size*, so they are not congruent. But they are all the same *shape*. They all have the same shape as the triangles in Figures 14.1 to 14.4. Triangles which

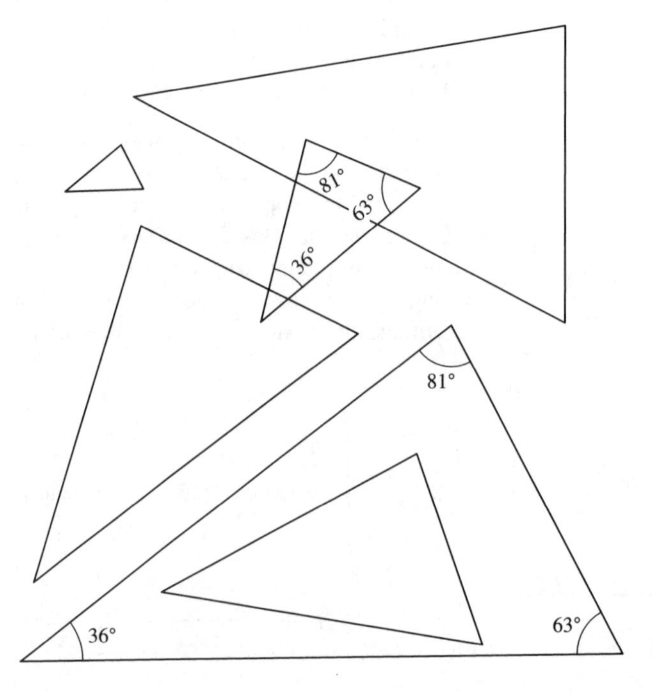

Figure 14.6

have the same shape are called **similar** triangles. There are millions more triangles that we could draw, all of different sizes but all similar to those in Figure 14.6.

Because the triangles in Figure 14.6 all have the same shape, the ratio between the lengths of their sides is the same for all of them (see page 135). If you measure any or all of them, you find that the sides are in the ratio 10:9:6. Other ratios produce triangles of different shapes (Figure 14.7). If the sides are in the ratio 1:1:1, the triangle is equilateral (Figure 14.7b). The triangle with the ratio 1:2:2 is an isosceles triangle (Figure 14.7d).

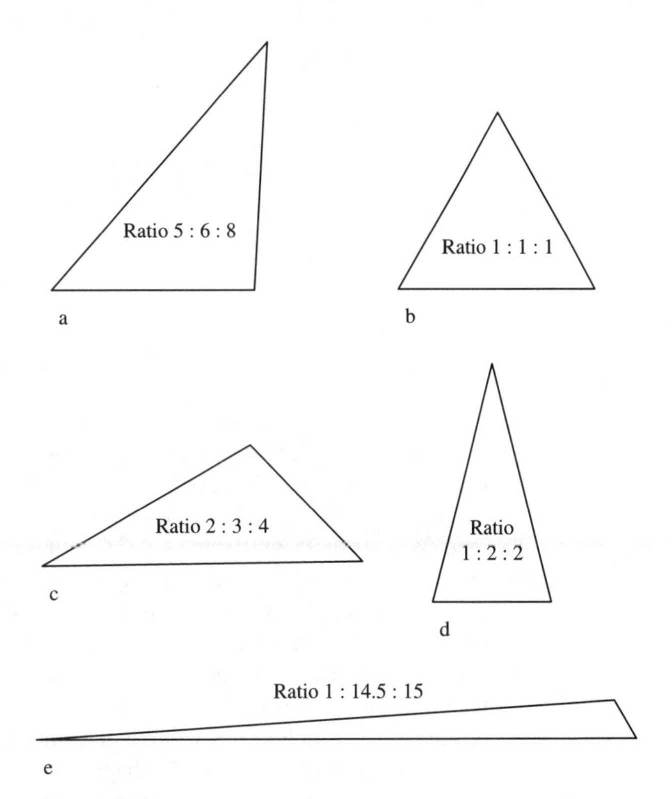

Figure 14.7

Figure 14.8 shows a set of similar right-angled triangles. The ratio of their sides is 1:1.6:1.9. The angle at vertex C is 58°. In a right-angled triangle, the sides are given special names:

Adjacent side:	next to angle C
Opposite side:	opposite to angle C
Hypotenuse:	the longest side, opposite to the right-angle.

The ratio of	adjacent : opposite : hypotenuse
is	1 : 1.6 : 1.9

We will ignore the hypotenuse for the moment and look at the ratio between the opposite side and the adjacent side.

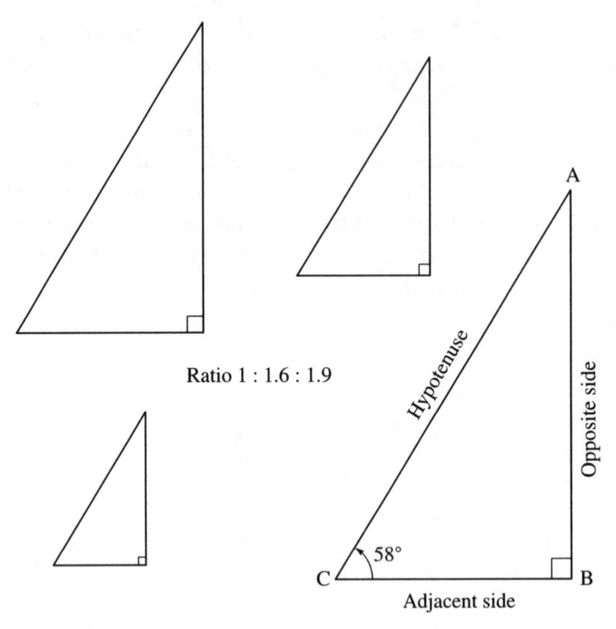

Figure 14.8

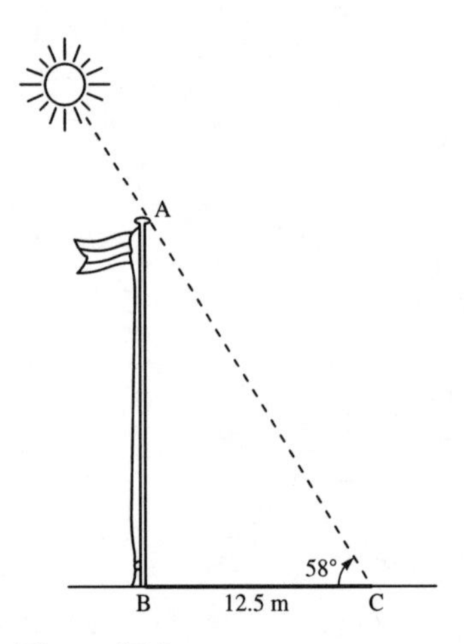

Figure 14.9

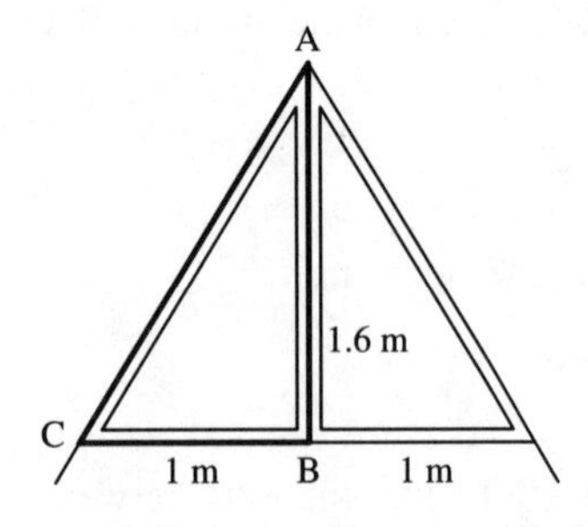

Figure 14.10

In all these triangles the opposite side is 1.6 times longer than the adjacent side. This is a useful property, as Figure 14.9 shows. The shadow of a flagpole falls on the ground. We measure the angle at which the Sun's rays slant down and find that this is 58°. We measure the length of the shadow: it is 12.5 m long. What is the height of the pole? Comparing the triangle in this figure with the triangles in Figure 14.7, we see that this is a similar triangle even if it is a very large one. This tells us that the height of the pole is 1.6 times the length of the shadow. The pole is $1.6 \times 12.5 = 20$ m tall.

Here is another example. A roof is constructed on a support which is 1.6 m high and 2 m wide (Figure 14.10). What is the slope of the roof? The left half of the frame, which is 1 m wide, is the side adjacent to $\hat{C}$. The opposite side is 1.6 times the adjacent side. So the triangle is similar to those in Figure 14.8. This means that the slope of the roof is 58°.

We have been lucky with these two examples because the triangles are similar to those in Figure 14.8 in which one angle is 58°. What we really need to know is the ratio of sides for *any* angle, not just for 58°. The ratio has a special name: it is called the **tangent** of the angle C. The symbol for tangent is **tan**:

$$\tan C = \frac{\text{opposite side}}{\text{adjacent side}} = \frac{AB}{BC}$$

In Figure 14.8:

$$\tan 58° = \frac{1.6}{1} = 1.6$$

Two kinds of tangent

The word **tangent** has two different meanings in maths:

- The ratio between the sides of a triangle, as explained in this chapter.
- A straight line which *just* touches a curve, as explained on page 203.

A tangent is a ratio, so it is a number without any units (page 133). The tangent of any angle *can* be found by drawing a right-angled triangle of a suitable shape, and finding the ratio of its sides. But the most usual way to find a tangent is to use a calculator.

Tangents on a calculator

Scientific calculators usually have a tangent key. But, before you use the key, check that the calculator is set to the right kind of angular unit. There are often three to choose from:

- degrees – used in this chapter and many other chapters
- radians – used later in this book (see Chapter 15)
- grads – not used in this book.

The display often shows 'DEG' or 'D' when it is in degree mode, and many go automatically to degrees when first switched on. If it is in radian mode, the display may show 'RAD' or 'R'.

Example

To find tan 37°. Check that the calculator is in degree mode. Then:

Keystrokes	*Display*
3 7	37.
tan	0.75355405

Usually a tangent to 4 dp is precise enough.

$\tan 37° = 0.7536$

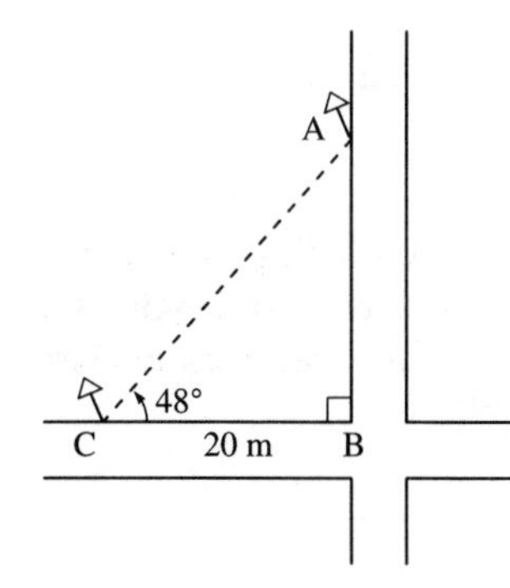

Figure 14.11

Here is a problem to solve, using the tangent ratio. One road crosses another at exactly 90° (Figure 14.11). There are traffic-signs at A and C. Sign C is 20 m from the cross-road. A surveyor finds that the angle between the road BC and the sign at A is 48°. How far is sign A from the cross-road? Begin with this equation:

$$\tan 48° = \frac{\text{opposite side}}{\text{adjacent side}} = \frac{AB}{BC}$$

We want to know AB so we must *change the subject* of the equation (page 168). Multiply both sides by BC:

$$BC \times \tan 48° = AB$$

Exchange sides:

$$AB = BC \times \tan 48°$$

To calculate AB we need to know the length BC, which is 20 m. We also need to know the tangent of 48°. A calculator tells us that tan 48° = 1.1106 (to 4 dp):

$$AB = 20 \times 1.1106 = 22.212$$

Sign A is 22.2 m from the cross-road (1 dp).

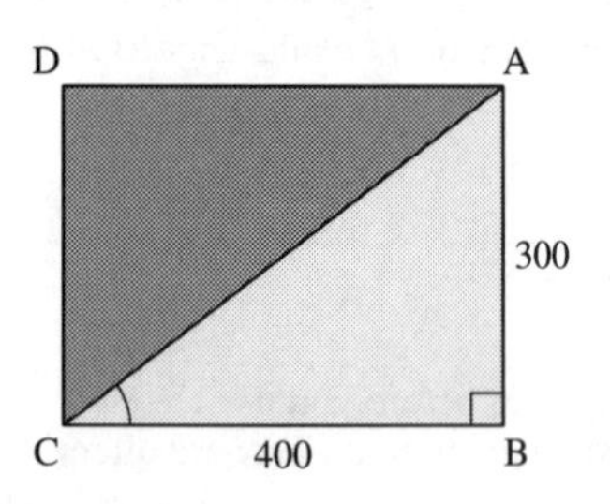

Figure 14.12

The next problem uses the tangent ratio in a different way. A rectangular flag measures 400 mm long and 300 mm wide (Figure 14.12), and is divided diagonally into two differently coloured areas. Find the angle ACB marked in the figure. Begin with this equation:

$$\tan ACB = \frac{\text{opposite side}}{\text{adjacent side}} = \frac{AB}{BC} = \frac{300}{400} = 0.75$$

Inverse tangents on a calculator

Inverse tangents are usually keyed as an 'inverse', 'shifted' or 'second' function.

Example

To find $\tan^{-1} 2.45$. Check that the calculator is in degree mode (see previous box). Then:

Keystrokes	*Display*
[2] [.] [4] [5]	2.45
[inv] [tan]	67.79652147

Usually an inverse tangent to 2 dp is precise enough.

$$\tan^{-1} 2.45 = 67.80°$$

We have to find the angle which has 0.75 as its tangent. Finding the *angle of a tangent* is the **reverse** of finding the *tangent of an angle*. The **angle of a tangent**, or **inverse tangent**, has the symbol $\mathbf{tan^{-1}}$, or sometimes **arctan**. With the calculator, we find that:

$$A\hat{C}B = \tan^{-1} 0.75 = 36.9°$$

The angle at C is 36.9° (1 dp).

Ups and downs

Two terms used to describe angles of view are:

Angle of elevation – when we look up at a point (example, looking up at the top of a tower from ground level), this is the angle between the horizontal and our angle of view.

Angle of depression – when we look down on point (example, looking down from a cliff-top to a person on a beach), this is the angle between the horizontal and our angle of view.

These angles are usually measured with a surveyor's instrument, such as a theodolite. In navigation, the angle of elevation of stars or the Sun is measured with a sextant.

Test yourself 14.2

1 For each of the triangles in Figure 14.7, draw two triangles that are similar.

2 Draw a right-angled triangle in which the opposite side AB is 60 mm and the adjacent side BC is 40 mm. Measure $\hat{C}$. What is the ratio AB:BC? What is the tangent of $\hat{C}$?

3 Draw a right-angled triangle in which the adjacent side is 55 mm and $\hat{C}$ is 37°. What is the length of the opposite side? What is the tangent of 37°?

4 Use a calculator to find to 4 dp:

a tan 72° **b** tan 31° **c** tan 13°

d tan 0° **e** tan 45° **f** tan 89°

5 Use a calculator to find to 2 dp:

a $\tan^{-1} 1.3$ **b** $\tan^{-1} 0.5$ **c** $\tan^{-1} 0.9$

d $\tan^{-1} 30$ **e** $\tan^{-1} 0.02$ **f** $\tan^{-1} 2.2$

When you answer the following questions, draw a *sketch* to help yourself understand the question, but do not make a scale drawing. Instead, calculate the result, obtaining tangents or inverse tangents with a calculator.

6 A flagpole which is 15 m tall casts a 11 m long shadow on level ground. What is the angle of elevation of the Sun?

7 From a point on a beach, 45 m away from the base of a vertical cliff, the angle of elevation of the top of the cliff is 42°. What is the height of the cliff?

8 The builder decides that the roof in Figure 14.10 is too steep. How high should the central post be to make the roof slope at an angle of 50°?

9 A walker starts at point A on a level sports field and walks 210 m due east to point B. She turns and walks 371 m due north to point C. What is the angle CAB?

10 Figure 14.13 is a partly-completed graph showing the tangent of an angle for different angles from 0° to 70°. Copy the graph on to graph paper and, using a calculator to find the remaining values, complete the graph. Is the line straight? If not, describe it.

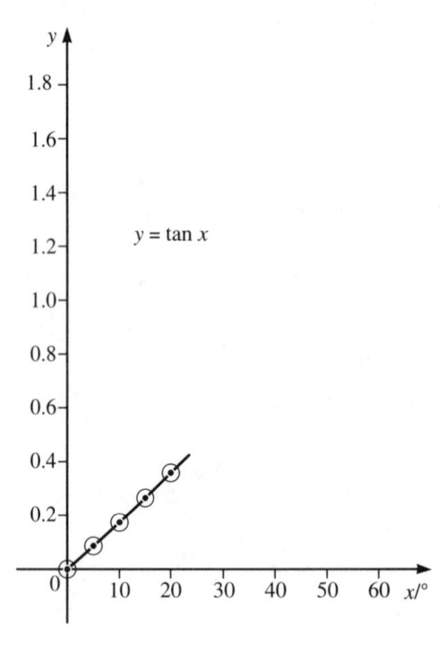

Figure 14.13

Another ratio

A triangle has three sides but, so far, we have considered only one of the possible ratios between them. Another useful ratio is:

$$\frac{\text{opposite side}}{\text{hypotenuse}}$$

This ratio is called the **sine** of the angle, symbol **sin**. Being a ratio, it has no unit. One thing to notice about sines is that the hypotenuse is always the longest side of the triangle, so a sine is always less than 1.

Sines are used for solving problems about triangles, when the size of the adjacent side is not relevant.

Sines and their inverses on a calculator

The procedure is the same as for tangents and inverse tangents, except that the sine key (sin) is used.

Examples

A person walks 56 m up a 23° slope (Figure 14.14). How much height has the person gained? Begin with an equation:

$$\sin 23° = \frac{\text{opposite side}}{\text{hypotenuse}} = \frac{AB}{CA}$$

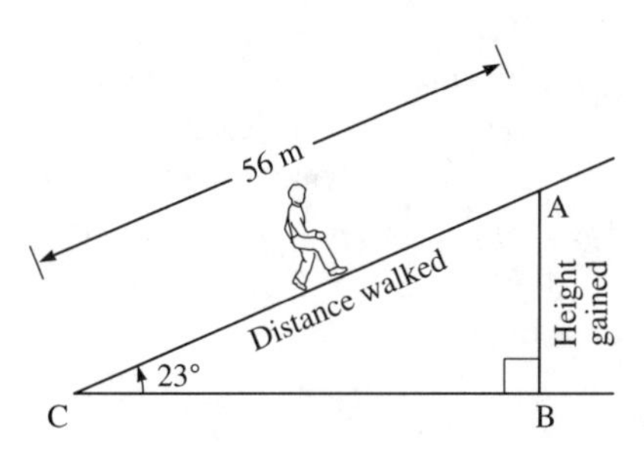

Figure 14.14

We want to know AB, so change the subject:

$$\Rightarrow \quad CA \sin 23° = AB$$
$$\Rightarrow \quad AB = CA \sin 23°$$
$$= 56 \times 0.3907$$
$$= 21.88$$

The height gained is 21.9 m (1 dp).

A radio mast is supported by wires 32 m long, attached 29 m up the mast (Figure 14.15). What is the angle x between the wires and the ground? Begin with an equation:

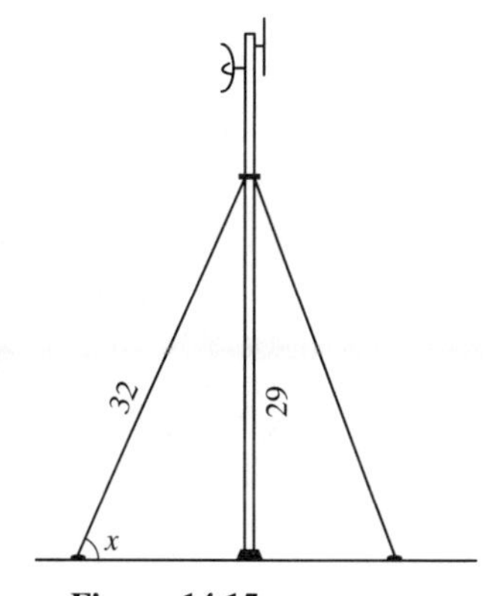

Figure 14.15

$$\sin x = \frac{\text{opposite side}}{\text{hypotenuse}} = \frac{29}{32} = 0.91$$

We have found the sine but need to know what angle has that sine. Use a calculator to find the inverse sine:

$$x = \sin^{-1} 0.91 = 65.5°$$

The angle between the wires and the ground is 65.5° (1 dp).

Test yourself 14.3

1 Use a calculator to find to 4 dp:

a sin 64° **b** sin 28° **c** sin 5°
d sin 0° **e** sin 45° **f** sin 90°

2 Use a calculator to find to 2 dp:

a $\sin^{-1} 0.4$ **b** $\sin^{-1} 0.5$ **c** $\sin^{-1} 0.1$
d $\sin^{-1} 0.86603$ **e** $\sin^{-1} 0.25$ **f** $\sin^{-1} 0.9$

When you answer the following questions, draw a *sketch* to help you understand the question. But do not make a scale drawing. Instead, calculate the result, obtaining sines or inverse sines with a calculator.

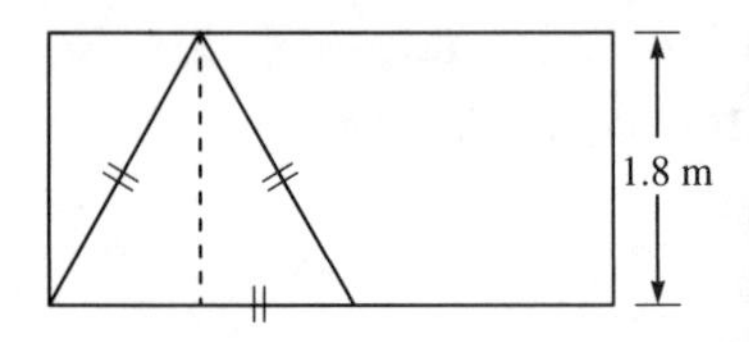

Figure 14.16

3 A person is cutting an equilateral triangle from a sheet of hardboard that is 1.8 m wide (Figure 14.16). What is the length of the sides of the largest triangle that can be cut?

4 A 2.2 m ladder leans against a vertical wall. It touches the wall 1.9 m above ground level. What is the angle between the ladder and the ground?

5 A kite is on a light string 27 m long. The angle between the string and the ground is 34°. Assuming that the string is straight, what is the height of the kite above ground?

6 Figure 14.17 is a partly-completed graph showing the sine of an angle for different angles from 0° to 90°. Copy the graph on to graph paper and, using a calculator to find the remaining values, complete the graph. Is the line straight? If not, describe it.

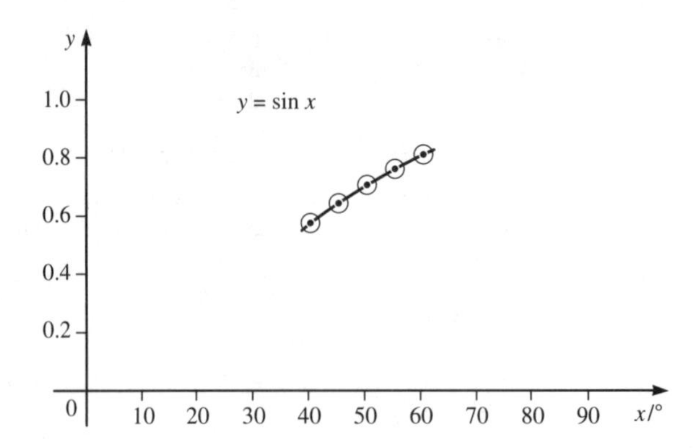

Figure 14.17

A third ratio

The third important ratio is the **cosine**, symbol **cos**. This is defined by:

$$\cos x = \frac{\text{adjacent side}}{\text{hypotenuse}}$$

Like the sine and tangent, the cosine is a ratio and has no unit. Like the sine, the cosine can never be greater than 1. The inverse cosine is $\mathbf{cos^{-1}}$. Cosines are used in problems when the size of the opposite side is not relevant.

Examples

What is the distance from A to C in Figure 14.11?

$$\cos 48° = \frac{\text{adjacent side}}{\text{hypotenuse}} = \frac{BC}{CA}$$

We need to find CA, so change the subject:

$$\Rightarrow \quad CA \cos 48° = BC$$

$$\Rightarrow \quad CA = \frac{BC}{\cos 48°}$$

$$= \frac{20}{0.6691}$$

$$= 29.89$$

The distance from A to C is 29.9 m (1 dp).

> **Cosines and their inverses on a calculator**
>
> The procedure is the same as for tangents and inverse tangents, except that the cosine key (cos) is used.

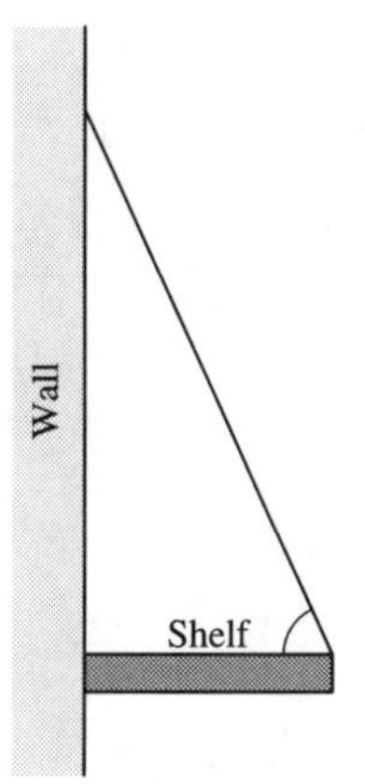

Figure 14.18

A shelf 225 mm wide is supported by a thin wire that is 470 mm long (Figure 14.18). What is the angle between the wire and the shelf?

$$\cos x = \frac{\text{adjacent side}}{\text{hypotenuse}} = \frac{225}{470} = 0.4787$$

$$\Rightarrow \quad x = \cos^{-1} 0.4787 = 61.4°$$

The angle between the wire and the shelf is 61.4° (1 dp).

> **Remember the ratios**
>
> $\text{Sin } x = \dfrac{\textbf{O}\text{pposite}}{\textbf{H}\text{ypotenuse}}$
>
> $\text{Cos } x = \dfrac{\textbf{A}\text{djacent}}{\textbf{H}\text{ypotenuse}}$
>
> $\text{Tan } x = \dfrac{\textbf{O}\text{pposite}}{\textbf{A}\text{djacent}}$
>
> Take the initial letters in order and remember how they sound:
>
> **SOHCAHTOA**

Test yourself 14.4

1 Use a calculator to find to 4 dp:

a cos 12° **b** cos 47° **c** cos 83°
d cos 0° **e** cos 45° **f** cos 90°

2 Use a calculator to find to 2 dp:

a $\cos^{-1} 0.7$ **b** $\cos^{-1} 0.5$ **c** $\cos^{-1} 0.05$
d cos–1 0.7071 **e** $\cos^{-1} 0.25$ **f** $\cos^{-1} 0.88$

3 On the same graph paper as your answer to *Test yourself 14.3*, question 6, draw the graph of $y = \cos x$. Is the line straight? If not, describe it. For what angle is $\sin x$ equal to $\cos x$?

The following problems may require tangents, sines, cosines or their inverses. The box helps you decide which to use.

4 A 2.5 m length of guttering is mounted with one end 0.1 m higher than the other. What is its angle of slope?

5 A rectangular paddock ABCD has the dimensions AB = CD = 100 m and AD = BC = 40 m. An electric fence is run straight across the paddock from corner A and meets the opposite side of the paddock DC at point E. Angle DAE = 35°. Find:

a the length of the fence, AE

b the distance DE

6 In the triangle ABC, $\hat{B} = 90°$, $\hat{C} = 26°$, and AB = 2.3 m. Find the lengths of the other sides of the triangle.

7 In the triangle XYZ, $\hat{Y} = 90°$, XZ = 15.7 mm, YZ = 10.4 mm. Find the angles $\hat{X}$ and $\hat{Z}$.

8 In the triangle PQR, $\hat{Q} = 90°$, PQ = 5 m, QR = 2 m. Find the angles $\hat{P}$ and $\hat{R}$.

9 To assess you progress in this chapter, work the exercises in *Try these first*, page 190.

15 Circles

Wheels, pulleys, gears, pistons, cake tins, bolts, control knobs – so many items used in technology are circular in one way or another. This chapter explains the properties of circles and relates them to other aspects of maths.

You need to know the Maths Essentials (Part 1), Ratios (Chapter 10), Formulae (Chapter 12), and Sines and Cosines (Chapter 14).

Try these first

Your success with this short test will tell you which parts of this chapter you already know.

1 A circle is 45 mm in diameter. How long is its circumference?

2 The circumference of a circle is 423 mm. What is its radius?

3 What is the area of a circle which has a circumference of 250 mm?

4 In a circle radius 50 mm, a sector is marked out by two radii, with an angle of 42° between them. What is the area of the sector? How long is the arc of the sector?

Describing circles

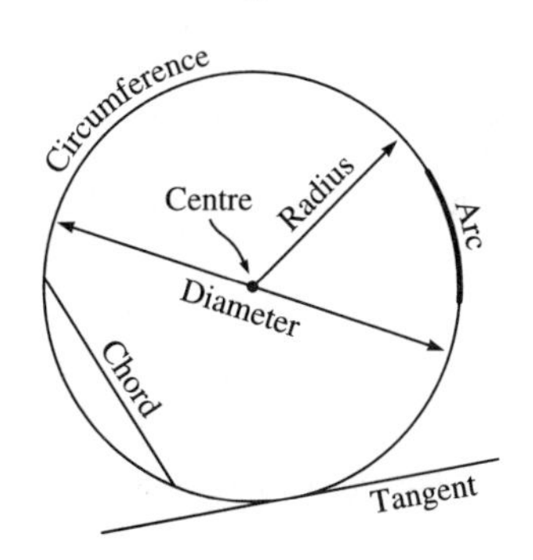

Figure 15.1

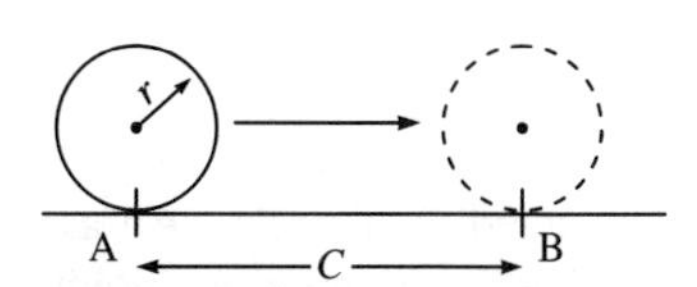

Figure 15.2

Figure 15.1 illustrates some of the terms used to describe circles. All points on the circle are the same distance from its **centre**. This distance is known as the **radius** of the circle. The word **radius** is also used to describe any straight line drawn from the centre to meet the circle. The plural of **radius** is **radii**.

The distance across the circle, along a straight line passing through the centre, is the **diameter** of the circle. The diameter is twice as long as the radius. The distance around the circle from any point on it and back to the same point is known as the **circumference**.

One way of measuring the circumference is to cut out a circle drawn on card, or to use a wheel (Figure 15.2). Make a mark on its edge. Draw a straight line on paper and make a mark A near one end of this. Place the circle edge-on against the line so that the marks coincide. Then roll the circle (wheel) along the line until its mark touches the paper again. Make a second mark B on the paper at this point.

Distance AB equals the circumference c of the circle. Measure this, and also the diameter, d. You find that c is slightly more than three times as long as d. There is a fixed ratio between them, whatever the size of the circle. With precise measurements, or with other even more precise techniques, we find that:

The ratio of circumference to diameter is 3.1416:1 (4 dp)

The ratio of circumference to diameter is so important in many branches of maths that the number 3.1416 has been given a special name. It is known as **pi**, and its symbol is the Greek letter pi, which is π. In the ratio above, pi is quoted to 4 dp, but it has many more decimal places than that. Pi has been calculated on a computer to over a thousand million decimal places. The sequence of numbers never repeats itself, and the end has not yet been reached. It is one of those numbers that it seems impossible to evaluate exactly. But 4 decimal places are enough for most calculations, though scientific calculators usually give the value to 9 dp.

Pi is used in all kinds of calculations involving the circumference and the diameter or radius or a circle.

> **Pi on a calculator**
>
> All except the simplest calculators have a special key for pi. Just press [π] and the value of pi appears on the display. On some calculators, pi may share the key with some other function, and you have to press the 'shift' or 'second function' key first. A 10-figure scientific calculator gives the value of pi as 3.141 592 654. This is the value used for the calculations in this book. If your calculator has fewer figures, or you are working on paper using approximate values such as 3.1416 or 3.142, you may find that your results differ slightly from ours in the last one or two significant figures.

Examples

A circular sports track is 125 m in diameter. What is the length of the track?

$$\text{Length of track} = \text{circumference} = \text{diameter} \times \pi$$
$$= 125 \times 3.1416 = 392.7$$

The track is 393 m long.

A model aeroplane is flown on a control line which is 5 m long. How far does it fly in one complete turn?

The aeroplane flies in a circle, the **radius** of which is the length of the control line. The **diameter** is twice the radius.

$$\text{Distance flown} = \text{circumference} = 2 \times \text{radius} \times \pi$$
$$= 2 \times 5 \times \pi = 31.42$$

Distance flown is 31.4 m (1 dp).

A garden roller, radius 0.3 m, is pushed across a lawn that is 20 m wide. How many times does the roller turn?

$$\text{Circumference of roller} = 2 \times 0.3 \times \pi = 1.885$$
$$\text{Number of turns} = 20/1.885 = 10.61$$

The roller turns 10.6 times (1 dp).

Other terms

Figure 15.1 also illustrates a few other terms connected with circles. An **arc** is any part of the circumference. A **chord** is a straight line joining two points on the circumference. A **tangent** is a straight line which touches the

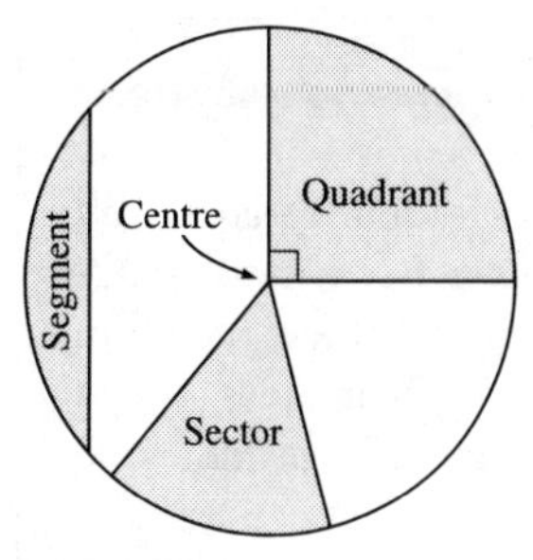

Figure 15.3

circumference at **one** point. The angle between the tangent and a radius drawn to that point is always a right-angle.

Figure 15.3 defines parts of the area inside the circle. A **sector** is the area enclosed by two radii and an arc. If the radii are at right-angles, the area is a **quadrant**. A **segment** of a circle is the area between a chord (see above) and an arc.

Checking calculations

As a rough check, use 3 as the value of pi. Often you can check your calculations in your head if you use this approximation.

Example

Find the circumference of a circle, diameter 56 mm.

Diameter $= 56 \times \pi \approx 56 \times 3 = 168$ mm

Test yourself 15.1

1 Calculate the circumference of these circles, given the diameter or the radius. Use a calculator, then check each result using the approximate method described in the box. Give the exact and approximate answer for each circle.

a diameter = 4.5 m **b** radius = 2.1 m

c radius = 1 mm **d** diameter = 1.3 km

e diameter = 6.7 m **f** radius = 145 mm

2 A circle has circumference c and radius r. Write a formula to use for calculating c, given r. Change the subject of the formula to r. [*Hint*: See page 168.]

3 Given the circumference of each of these circles, calculate the radius.

a 14.5 m **b** 3.2 mm

c 5.5 m **d** 5.46×10^3 km

4 The diameter of the tyres of a car is 0.52 m, to the outside of the tread. How many times do the wheels turn during a journey of 1.5 km?

5 A coil of 120 turns of fine wire is wound round a former that is 35 mm in diameter. Adjacent turns are adjacent but not overlapping, so we can assume that each turn is a circle 35 mm in diameter. Allowing 100 mm of wire for connections at each end, how much wire is required?

Explore this

Draw a circle 20 mm radius. Decide on a way of finding its area. You may decide to use the square-counting technique from page 81. A quicker way is to draw the circle on thin card, cut it out and weigh it. Then weigh a rectangle cut from the same card. Their areas are in proportion to their weights, so you can calculate the area of the circle using the techniques on page 134.

Repeat this for five or six circles of different sizes and set out your results in a table (leave space for two more columns on the right):

Radius (mm), r	*Area* (mm^2)*, A*
20	
30	
and more	

Plot a graph of r (on the x-axis) against A (on the y-axis). Is A directly proportional to r? [*Hint*: Is the graph a straight line?] At this stage you may prefer to continue your work on a spreadsheet (see next section).

Add another column to the table, headed 'r^2'. Calculate and enter the *square* of r for each circle. Now plot a graph of r^2 (x-axis) against A (y-axis). Is A proportional to r^2?

In the fourth column of the table, calculate A/r^2 for each circle. Allowing for small errors of measurement, is the ratio the same for all the circles?

Explore this too

The calculations above can be done on a spreadsheet. Here is the format:

```
  A....... B....... C....... D.......
1 Area of a circle
2
3   Radius      Area        r2      A/r2
4
5
6
7
8
9                        Average
```

This has space for details of five different circles. These are the formulae:

```
C4=A4*A4
D4=B4/C4
C5=A5*A5
D5=B5/C5
C6=A6*A6
D6=B6/C6
C7=A7*A7
D7=B7/C7
C8=A8*A8
D8=B8/C8
D9=D4#D8/5
```

Enter the radius of each circle in cells A4 to A8, and your estimates of area in cells B4 to B8. Update the spreadsheet. It calculate r^2 and A/r^2 for each circle. These are the values that you plot as graphs, as described in the previous section. The spreadsheet also calculates the average value of A/r^2 for all the circles.

Length of an arc

In Figure 15.4a there are two radii, length r, with an angle of 70° between them. They define a segment of the circle. The length of the arc AB of this segment is directly proportional to the angle. We are going to calculate the length of this arc, measured *along* the arc, not straight across from one end to the other.

In Figure 15.4b we have a segment which is really the whole circle. Its angle is 360°. Its length is the whole circumference of the circle, which is $2\pi r$. If length of arc is proportional to angle, we can work this in the same way as the examples on page 134:

When the angle is 360° the length of the arc is $2\pi r$

When the angle is 1° the length of the arc is $\frac{2\pi r}{360}$

When the angle is 70° the length of the arc is $\frac{70}{360} \times 2\pi r$

If the radius is 45 mm, we evaluate the arc:

$$\text{arc} = \frac{70}{360} \times 2\pi r = 55\,\text{mm}$$

Another example: How long is the arc between radii 127° apart in a circle, radius 140 mm?

$$\text{arc} = \frac{127}{360} \times 2\pi \times 140 = 310\,\text{mm}$$

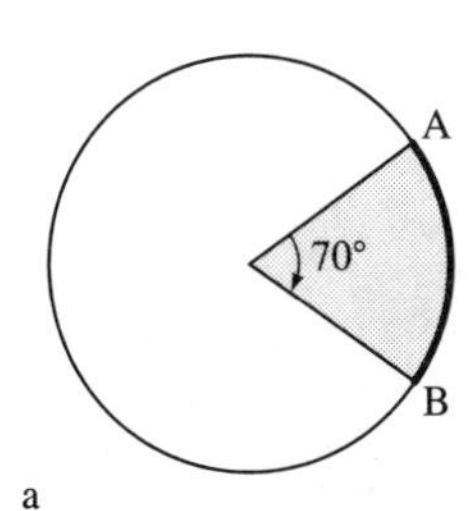

a

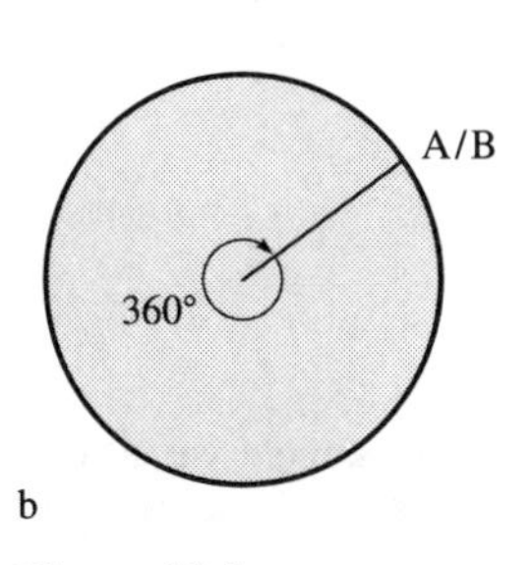

b

Figure 15.4

Area of a circle

In Figure 15.5 we have taken a circle, radius r, cut it into 20 sectors and laid the sectors out, side by side, but touching. They form the rough outline of a parallelogram (page 72). The area of the parallelogram is very close to the area of the circle, because it is made up from the same set of sectors.

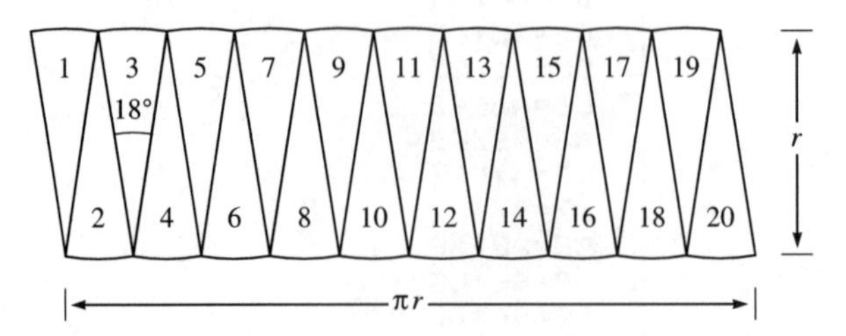

Figure 15.5

The width of the parallelogram is about the same as the length of a sector, which is r.

The length of the parallelogram is made up of 10 arcs of the circle. The circle was cut into 20 sectors, so the length on one arc is 1/20 of the circumference. The length of 10 arcs is half the circumference. We already know that the circumference is $2 \times r \times \pi$. So half the circumference is $r \times \pi$. This is the length of the parallelogram. The rule on page 79 tells us how to find its area.

$$\text{The area of the parallelogram} = \text{length} \times \text{width} = (r \times \pi) \times r = \pi r^2$$

This is the same as the area of the circle, so:

The area of a circle is πr^2

> **Approaching perfection**
>
> In Figure 15.5 the area has wavy edges, so it is not *exactly* a rectangle and its area is only approximately πr^2. But, if we cut the circle into many more sections (say 2000), they would line up to make an almost perfect rectangle with area very much closer to πr^2. With an infinitely large number of sections, the rectangle is perfect and its area is exactly πr^2.

Examples

What is the area of a circle, radius 72 mm?

$$\text{Area} = \pi \times 72^2 = 16\,286\ \text{mm}^2$$

A circle has an area of 6700 mm^2. What is its radius? In this example we need to change the subject of the formula:

$$\text{area} = \pi r^2$$

Divide both sides by π:

$$\Rightarrow \qquad \frac{\text{area}}{\pi} = r^2$$

Take square root of both sides:

$$\sqrt{\frac{\text{area}}{\pi}} = r$$

Exchange sides:

$$r = \sqrt{\frac{\text{area}}{\pi}}$$

Now we are ready to calculate the radius:

$$r = \sqrt{\frac{6700}{3.1412}} = \sqrt{2132.7} = 46.2\,\text{mm}$$

The final stage, working out the square root, is done with a calculator (see box on page 180).

Area of a segment

On page 207, we saw how the length of the arc of a segment is directly proportional to the angle. The same reasoning applies to the area. We use a similar formula:

$$\text{area} = \frac{\text{angle}}{360} \times \pi r^2$$

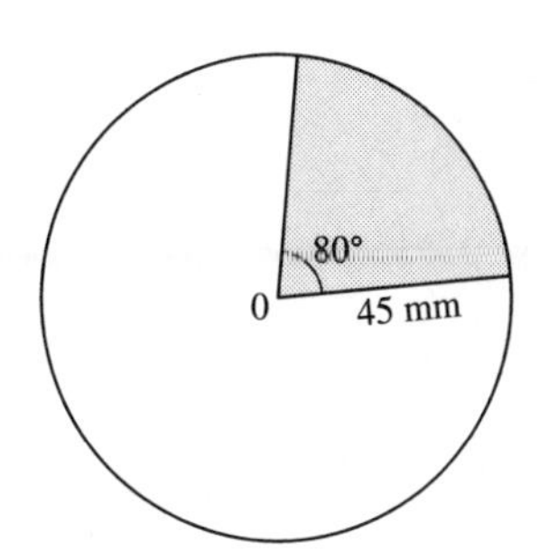

Figure 15.6

Example

What is the area of an 80° segment of a circle, radius 45 mm? (Figure 15.6)

$$\text{Area} = \frac{80}{360} \times 2\pi \times 45^2 = \frac{80}{360} \times 12723$$

$$\text{Area} = 2827\,\text{mm}^2$$

Test yourself 15.2

1 Calculate the lengths of these arcs:

a radius 60 mm angle 20°.

b radius 1.2 m, angle 60°.

c radius 102 mm, angle 210°.

2 A belt passes around a drum, 0.3 m diameter, and is in contact with it over an angle of 200° (Figure 15.7). What length of the belt is in contact with the drum? The belt is 200 mm wide. What area of the belt is in contact with the drum?

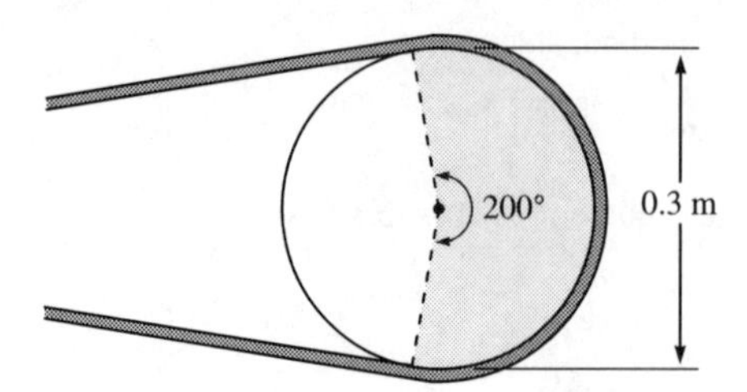

Figure 15.7

3 Calculate the areas of these circles, given the following data:

a radius = 35 mm **b** radius = 6.7 m

c diameter = 50 mm **d** diameter = 3 mm

e circumference = 300 mm **f** radius = 4.5×10^{-3} mm

4 A circular lawn has circular flower beds in it, as in Figure 15.8. What is the area of the lawn?

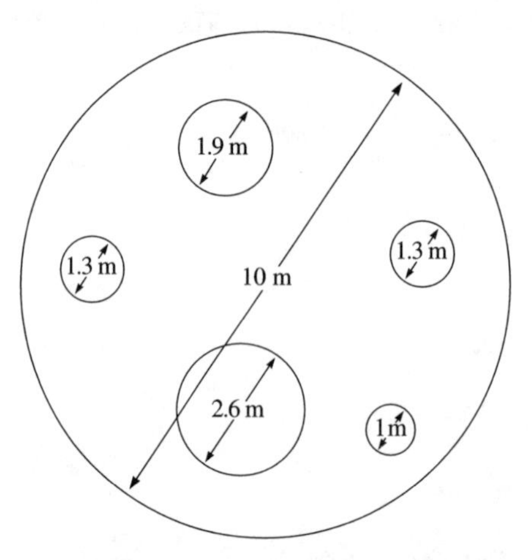

Figure 15.8

5 A rectangle of raw pastry measures 350 mm by 210 mm. How many pieces of pastry can be cut from it using a circular pastry cutter 70 mm in diameter? What area of pastry is left? The left-over pastry is rolled into a strip 70 mm wide, and of the same thickness. What is the length of the strip? How many pieces of pastry can be cut from it with the same pastry-cutter? What area of pastry is left over? Express this as a percentage of the original piece.

6 A can of spray paint holds enough to cover $2\,m^2$. What diameter circle can just be covered by a can of this paint?

7 A lampshade is made from a piece of sheet plastic shaped as shown in Figure 15.9. What is the area of the plastic, to the nearest $100\,mm^2$?

8 To assess your progress in this chapter, work the exercises in *Try these first*, page 203.

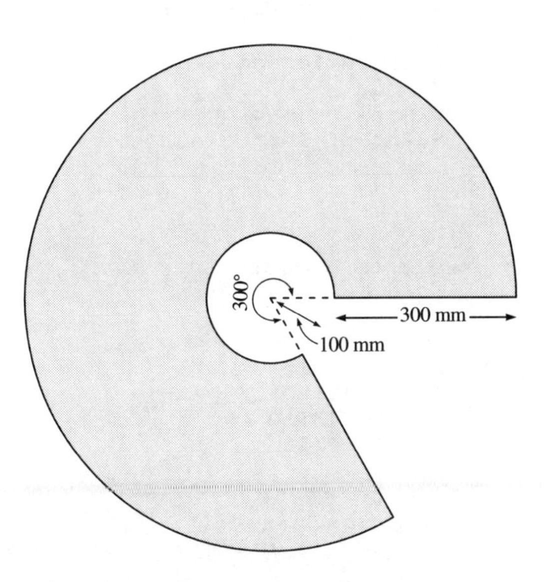

Figure 15.9

Circle formulae

Given a circle, radius r:
Diameter, $d = 2r$
Circumference, $c = \pi d = 2\pi r$
Area, $A = \pi r^2$

Changing the subject of some of these formulae gives:
Radius $= d/2$
Radius $= c/\pi$
Radius $= \sqrt{A/\pi}$

Given a sector, radius r, angle x (in degrees):
Arc $= 2\pi rx/360$
Area $= \pi r^2x/360$

Explore this

Use this spreadsheet to calculate facts about circles and sectors, given the radius and angle.

The format is:

```
  A.......  B.......  C.......
1 CIRCLE DATA SHEET
2
3 Radius =
4 Angle of sector =
5 Diameter =
6 Circumference =
7 Area =
8 Area of sector =
9 Arc of sector =
```

The formulae are:

```
C5=2*C3
C6=3.142*C5
C7=3.142*C3*C3
C8=C7*C4/360
C9=C6*C4/360
```

Enter the radius in cell C3. If you need information about a sector, enter the angle (in degrees) in cell C4. Update the spreadsheet; it calculates all the other data. Here is an example for a circle, radius 5.72, with sector angle 62°:

```
CIRCLE DATA SHEET

Radius =                   5.72
Angle of sector =         62.00
Diameter =                11.44
Circumference =           35.94
Area =                   102.80
Area of sector =          17.70
Arc of sector =            6.19
```

16 Pythagoras

Pythagoras was a famous thinker who lived and worked in Greece about 2500 years ago. He is said to have discovered a useful fact about right-angled triangles. This fact is still important today in technology.

You need to know the Maths Essentials (Part 1), sines, cosines and tangents (Chapter 14), and squares and square roots (Chapter 13).

Try these first

Your success with this short test will tell you which parts of this chapter you already know.

1 In a $\triangle ABC$, $\hat{B} = 90°$, AB = 5.4 m, BC = 8.3 m. Use Pythagoras' theorem to find AC, to 2 sf.

2 In a $\triangle XYZ$, $\hat{Y} = 90°$, XZ = 20 m, YZ = 12 m. Use Pythagoras' theorem to find XY, to 2 sf.

3 Without using a calculator or tables, find

a sin 45° **b** cos 60° **c** tan 30°

Pythagoras' theorem

A theorem is a statement that we can prove from facts we already know to be true. This theorem is based on a right-angled triangle (Figure 16.1) and says that:

The square of the length of the longest side (hypotenuse) equals the sum of the squares of the other two sides

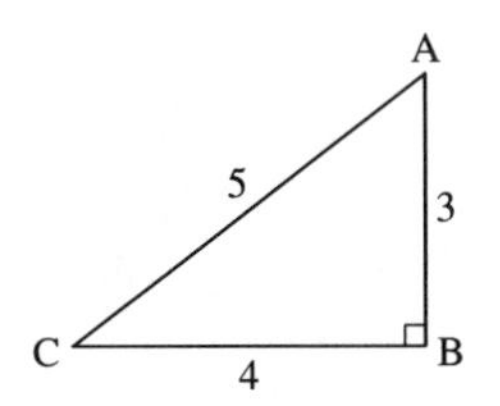

Figure 16.1

In Figure 16.1:

$$AC^2 = AB^2 + BC^2$$

Pythagoras' theorem applies to right-angled triangles of any size and shape. In a few cases the lengths of the sides are whole numbers.

In Figure 16.1, AC = 5, AB = 3, BC = 4:

$$AC^2 = 5^2 = 25$$

also $AB^2 + BC^2 = 4^2 + 3^2 = 16 + 9 = 25$

This shows that the theorem is true for this triangle at least. In this triangle, the sides have whole numbers. An accurate way of making a right-angled frame is to nail together three strips of wood to form a triangle with sides 300 mm, 400 mm and 500 mm (Figure 16.2); or make it any other size in which the sides are in the ratio 3:4:5, for example, 150 mm, 200 mm and 250 mm.

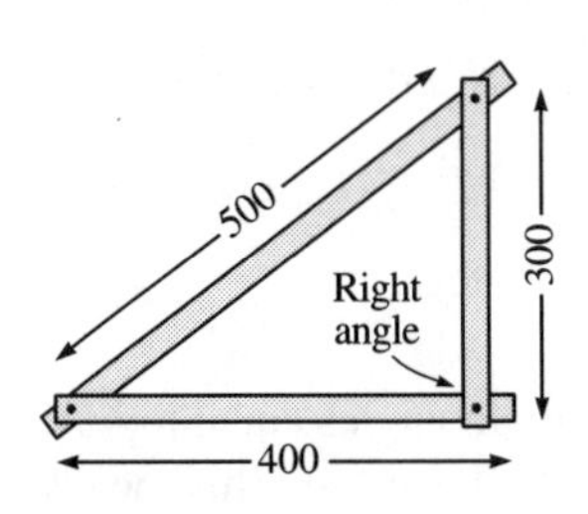

Figure 16.2

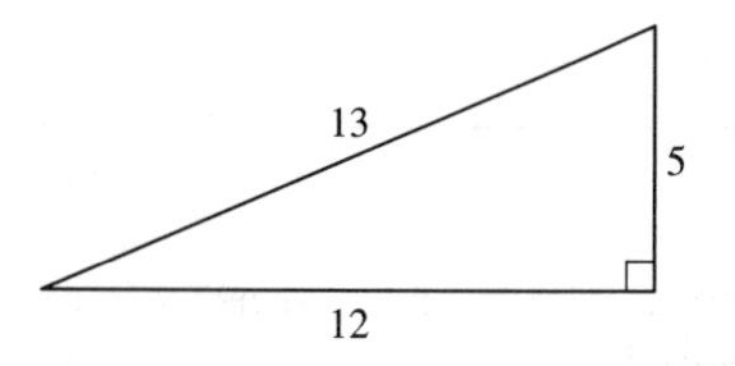

Figure 16.3

Figure 16.3 shows another right-angled triangle with sides that are whole numbers. But Pythogoras' theorem applies to right-angled triangles of any size and shape.

Using Pythagoras

We use Pythagoras (we should say 'Pythagoras' theorem' but most people just call it 'Pythagoras') to find the length of one side of a right-angled triangle when we know the lengths of the other two sides.

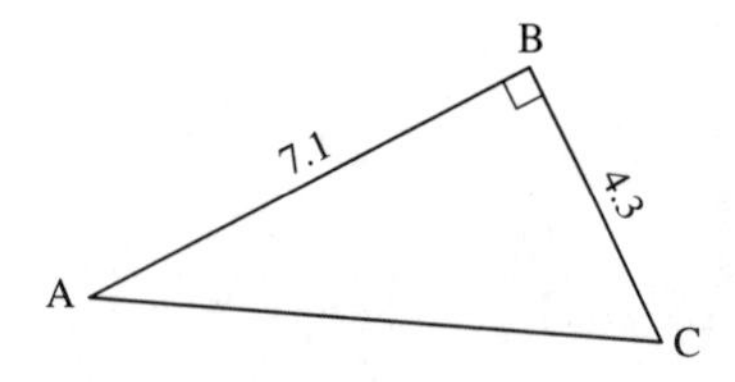

Figure 16.4

Examples

Find AC in Figure 16.4.

From Pythagoras:
$$\begin{aligned} AC^2 &= AB^2 + BC^2 \\ &= 7.1^2 + 4.3^2 \\ &= 50.41 + 18.49 \\ &= 68.9 \end{aligned}$$

Use a calculator to find the square root

$$\underline{AC = \sqrt{68.9} = 8.30}$$

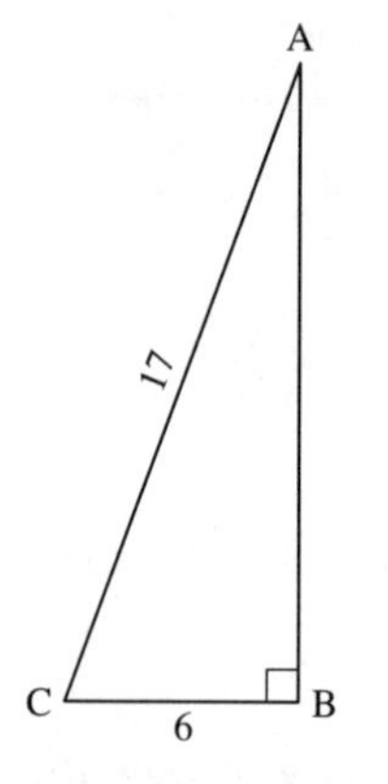

Figure 16.5

In the example above, we found the length of the hypotenuse (the longest side) by adding the squares of the other two sides. In the next example (Figure 16.5), we are told the hypotenuse and one other side and asked to find the remaining side:

$$\begin{aligned} AB^2 &= AC^2 - BC^2 \\ &= 17^2 - 6^2 \\ &= 289 - 36 \\ &= 253 \end{aligned}$$

$$\underline{AB = \sqrt{253} = 15.91\ (2\,\text{dp})}$$

We square the longest side, then *subtract* the square of the other known side. The square root of the difference is the unknown side.

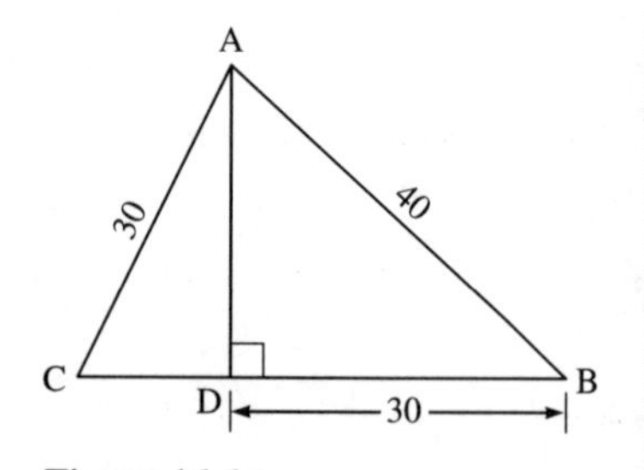

Figure 16.6

Test yourself 16.1

Round the results to 3 sf.

1 In $\triangle JKL$, $\hat{K} = 90°$, JK = 6 m, KL = 8 m. Find JL.

2 In $\triangle XYZ$, $\hat{Y} = 90°$, XY = 11 m, YZ = 13 m. Find XZ.

3 In $\triangle PQR$, $\hat{Q} = 90°$, PQ = 5 mm, QR = 3 mm. Find PR.

4 In $\triangle ABC$, $\hat{B} = 90°$, AC = 10 m, AB = 7 m. Find BC.

5 In $\triangle RST$, $\hat{S} = 90°$, RT = 13.1 mm, ST = 3.6 mm. Find RS.

6 Find AD and CD in Figure 16.6.

Standard triangles

The right-angled triangle in Figure 16.7, has its other two angles equal to 45°. Since it has two equal angles, it is an isosceles triangle (page 68). This means that sides AB and BC are equal in length.

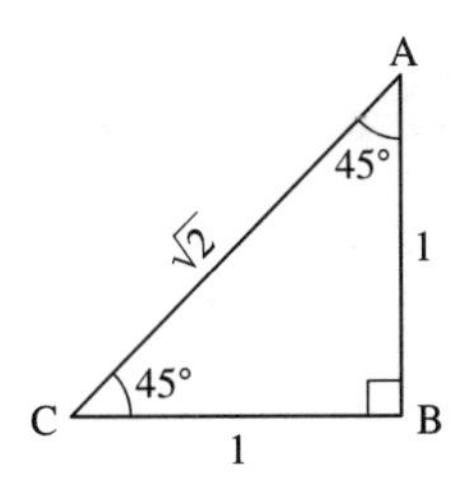

Figure 16.7

Suppose that AB and AC are 1 unit long. Then, by Pythagoras,

$$\begin{aligned} AC^2 &= AB^2 + BC^2 \\ &= 1^2 + 1^2 \\ &= 1 + 1 \\ &= 2 \\ AC &= \sqrt{2} \end{aligned}$$

If the angles of a triangle are 45°, 45° and 90°, its sides are in the ratio 1:1:√2

We use this information to write formulae for the trig ratios (sine, cosine and tangent) for the angle 45°. Base these on C in Figure 16.7:

$$\sin 45° = \frac{\text{opposite}}{\text{hypotenuse}} = \frac{1}{\sqrt{2}}$$

$$\cos 45° = \frac{\text{adjacent}}{\text{hypotenuse}} = \frac{1}{\sqrt{2}}$$

$$\tan 45° = \frac{\text{opposite}}{\text{adjacent}} = \frac{1}{1} = 1$$

Note that the sine and cosine of 45° are equal. Knowing these values is useful because the angle 45° is often found in frameworks and other structures.

Another standard triangle

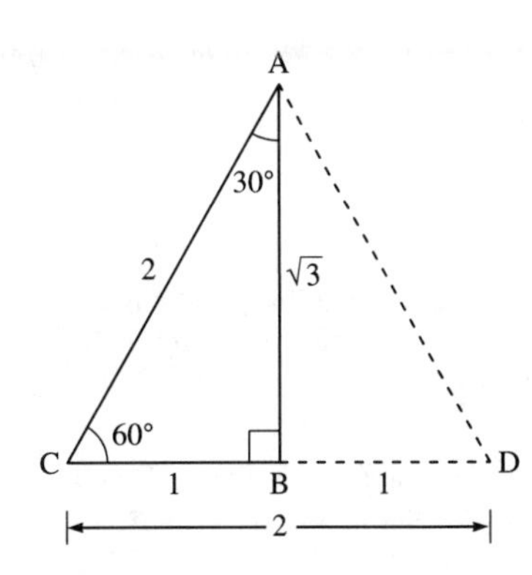

Figure 16.8

Triangle ABC in Figure 16.8 has angles 30°, 60° and 90°. As the figure shows, if the triangle is repeated as a mirror image to the right of AB, we obtain an equilateral triangle, ADC.

Triangle ABC is half of an equilateral triangle, so BC is half the length of AC. If BC is 1 unit long, AC is 2 units long. By Pythagoras:

$$\begin{aligned} AB^2 &= AC^2 - BC^2 \\ &= 2^2 - 1^2 \\ &= 4 - 1 \\ &= 3 \\ AB &= \sqrt{3} \end{aligned}$$

If the angles of a triangle are 30°, 60° and 90°, its sides are in the ratio 1:2:√3

As with the previous triangle, we can use the ratios of the sides to write formulae for trig ratios. First, based on CAB:

$$\sin 30° = \frac{\text{opposite}}{\text{hypotenuse}} = \frac{1}{2} = 0.5$$

$$\cos 30° = \frac{\text{adjacent}}{\text{hypotenuse}} = \frac{\sqrt{3}}{2}$$

$$\tan 30° = \frac{\text{opposite}}{\text{adjacent}} = \frac{1}{\sqrt{3}}$$

Also, based on ACB:

$$\sin 60° = \frac{\text{opposite}}{\text{hypotenuse}} = \frac{\sqrt{3}}{2}$$

$$\cos 60° = \frac{\text{adjacent}}{\text{hypotenuse}} = \frac{1}{2} = 0.5$$

$$\tan 60° = \frac{\text{opposite}}{\text{adjacent}} = \frac{\sqrt{3}}{1} = \sqrt{3}$$

Note that sin 30° = cos 60° and cos 30° = sin 60°. These formulae may be used in solving problems when angles of 30°, 45° or 60° are involved.

Example

What is the height of the coping stone shown in section in Figure 16.9?

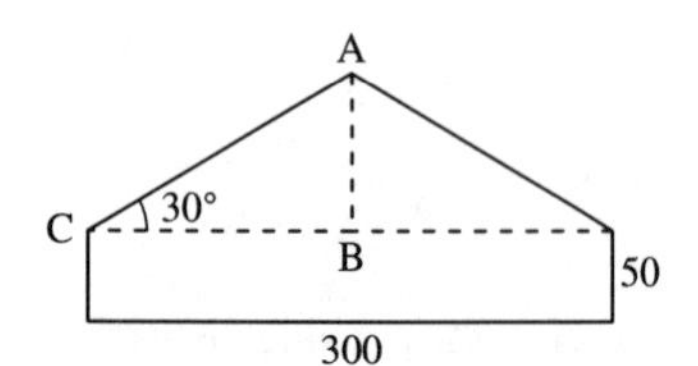

Figure 16.9

In $\triangle$ABC, BC = 150 mm.

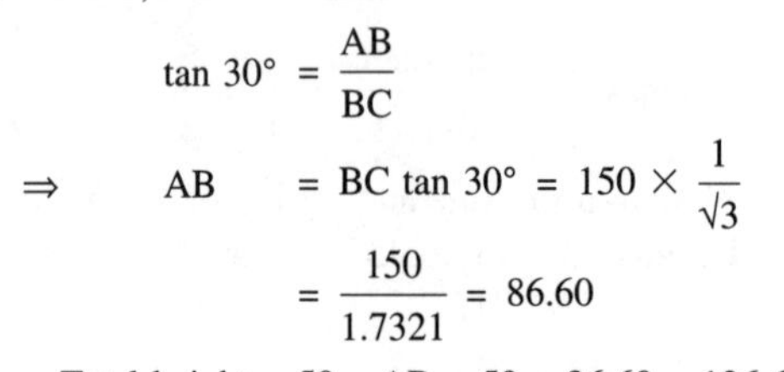

$$\tan 30° = \frac{AB}{BC}$$

$$\Rightarrow \quad AB = BC \tan 30° = 150 \times \frac{1}{\sqrt{3}}$$

$$= \frac{150}{1.7321} = 86.60$$

Total height = 50 + AB = 50 + 86.60 = 136.60

Height of stone = 136.6 mm (1 dp)

Constructing angles

Use of a pair of compasses to construct angles without a protractor.

60° angle Point of compasses at A. Draw arc, cutting the line at B (Figure 16.10a). Do not alter the setting of the compasses and place the point at B. Draw a short arc to cut the other arc at C (Figure 16.10b). Join AC. (*Note*: $\triangle$ABC is equilateral so all its angles are 60°).

Right-angle Point of compasses at A. Draw short arcs to cut line at B and C (Figure 16.10c). Open up compasses slightly. Point at B. Draw short arc (Figure 16.10d). Do not alter setting of compasses. Point at C. Draw short arc to cross other arc at D. Join AD.

30° angle Construct a 60° angle as above (Figure 16.10e). Point at A. Draw short arcs to cut line of 60° angle at B and C. Do not alter setting. Point at B. Draw short arc. Point at C. Draw short arc to cross other arc at D. Join AD.

This technique can be used to divide into two (*bisect*) any other angle. The right-angle above is obtained by bisecting 180° (a straight line).

45° angle Bisect a right-angle, as in Figure 16.10f. Draw right angle at A, as above. Close compasses slightly. Point at A, draw short arcs to cut lines of right-angle at E and F. Do not alter setting. Draw arcs with point at E and at F, to cross at G. Join AG.

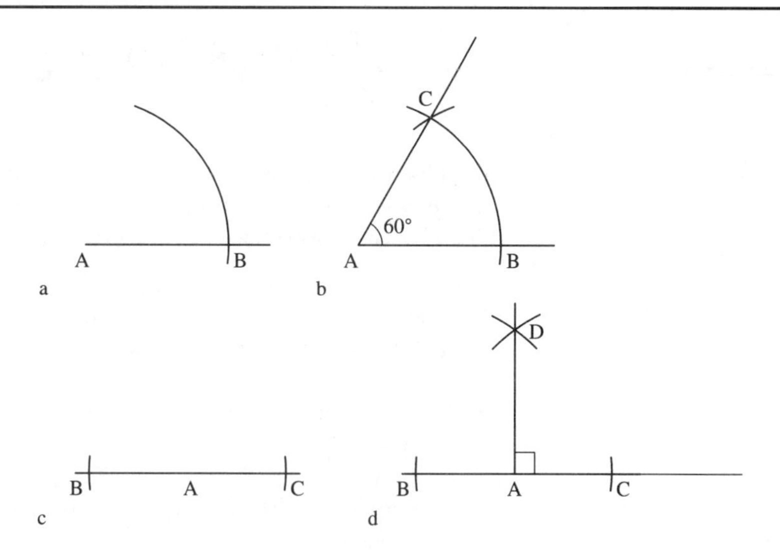

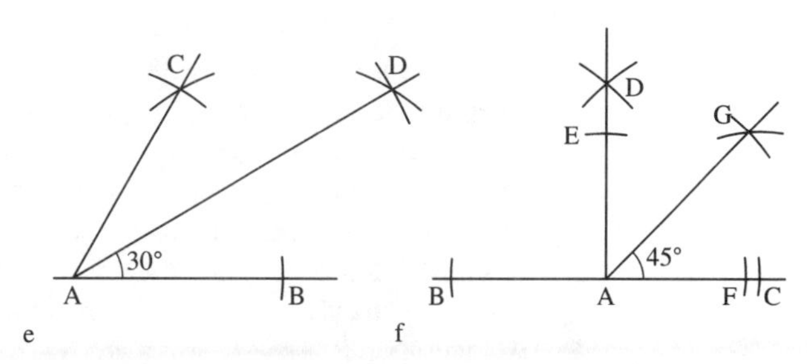

Figure 16.10

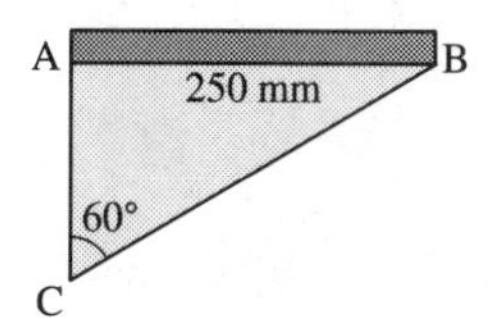

Figure 16.11

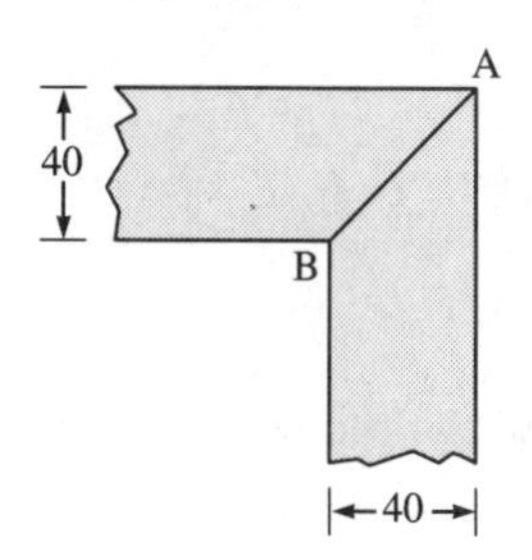

Figure 16.12

Test yourself 16.2

Solve these problems without using a calculator or tables for finding the trig ratios. Calculate results to 3 sf.

1 A shelf is supported by a bracket (Figure 16.11). Find BC and AC.

2 A piece of sheet metal for a road sign is cut as an equilateral triangle with sides 0.5 m. What is the area of the sheet?

3 A picture frame is mitred at its corners, as in Figure 16.12. The moulding is 40 mm wide. Find the length AB?

4 Using the techniques described in the box, construct these angles:

a 90° **b** 30° **c** 120° **d** 135° **e** 300°

5 To assess if you have covered this chapter fully, answer the questions in *Try these first*, page 213.

17 Interpreting graphs

Graphs are a way of showing how two quantities relate to each other. In this chapter we look at many kinds of graph to see how they help us discover the connections between two variables.

You need to know the Maths Essentials (Part 1), especially graphs (Chapter 7). You also need to know about proportions (page 143), balancing equations (page 168) and inequalities (pages 171–3).

Try these first

Your success in this short test will show you which parts of this chapter you already know.

1 Without plotting the line, find the gradient, x-intercept, and y-intercept of the graph of the equation $2y - 7x = -14$.

3 Two rooms each have a wall area of $30\,m^2$. Two friends A and B start to paint the walls at the same time, one in each room. A covers $0.3\,m^2$ of wall in 1 minute. B covers $0.24\,m^2$ in 1 minute, but $5\,m^2$ is already newly painted in B's room. Using the same set of axes, plot graphs to show the newly painted area in each room, from the starting time until both rooms are finished. Use the graphs to find when the same area has been newly painted in both rooms, and which friend finishes first.

3 By plotting graphs, find the values of x and y which satisfy *both* of these equations:

$$y = x^2 - 3$$
$$y = 2x + 5$$

4 Plot the graphs of these inequalities on the same axes:

$$y \leqslant 5 - x$$
$$y < 2 + 2x$$

What is the greatest integer value that y can have that satisfies both these inequalities?

5 Graph the data below, identify the relationship between x and y and find the values of the constants in the equation.

x	1	2	3	4	5	6	7	8
y	4.05	5.47	7.38	9.96	13.45	18.15	24.50	33.07

Straight-line graphs

In Chapter 7, we plotted many straight-line graphs and noticed that they all have the same basic equation:

$$y = mx + c$$

$y = mx + c$
Learn this equation: we use it over and over again.

In this equation:

- x is the independent variable, plotted along the horizontal axis (x-axis).
- y is the dependent variable (its value depends on the value chosen for x), plotted along the vertical axis (y-axis).
- m, the coefficient of x, decides the steepness of the line. We call m the **gradient** of the line.
- c, the constant, tells us where the line cuts the y-axis (the y-intercept).

These features are illustrated in Figure 17.1.

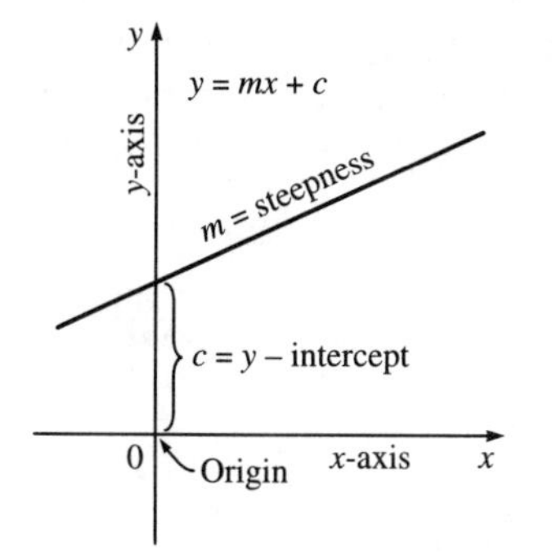

Figure 17.1

In Chapter 7, we described one way to plot straight-line graphs. Make a table listing a range of values of x, then calculate the corresponding values of y. Using the pairs of values (x,y) as coordinates, plot the points of the graph and join them with a straight line. This method is suitable for plotting graphs of other shapes, as we shall see later in this chapter. But first we will look at two more methods of plotting straight-line graphs.

Intercepts method

This technique locates the points where the line cuts across the axes. These points are the y-intercept on the y-axis (Figure 17.1) and the x-intercept on the x-axis.

The line cuts the y-axis when $x = 0$, so we can find the y-intercept by making x equal to zero in the equation.

Example

Find the y-intercept for the graph of the equation $y = 2x + 3$.

Make $x = 0$ in the equation:

$$\begin{aligned} y &= 2x + 3 \\ &= 2 \times 0 + 3 \\ &= 0 + 3 \\ &= 3 \end{aligned}$$

The y-intercept is 3.

To continue with this example, we find the x-intercept too. This is where the line cuts the x-axis. It cuts the x-axis when $y = 0$.

Make $y = 0$ in the equation:

$$\begin{aligned} 0 &= 2x + 3 \\ -2x &= 3 \\ -x &= 3/2 = 1.5 \\ x &= -1.5 \end{aligned}$$

The x-intercept is -1.5.

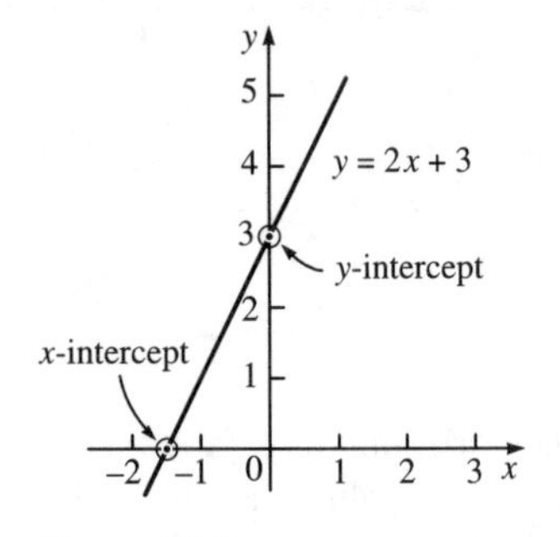

Figure 17.2

In Figure 17.2 we have plotted the y-intercept at (0,3) and the x-intercept at (−1.5,0). The straight line running through these two points is the graph of $y = 2x + 3$.

Note that this technique can be used only for graphs which we *know* are straight lines (they have the basic equation $y = mx + c$). Some other graphs might *possibly* have the same two intercepts but may curl round in some quite different way between the two points.

Gradient and *y*-intercept method

In this method we first of all find the y-intercept. Taking the same example as above, we have already found that this is at (0,3). This gives us one point on the line.

To get the next point, we make use of the gradient, m. The equation shows that, as x increases by 1, y increases by m. In this equation, $m = 2$. As we step one unit to the right, we step 2 units up. Then we reach another point on the graph. Figure 17.3 shows the result of stepping 'along 1, up 2' several times. Joining the points gives us the graph. Once we have plotted a few points in this way, we can extend the graph in either direction by continuing the straight line with a ruler.

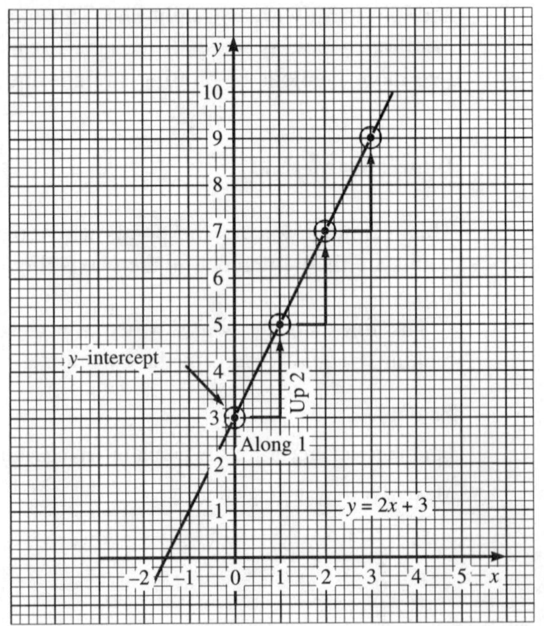

Figure 17.3

If m is negative, as in the equation $y = 6 - 3x$, the line slopes down toward the right. We have a negative gradient. To plot this line, first calculate the y-intercept:

$$
\begin{aligned}
y &= 6 - 3 \times 0 \\
&= 6 - 0 \\
&= 6
\end{aligned}
$$

Begin at point (0,6) in Figure 17.4. Then, since $m = -3$, we step 'along 1, *down* 3', a few times. This gives the points on the line.

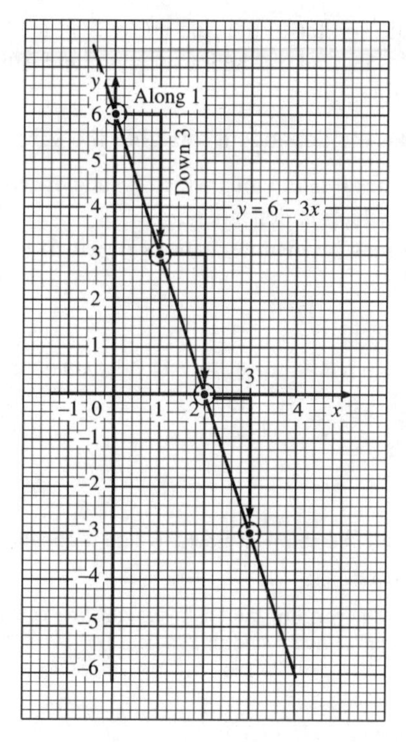

Figure 17.4

Other forms of the basic equation

Look at these equations:

$$y = 5x + 3$$
$$y - 3 = 5x$$
$$y - 5x = 3$$
$$-5x = 3 - y$$
$$5x = y - 3$$
$$y - 5x - 3 = 0$$

All of these are obtained from the first one simply by using the rules for keeping equations balanced. They are all versions of the same equation. They all give the same graph. But it makes it much easier to identify m and c, if we put the equation into the form that has just 'y =' on the left. In other words:

Make y the subject of the equation

Example

Plot a graph of $3x = 6 - y$.

This has x as the subject but we want make y the subject:

$$3x = 6 - y$$
$$\Rightarrow \quad y + 3x = 6$$
$$\Rightarrow \quad y = 6 - 3x$$

This is the equation which is plotted in Figure 17.4.

A rule similar to the rule above also applies when we have other pairs of variables instead of x and y, such as a and b or p and q. A more general version of the rule is:

Make the dependent variable the subject of the equation

Sometimes it is obvious which is the dependent variable; sometimes you will be told which is dependent.

You may be given equations in which y has a coefficient.

Example

Plot the graph of $6y = 3x + 15$.

We have to reduce the $6y$ to y. Divide *both* sides of the equation by 6:

$$\frac{6y}{6} = \frac{3x}{6} + \frac{15}{6}$$

$$\Rightarrow \qquad y = \frac{x}{2} + \frac{5}{2}$$

Converting the common fractions into decimals:

$$y = 0.5x + 2.5$$

This has the form of the basic straight-line equation. We can tell at a glance that its y-intercept (c) is 2.5 and its gradient (m) is 0.5.

Test yourself 17.1

1 For each of the graphs in Figure 17.5, find the y-intercept (c) and the gradient (m).

2 Without plotting the graphs, find the y-intercept (c) and the gradient (m) of the graphs of these equations. In each case say whether the graph slopes up or down to the right.

a $y = 8x + 2$ **b** $y = 8 - 7x$

c $y + 5x = 7$ **d** $6x = 10 - y$

e $4y = 12x - 10$ **f** $15 - 6x = 5y$

3 Use the intercepts method to plot the graphs of these equations. The dependent variable is y, except where a different letter is given in brackets after the equation.

a $y = 2x + 5$ **b** $y - x = -7$

c $3x + 10 + y = 0$ **d** $10x + 12 = 4y$

e $2b = 4 - 3a \quad (b)$ **f** $j - 2k = 8 \quad (k)$

4 Use the intercept and gradient method to plot the graphs of these equations. The dependent variable is y, except where a different letter is given in brackets after the equation.

a $y = x + 3$ **b** $y = 2x - 6$

c $y + 4x = 20$ **d** $2x = 7 - y$

e $2q = 6 + 3p \quad (q)$ **f** $4v + 2u - 8 = 0 \quad (v)$

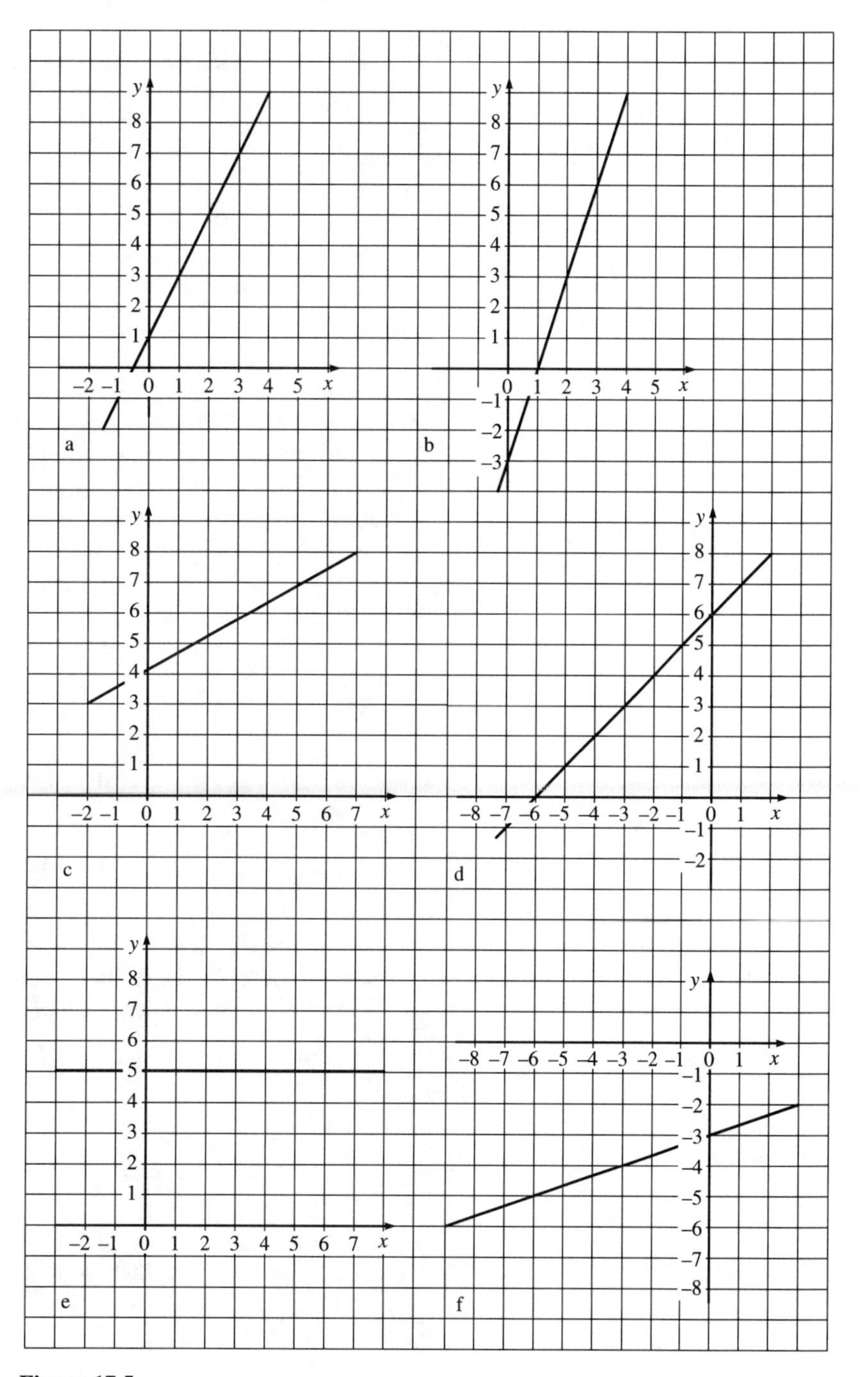

Figure 17.5

Explore this

Use a graphic calculator or computer graphic package to plot graphs quickly, so that you can investigate the effects of altering m and c. For instructions for operating the calculator or software, see its Instruction Manual.

Before you start, set the calculator to *rectangular mode*. Polar mode draws graphs of a different type. If possible, set the scale on both x and y axis to 0.5 (markings spaced 0.5 apart).

For the first graph, key in the equation:

$$y = x + 1$$

Note where the y-intercept comes, and observe the slope of the line. Use the Trace function to follow along the line to check exactly where the line crosses the y-axis. When the coordinate display shows $x = 0$, then y equals the y-intercept. Try finding the x-intercept. You may have to shift the axes slightly to the right, by pressing the [→] cursor key. When the display shows $y = 0$, then x equals the x-intercept.

The line you have just plotted has $m = 1$ and $c = 1$. Without clearing the screen, plot a whole *family* of lines all with $m = 1$, but with different values of c, positive or negative, whole numbers or fractions. For example, try:

$$y = x + 2$$
$$y = x + 5$$
$$y = x + 4.6$$
$$y = x - 2$$

and many more. These lines all have the same m. What do you notice about these lines. For some of them, find the x and y intercepts as described above. Plot a few more lines, but with $m = 2$.

Clear the screen. Another family to investigate is those which have the same c but different values of m. For example, try:

$$y = 2x + 2$$
$$y = 3x + 2$$
$$y = 10x + 2$$
$$y = 0x + 2$$
$$y = -3x + 2$$

and many more. What do you notice about this family? How can you pick out the lines with the highest value of m?

Continue in this way until you can predict in advance what a line will look like when it is plotted.

Work with a friend. Your friend keys in a straight-line equation without telling you what it is. When the graph is plotted, you look at it and have to discover the values of c and m.

Practical graphing

A cyclist travels along a level road. There are trees beside the road, spaced exactly 20 m apart. These are the times at which the cyclist passes each tree:

Tree	1	2	3	4	5	6	7
Time (s)	3.4	5.6	7.8	10.0	12.2	14.4	16.6

It makes more sense to replace the tree numbers with actual distances travelled:

Distance (m)	0	20	40	60	80	100	120
Time (s)	3.4	5.6	7.8	10.0	12.2	14.4	16.6

Figure 17.6 is the graph of this data. When time is a variable, it is often taken to be the independent variable, because time passes steadily as events happen and is not affected by them. So we plot time (t) along the x-axis and distance (d) along the y-axis.

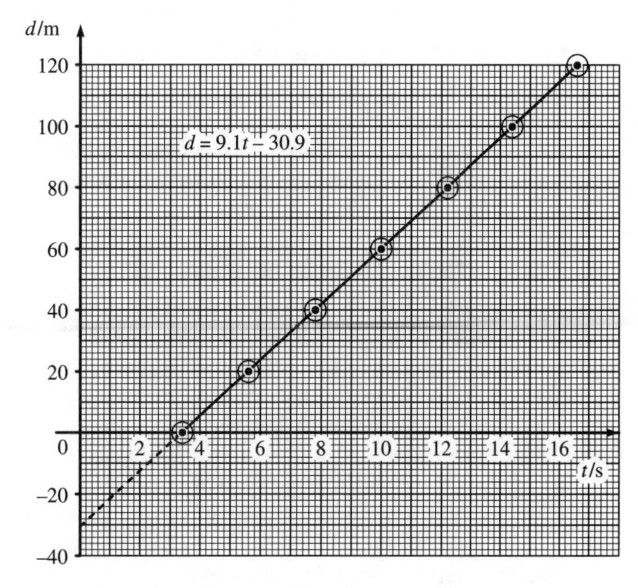

Figure 17.6

The y-intercept of the line is $d = -30.9$. This might tell us the distance of the cyclist from the first tree when the stop-watch was started. This distance is 30.9 m back along the road, before the cyclist reaches the first tree. But we must be careful in interpreting the graph in this way. The *first* thing we actually know about the position of the cyclist is that he passed the first tree at 3.4 s. Maybe the cyclist did not start until a second or two *after* the watch was started. The truth is, we just do not know what happened before 3.4 s. This illustrates a point about practical graphs. If a graph is based on a maths equation, we can usually extend the line as far as we like in either direction. But, if the graph is dealing with something from real life, we must not always assume that we can extend it beyond the range of the data supplied.

For the same reason, we cannot reliably extend the line beyond $t = 16.6$. The cyclist may have become tired and dismounted after 17 s.

Now to look at the x-intercept, which is $t = 3.4$. This is the time that the cyclist passed the first tree. We are certain of this because this is one of the data points.

Finally, there is the gradient. On this graph, it is not easy to 'count squares' along and up or down. Instead, we will calculate distances along and up between two points. We could choose any two points for this, let us take the first and the last:

The time taken from tree 1 to tree 7 is: $16.6 - 3.4 = 13.2$
The distance from tree 1 to tree 7 is: $120 - 0 = 120$

In terms of stepping from part of the line to another we have:

13.2 along and 120 up

In proportion, this is equal to:

1 along and 120/13.2 up
$\Rightarrow$ 1 along and 9.1 up (1 dp)

The gradient, $m = 9.1$

Having found the y-intercept (or rather, the d-intercept) and the gradient, we can now write the equation for this line:

$$d = 9.1t - 30.9 \qquad 3.4 < t < 16.6$$

The inequality on the right specifies that the graph reliably extends only from $t = 3.4$ to $t = 16.6$. It is important to place such limits on an equation if there are practical reasons for doing so.

The meaning of the gradient

We will continue with the example of the cyclist, but suppose that a steady following wind begins to blow just as the cyclist passes tree 3. We might expect that the cyclist goes faster now, and reaches trees 4 to 7 earlier than before. The new data might be:

Distance (m)	0	20	40	60	80	100	120
Time (s)	3.4	5.6	7.8	9.4	11.0	12.6	14.2

Figure 17.7 shows the graph. It is obvious that the gradient is greater from tree 3 onward. A steeper gradient results from an increased speed. The sharp bend in the line is not completely realistic. The cyclist could not change speed *instantly*. The bend should be gradual but, as an approximation, the sharp bend is acceptable.

In the previous section, we calculated m by dividing 120 by 13.2. Putting units to these values, we divided 120 m by 13.2 s.

$$m = \frac{120}{13.2} = 9.1 \text{ metres per second}$$

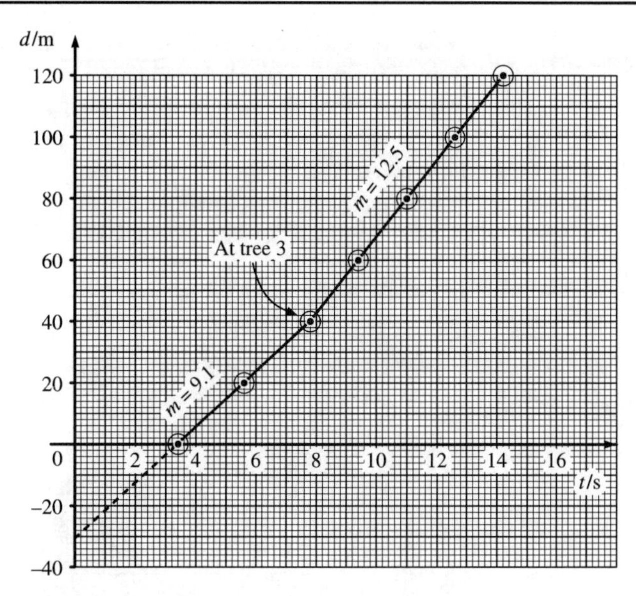

Figure 17.7

In this graph, m represents the cyclist's *speed*. It is the *rate of change* of distance with respect to time. On any graph:

***m* is the rate of change of the dependent variable with respect to the independent variable**

Just as we calculated m for Figure 17.6, we can calculate m for the right-hand section of Figure 17.7:

The time taken from tree 3 to tree 7 is: $14.2 - 7.8 = 6.4$
The distance from tree 3 to tree 7 is: $120 - 40 = 80$

In terms of stepping from one part of the line to another, we have:

6.4 along and 80 up

In proportion, this is equal to:

1 along and 80/6.4 up
$\Rightarrow$ 1 along and 12.5 up
Gradient, $m = 12.5$

The meaning of ms^{-1}

When used with *units*, the negative index -1 is read as '*per*'. For example, $12.5\,ms^{-1}$ is read as '12.5 metres *per* second'.

The cyclist's speed assisted by the wind is $12.5\,ms^{-1}$.

An example from food technology

One recommendation for the cooking time for a joint of sirloin in a microwave is:

8 min per 450 g, plus 20 min standing time

We are to plot a graph to show total cooking time for joints of various weights from 200 g to 1000 g. We could then use this graph to read off cooking times instantly, given the weight of the joint.

If w is the weight of the joint (in g) and t is the total time (in min), the formula is:

$$t = \frac{8w}{450} + 20 \qquad 200 < w < 1000$$

In this example, time is the *dependent* variable (compare with the example above). Here, time is not something that passes steadily, but is a setting on the display of the microwave, *depending on* the weight of the joint. So we plot t on the y-axis and w on the x-axis (Figure 17.8).

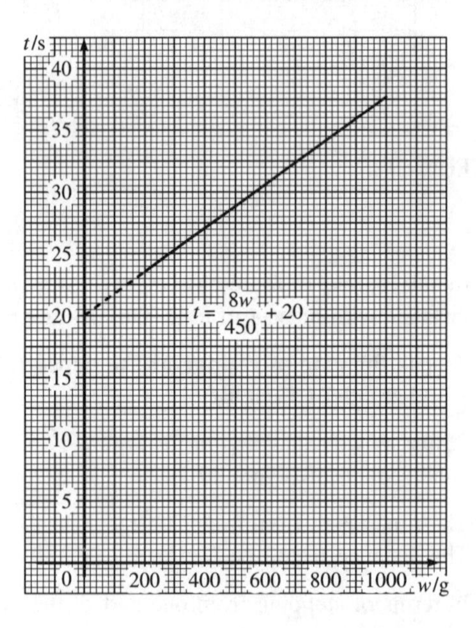

Figure 17.8

The y-intercept, taken from the formula, is 20. This is the standing time. We have dotted in the left-hand section of the line because weights less than 200 g are not practicable. A very small joint would dry out, and it might damage the oven to run it with no joint at all. Beyond 1000 g there might be problems in fitting the joint into the oven.

There is no x-intercept in Figure 17.8. To find this we would have to continue the line toward the left. Cooking a negative amount of sirloin is meaningless.

The gradient, taken from the formula, is 8/450. This is the *rate of change* of heating time with respect to weight. What it means is that we increase the time by 8 minutes for every 450 g of meat.

The timings are based on a 700 W microwave oven. With a higher-powered oven of 750 W, we would expect heating times to be shorter, but standing time would be unchanged. The rate of change of heating time to weight is reduced because the 750 W oven delivers heat energy to the joint at a greater rate. The gradient of the graph is correspondingly reduced, giving shorter total cooking times than in the 700 W oven. Conversely, the slope of the graph for a 650 W oven is steeper.

Test yourself 17.2

1 An 800-litre water tank contains 250 ℓ of water. A filling tap is turned on and, after 12 min, the tank contains 382 ℓ. Plot a graph to show how the volume of water in the tank increases with time. Find the y-intercept, and gradient and state what they represent. Why is there no point in extending the graph beyond $t = 50$?

2 The time taken to weave a 200 mm length of fabric is 80 min. The time taken for 500 mm of fabric is 140 min. These times include a fixed time for setting up the loom, which is not affected by the length of fabric woven. Up to 1 m of fabric can be woven in one run. Plot a graph to show how the time taken relates to the length of fabric woven. Find the y-intercept and gradient and state what they represent. How long is the setting-up time? How long does it take to weave 1 m of fabric?

3 The input voltage (x) and output voltage (y) of an amplifier are related by the equation:

$$y = -12x - 1.5$$

y can vary between −13 V and +13 V. Plot a graph of this equation for values of y between −13 and +13. Find the x- and y-intercepts and the gradient and state what they represent.

Graphs which cross

Figure 17.9 has two graphs plotted on the same set of axes. Their equations are:

$$y = 18 - 3x \qquad \text{Equation 1}$$
$$y = 2.5x - 4 \qquad \text{Equation 2}$$

The lines cross at point A, which has coordinates (4,6).

The line for Equation 1 joins all the points in which the values of x and the corresponding values of y are consistent with (or **satisfy**) Equation 1.

For example, when $x = 1$, then $y = 18 - 3 = 15$
when $x = 2$, then $y = 18 - 6 = 12$
when $x = 4$, then $y = 18 - 12 = 6$
when $x = 7$, then $y = 18 - 21 = -3$

Similarly, the points on the line for Equation 2 all have values of x and y which satisfy Equation 2.

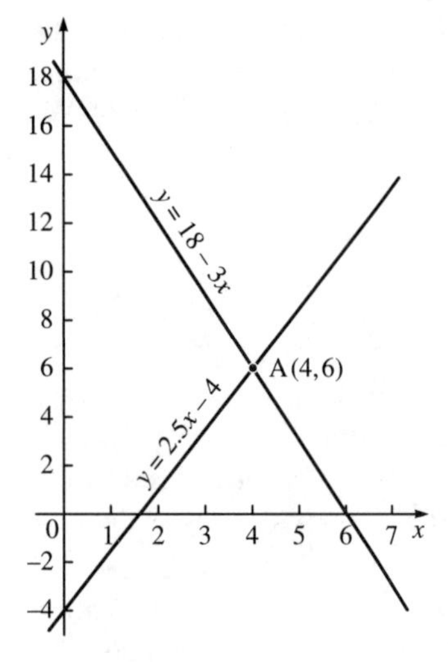

Figure 17.9

For example, when $x = 1$, then $y = 2.5 - 4 = -1.5$
when $x = 2$, then $y = 5 - 4 = 1$
when $x = 4$, then $y = 10 - 4 = 6$
when $x = 7$, then $y = 17.5 - 4 = 13.5$

Looking through the two lists of examples, we find that both lists have just one pair of values of x and y that are the same in both lists. In both lists, when $x = 4$, then $y = 6$. These are the coordinates of the point A, at which the lines cross.

Point A is where the value of x and the corresponding value of y satisfy *both* equations at the same time. Plotting the graphs and noting where the lines cross is a way of finding values to satisfy two different equations.

Example

In a water-garden, the water from one pool drains into another pool. The volume of water v in one pool at time t is given by the equation:

$$v = 25 - 2t$$

In the other pool, the volume is:

$$v = 1.5t + 4$$

v is in cubic metres and t is in hours. At what time do the pools contain the same amount of water, and how much do they hold at that time? In other words, what values of t and v satisfy both equations?

Plotting the graphs (Figure 17.10) show one pool emptying and the other pool filling. The lines cross at (6, 13). The time $t = 6$ is the same for both pools, and the volume $v = 13$ is the same for both pools.

The pools both hold 13 m^3 after 6 h.

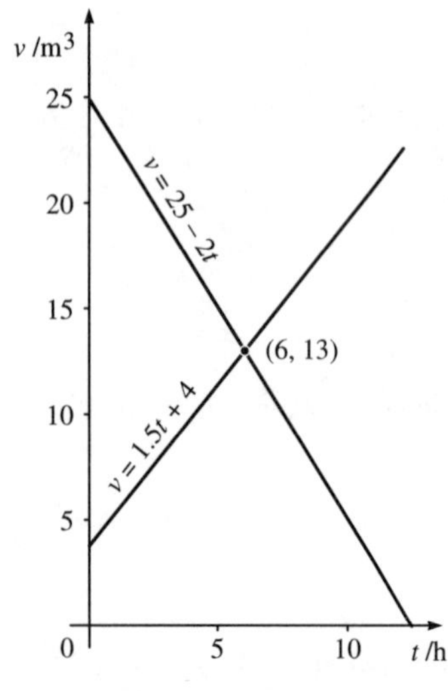

Figure 17.10

As we shall see in Chapter 20, it is easy to solve a problem such as this without plotting graphs. But this is an illustration of a method which can be very useful when the equations are more complicated. Such an example is given on page 242.

Test yourself 17.3

For each of these pairs of equations, draw their graphs and find values of x and y that satisfy both equations.

1 $y = 3x - 1$
$y = x + 3$

2 $y = 2x - 9$
$y = 15 - x$

3 $y = 6 + x$
$y = 2 - x$

Graphing inequalities

In Chapter 12 we saw that a statement such as:

$$y \leqslant 2$$

means that the value of y is less than 2 or equal to 2, but is not greater than 2. This type of statement is known as an **inequality**. It is possible to add an unknown amount to such a statement. For example, we might have:

$$y \leqslant x + 2$$

This means that y is less than or equal to $x + 2$. The values that y can take now depend on the value of x.

A graph makes the situation much easier to understand. We have said that y is less than OR equal to $x + 2$. First we will plot the part of the statement which says that y is *equal* to $x + 2$. This is just a straight-line graph of the kind we have plotted so often (Figure 17.11). Depending on the value of x, the line tells the values y may have. But the statement says that y can also have values *less than* these. For any given point on the line, it may also have values *below* that point. In Figure 17.11 we have plotted six points from the many possible ones below the line. When $x = 1$, for example, the value of y on the line is:

$$y = x + 2 = 1 + 2 = 3$$

At a chosen point A, $x = 1$, but y is only 1.5, which is less than 3. So point A satisfies the inequality. Similarly, when $x = 5$, then y on the line is $x + 2 = 7$. At a chosen point B, $x = 5$, but y is only 4, which is less than 7.

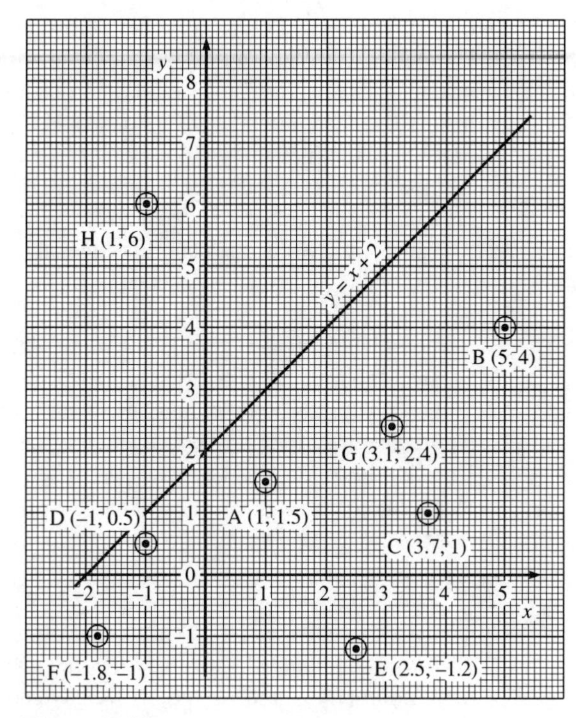

Figure 17.11

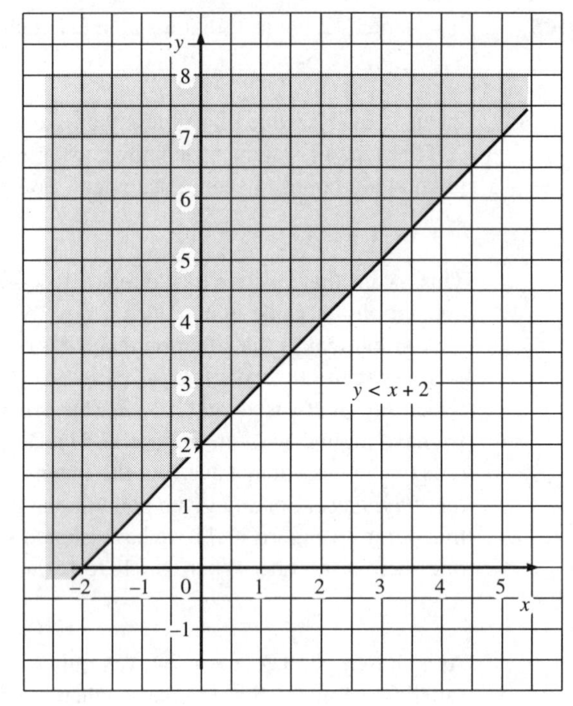

Figure 17.12

If you check all the points A to F, you will find that y is always less than $(x + 2)$. These points, and all other points on the same side of the line, satisfy the inequality. Without any further calculations, we can use the graph to tell instantly that point G, (3.1,2.4), for example is one which satisfies the inequality. By contrast, a point such as H,(1,6), which is above the line, does not satisfy the inequality. To make it clearer, we shade in the area containing the non-satisfying points, as in Figure 17.12. Any point on the line or in the unshaded area satisfies the inequality.

Figure 17.13 illustrates another inequality:

$$y \geqslant 2x - 2$$

Comparing this with Figure 17.12, this is shaded on the other side of the line, because the inequality is 'greater than . . .' instead of 'less than . . .'.

It is always a good idea to check which side of the line should be shaded, *before* shading. Take any point which is easy to compare with the inequality. Usually the point (1,1) is suitable. Then make $y = 1$ and $x = 1$ in the inequality:

$$1 \geqslant 2 \times 1 - 2$$
$$\Rightarrow \quad 1 \geqslant 0$$

This is obviously *true*, so point (1,1) is on the *un*-shaded side of the line. It is one of those points which satisfy the inequality. It is not necessary to actually mark point (1,1) on the graph, as we have done in Figure 17.13. Just imagine it.

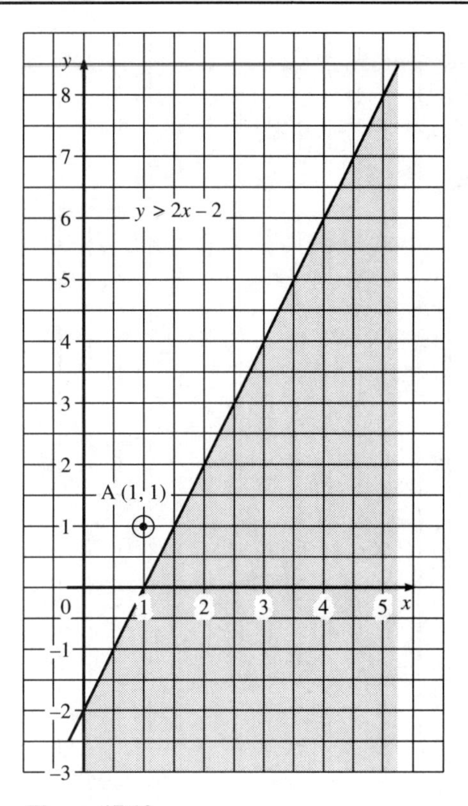

Figure 17.13

> **Test points**
>
> Another good point for testing inequalities is (0,0). The main thing about a test point is that it must not be *directly on* the equality line, but to one side of it.

In Figure 17.14, the inequality is:

$y > 7 - 2x$

This states that y is greater than, but is not equal to $7 - 2x$. We begin by plotting the equation $y = 7 - 2x$. But, to indicate that y does not actually include the values *on* this line, we draw a dashed line. Now to decide which side of the line to shade. Select point (1,1) for testing. Substituting $x = 1$, $y = 1$ in the inequality:

$$1 > 7 - 2 \times 1$$

$$\Rightarrow \quad 1 > 5$$

This is definitely *not* true, so the shading must be on the *same* side of the line as point (1,1). This is the side of points that do not satisfy the inequality.

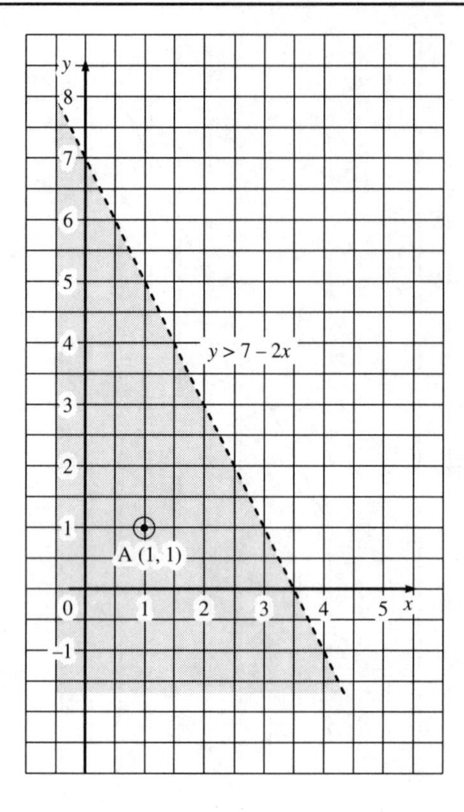

Figure 17.14

A practical example

A shelf is stacked with cans of soup and cans of fruit, all cans being the same size. The shelf can hold up to 40 cans. If s is the number of cans of soup and f is the number of cans of fruit, the total number of cans is $s + f$. If the shelf is full:

$$s + f = 40$$

But the shelf need not be full, so we have an inequality:

$$s + f \leqslant 40$$

This can be plotted as a graph. In this example we cannot say which is the dependent variable; the variables depend on each other. Let us make s the subject of the inequality:

$$s \leqslant 40 - f$$

Figure 17.15 shows the inequality plotted and shaded to show the unacceptable values of s and f. These are pairs of values where there are more cans than will fit on to the shelf.

In the unshaded area, only integer values are allowed, as it is not possible to have fractions of cans. The integer values are represented by the points where the grid lines cross. With values on the equality line itself, the shelf is

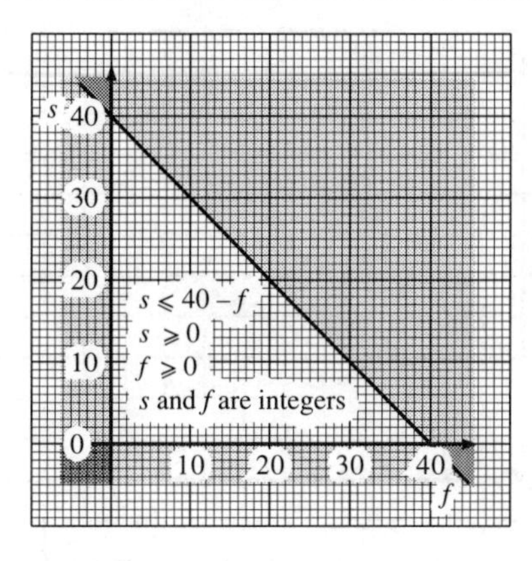

Figure 17.15

full. All other values in the unshaded area are possible, but the shelf is not full. The allowable values include point (0,0) when the shelf is completely empty.

We have also shaded the areas to the left of the y-axis and below the x-axis. The reason for this is that negative cans of soup or fruit are not possible. This practical fact is expressed in two more inequalities:

$$s \geqslant 0 \quad \text{and} \quad f \geqslant 0$$

All three inequalities are required to specify the situation completely.

Test yourself 17.4

1 Plot graphs of these inequalities or pairs of inequalities. In each case state whether the point (4,3) satisfies the inequality or not.

a $y \geqslant 2x + 3$ **b** $y < 0.5x + 2$

c $y > 8 - 4x$ **d** $y \geqslant 7 - 1.5x$
$y \geqslant 2$ $x < 4$

2 What inequalities are represented by the graphs in Figure 17.16?

3 A shelf can hold up to 10 ammeters or voltmeters. Both types of meter are the same size. Write an inequality to relate possible numbers of ammeters (a) and voltmeters (v). Plot a graph of the inequality.

4 Each poster on a display board is fixed by at least 3 drawing pins and there are 5 spare pins pushed into the board at one side. There are no more than 50 pins on the board at any time. Write inequalities to relate the number of pins (y) to the number of posters (x), for up to 0 posters. Plot the graph of the inequality.

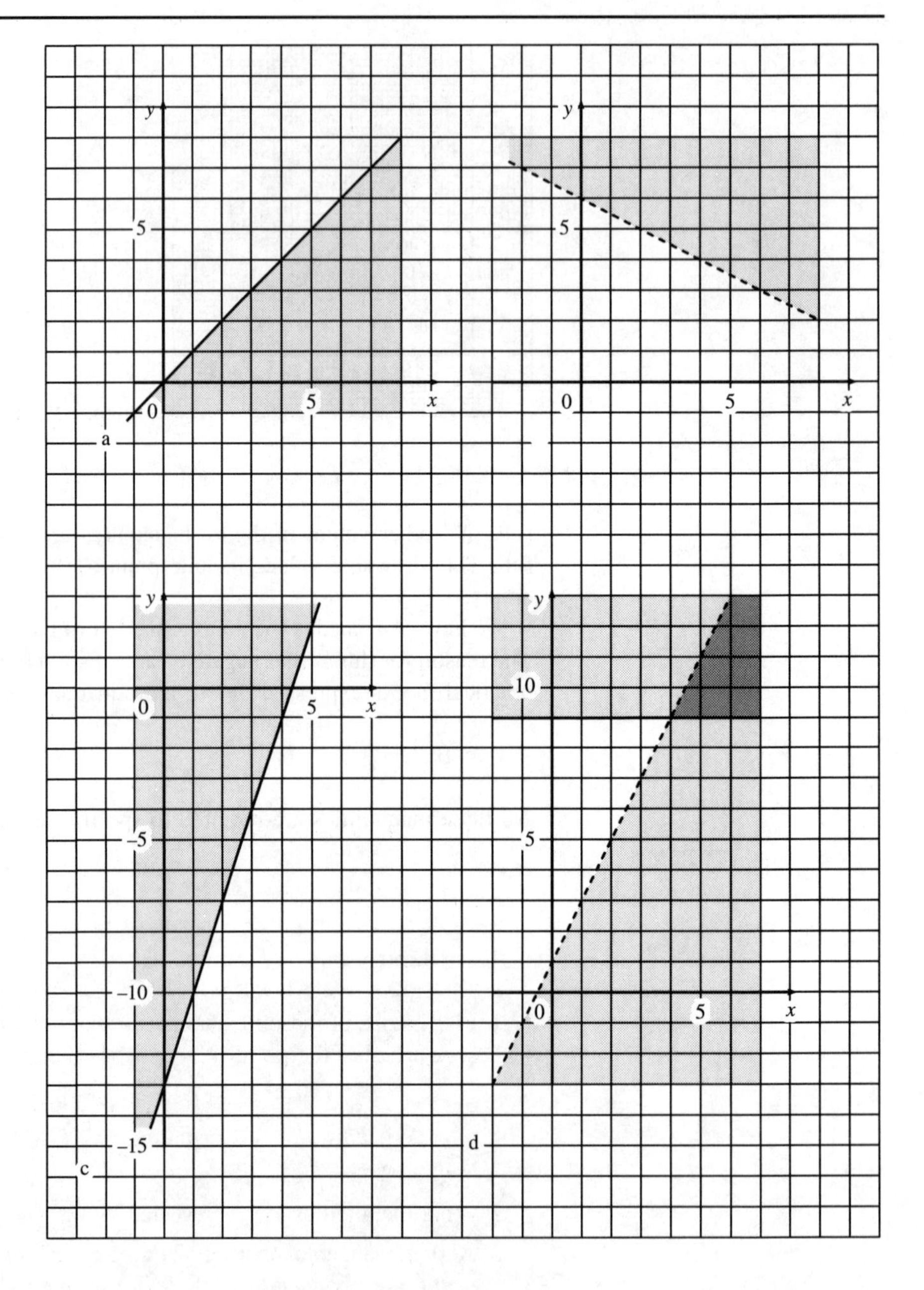

Figure 17.16

Straight lines and curves

An equation of the form $y = mx + c$ always produces a straight-line graph. In Figure 17.17 we see what happens when we have x^2 in the equation. The equation of this graph is the simplest possible one containing x^2:

$$y = x^2$$

To plot this graph we have to calculate a set of values of x and y:

x	−5	−4	−3	−2	−1	0	1	2	3	4	5
y	25	16	9	4	1	0	1	4	9	16	25

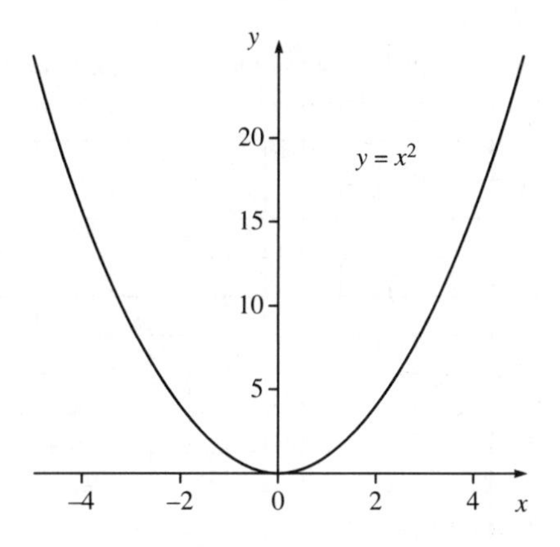

Figure 17.17

Note some of the features of this curve:

- There are no negative values of y. This is because the squares of all real numbers are positive (page 60).
- The curve is symmetrical about the y-axis. This is because $(-2)^2 = 2^2$, $(-4)^2 = 4^2$ and so on.
- It slopes up more and more steeply to the left (negative-going direction) and right (positive-going direction).

This type of curve is called a **parabola**.

Shifting the parabola

Figure 17.18a shows the parabola $y = x^2$ plotted several times on a smaller scale.

Try adding a constant to the y, for example:

$$y + 3 = x^2$$

Change the subject: $y = x^2 - 3$

This has the effect of subtracting 3 from all the values of y in the table above. It also has the effect of shifting the parabola three units down (Figure 17.18b). Conversely, subtracting a constant from the y shifts the parabola up (Figure 17.18c).

Try adding a constant to the x, for example:

$$y = (x + 3)^2$$

The table of values is now:

x	–5	–4	–3	–2	–1	0	1	2	3	4	5
$(x + 3)$	–2	–1	0	1	2	3	4	5	6	7	8
y	4	1	0	1	4	9	16	25	36	49	64

If you compare this table with the previous one, you find that the sequence of values of y is the same, but shifted 3 places to the left. Similarly, the parabola is shifted 3 units to the left (Figure 17.18d). Conversely, subtracting a constant from x shifts the parabola to the right.

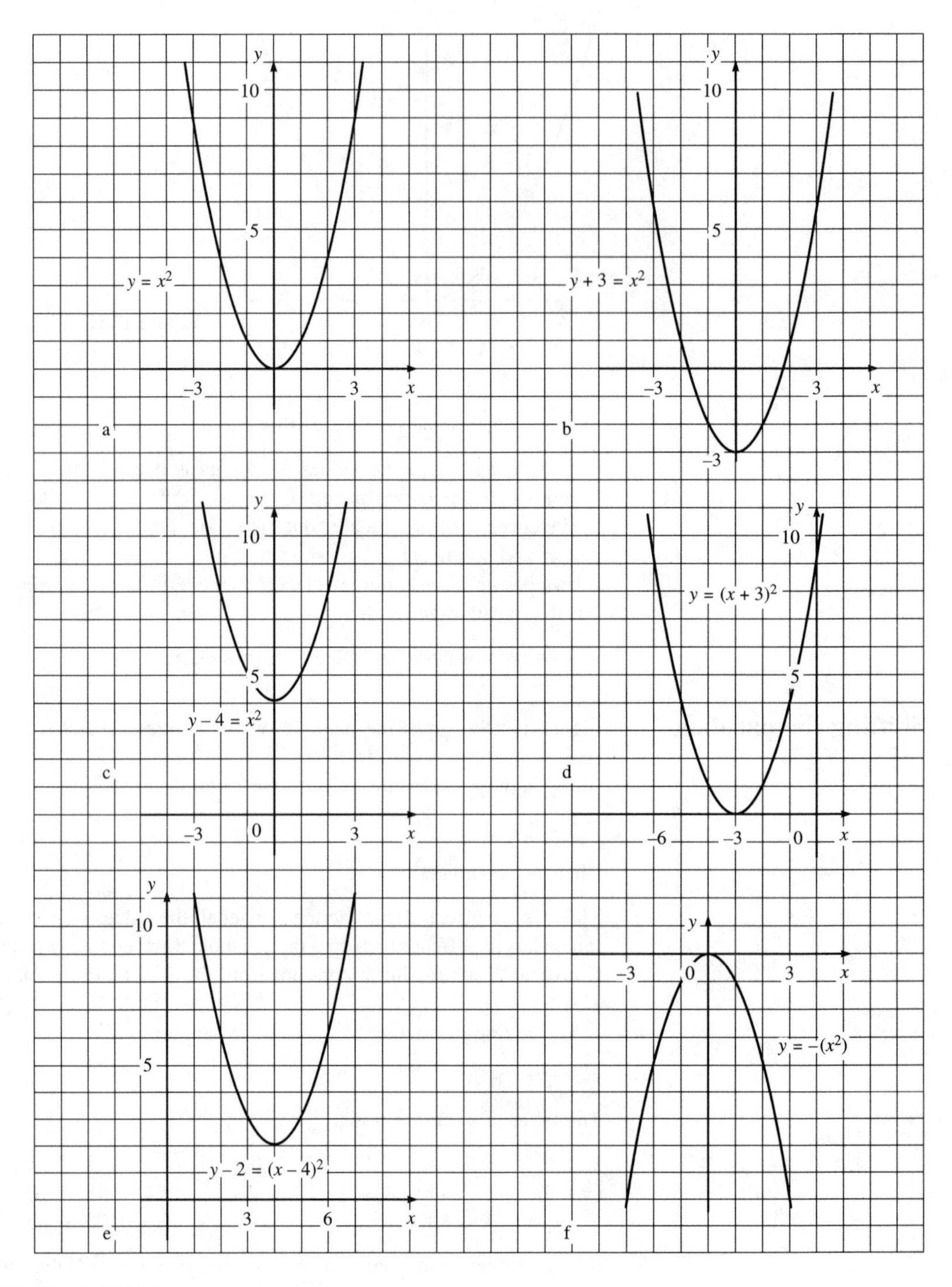
y
10
5
y = x²
−3
3
x
a
y
10
5
y + 3 = x²
−3
3
x
−3
b
y
10
5
y − 4 = x²
−3
0
3
x
c
y
10
y = (x + 3)²
5
−6
−3
0
x
d
y
10
5
y − 2 = (x − 4)²
3
6
x
e
y
−3
0
3
x
y = −(x²)
f

Figure 17.18

Watch the brackets

Note the difference between $(-x)^2$ and $-(x^2)$.

$(-x)^2$ means negate x first, *then* square it.
$-(x^2)$ means square x first, *then* negate it.

The brackets are not needed for $-(x^2)$; write it as $-x^2$

Values of $(-x)^2$ are always positive.

Examples

$$(5)^2 = 25$$
$$(-5)^2 = 25$$

Values of $-(x^2)$ are always negative.

Examples

$$-(5^2) = -25$$
$$-(-5^2) = -25$$

We can shift the parabola in two directions at once:

$$y - 2 = (x - 4)^2$$

Figure 17.18e shows the result: the parabola is shifted 2 units up and 4 units to the right.

Finally, make the x^2 negative:

$$y = -(x^2)$$

As might be expected, this turns the parabola upside down (Figure 17.18f).

Expanding the equation

So that we can show what has been added to or subtracted from the x and the y, we have written the parabola equation in the form:

$$y - 2 = (x - 4)^2$$

But we could expand and then simplify this:

$$\Rightarrow \quad y - 2 = x^2 - 8x + 16$$
$$\Rightarrow \quad y = x^2 - 8x + 14$$

The equation remains balanced, so the graph is exactly the same.

An expression such as $x^2 - 8x + 14$ is called a **polynomial**. It has several terms, involving different *whole-number* powers of x. Here the highest power is x^2 so this is called a **second-order** polynomial. The graph of any second-order polynomial is a parabola.

Explore this

Use a graphic calculator or computer graphic package to investigate the effects of varying the basic parabola equation. First put the calculator in rectangular mode. If the calculator does not have a special $\boxed{x^2}$ key, use the $\boxed{x^y}$ key when entering the formula.

Build up families of curves on the screen:

- Add or subtract different values to the y
- Add or subtract different values to the x (before squaring)
- Multiply or divide the x by different values.

Observe how these changes affect the position and the shape of the parabola.

Plot the parabola $y = x^2$. Then, without clearing the screen, plot the straight line $y = x + 2$. Use the Trace function to find where these two lines cross. There are two crossing points, which represent the two sets of x's and y's which satisfy these two equations. Try this for other pairs of parabola and straight-line equations.

Identifying parabolas

A variable current (I) is passed through a heater coil and the rate of heat production (h) is measured. Here are the results:

Current, I	0	0.5	1.0	1.5	2.0	2.5	3.0
Heat, h	0	12	53	117	205	302	465

The graph of these results (Figure 17.19) is a curve which looks very much like a parabola. If this is so, we could say that heating is proportional

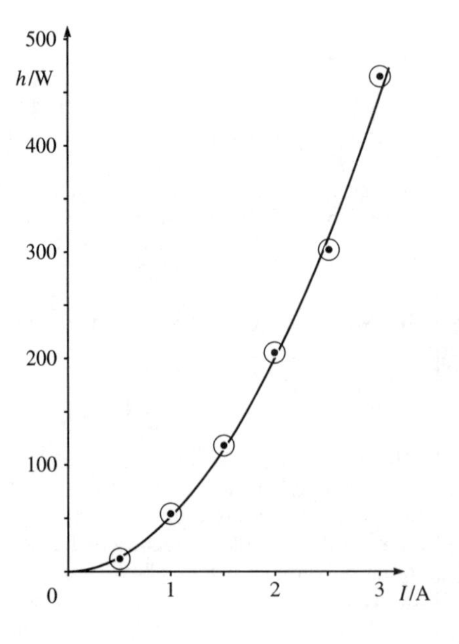

Figure 17.19

Negative current

Figure 17.19 shows only the positive half of the parabola. This is because the data represents only positive current. Although negative values are not possible in some examples (for example, negative cans of fruit) negative current is simply current flowing through the heater in the opposite direction. The heating effect of a current is just the same, whichever way the current flows. So, by measuring the heating effect with reversed current, we would be able to complete both sides of the parabola.

to the square of the current. The difficulty is knowing if this curve really is a parabola, for there are possibly curves for other equations but similar shape. In any case, there are experimental errors which make it difficult to judge the exact shape of the curve.

When two quantities are directly proportional, the graph of one plotted against the other is a straight line (page 144). If we believe that h is proportional to I^2, we can test this idea by plotting I^2 against h. These are the figures:

Current², I^2	0	0.25	1.0	2.25	4.0	6.25	9.0
Heat, h	0	12	53	117	205	302	465

Figure 17.20 shows the graph. There is no problem in recognizing a straight line, even when a few of the points are slightly off it. Because the graph is a straight line we can say that:

The heating effect is proportional to the square of the current.

We can say more than this, for Figure 17.20 is a straight-line graph with zero y-intercept (so $c = 0$) and a gradient that we can measure. Taking two

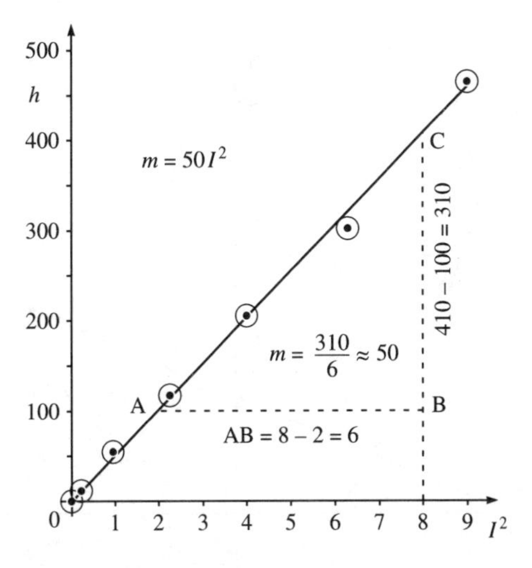

Figure 17.20

convenient points on the graph, at which x and y have integer values, we measure:

Horizontally AB $= 8 - 2 = 6$
Vertically BC $= 410 - 100 = 310$

To step from A to C we move '6 across the 310 up' (compare page 226). The gradient is:

$$m = \frac{BC}{AB} = \frac{310}{6} \approx 50$$

Solving equations

On page 230 we showed how to solve a pair of equations by plotting their graphs and finding where the lines cross. This can also be done when one or both curves are parabolas. Often equations with x^2 can be solved by factorizing, as explained on page 299, but the graphing technique works well when it is not possible to factorize the equations. In Figure 17.21 we have plotted the lines for:

$y = x + 4$
$y = x^2 - 2$

$(x^2 - 2)$ does not factorize. The line cross at *two* places, so there are two sets of solutions:

$x = -2$ and $y = 2$

and

$x = 3$, and $y = 7$

Check by substituting $x = -2$ and $y = 2$ into both equations. Then check the other pair of values.

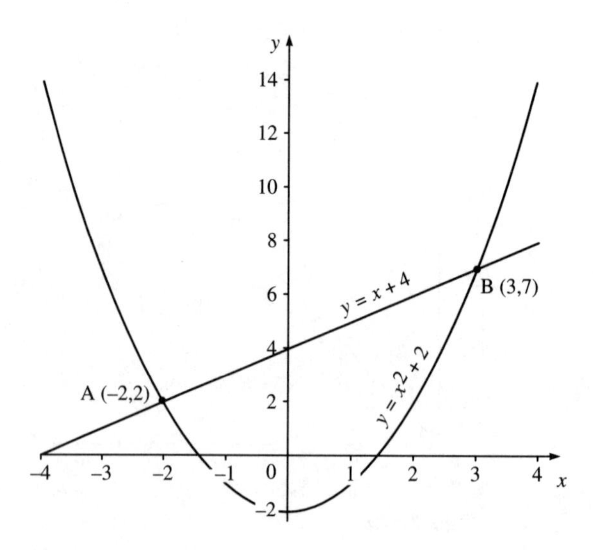

Figure 17.21

It is sensible to use an approximate result, since we have had to estimate where the line should be drawn through the irregular row of points.

Since $c = 0$ and $m = 50$, the equation for the line is:

$$h = 50I^2$$

This technique of plotting squares can be used in other instances when we suspect that one variable is proportional to the square of another.

Test yourself 17.5

1 Plot the graphs of these equations for $x = -4$ to $x = 4$:

a $y - 1 = x^2$ **b** $y + 2 = x^2$

c $y + 3 = (x + 2)^2$ **d** $y = -(x - 1)^2$

e $y = x^2 - 4x + 4$ **f** $y = x^2 + 3x - 1$

2 The minimum breaking distance (b) of a vehicle is measured at different speeds (v), b is in metres, v is in kilometres per hour. The results of the measurements are:

Speed (v)	30	50	70	90	110	130
Distance (b)	10	24	50	79	125	161

Plot the graph of this data and draw the best line through the points. These figures are the *minimum* breaking distances; in other words, actual breaking distance may be equal to or greater than b. Shade the graph, leaving clear the area which includes all possible breaking distances.

Plot the graph which confirms that b is directly proportional to v^2, and write the equation connecting v and b.

Exponential curves

These are curves in which x is an index or **exponent** (page 185). For example:

$$y = 2^x$$

Figure 17.22 shows this curve and also curves for other numbers to the power of x. Note that all the curves have their y-intercept at $y = 1$. This is because any number to the power 0 is equal to 1 (page 177). To the right they slope upward with increasing gradient. To the left their gradient decreases more and more slowly; the curve approaches but never quite touches the x-axis.

In electronics, we often find equations in which y equals a power of e, the exponential constant (page 187):

$$y = e^x$$

The graph has the same form as those in Figure 17.22. If constants are introduced into the equation, the graph may be shifted or modified in other

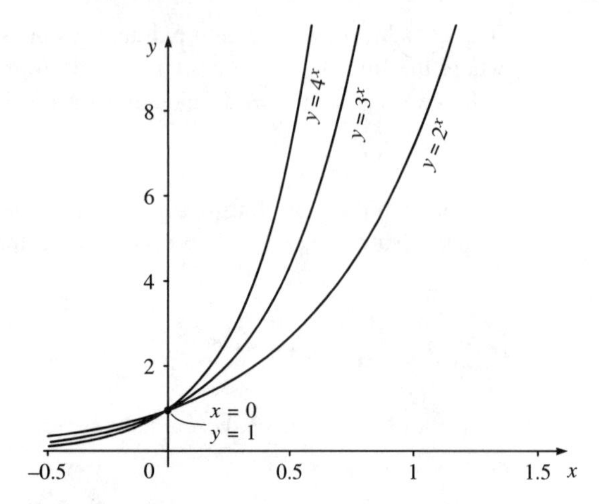

Figure 17.22

ways (we saw this happening with parabolas too, on page 238). Here is an exponential equation with various constants included:

$$y = 2 - 1.5e^{0.7x}$$

Figure 17.23 shows the graph. The negative sign has turned the curve upside down, and the 2 has shifted it 2 units up. The 0.7 has stretched the curve out in the x-direction.

If you take test measurements and obtain a curve which you suspect is exponential, it is difficult to be sure about it. Exponential curves are not identified for certain by just looking at them. Instead of being exponential, your curve might be part of a parabola, or possibly something different again. To check on an exponential relationship, we plot the inverse of the exponential, the natural logarithm (page 187).

As an example, suppose we have this data:

x	1	2	3	4	5	6	7
y	4.6	8.3	15.1	27.6	50.2	91.5	166.7

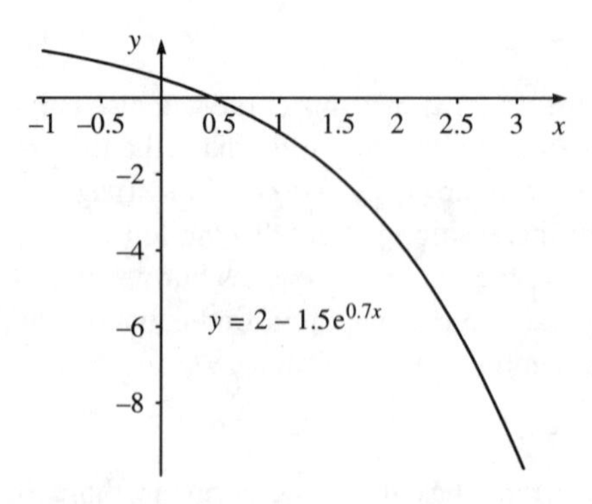

Figure 17.23

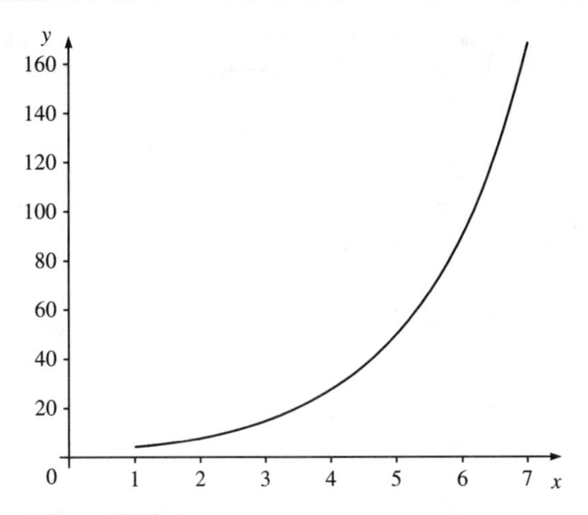

Figure 17.24

When it is plotted (Figure 17.24), it looks as if the line passes through the point (0,1). It curves upward to the right at an increasing rate. It *might* be an exponential curve, with an equation of this form:

$$y = ae^{bx}$$

We want to know if it really is exponential and, if so, the values of a and b. In the previous paragraph, we mentioned taking natural logarithms, logs to base e. The rules for working with logs are in the box on page 187.

First we find the log of e^{bx} by applying the power rule:

$$\Rightarrow \qquad \ln e^{bx} = bx \ln e$$

But ln e = 1 so this simplifies to:

$$\ln e^{bx} = bx$$

Next we must allow for the fact that e^{bx} is multiplied by a. Apply the multiplication rule:

$$\ln (ae^{bx}) = \ln a + bx$$

Applying the log rules

The rule	*As on page 187*	*As applied in this example*
Powers	$\log m^n = n \log m$	$\ln e^{bx} = bx \ln e$ $= bx$ (because $\ln e = 1$)
Multiplication	$\log (m \times n) = \log m + \log n$	$\ln (ae^{bx}) = \ln a + \ln (e^{bx})$ $= \ln a + bx$

This completes taking the log of the right side of the equation. The log of the left side is simply ln *y*, and so:

$$\ln y = \ln a + bx$$

Looking at each term in this equation, we find that it has the same form as the basic straight-line equation:

$$y = c + mx$$

Instead of *y*, we have ln *y*
Instead of *c*, we have ln *a* – the *y*-intercept
Instead of *m*, we have *b* – the gradient

This means that, if we plot ln *y* against *x*, we get a straight line with ln *a* as the *y*-intercept and *b* as the gradient.

When we plot the logs we have a choice of two techniques. One technique is to use a calculator to find the natural log (ln) of each value of *y*, then plot ln *y* against *x*. Here is the table of values again, with a third line added for ln *y*:

x	1	2	3	4	5	6	7
y	4.6	8.3	15.1	27.6	50.2	91.5	166.7
$\ln y$	1.53	2.12	2.71	3.32	3.92	4.52	5.12

Figure 17.25 is the graph of ln *y* against *x*.

A much easier technique is to use special graph paper which works out the logs for you. On the *x*-axis the scale is normal. On the *y*-axis the scale is such that the distance of any point above the *x*-axis is equal to its log, Figure 17.26 shows the same data plotted on this kind of graph paper. Instead of plotting the *log* of *y* on an ordinary scale, as in Figure 17.25, we plot *y* on a *log* scale.

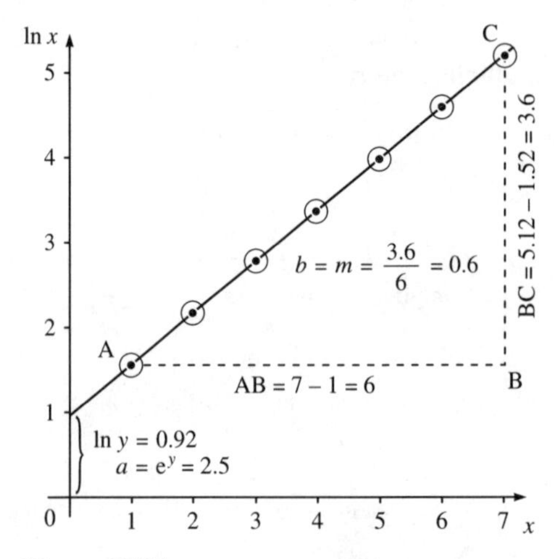

Figure 17.25

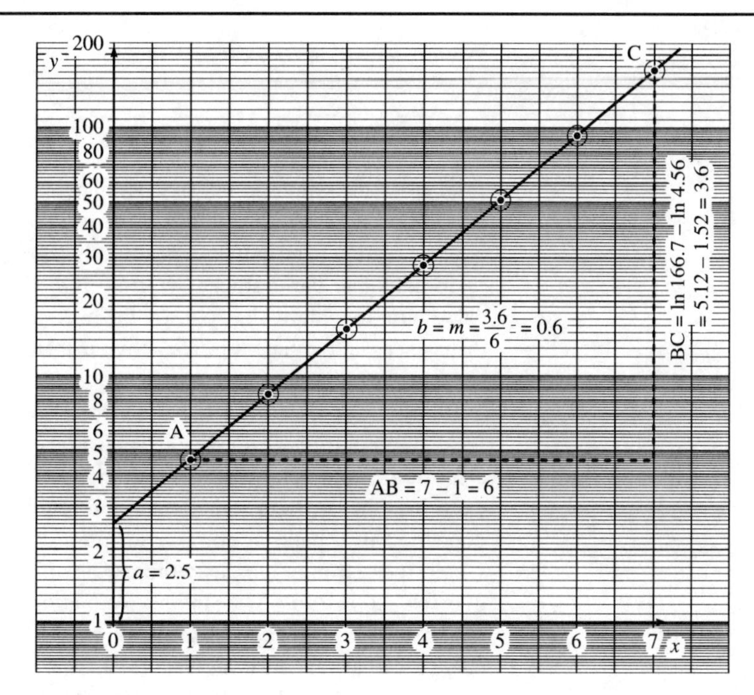

Figure 17.26

Either way, the graph is the same, as can be seen by comparing the figures. But using log graph paper makes everything much easier. This kind of graph paper is called *semi-log* paper because it has a log scale on the y-axis only. Another type of graph paper, *log-log* paper has log scales on both axes. It is used for analysing certain other types of equations.

Now we interpret the graph of Figure 17.25. First of all, it is a straight line, confirming that Figure 17.24 is an exponential curve. Looking at the y-intercept, and reading its value from the graph, we find that $\ln y = 0.92$. With a calculator, we find that $e^{0.92} = 2.5$. This tells us that $a = 2.5$.

We find the gradient by measurements on a part of the graph. As the whole graph is a straight line, and has the same gradient all along it, we can take the first and last points. The horizontal distance AB is 7 – 1 = 6. The vertical distance BC is 5.12 – 1.52 = 3.6. The gradient is:

$$m = 3.6/6 = 0.6$$

This is also the value of b, so $b = 0.6$.

Have you noticed?

The y-axis in Figure 17.26 starts from $y = 1$.

We might have started it from 0.1, 0.001 or even smaller *positive* values, but it can *never* start from zero or from a negative numbers. This is because we can not have the log of zero or a negative number.

We are now able to write out the equation for the data:

$$\underline{y = 2.5e^{0.6x}}$$

Figure 17.26 is interpreted in a similar way. It is a straight line, confirming that the relationship is exponential. Since the scale on the y-axis is a log scale, we can read off the y-intercept directly: $a = 2.5$.

To calculate the gradient, we first find AB as above. To find BC we use a calculator to obtain the logs of the y-coordinates of A and B. BC is their difference, and equals 3.6. As above the gradient is $b = 0.6$. We arrive at the same equation as before.

Plotting curves with a log scale for y is a useful technique for analysing measurements, particularly in electronics, where exponential relationships are often found. The exponential equation is also found in other fields, for example the growth of plants and animals, or the growth of money deposited in a bank at compound interest. Because it expresses the way in which these things increase, the exponential equation is often known as the **growth equation**. The inverse equation, in which the index of e is negative, is the **decay equation**. Figure 17.27 is the curve for $y = e^{-x}$. The charge on a capacitor decreases in this way. Another example is radioactive decay by which the amount of radioactivity in a sample decreases exponentially.

As Figure 17.27 shows, the charge on a capacitor falls rapidly when it is first being discharged. Then it falls more and more slowly. The curve approaches *but never reaches* zero. Similarly, the charge becomes very small, and for practical purposes becomes zero, but actually the capacitor is never *completely* discharged. The same applies to radioactivity, which decreases relatively rapidly at first but never disappears altogether.

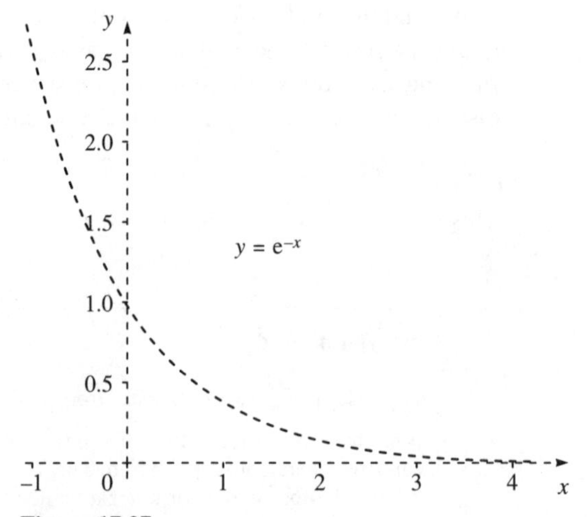

Figure 17.27

Explore this

With a graphic calculator or computer graphic software, try plotting families of exponential curves. Note their features and how their shape changes when different constants are introduced into the equation.

Begin with $y = 1^x$, $y = 2^x$, $y = 3^x$ and others like this.

Then add a constant to the right-hand side of the equation. Use the e^x function to plot graphs of:

$y = e^x$ $\quad y = e^{0.5x}$

$y = e^{2x}$ $\quad y = e^{-x}$

and other equations in which the x has positive or negative coefficients. The e^x function is usually obtained by using the shift, or inverse function key followed by the [ln] key.

e^x on a calculator

A log expresses a number as the power of another number (page 186). As an equation, this statement about natural logs expresses numbers as powers of e:

if $\quad a = e^b$

then we say $\quad \ln a = b$

When we are finding the log of a number on a calculator, we are given a and find b. This is done by using the [ln] key.

If we are given a log and want to find the corresponding number (the antilog), we need the inverse operation. On most calculators we use the same key but press the inverse, shift, or second function key first. The calculator panel is usually marked e^x beside the ln key, to indicate this operation.

Example

Find the number of which 0.45 is the natural log.

Key-strokes	*Display*
[.] [4] [5]	0.45
[INV] [e^x]	1.568312185

Calculators usually have the same pair of functions for logs to the base 10. The [log] key produces logs to base 10. The [log] key is also marked 10^x. Using the same key after pressing the inverse key gives the antilogarithm.

More curves

There are several other curves that you might come across in technology. It is useful to be able to recognize them by their shapes and other features.

Cubic curves

Figure 17.28 shows the basic cubic curve:

$$y = x^3$$

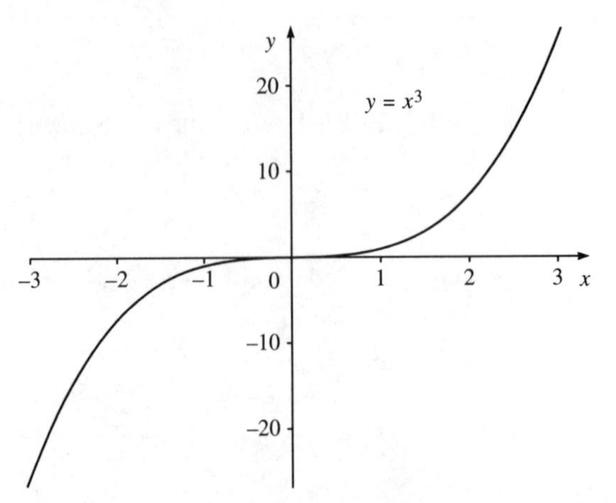

Figure 17.28

This curves up for positive values of x but, since the cube of a negative number is negative, it curves down toward the left. The curve has a change of curvature at the origin. If we include constants in the equation as well as terms in x^2 and x, the curve shows a double change of direction, usually around the origin, as in Figure 17.29. It may also be shifted to the left or right, and up or down, as was the case with the parabola. If x^3 is negated, the curve is upside down, going steeply down to the right and up to the left.

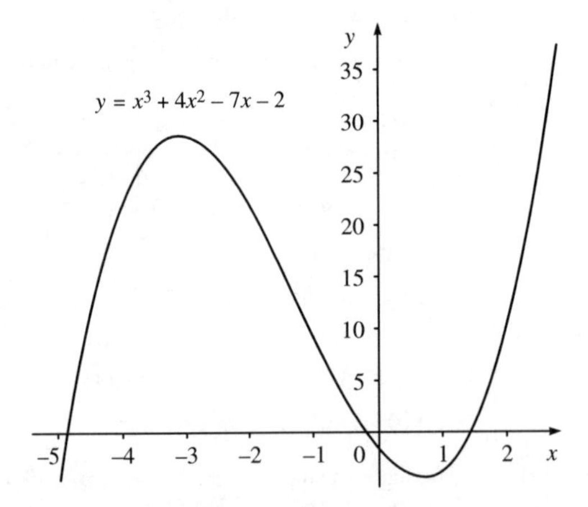

Figure 17.29

Hyperbola

This is a double curve, resulting from an equation such as:

$$y = \frac{2}{x} + 4$$

Figure 17.30 is the graph of this equation. The equation has the reciprocal of x ($1/x$) instead of x, and this makes the value of y sweep up to very high and down to very low values when x is close to 0. When $x = 0$, y has no value.

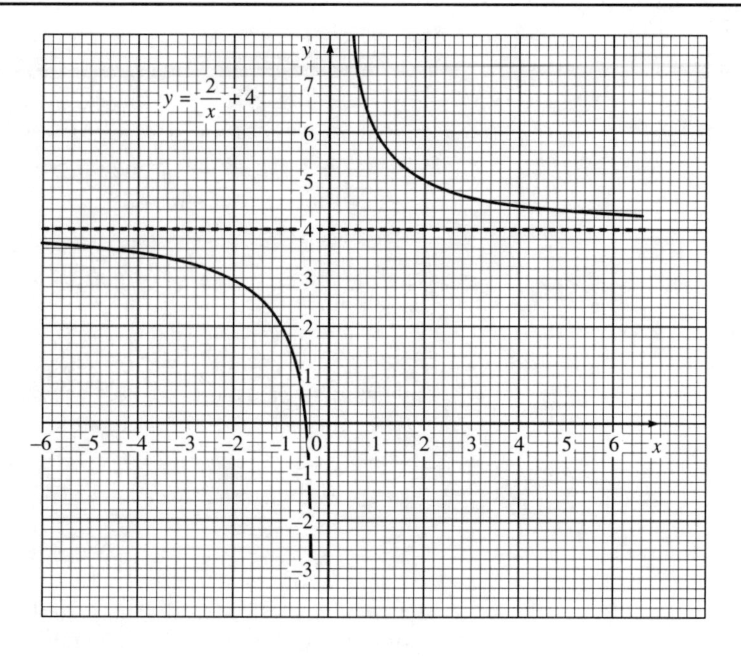

Figure 17.30

Circle

The equation of a circle is:

$$y^2 = x^2 + r^2$$

In this equation, r is the radius of the circle. Figure 17.31 is an example. Here the centre of the circle is at the origin, but it is shifted to other positions if constants are added to or subtracted from the equation.

Trig ratios

The curves of the trig ratios are so important in several branches of technology that they deserve a chapter to themselves. Refer to Chapter 19.

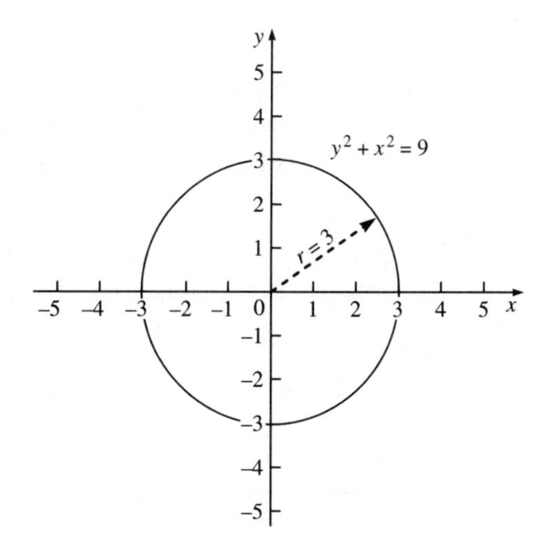

Figure 17.31

Test yourself 17.6

1 Plot the curves of these equations from $x = -1$ to $x = 3$ using a graphic calculator if you prefer.

a $y = 0.01^{2x}$ **b** $y = e^{0.5x}$ **c** $y = 2e^{-0.4x}$

For each curve, state the y-intercept.

2 Plot the graph of this set of data:

x	1	2	3	4	5	6
y	3.7	4.5	5.5	6.7	8.2	10.0

By plotting a graph of ln y against x, confirm that it is an exponential curve with the equation $y = ae^{bx}$, and find the values of a and b.

3 Plot the graph of this set of data:

x	1	2	3	4	5	6
y	1.2	0.74	0.45	0.27	0.16	0.10

By plotting a graph of ln y against x, confirm that it is an exponential curve with the equation $y = ae^{bx}$, and find the values of a and b.

4 To assess if you have covered this chapter fully, answer the questions in *Try these first*, page 218.

Explore this

With a graphic calculator or computer graphic software, try plotting families of cubic curves. Note their features and how their shape changes when different constants are introduced into the equation.

Plot graphs of cubic equations in factor form, such as:

$$y = (x - 1)(x + 2)(x + 4)$$

Your calculator should allow you to enter such equations directly. Note the x-intercepts and how these relate to the constants in the equation. Vary this by making two factors the same:

$$y = (x - 1)(x + 2)(x + 2)$$

What effect does this have on the shape of the curve and on the number of x-intercepts? Also try all three factors the same:

$$y = (x + 1)(x + 1)(x + 1)$$

You can enter this as:

$$y = (x + 1)^3$$

18 More areas and volumes

In this chapter we extend the work of Chapter 6 and make this an opportunity to revise many of the topics of Chapters 9 to 17. If there are any references in this chapter to topics which you do not remember or understand, check back to the earlier chapters immediately.

You need to know the Maths Essentials (Part 1) and all the main topics in the chapters of Part 2.

Try these first

Your success with this short test will tell you which parts of this chapter you already know. Express all answers to 3 sf.

1 Find the area of Figure 18.1.

2 2000 spherical glass beads, diameter 4 mm, are placed in a cylindrical container, radius 50 mm and melted. Assuming no loss or expansion, what is the depth of molten glass in the container?

3 The barrel of a plastic ball-point pen is 120 mm long and is hexagonal in section, with 4 mm sides. A cylindrical hollow, 3 mm radius runs down the length of the barrel. What is the volume of plastic?

4 A boiler is cylindrical, 0.5 m radius and 2 m long. One end is flat, the other is hemispherical (Figure 18.2). Find the surface area.

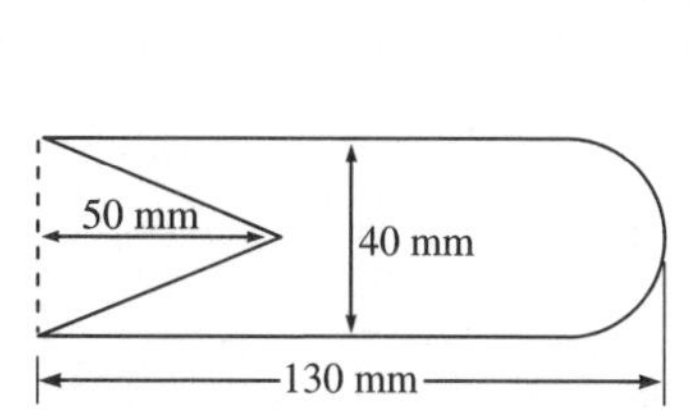

Figure 18.1

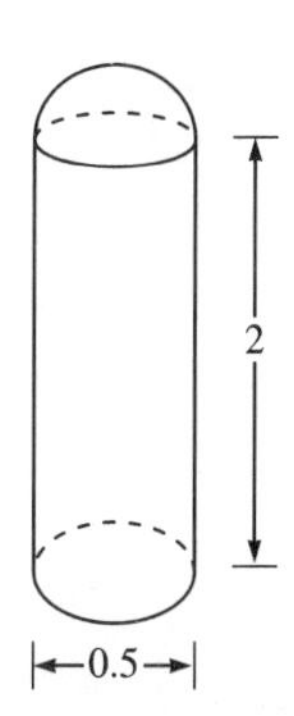

Figure 18.2

Areas of plane figures

Figure	*Examples*	*Formula*
Triangle	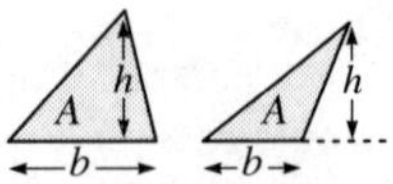	$A = \frac{1}{2}bh$ b = base h = height
Quadrilaterals (Square, rectangle, parallelogram, rhombus)	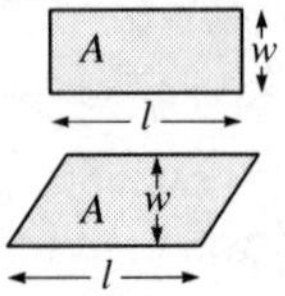	$A = lw$ l = length w = width
Other quadrilaterals and polygons	Trapezium, kite, hexagon, octagon	Divide into triangles and rectangles; sum their areas
Circle	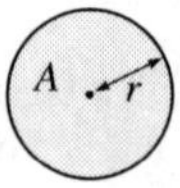	$A = \pi r^2$ r = radius
Sector	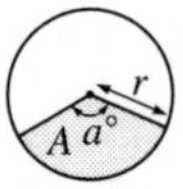	$A = \frac{a}{360} \times \pi r^2$ a = angle (degrees), r = radius
Segment	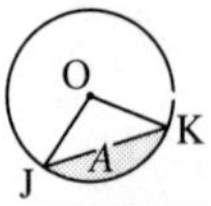	A = area of sector − area of triangle OJK

Figure 18.3

Volumes of solids

Solid	*Examples*	*Formula*
Prism	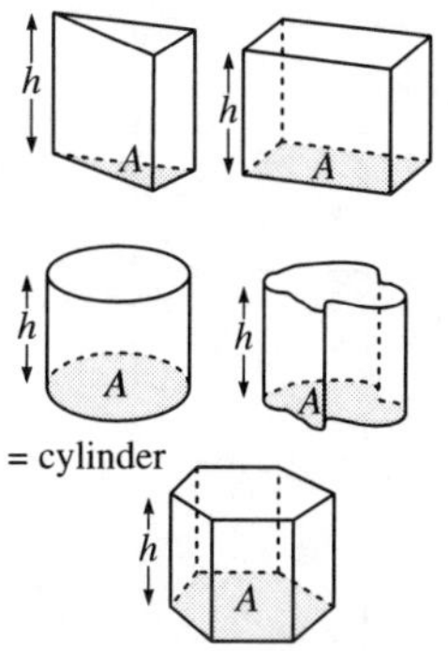 h, A; = cylinder	$V = Ah$ A = base area h = height
Pyramid	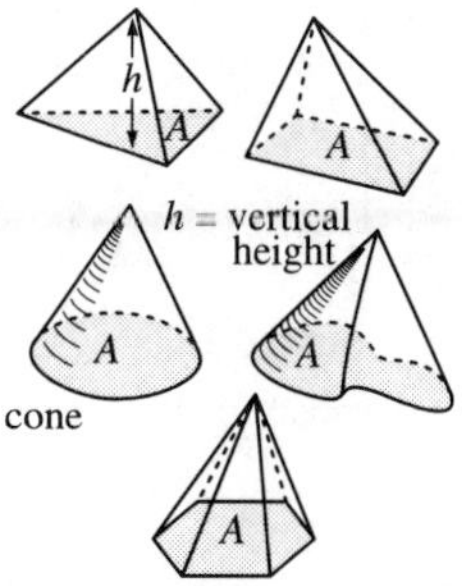 h = vertical height; A; cone	$V = \frac{Ah}{3}$ A = base area h = vertical height
Sphere	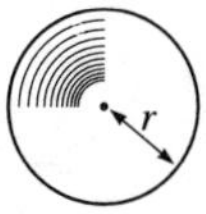 r	$V = \frac{4\pi r^3}{3}$ r = radius

Figure 18.4

Surface areas of solids

Solid	*Examples*	*Formula*
Rectangular prism		$A = 2(wl + wh + lh)$ w = width l = length h = height
Cylinder		Curved surface: $A = 2\pi rh$ Whole: $A = 2\pi r(h + r)$ r = radius h = height
Cone		Curved surface: $A = \pi rl$ Whole: $A = \pi r(l + r)$ r = radius l = slant height
Sphere		$A = 4\pi r^2$ r = radius

Figure 18.5

Formulae for areas and volumes

The first three boxes in this chapter summarize the formulae from Chapters 6 and 15, and include several new ones.

Areas of plane figures

All of these have been explained earlier, except for the area of a **segment** of a circle.

To find this, first find the area of the **sector**, using the formula given in the table. For this you need to know the radius r and the angle a. The area of the triangle can be found by using the geometry and trig from Chapter 14, as explained below.

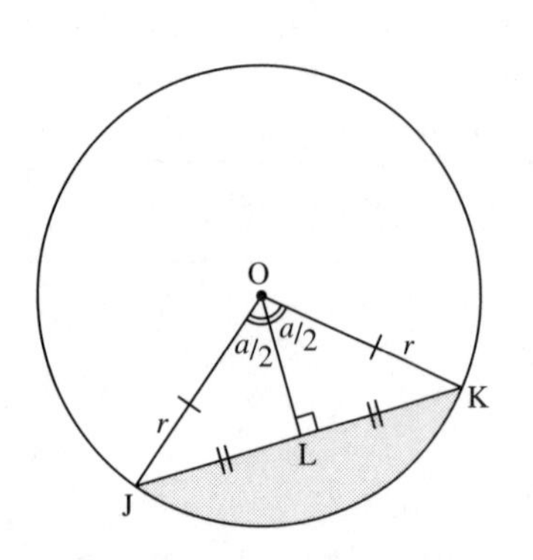

Figure 18.6

Imagine a line OL drawn from the centre O, at right-angles to the chord (Figure 18.6). Triangle OJK is isosceles, because OK = OJ (two radii). Therefore $O\hat{K}J = O\hat{J}K$. In the triangles OKL and OJL, OK = OJ (two radii), $O\hat{K}J = O\hat{J}K$ (angles of the isosceles triangle) $O\hat{L}K = O\hat{L}J$ (right-angles), and side OL is common to both triangles. Therefore, triangles OKL and OJL are congruent. This means that JL = LK, and $K\hat{O}L = a/2$.

In $\triangle$OKL, $O\hat{L}K = 90°$ and $\sin K\hat{O}L = \sin(a/2) = LK/r$

$$\Rightarrow \quad LK = r \sin(a/2)$$

Similarly $\cos K\hat{O}L = \cos(a/2) = OL/r$

$$\Rightarrow \quad OL = r \cos(a/2)$$

The area of $\triangle$OLK $= \frac{1}{2}$ base $\times$ height $= \frac{1}{2}$ LK $\times$ OL
$= \frac{1}{2}r^2 \sin(a/2) \times \cos(a/2)$

But $\triangle$OJK has twice this area:

Area of $\triangle$OJK $= r^2 \sin(a/2) \times \cos(a/2)$

Calculate this area and subtract it from the area of the sector to find the area of the segment.

Example

Find the area of a segment of a circle when $r = 1.5$, and $a = 120°$.

$$\text{Area of sector} = \frac{a}{360} \times \pi r^2$$

$$= \frac{120}{360} \times 3.1416 \times 1.5^2$$

$$= 2.356$$

$$\text{Area of } \triangle\text{OJK} = 1.5^2 \times \sin 60° \times \cos 60°$$

Using the standard triangle from page 215, $\sin 60° = \sqrt{3}/2$ and $\cos 60° = 1/2$:

$$\Rightarrow \quad \text{Area of } \triangle\text{OJK} = 1.5^2 \times \frac{\sqrt{3}}{4} = 0.974$$

$$\underline{\text{Area of segment} = 2.356 - 0.974 = 1.382}$$

Volumes of solids

The box emphasizes that the same formula applies to prisms of any kind. A cylinder is a circular prism. All the prisms shown as examples are **right prisms** in which the upper surface is directly above the base. The formula also applies to **oblique prisms** (Figure 18.7) in which the upper surface is parallel to the base but is not directly over it. In oblique prisms the height h is the vertical height.

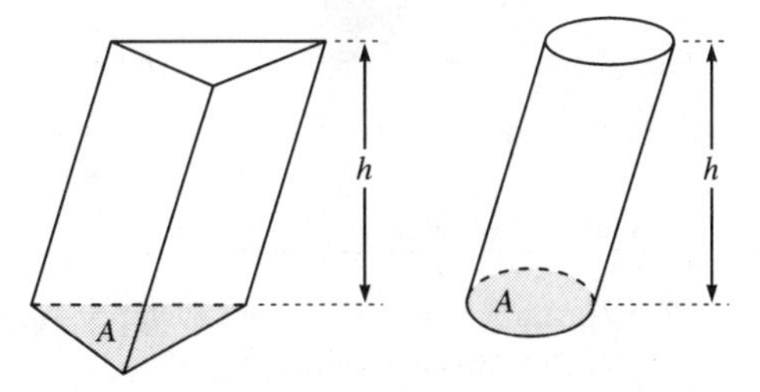

Figure 18.7

Pyramids are another class of solid which share a common volume formula. The circular pyramid is more often called a **cone**. The formula applies to right pyramids, in which the apex is directly above the centre of the base and to oblique pyramids (all in the diagram are slightly oblique) in which the apex is displaced to one side.

The volume of a solid of more complicated shape may often be found by considering it to be made up of two or more of the solids in the table, or it may be the *difference* between two of these solids.

Example

The shank of a rivet is 10 mm long, radius 1.5 mm. It has a hemispherical head, radius 2 mm. Find its volume, expressed in standard form to 3 sf.

The rivet is made up of a cylinder and a hemisphere (half-sphere). Volume of shank:

$$\text{Area of cross-section} = A = \pi r^2$$
$$= 3.1416 \times 1.5^2 = 7.0686$$
$$V = Ah = 7.0686 \times 10 = 70.686$$

Volume of head:

$$V = \tfrac{1}{2} \times \frac{4\pi r^3}{3} = \frac{2 \times 3.1416 \times 2^3}{3}$$
$$= 16.755$$

$$\text{Total volume} = 70.686 + 16.755 = 87.441$$

$$\underline{\text{Volume of rivet} = 8.74 \times 10\,\text{mm}^3 \text{ (3 sf).}}$$

Surface areas of solids

The formula for a prism also applies to a prism which is a parallelogram or rhombus in section, in which case h is the vertical height. With a cube, side ℓ, $w = h = \ell$. The formula simplifies to:

$$A = 6\ell^2$$

Note that, in the equation for the cone, we refer to the *slant* height (ℓ), not the vertical height. If you are given the vertical height, use trig find the slant height.

Example

Find the surface area of a right cone, radius 10 and *vertical* height 25. Figure 18.8 is a vertical section through the cone. The dashed line is the axis of the cone from the apex A to the centre of the base D. $\triangle$ACD is right-angled so, by Pythagoras:

$$AC^2 = AD^2 + DC^2$$
$$= 25^2 + 10^2 = 725$$

$$\text{Slant height, } \ell = AC = \sqrt{725} = 26.93$$

$$A = \pi r(\ell + r) = 3.1416 \times 10\,(26.93 + 10)$$
$$= 1160.2$$

$$\underline{\text{Total area of cone} = 1160 \text{ (3 sf).}}$$

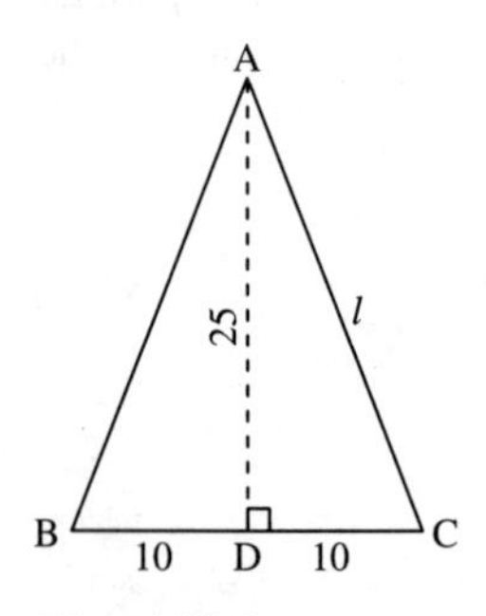

Figure 18.8

The formula for surface areas apply only to right prisms, cylinders and cones, not to oblique solids.

Test yourself 18.1

1 Find the areas of these figures, expressing the results in standard form to 3 sf.

a Triangle with base 13, vertical height 14

b Circle, radius 30

c Hexagon, sides 10

d Sector, angle 250°, radius 7

e Parallelogram, length 24, width 17

f Quadrant, radius 39

g Segment of a circle, radius 5, angle 50°

2 Find the volumes of these solids, to 4 sf.

a Rectangular pyramid with base length, 23, base width 12, vertical height 31

b Cone, base radius 5, vertical height 52

c The planet Mars, taking it as a sphere with average radius 3385 km. Express the result in standard form

d Cylinder, radius 72, height 3

3 Find the surface areas of these right solids, rounding the results to 1 dp.

a Cylinder, radius 5, height 15

b Sphere, radius 6

c A rectangle of paper rolled to make an open-ended cylinder, radius 12, height 7

d Hemisphere, radius 11

e Cone, base radius 4, slant height 17

f Cone, base radius 5, vertical height 4

Changing the subject

A bar of modelling clay measures 60 mm × 40 mm × 15 mm. It is rolled to make a sphere. Find the diameter of the sphere (3 sf).

The volume of clay in the bar is $60 \times 40 \times 15 = 36\,000\,\text{mm}^3$. To find the radius of a sphere of this volume we need to change the subject of the formula for the volume of a sphere:

$$V = \frac{4\pi r^3}{3}$$

$$\Rightarrow \quad 3V = 4\pi r^3$$

$$\Rightarrow \quad \frac{3V}{4\pi} = r^3$$

$$\Rightarrow \quad r^3 = \frac{3V}{4\pi}$$

$$\Rightarrow \quad r = \sqrt[3]{\frac{3V}{4\pi}}$$

We have changed the subject of the formula from V to r. Substituting the value, $V = 36\,000$, in the new formula:

$$r = \sqrt[3]{\frac{3 \times 36\,000}{4\pi}} = \sqrt[3]{\frac{27\,000}{\pi}}$$

$$= \sqrt[3]{8594.4} = 20.48$$

The question asks for the diameter:

Diameter of sphere = 41.0 mm.

Explore this

Use a spreadsheet program to produce a data sheet for rectangular prisms. Here is a suggested layout, though you may think of things to add to it and other ways to improve it.

```
   A....... B....... C....... D.......
 1 RECTANGULAR PRISM DATA SHEET
 2
 3 Length =
 4  Width =
 5 Height =
 6
 7 Area of side          =
 8 Area of end           =
 9 Area of top/bottom    =
10           Total area  =
11 Volume =
```

The formulae are:

```
D7=B3*B5
D8=B4*B5
D9=B3*B4
D10=D7#D9*2
B11=B3*B4*B5
```

Key in the width, length and height (all in the same units). Update the spreadsheet to find the volume and the various surface areas. Here is an example of the spreadsheet in use:

```
RECTANGULAR PRISM DATA SHEET

Length =      7.20
 Width =      3.10
Height =      4.90

Area of side       =          35.28
Area of end        =          15.19
Area of top/bottom =          22.32
        Total area =         145.58
Volume =    109.36
```

The remainder of this chapter consists of problems about areas and volumes. Results are given to 3 sf unless otherwise stated. Results greater than 1000 are expressed in standard form. There are questions for you to solve on page 269.

Signal cone

A signal cone is made from sheet metal to mount on top of a marker buoy. Figure 18.9 is a vertical section of the cone. Ignoring overlaps, what area of sheet metal is required?

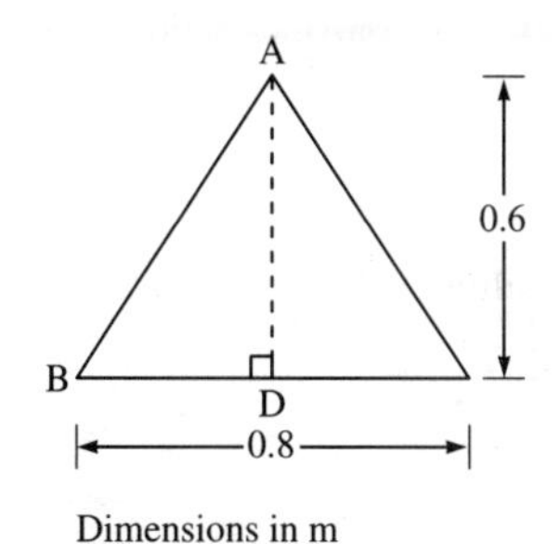

Figure 18.9

The area of metal is the surface area of the cone. We need to know the slant height. By Pythagoras, in the right-angled triangle ABD, AB is the slant height:

$$AB^2 = AD^2 + BD^2 = 0.6^2 + 0.4^2 = 0.52$$

$$\Rightarrow \quad AB = \sqrt{0.52} = 0.721$$

$$\text{Area of cone} = \pi r\,(\ell + r) = \pi \times 0.4\,(0.721 + 0.4)$$

$$= 1.4087$$

$$\underline{\text{Area of sheet metal} = 1.41\ \text{m}^2.}$$

Drainpipes

A plastic drain-pipe is 80 mm diameter externally, with walls 3 mm thick. What volume of plastic is needed to mould a pipe 1.5 m long?

The volume of plastic is the difference between the external and internal volumes of the pipe. The external diameter is given as 80 mm, so the external radius is 40 mm. The internal radius is one wall-thickness less than this:

Internal radius = 40 – 3 = 37

In millimetres, the length of pipe, h = 1500

$$\text{External volume} = \pi r^2 h = \pi \times 40^2 \times 1500 = 7\,539\,822$$

$$\text{Internal volume} = \pi \times 37^2 \times 1500 = 6\,451\,261$$

$$\text{Volume of plastic} = 7\,539\,822 - 6\,451\,261 = 1\,088\,561$$

Volume of plastic = $1.09 \times 10^6\,\text{mm}^3$ (3 sf).

Ice cream cones

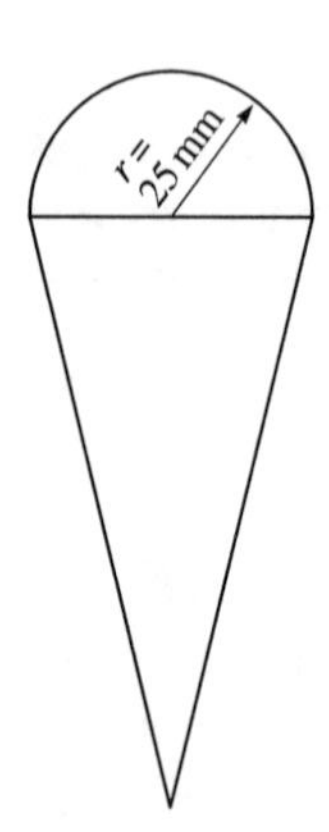

Figure 18.10

Write a formula for the volume of ice cream in Figure 18.10. The internal radius of the cone is r, and this is also the radius of the hemisphere of ice-cream on top of the cone. The internal radius and length of the cone are in the ratio 1:4. Assume that the cone tapers to a point at the bottom and that it is completely filled. Find the volume of ice cream if r = 25 mm.

$$\text{The volume of the hemisphere is } \tfrac{1}{2} \times \frac{4\pi r^3}{3} = \frac{2\pi r^3}{3}$$

The area of the mouth of the cone is $A = \pi r^2$

The vertical height h of the cone is 4 times the radius = $4r$

$$\text{The volume of the cone is } \frac{Ah}{3} = \frac{\pi r^2 \times 4r}{3} = \frac{4\pi r^3}{3}$$

The total volume V of ice cream is the sum of these two volumes:

$$V = \frac{2\pi r^3}{3} + \frac{4\pi r^3}{3}$$

Put the HCF of these two terms in front of the brackets:

$$= \frac{2\pi r^3}{3}(1 + 2) = \frac{2\pi r^3}{3} \times 3$$

$$\Rightarrow \quad V = 2\pi r^3$$

If r = 25 mm:

$$V = 2\pi 25^3 = 98\,175$$

When r = 25 mm, $V = 9.82 \times 10^4\,\text{mm}^3$

Fish tank

A rectangular fish tank with vertical sides measures 0.6 m long and 0.3 m wide. It is about half filled with water. Then some rocks are placed in the tank, and are completely submerged in the water. The water level in the tank rises 80 mm. Find the total volume of the rocks.

Let the original depth of water be d m. If ℓ and w are the length and width of the tank, the volume of water is:

$$\ell \times w \times d$$

If all dimensions are in mm, the volume is in mm^3. When the rocks are submerged the total volume of water *and* rocks is:

$$\ell \times w \times (d + 80)$$

The volume of the rocks, V, is the difference of these:

$$V = \ell wd - \ell w(d + 80)$$

Bringing out common factors ℓ and w and placing them before the brackets:

$$V = \ell w(d - d + 80) = \ell w \times 80$$

With the dimensions given, working in mm:

$$V = 600 \times 300 \times 80 = 1.44 \times 10^7$$

The volume of rocks is 1.44×10^7 mm^3.

Tennis ball packs

Tennis balls, diameter 62 mm, are packed in sixes in cylindrical card tubes (Figure 18.11). Find the volume of the six balls, and the internal volume of the tube. Find the volume of unfilled space in the tube and express this as a percentage of the volume of the tube.

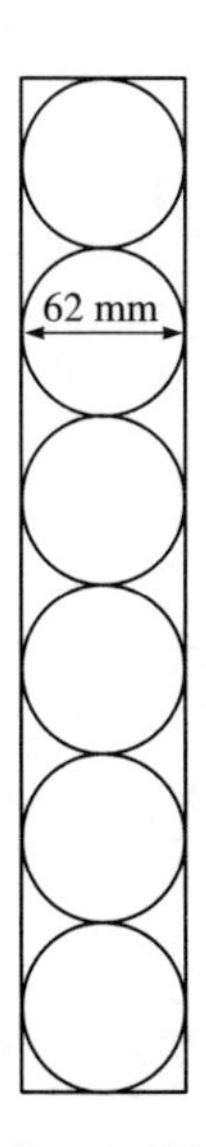

Figure 18.11

Radius of the balls and tube is half the diameter: $r = 31$ mm.

Volume of 1 ball is $\dfrac{4\pi r^3}{3} = \dfrac{4\pi 31^3}{3} = 124\,788$

Volume of 6 balls is $6 \times 124\,788 = 748\,728$
Height of tube h is 6 times the diameter of a ball $= 12r = 372$

Volume of tube is $\pi r^2 h = \pi \times 31^2 \times 372 = 1\,123\,094$

Volume of unfilled space is $1\,123\,094 - 748\,728 = 374\,366$

Space as a percentage of the volume of the tube:

$$\frac{374\,366}{1\,123\,094} \times 100 = 33.33$$

Volume of unfilled space is 33.3% of tube (3 sf).

Lamingtons

Lamington cakes traditionally have the shape of a cone with the apex cut off. Technically, we describe this as a **frustum**. A batch of Lamingtons has the dimensions shown in the vertical section of Figure 18.12a. The top and sloping surface are coated with jam, 2 mm thick. What volume of jam is needed for each cake? How many teaspoonfuls is this?

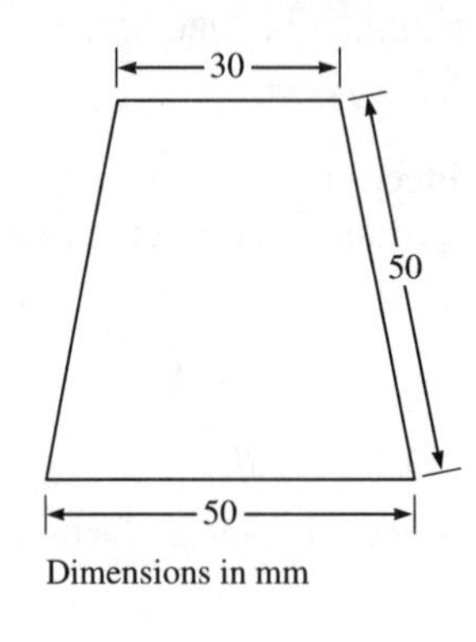

a

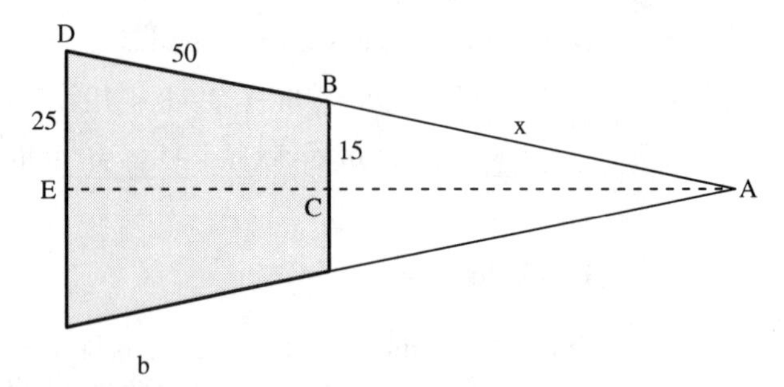

b

Figure 18.12

The frustum is considered to be the difference between two cones. In Figure 18.12b we see a section through these. Triangles ABC and ADE have the same angles, so they are similar triangles. This means that the ratio between the sides is the same in both triangles. If AB = x:

$$\frac{x}{15} = \frac{x + 50}{25}$$

$$\Rightarrow \quad \frac{25x}{15} = x + 50$$

$$\Rightarrow \quad 25x = 15\,(x + 50)$$

$$\Rightarrow \quad 25x = 15x + 750$$

$$\Rightarrow \quad 10x = 750$$

$$\Rightarrow \quad x = 75$$

Volumes of fluids

Units for measuring the volumes of fluids (liquids and gases) are:

Litre: the volume of a cube measuring 100 mm × 100 mm × 100 mm. Symbol l, but often printed ℓ to avoid confusion with I and 1.

Millilitre: a thousandth of a litre. Symbol ml. 1 ml = 1000 mm^3.

Cubic centimetre: The volume of a cube measuring 1 cm × 1 cm × 1 cm. Symbol cm^3. 1 cm^3 = 1 ml = 1000 mm^3.

Litres are used for measuring relatively large volumes, such as petrol, milk, beer, and the cylinder capacities of powerful petrol engines.
Millilitres and cubic centimetres are used for measuring small volumes, including medicines and the cylinder capacities of small petrol engines. A handy household measure is the teaspoon, which is standardized to 5 cm^3 = 5 ml = 5000 mm^3.

The frustum is the difference between a larger cone, slant height 125 mm, base radius 25 mm, and a smaller cone, slant height 75 mm, base radius 15 mm.

Curved surface of the larger cone $= \pi r\ell = \pi \times 25 \times 125$
$= 9817.5$

Curved surface of the smaller cone $= \pi \times 15 \times 75$
$= 3534.3$

Difference between the cones = curved surface area of Lamingtons
$= 9817.5 - 3534.3 = 6283.2$

Area of top surface of Lamington $= \pi \times 15^2 = 706.9$
Total area $= 6283.2 + 706.9 = 6990.1$

The jam coating can be thought of as having the same area, and is 2 mm thick. Its volume is:

$$6990.1 \times 2 = 13\,980.2\text{ mm}^3$$

Given that a standard teaspoonful is 5000 mm^3, the number of teaspoonfuls is:

$$\frac{13\,980.2}{5000} = 2.796$$

Each cake needs 1.40×10^4 mm^3 jam, approximately 3 teaspoonfuls.

Making a funnel

A funnel is to be made from sheet metal, with the dimensions shown in the vertical section of Figure 18.13a. Draw a diagram to show the shape and dimensions of the two pieces of metal required, ignoring overlaps. What is the area of each piece?

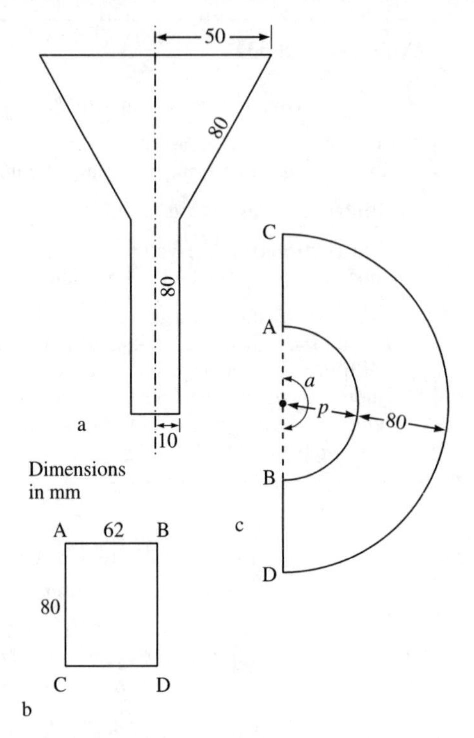

Figure 18.13

The stem of the funnel is a cylinder, radius 10 mm. It is made from a rectangle, length 80 mm and width equal to the circumference of a circle 10 mm radius.

Width $= 2\pi \times 10 = 62.832$

Area $=$ width $\times$ length $= 62.832 \times 80 = 5027$

Area of the stem piece $= 5.027 \times 10^3\,\text{mm}^2$

The rectangle is drawn in Figure 18.13b; it is rolled so that A meets B and C meets D.

The bell of the funnel is made from a piece of metal shaped as in Figure 18.13c. The circumference of the narrow end of the bell is:

$2\pi r = 20\pi$

This is to be formed when A and B of the sector are brought together. In other words, the arc AB = 20π.

But the arc is also calculated from angle a and the radius p:

$$\frac{a}{360} \times 2\pi p = 20\pi$$

$$\Rightarrow \qquad ap = 3600 \qquad (1)$$

Similarly, for the wide end of the funnel and the arc CD:

$$\frac{a}{360} \times 2\pi(p + 80) = 100\pi$$

$$\Rightarrow \quad a(p + 80) = 18\,000$$

$$\Rightarrow \quad ap + 80a = 18\,000$$

Substituting the value of ap from Equation 1:

$$80a = 18\,000 - 3600 = 14\,400$$

$$a = 180$$

The angle is 180°. Substituting $a = 180$ in equation 1:

$$180p = 3600$$

$$\Rightarrow \quad p = 20$$

In Figure 18.13c, $a = 180°$, $p = 20$ mm.

The bell is made from a semicircle radius 100 from which a semicircle radius 20 is cut away. The area of a semicircle radius 100 is:

$\frac{1}{2} \times \pi \times 100^2 = 15\,708$

The area of a semicircle radius 20 is:

$\frac{1}{2} \times \pi \times 20^2 = 628$

The area of the bell is $15\,708 - 628 = 15\,080$

Area of bell piece is 1.51×10^4 mm^2.

Cans of paint

Paint is sold in cans of two sizes:

Large: 85 mm diameter, 88 mm high
Small: 60 mm diameter, 71 mm high

It takes 4 litres to paint a machine. Draw a graph to show the different combinations of large cans and small cans that will provide enough paint for the job, with no more than 10% left over.

Volume of large can $= \pi \times 42.5^2 \times 88 = 499\,356$ mm^3
Volume of small can $= \pi \times 30^2 \times 71 = 200\,748$ mm^3

It is more convenient to work in litres:

$499\,356$ mm^3 $= 499$ ml ≈ 0.5 litres
$200\,748$ mm^3 $= 201$ ml ≈ 0.2 litres

If we have x large cans and y small cans, the least number of cans required for 4 litres is given by the inequality:

$$0.5x + 0.2y \geqslant 4$$

$$\Rightarrow \quad 0.2y \geqslant 4 - 0.5x$$

$$\Rightarrow \quad y \geqslant \frac{4 - 0.5x}{0.2}$$

$$\Rightarrow \quad y \geqslant 20 - 2.5x$$

Up to 10% may be left over, but no more. The maximum amount of paint is 110% of 4 litres, which is 4.4 litres. The greatest number of cans for 4.4 litres is given by the equality:

$$0.5x + 0.2y \geqslant 4.4$$
$$\Rightarrow \quad y \geqslant 22 - 2.5x$$

Figure 18.14 is the graph of the inequalities. The line for $y = 20 - 2.5x$ is plotted by using the two-intercepts method:

Making $x = 0$ gives $y = 20 - 0 = 20$
Making $y = 0$ gives $0 = 20 - 2.5x$
$\Rightarrow$ $y = 0$ gives $x = 20/2.5 = 8$

Points (0,20) and (8,0) are plotted and joined with a straight line. The line for $y = 22 - 2.5x$ is plotted in a similar way.

Because only whole numbers of cans can be bought, the permissible numbers of cans are indicated by the points where the grid lines cross in the unshaded zone. For example, we can have:

20, 21 or 22 small cans and no large cans
18 or 19 small cans and 1 large can
. . . and so on . . .
0, 1, or 2 small can and 8 large cans.

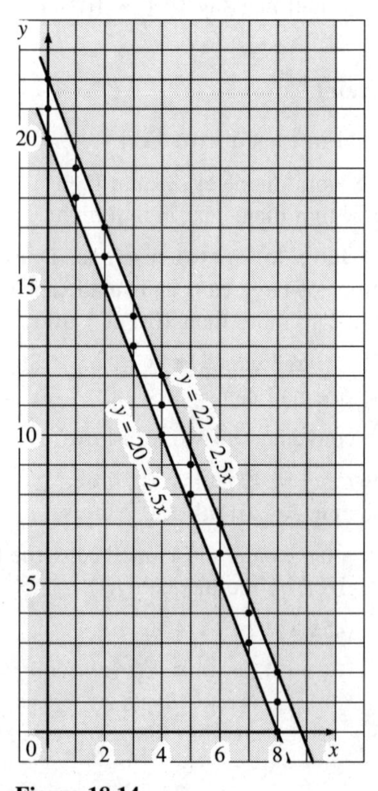

Figure 18.14

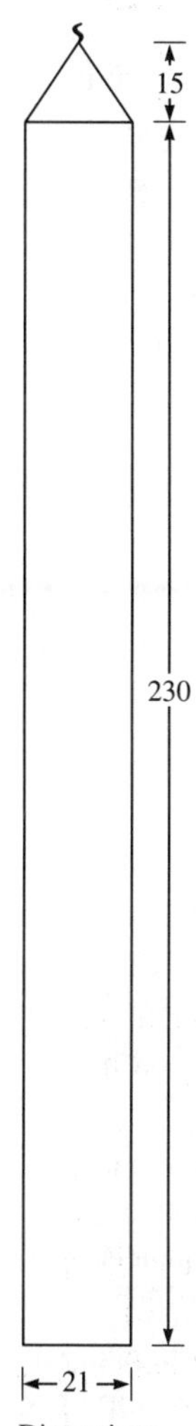

Figure 18.16

Test yourself 18.2

Express answers to 3 sf. Express answers in standard form if they are greater than 1000.

1 Find the volume of the objects shown in Figure 18.15.

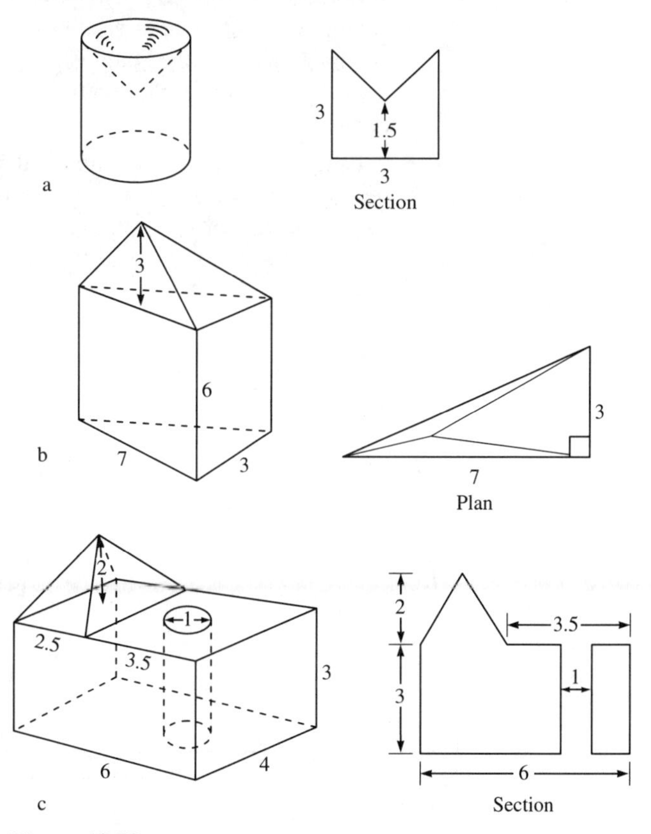

Figure 18.15

2 The base of a rectangular milk carton is 90 mm long and 65 mm wide. If the carton holds 500 ml of milk, find the depth of the milk.

3 The cylinder of a petrol engine is 75 mm in diameter, and takes in 400 cc of fuel/air mixture at each stroke. How far does the piston travel?

4 The candle shown in Figure 18.16 is melted down and cast in a mould shaped as a square pyramid with vertical height equal to twice the length of one side of the base. Find the side of the base and the vertical height of the pyramid candle.

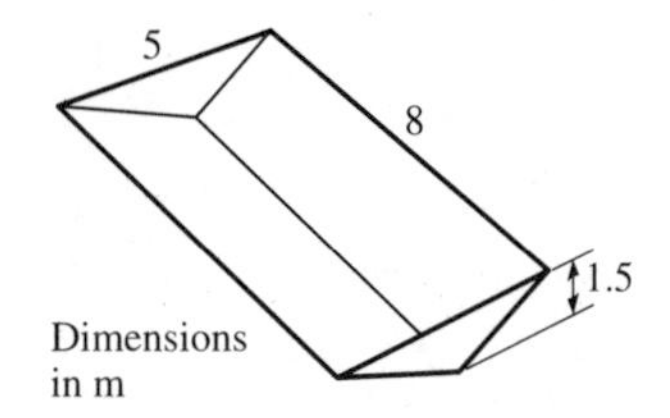

Figure 18.17

5 A water trough as sloping sides making it triangular in section. (Figure 18.17). Its ends are vertical. Find the volume of the trough. Find how deep the water is when the trough holds half its maximum volume.

6 A regular tetrahedron is a triangular pyramid with all sides equal; all four faces are equilateral triangles. If a side is 50 mm long, find the vertical height, the surface area and volume of the tetrahedron.

7 A flower-pot of circular section is 130 mm in diameter at the top, 90 mm diameter at the bottom and its vertical height is 115 mm. It is three-quarters filled with potting compost. Find the volume of the compost.

8 A lampshade is designed with the dimensions shown in the vertical section of Figure 18.18. Its top and bottom are open. It is made by rolling a single piece of plastic into a cone. Ignoring overlaps, draw a pattern for the plastic and find its area.

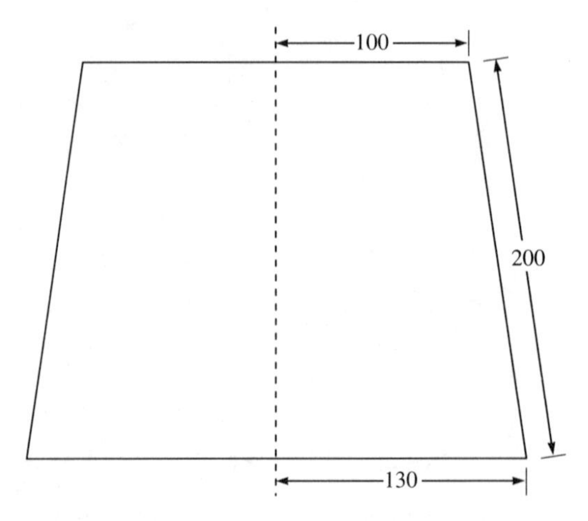

Figure 18.18

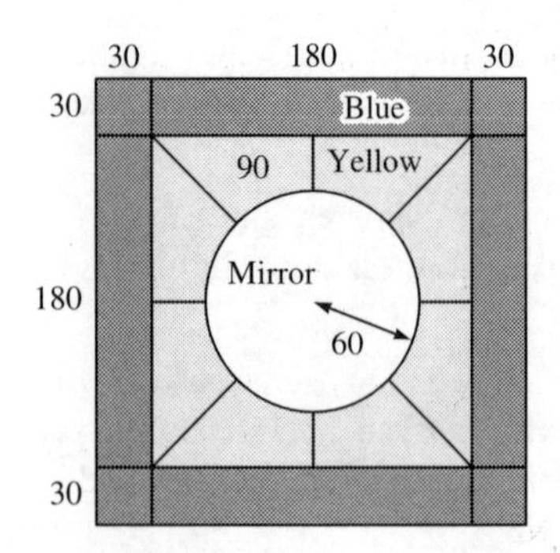

Figure 18.19

9 An unsharpened hexagonal pencil, length 180 mm, sides 4 mm, has a 'lead' 2.5 mm in diameter. Express the volume of the lead as a percentage of the volume of the pencil.

10 A circular Tiffany mirror is surrounded by panels of yellow glass, and bordered with blue glass, as in Figure 18.19. Find the area of the mirror, a panel of yellow glass, and the long and square panels of blue glass. Find the ratio of the total areas, mirror:yellow:blue.

11 A lozenge is moulded in plastic to decorate an exhibition of heraldry (Figure 18.20). The base is a prism shaped as a rhombus, on top of which is a rhombus-shaped pyramid. Find the volume of plastic needed for one lozenge.

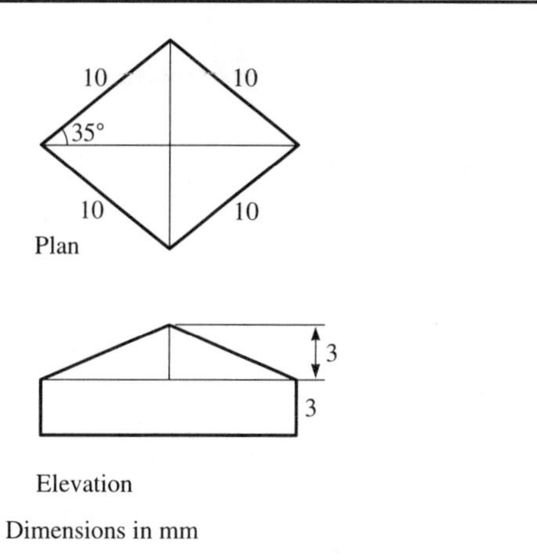

Figure 18.20

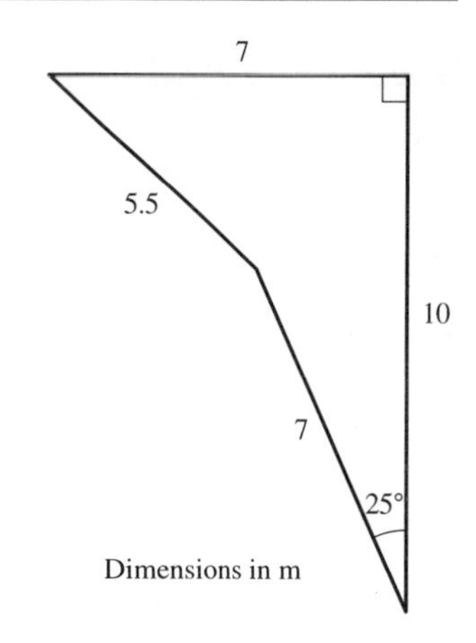

Figure 18.21

12 A garden bed, shaped as in Figure 18.21, is surrounded by a wall 0.5 m high to make a raised bed. Find the volume of soil needed to raise the bed to be level with the top of the wall.

13 On the same pair of axes, plot graphs to show how

a the surface **b** the volume

of a cube varies with the length x of the side, for $x = 0$ to $x = 7$. Find x when the surface area and the volume have the same value. [*Hint*: Use Rectangular Prism Data sheet, page 260]

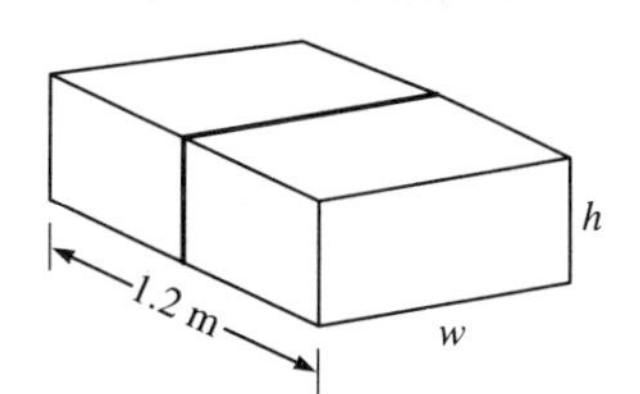

Figure 18.22

14 Goods are packed in cartons 1.2 m long (Figure 18.22), with a range of widths and heights. The cartons are secured by straps just over 6 m long, which limit the total of width and height to 3 m. The minimum width is 0.5 m. The minimum height is 0.3 m. The width of a carton must be equal to or greater than the height. Express these conditions in the form of four inequalities. Plot a graph of the equalities, shading the disallowed areas. The unshaded area shows which combinations of width and height are permitted. From the graph find:

a the maximum width **b** the maximum height

Calculate the corresponding volume of a carton for **a** and **b**.

15 Work the exercises in *Try these first*, page 253.

Part 3 – Maths Topics

This part contains a selection of topics from which you can choose those you need for your branch of technology. It is assumed that you are completely familiar with the Maths Essentials of Part 1, but we shall tell you which Chapters of Part 2 you need to know for each topic.

19 Exploring trig ratios

In this chapter we look into trig ratios in more detail, including the ratios for angles greater than 90°. We also learn more ways of finding the sides and angles of triangles.

You need to know the basic facts about sines, cosines and tangents, as presented in Chapter 14, and also the standard triangles (page 214).

Try these first

1 Use a calculator to find these trig ratios (4 dp).

a sin 100° **b** cos 357°

c tan 145° **d** cos 2 (angle in rad)

2 Use a calculator to find the angles which have these trig ratios, in degrees (2 dp).

a $\tan^{-1} 2.5$ **b** $\sin^{-1} -0.4$

3 Express in radians.

a 90° **b** 234°

4 Express in degrees.

a 1.5 rad **b** $3\pi/2$

5 In a triangle ABC, AB = 4.5, $\hat{A} = 55°$, $\hat{B} = 34°$. Find AC.

6 In a triangle PQR, PQ = 5.6, QR = 8.1, RP = 6.4. Find $\hat{R}$.

7 In a triangle XYZ, XZ = 12, XY = 10, $\hat{X} = 12°$. Find YZ, and the area of the triangle.

Bigger angles

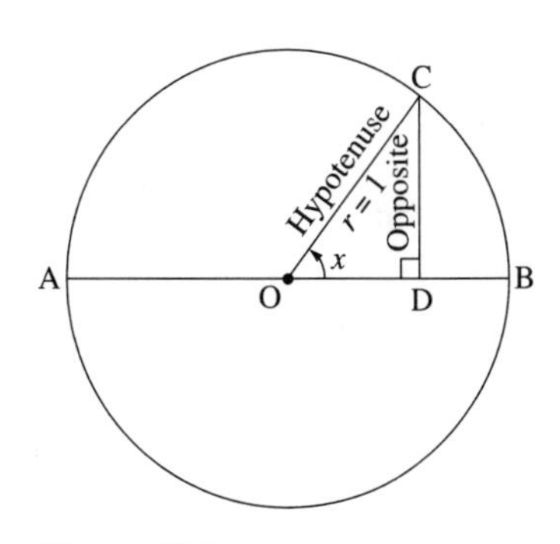

Figure 19.1

In Chapter 14, we saw that the sine of an angle increases from 0 to 1 as the angle increases from 0° to 90°. Figure 19.1 shows a slightly different way of looking at sines. The circle has unit radius (1 mm, 1 m, 1 km) and AB is a diameter. OC is a radius and we drop a line from C to cut AB at D. The angle $C\hat{D}O$ is a right-angle.

In the triangle OCD, OC is the hypotenuse, length 1, and CD is the side opposite the angle $C\hat{O}D$, size $x°$

$$\sin x = \frac{\text{opposite}}{\text{hypotenuse}} = \frac{\text{CD}}{1} = \text{CD}$$

Because the circle has unit radius, the sine of x equals the length of CD. When $x = 0°$, OC lies along OB and CD = 0. This agrees with what we already know, that sin 0° = 0. As OC begins to swing anti-clockwise around the circle, x

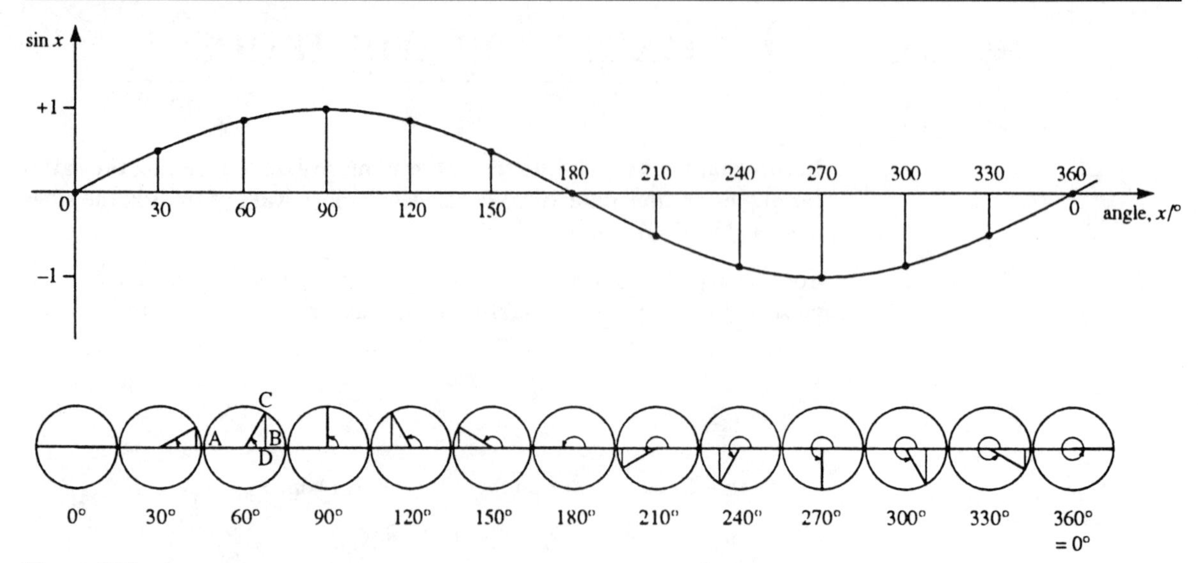

Figure 19.2

increases. So does CD. When $x = 90°$, CD lies along CO. CD = CO = 1. This also agrees with what we already know, that sin 90° = 1.

The lower part of Figure 19.2 shows what happens as x increases from 0° to 90° and beyond. From 0° to 90° the radius swings from the horizontal position to the vertical position, as we have already described. The sine of x (= CD) increases from 0 to 1. In the upper part of Figure 19.2, the sine is plotted against angle. The vertical lines are drawn twice the length of CD to make the curve more pronounced.

Between 90° and 180° the length of CD, and consequently sin x decreases from 1 to 0. Beyond 180° the radius is in the lower half of the circle. CD is negative. This makes sin x negative. The value of sin x decreases from 0 at 180° to −1 at 270°, then returns to 0 at 360°. At 360° we have completed a whole turn, and are back to the starting point:

$$\sin 360° = \sin 0° = 0.$$

The maximum value of sin x is 1, and its minimum value is −1.

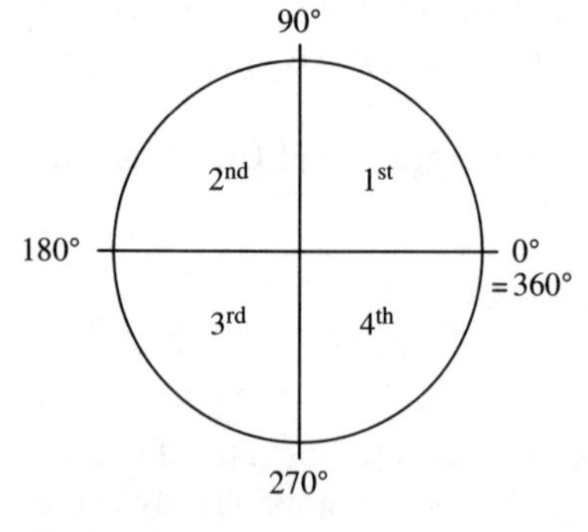

Figure 19.3

Quadrants

Angles are usually measured from a line directed to the right, and increase in an anticlockwise direction. The angular scale is divided into four **quadrants** (Figure 19.3):

Quadrant	*Degrees*	
	From	*To*
1st	0	90
2nd	90	180
3rd	180	270
4th	270	360

Each half of the sine curve is symmetrical. Considering the curve between 0° and 180°, the rise from 0° to 90° is symmetrical with the fall from 90° to 180°. In the lower part of Figure 19.2, the diagram for 30° is the mirror image of that for 150°. This means that sin 30° = sin 150°. Similarly, sin 60° = sin 120° and sin 0° = sin 180°. In general:

$$\sin x \equiv \sin(180° - x) \qquad 0° \leqslant x \leqslant 180°$$

This fact is important when finding inverse sines with a calculator. If you find $\sin^{-1} 0.64$, for example, the calculator displays 39.79 (rounding to 2 dp). The sine of 39.69° is 0.64. But, as can be seen from Figure 19.4 there is an angle in the second quadrant which also has 0.64 as its sine. The calculator gives us the angle in the first quadrant. Subtract this from 180° to obtain the angle in the second quadrant.

$\sin^{-1} 0.64 = 39.69°$ or $140.21°$

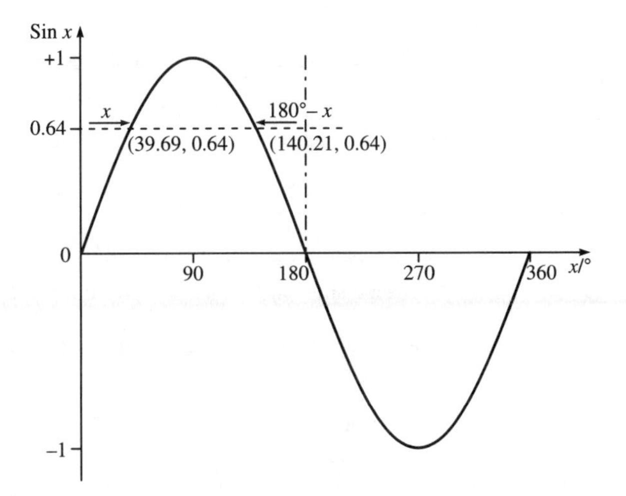

Figure 19.4

In most problems it is easy to know which angle is the right one. Drawing a sketch of the problem often helps to sort it out.

The same applies to the sines of angles in the third and fourth quadrants. To obtain a symmetrical curve we have to extend the curve to 540° (Figure 19.5). Then:

$$\sin x \equiv \sin(540° - x) \qquad 180° \leqslant x \leqslant 360°$$

Figure 19.2 also shows that the curve from 180° to 360° is a repeat of the curve from 0° to 180° but upside down. In other words:

$$\sin x \equiv -\sin(x \pm 180°)$$

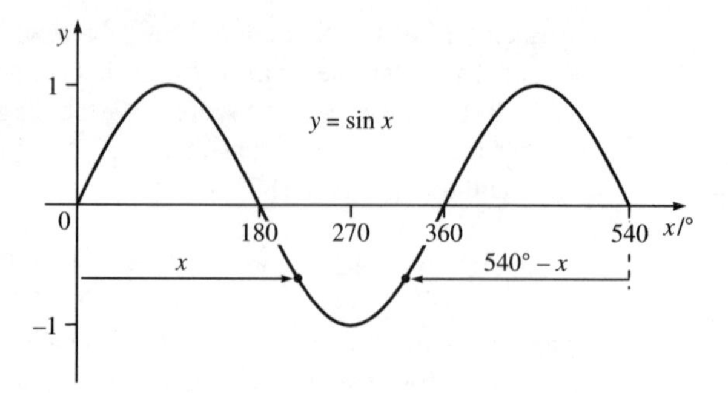

Figure 19.5

Using the relationships above and the formulae on page 215, we can find sines of standard angles in the second to fourth quadrants, without using a calculator.

Examples

$$\sin 135° = \sin(180 - 135°) = \sin 45° = 1/\sqrt{2}$$

$$\sin 240° = -\sin(240 - 180°) = -\sin 60° = -\sqrt{3/2}$$

$$\sin 330° = -\sin(330° - 180°) = -\sin 150° = -\sin(180 - 150°)$$
$$= -\sin 30° = -0.5$$

Inverse sines in the 3rd and 4th quadrants

Some calculators return a *negative* value when the angle is in the 3rd or 4th quadrants.

Example

$$\sin^{-1} - 0.6 = -36.87°$$

A **negative** angle, such as –36.87°, is measured from zero in a **clockwise** direction.

To convert this to a positive angle, add 360° to it:

$$-36.87° + 360° = 323.13°$$

The fourth-quadrant angle with a sine of –0.6 is 323.13°.

Next, subtract 540°:

$$323.13° - 540° = -216.87°$$

Ignore the negative sign; read the display as a positive number:

The third-quadrant angle with a sine of –0.6 is 216.87°.

All of this can be quickly done as a continuous chain of operations on the calculator. This situation does not occur with cosines.

Test yourself, 19.1

1 Use a calculator (page 277) to find these sines (4 dp).

a 25° **b** 212°

c 312° **d** 167°

2 Use a calculator to find the angles which have these sines (2 dp).

a 0.6 **b** –0.75

c –0.27 **d** 0.005

3 Write the sines of these angles without using a calculator.

a 315° **b** 120°

c 270° **d** 225°

Explore these

Spreadsheet for sines

Set up a spreadsheet like this:

```
     A.......  B.......  C.......  D.......
  1 SINE GRAPH
  2
  3 Radius =    5.0000
  4
  5     Angle   Height      Sine
  6         0
  7        15
  8        30
  9        45
 10        60                        Mean =
 11        75
 12        90
 13       105
 14       120
 15       135
 16       150
 17       165
 18       180
 19       195
 20       210
 21       225
 22       240
 23       255
 24       270
 25       285
 26       300
 27       315
 28       330
 29       345
 30       360
```

Draw a circle with radius between 50 mm and 100 mm, as in Figure 19.1. Enter the length of the radius in cell B3.

For each angle from 0° to 360° in steps of 15°, draw OC and CD, then measure CD. Enter the lengths of CD in column B of the spreadsheet. When the spreadsheet is updated, it calculates and displays the sine of every angle. It also calculates the mean (or average) of the sines. What do you expect this to be?

Transfer the data of cells B6 to B30 and C6 to C30 to a graphic display program. Instruct it to display a line graph of the data. This is a sine curve.

Investigating sine graphs

Use a graphic calculator or software to display graphs of sines. Make sure that it is in 'degree' mode before you begin.

Set the range for x: 0 to 720. Set the range for y: –10 to +10. Now try these equations, all plotted on the same area without clearing the screen:

$y = \sin x$ $y = 2 \sin x$ $y = 4 \sin x$
$y = 8 \sin x$ $y = 2 + \sin x$ $y = -4 + 2 \sin x$

Set the range x: 0 to 1080. Set the range for y: –2 to +2. Clear the screen, then try variations such as these:

$y = \sin 2x$ $y = \sin 3x$ $y = \sin 4x$
$y = \sin (x/2)$

Cosines beyond 90°

Figure 19.6 is very similar to Figure 19.1, except that it is rotated 90° anticlockwise for a reason which will be explained in a moment. We find the cosine of x by measuring OD:

$$\cos x = \frac{\text{adjacent}}{\text{hypotenuse}} = \frac{\text{OD}}{1} = \text{OD}$$

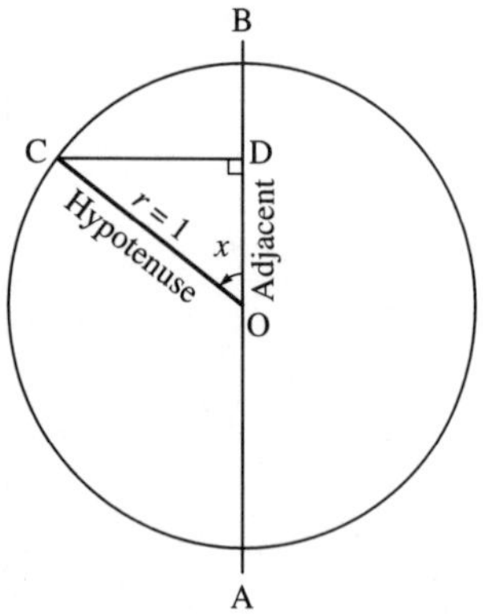

B
C
D
$r = 1$
Hypotenuse
x
Adjacent
O
A

Figure 19.6

In Figure 19.7 the lengths of OD are transferred to the graph (with doubling to make the curve more pronounced). Now we can see that Figure 19.6 was rotated so that the direction of OD is the same as the y-axis of the graph. The result is a curve which shows $\cos x = 1$ when $x = 0$, and falls to zero when $x = 90°$.

As the rotating radius moves into the second quadrant, OD becomes negative. Cosines are negative from 90° to 270°. They are positive again in the fourth quadrant.

The curve for $\cos x$ is symmetrical from 0° to 360° and so:

$$\mathbf{\cos x = \cos(360 - x)}$$

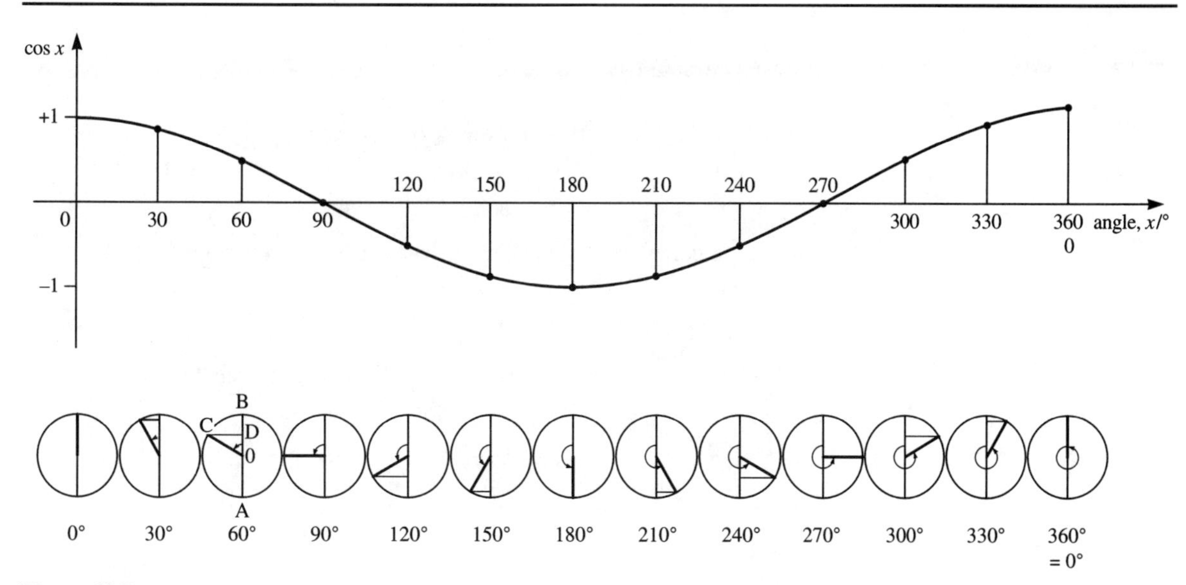

Figure 19.7

The values change from positive to negative or from negative to positive every 180°, as they did on the sine curve. This means that:

$$\cos x \equiv -\cos(x \pm 180°)$$

Sine-cosine relationships

Plotting the sine and cosine curves on the same axes gives Figure 19.8. Now it is clear that the two curves are alike, except that the cosine curve is shifted 90° to the left compared with the sine curve. This is not surprising; we rotated the circle 90° anti-clockwise when plotting Figure 19.7.

We express this relationship in these equations:

$$\sin x \equiv \cos(x - 90°)$$
$$\cos x \equiv \sin(x + 90°)$$

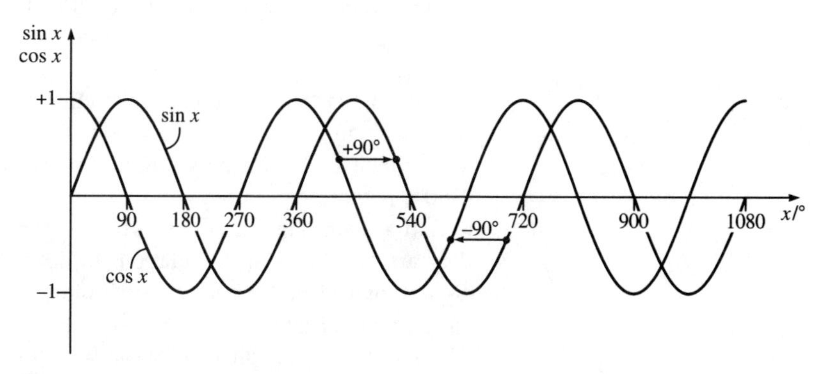

Figure 19.8

Test yourself 19.2

1 Use a calculator to find these cosines (4 dp).

a 275° **b** 181°

c 12° **d** 132°

2 Use a calculator to find the angles which have these cosines (2 dp).

a 0.7 **b** 0.3

c –0.88 **d** –0.12

3 Write the cosines of these angles without using a calculator.

a 135° **b** 330°

c 150° **d** 240°

Explore these

Spreadsheet for cosines

Follow the suggestions on page 279, but adapt them to making a spreadsheet for cosines and plotting the cosine graph.

Investigating cosine graphs

Use a graphic calculator or software to plot cosine graphs just as you did with sine graphs on page 280. Also plot sine and cosine graphs on the screen at the same time.

Units for angles

The most commonly used unit for practical measurement of angles is the degree. 360 degrees make one complete revolution. The other important unit is the **radian** (symbol, rad). If we take an arc AB of a circle (Figure 19.9) equal in length to the radius of the circle, the angle between radii OA and OB is 1 rad.

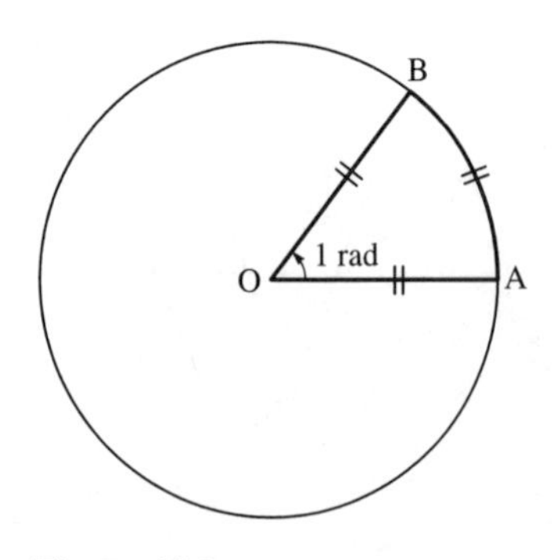

Figure 19.9

If the arc is twice as long as the radius, the angle is 2 rad. If the arc is three times as long as the radius, the angle is 3 rad. The circumference of a circle is $2\pi r$, which is 2π times as long as the radius. The angle of a whole circle (360°) is therefore 2π rad.

Radians are rarely used when finding the sides and angles of triangles, but they are often used in calculations to do with rotary motion, and with audio signals, as we shall see later. Using radians often helps to make the formulae simpler. One example we can quote now is the area of the sector of a circle. Think of this as the area swept out by a lever on a rotating shaft, or the area of windscreen swept clean by a windscreen wiper blade.

Converting angles

Degrees to radians

Multiply by π, then divide by 180.

Example

$78° \times \pi/180 = 1.361$ rad (4 sf)

Radians to degrees:

Multiply by 180, then divide by π.

Example

$2.4\,\text{rad} \times 180/\pi = 137.5°$ (4 sf)

The usual formula (page 254) for A, the area of a sector, angle a, radius r is

$$A = \frac{a}{360} \times \pi r^2$$

The fraction $a/360$ is in the formula because the sector is a fraction of the total area of the circle. The fraction is:

$$\frac{\text{number of degrees in sector}}{\text{number of degrees in whole circle}}$$

If we put the angles in radians instead of degrees, the fraction becomes:

$$\frac{\text{number of radians in sector}}{\text{number of radians in whole circle}} = \frac{a}{2\pi}$$

When we multiply the fraction by the area of the circle, the π's cancel out:

$$A = \frac{a}{2\pi} \times \pi r^2 = \frac{ar^2}{2}$$

This formula is easier to evaluate than the one based on degrees.

When a trig ratio is based on degrees, the degree symbol is usually written in, for example sin 5°. However, if there is no degree symbol, the angle should be taken to be in radians. For example, sin 5 means the sine of 5 rad, *not* the sine of 5°. When you are working with a calculator, it is essential to put it in the correct mode, degrees or radians, before you start the calculations.

Test yourself 19.3

1 Convert into radians (2 dp)

a 20° **b** 235°

c 270° **d** 343.7°

2 Convert into degrees (2 dp)

a 1 rad **b** 2.5 rad

c $\pi/3$ rad **d** $3\pi/5$ rad

3 Find these ratios for angles expressed in rad (4 dp)

a cos 3 **b** sin 2π

c tan 2 **d** cos π

The sine rule

In Chapters 14 and 16 we looked at methods for solving right-angled triangles. Sometimes we want to solve a triangle that is *not* right-angled. The sine rule may help in such cases.

The sine rule is proved in this way. Take any triangle, such as △ABC (Figure 19.10) and draw a line from apex C to meet the base AB at right angles. The line from C meets the base at D. CD divides △ABC into two right-angled triangles.

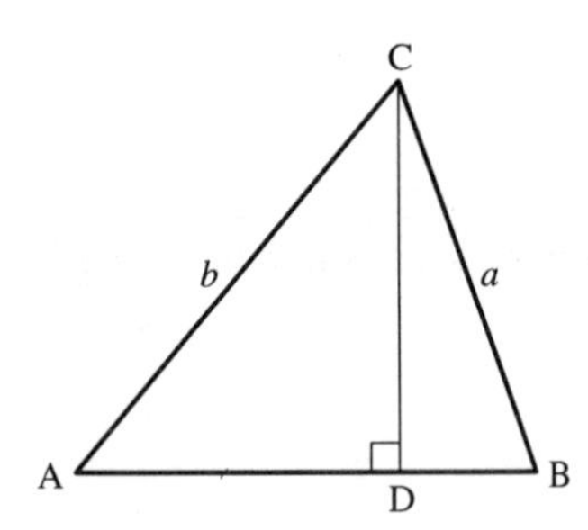

Figure 19.10

In △BDC, $\sin B = \dfrac{CD}{CB}$

Changing the subject: $CD = CB \sin B$ (1)

In △ADC, $\sin A = \dfrac{CD}{AC}$

Changing the subject: $CD = AC \sin A$ (2)

CD is on the left side of both equations (1) and (2) so their right sides must be equal:

$$CB \sin B = AC \sin A \qquad (3)$$

Instead of writing CB and AC to represent the sides, we call their lengths a and b. Side a is opposite angle $\hat{A}$ and side b is opposite angle $\hat{B}$. Rewriting equation (3):

$$a \sin B = b \sin A$$

Dividing by sin A:

$$\frac{a \sin B}{\sin A} = b$$

Dividing by sin B:

$$\frac{a}{\sin A} = \frac{b}{\sin B} \qquad (4)$$

This equation is saying that the ratio of side/sine is the same for both angles $\hat{A}$ and $\hat{B}$. We can also take any other pair of angles, such as $\hat{A}$ and $\hat{C}$, or $\hat{B}$ and $\hat{C}$ and prove the same thing:

$$\frac{a}{\sin A} = \frac{c}{\sin C} \qquad (5)$$

$$\frac{b}{\sin B} = \frac{c}{\sin C} \qquad (6)$$

Combining these equations (4)–(6) into one, we have the full sine rule:

$$\frac{a}{\sin A} = \frac{b}{\sin B} = \frac{c}{\sin C} \qquad (7)$$

Using the sine rule

Equation 7 is useful for remembering the rule, but we actually use the simpler rules (4)–(6) more often. Equation (4) has four quantities in it:

length a
length b
sin A
sin B

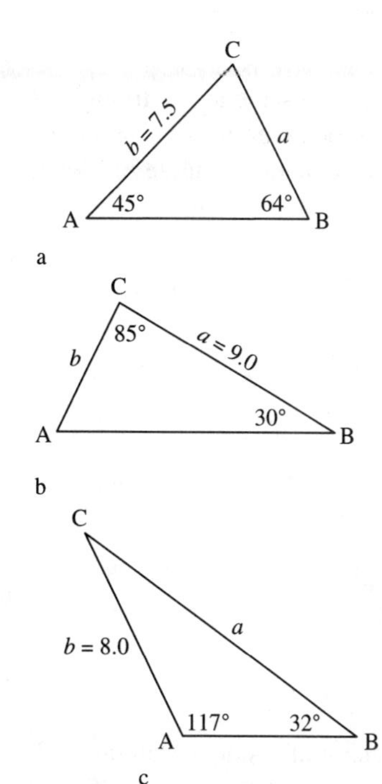

Figure 19.11

To find any *one* of these quantities we need to know the other three. For example, in the triangle of Figure 19.11a we know length b and the two angles $\hat{A}$ and $\hat{B}$, so we can calculate length a. The first thing to do is to set out what we *know* about the triangle:

$b = 7.5$
$\hat{A} = 45° \quad \Rightarrow \quad \sin A = 0.7071$
$\hat{B} = 64° \quad \Rightarrow \quad \sin B = 0.8988$

Enter these in equation (4):

$$\frac{a}{0.7071} = \frac{7.5}{0.8988}$$

$$\Rightarrow \qquad a = \frac{7.5 \times 0.7071}{0.8988} = 5.9004$$

Side CB = 5.9 (1 dp)

In Figure 19.11b we are given side a, but are not told the value of $\hat{A}$.

We are given the other two angles, so finding $\hat{A}$ is easy:

$$\hat{A} = 180° - 85° - 30° = 65°$$

List what we know:

$$\begin{aligned} a &= 9.0 \\ \hat{A} &= 65° \quad \Rightarrow \quad \sin A = 0.9063 \\ \hat{B} &= 30° \quad \Rightarrow \quad \sin B = 0.5000 \end{aligned}$$

Enter the known quantities in equation (4):

$$\frac{9.0}{0.9063} = \frac{b}{0.5}$$

$$\Rightarrow \quad b = \frac{9.0 \times 0.5}{0.9063} = 4.9652$$

$$\underline{\text{Side AC} = 5.0 \text{ (2 sf)}}$$

> **Have you noticed?**
>
> In any triangle:
>
> The *longest* side is opposite the *largest* angle.
> The *shortest* side is opposite the *smallest* angle.
>
> This is a useful fact for checking calculations.

Figures 19.11a and b are acute-angled triangles, the same as the triangle of Figure 19.10. A similar proof of the sine rule can be given for obtuse-angled triangles and the formulae are the same. The obtuse-angled triangle of Figure 19.11c is solved as follows.

List what we know:

$$\begin{aligned} b &= 8.0 \\ \hat{A} &= 117° \quad \Rightarrow \quad \sin A = 0.8910 \\ \hat{B} &= 32° \quad \Rightarrow \quad \sin B = 0.5299 \end{aligned}$$

Enter the known quantities in equation (4):

$$\frac{a}{0.8910} = \frac{8.0}{0.5299}$$

$$\Rightarrow \quad a = \frac{8.0 \times 0.8910}{0.5299} = 13.4516$$

$$\underline{\text{Side BC} = 13.5 \text{ (3 sf)}}$$

These examples show how to calculate the length of a side, given *one side and any two* angles. We have always used equation (4), but equations (5) and (6) can be used, given other sides and angles.

Sine rule on a calculator

In the examples in the text, we find the sine of the angle and write it down, to be used later. Once you are sure of the rule, you can do the whole calculation in one operation:

Example

Solve: $\dfrac{a}{\sin 46°} = \dfrac{3.5}{\sin 54°}$

Key-strokes	*Display*	*Notes*
[3] [.] [5]	3.5	
[×] [4] [6]	46.	
[sin]	0.7193398	sin 46
[/]	2.517689301	3.5 × sin 46
[5] [4]	54.	
[sin]	0.809016994	sin 54
[=]	3.112035122	= *a*

$a = 3.11$ (3 sf)

You may need to modify this routine to suit your own calculator. Some calculators have the sine rule as a built-in routine.

Finding an angle

Equations (4) and (6) can be solved if we know three of the four quantities. In the previous section, we used it with one side and two angles. Now we will use it with two sides and one angle. But there is a limitation. In Figure 19.12a,

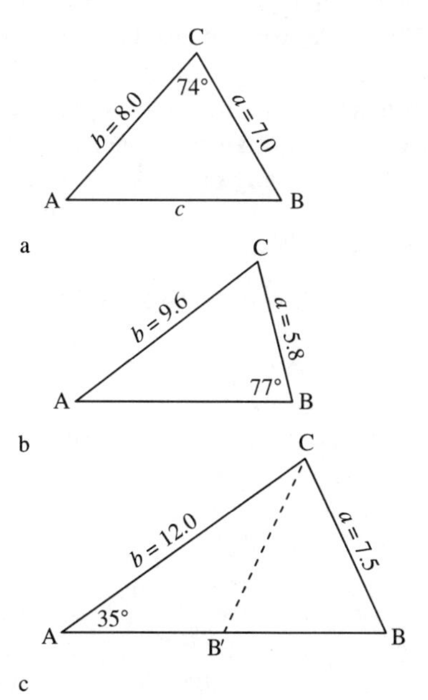

Figure 19.12

we know side a, side b, and angle C. But these three quantities provide only two items of information for each of equations (4) to (6). They provide three items for equation (7) but, for any part of it, we know only the side *or* the angle; we do not know both. We can never find the ratio and solve the triangle. The limitation is summarized by saying:

The sine rule cannot be used with two sides and the angle *between* them

In Figure 19.12b, we have two sides an an angle that is *not* between them. List what we know:

$$a = 5.8$$
$$b = 9.6$$
$$\hat{B} = 77° \quad \Rightarrow \quad \sin B = 0.9744$$

In this application of the sine rule we are finding an angle so we usually turn the equations upside down. Equation (4) becomes:

$$\frac{\sin A}{a} = \frac{\sin B}{b}$$

Substitute the known values:

$$\frac{\sin A}{5.8} = \frac{0.9744}{9.6}$$

$$\Rightarrow \quad \sin A = \frac{0.9744 \times 5.8}{9.6} = 0.5887$$

$$\Rightarrow \quad \hat{A} = \sin^{-1} 0.5887 = 36.1°$$

The calculator gives the result as 36.1° but there is another possible answer (page 277). Angle $\hat{X}$ might also be 180° – 36.1° = 143.9. It is easy to see that this is impossible. We know that B = 77° and, if $\hat{A}$ = 143.9, their total is 220.9°. If this were so, these two angles alone total more than 180°, which is more than the total of all three angles of any triangle.

We are left with only one possible result:

$$\underline{\text{Angle } \hat{A} = 36.1°}$$

In some triangles there may be two solutions. For example, in Figure 19.12c:

$$a = 7.5$$
$$b = 12.0$$
$$\hat{A} = 35° \quad \Rightarrow \quad \sin A = 0.5736$$

$$\frac{0.5736}{7.5} = \frac{\sin B}{12.0}$$

$$\Rightarrow \quad \sin B = \frac{0.5736 \times 12.0}{7.5} = 0.9178$$

$$\Rightarrow \quad \hat{B} = \sin^{-1} 0.9178 = 66.6° \text{ or } 113.4°$$

There are two possibilities:

If $\hat{B} = 66.6°$, $C = 180° - 66.6° - 35° = 78.4°$
The triangle is △ABC of Figure 19.12c.

If $\hat{B} = 113.4°$, $C = 180° - 113.4° - 35° = 31.6°$
The triangle is △AB'C of Figure 19.12c. The side B'C (dashed line) is equal in length to BC.

Without further information, either solution is possible.

Test yourself 19.4

Solve these triangles. Drawing a sketch (not a scale diagram) helps to make the problem clearer. Round answers to 2 dp.

1 In △ABC, $\hat{A} = 63°$, $\hat{B} = 52°$, $b = 9.5$. Find a.
2 In △ABC, $\hat{B} = 47°$, $\hat{C} = 72°$, $a = 5.2$. Find b.
3 In △PQR, $\hat{P} = 31°$, $\hat{Q} = 86°$, $q = 12.5$. Find p.
4 In △XYZ, $\hat{X} = 105°$, $\hat{Y} = 25°$, $x = 6.2$. Find y.
5 In △ABC, $\hat{A} = 84°$, $a = 8.2$, $b = 5.0$. Find $\hat{B}$.
6 In △LMN, $\hat{M} = 80°$, $l = 4.4$, $m = 12.0$. Find $\hat{L}$.
7 In △JKL, $\hat{J} = 42°$, $j = 5.4$, $k = 7.0$. Find $\hat{K}$.
8 In △RST, $\hat{S} = 86°$, $r = 11.7$, $s = 11.9$. Find $\hat{T}$.

Cosine rule

If you are given two sides and the angle between them (the **included** angle), the sine rule is no help. Instead, use the cosine rule. To prove the rule, we use a triangle the same as that in Figure 19.10, but with some of the sides differently labelled (Figure 19.13). Side AB has length c so, if we make AD equal to an unknown length x, then DB = $c - x$.

In the right-angled triangle ADC, Pythagoras' theorem tells us that:

$$b^2 = x^2 + y^2 \qquad (8)$$

Similarly, in the right-angled triangle BDC:

$$a^2 = y^2 + (c - x)^2$$

Multiplying out $(c - x)^2$:

$$a^2 = y^2 + c^2 - 2cx + x^2 \qquad (9)$$

The next step is to subtract equation (8) from equation (9). We subtract the left side of (8) from the left side of (9); we subtract the right side of (8) from the right side of (9). Because we are subtracting equal amounts from both sides of (9) the result is still a balanced equation:

$$a^2 - b^2 = y^2 + c^2 - 2cx + x^2 - x^2 - y^2$$

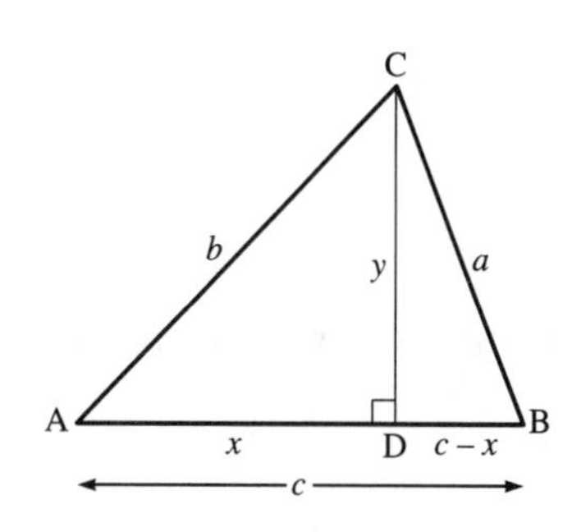

Figure 19.13

The x^2 and y^2 terms disappear by subtraction, leaving:

$$a^2 - b^2 = c^2 - 2cx$$

$$\Rightarrow \quad a^2 = b^2 + c^2 - 2cx \qquad (10)$$

From △ADC:

$$\cos A = \frac{x}{b}$$

$$\Rightarrow \quad x = b \cos A$$

Substitute this value of x in equation (10):

$$\boldsymbol{a^2 = b^2 + c^2 - 2bc \cos A} \qquad (11)$$

This the cosine rule. We could also prove it with a line drawn from B to meet AC, or from A to meet CB, giving two other versions of the rule:

$$\boldsymbol{b^2 = a^2 + c^2 - 2ac \cos B} \qquad (12)$$

$$\boldsymbol{c^2 = a^2 + b^2 - 2ab \cos C} \qquad (13)$$

All versions of the rule end with an expression which contains all three letters of the alphabet, two as sides, one as an angle. The letter that is the angle is the same letter as the equation begins with.

Now we can solve the triangle of Figure 19.12a, in which we are given the included angle:

$$a = 7.0$$
$$b = 8.0$$
$$\hat{C} = 74° \quad \Rightarrow \quad \cos C = 0.2756$$

Substituting in equation (13):

$$c^2 = 7.0^2 + 8.0^2 - 2 \times 7.0 \times 8.0 \times 0.2756$$
$$= 49 + 64 - 30.8672$$
$$= 82.1328$$
$$\Rightarrow \quad c = \sqrt{82.1328} = 9.0627$$

Side AB = 9.06 (2 dp)

Finding the angles

If we change the subject of equation (11) from a to cos A, we have another useful form of the rule:

$$\boldsymbol{\cos A = \frac{b^2 + c^2 - a^2}{2bc}}$$

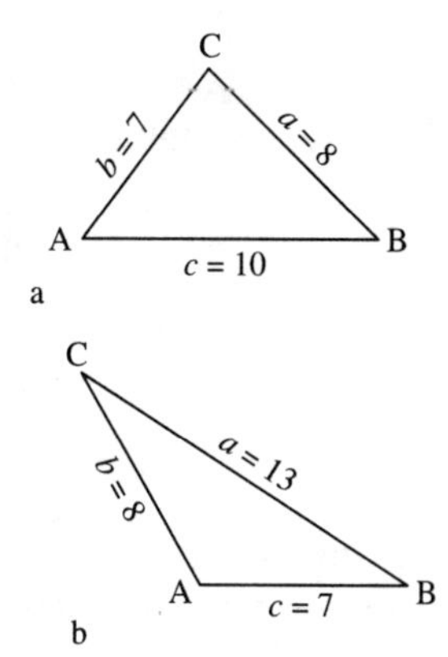

Figure 19.14

This also has two other versions, for cos B and cos C. The cosine rule is used to find the angles of a triangle when we know all three sides. For example, in the triangle of Figure 19.14a:

$$\cos A = \frac{7^2 + 10^2 - 8^2}{2 \times 7 \times 10} = 0.6071$$

$$\Rightarrow \qquad \hat{A} = \cos^{-1} 0.6071 = 52.6°$$

Note that there is no possibility of two solutions here. Angles with positive cosines are either in the 1st or 4th quadrants. But angles in the 4th quadrant are between 270° and 360° so are too big to be angles in a triangle.

Similarly:

$$\cos B = \frac{a^2 + c^2 - b^2}{2ac}$$

$$= \frac{8^2 + 10^2 - 7^2}{2 \times 8 \times 10} = 0.7188$$

$$\hat{B} = \cos^{-1} 0.7189 = 44.0°$$

The third angle is found from:

$$\hat{C} = 180° - 52.6° - 44.0° = 83.4°$$

The angles are: $\hat{A} = 52.6°$, $\hat{B} = 44.0°$, $\hat{C} = 83.4°$

In the triangle of Figure 19.14b, we have:

$$\cos A = \frac{8^2 + 7^2 - 13^2}{2 \times 8 \times 7} = -0.5$$

A negative sign means that $\hat{A}$ is in the second or third quadrants. But angles in the third quadrant are too big to be angles in a triangle.

$$\hat{A} = \cos^{-1}(-0.5) = 120°$$

Check the angles and sides (see box on page 286).

Area of a triangle

The area of the triangle ABC in Figure 19.13 is:

$$\tfrac{1}{2} \text{ base} \times \text{height} = \tfrac{1}{2} \times c \times y \qquad (16)$$

In the right-angled triangle ADC:

$$\sin A = \frac{y}{b}$$

$$\Rightarrow \qquad y = b \sin A$$

Substituting this value in equation (16):

Area ΔABC = ½*bc* sin A

In words:

Area = half of the product of two sides multiplied by the sine of the included angle

As an example, the area of the triangle in Figure 19.12a is:

$$\text{Area} = \tfrac{1}{2} \times 8.0 \times 7.0 \times \sin 74° = 28 \times \sin 74° = 26.9153$$

$$\underline{\text{Area} = 26.9 \text{ (3 sf)}}$$

Another useful formula for the area of a triangle was discovered by Hero, a Greek mathematician living in the 1st Century. Hero's formula states that:

$$\textbf{Area} = \sqrt{s(s-a)(s-b)(s-c)}$$

$$\text{Where } s = \frac{a+b+c}{2}$$

$$\text{In Figure 19.14a, } s = \frac{7+8+10}{2} = 12.5$$

$$\text{Area} = \sqrt{12.5 \times 5.5 \times 4.5 \times 2.5} = \sqrt{773.4} = 27.8107$$

$$\underline{\text{Area} = 27.8 \text{ (3 sf)}}$$

Test yourself 19.5

1 Find the third side of each of these triangles, given two sides and the included angle (2 dp). Find the area of each triangle.

a Sides 3.4 and 7.2, angle 50°.

b Sides 12.6 and 4.7, angle 106°.

c Sides 7.1 and 9.3, angle 12°.

2 Find the angles of these triangles, given the three sides. Find the area of each triangle.

a $a = 5, b = 4, c = 7$.

b $p = 12, q = 8, r = 5$.

c $x = 7.5, y = 7.3, z = 7.9$.

3 To assess your progress in this chapter, work the exercises in *Try these first*, page 275.

20 Equations in pairs

Given two equations, each with the same two unknown variables, there are ways of solving them to find both variables. This chapter describes the techniques.

You need to know about handling equations (Chapter 12), and interpreting graphs (Chapter 17).

Try these first

Solve these pairs of equations.

1 $2x + 7y = 34$
$3y = 12$

2 $5a - 2b = 26$
$3a + 2b = 6$

3 $-9s + t = 23$
$3s - 4t = -26$

Using graphs

We have already had examples of solving pairs of equations by plotting graphs, as illustrated by Figures 17.9 and 17.10. The point where the lines cross gives the values of x and y. This technique is very useful when the equations are complicated. In such cases, it may be the only practicable way of finding the variables. But graphs take a long time to plot and there may be difficulties in deciding the exact values of the variables at the crossing-point. For simple equations, it is quicker and more accurate to use algebra to find the answers.

Substituting

A straight path 10 m long is made from 20 grey paving slabs, laid end-to-end. Then it is decided that the path would look better if every third slab is red. The red slabs have the same width as the grey slabs, but are not the same length. If 14 grey slabs and 7 red slabs are laid end-to-end, the path is 9.8 m long. How long are the grey and red slabs?

To solve this problem we express the facts as two equations, with the lengths of the two types of slabs as the unknown variables. If the length of the grey slabs is g, the equation for the all-grey path is:

$$20g = 10 \qquad (1)$$

If the length of red slabs is r, the equation for the grey and red path is:

$$14g + 7r = 9.8 \qquad (2)$$

We now have to find a value for g and a value for r that satisfy both equations *at the same time.* Pairs of equation that have to be satisfied at the same time are called **simultaneous equations**.

Equation (1) has only one unknown, g:

$$\Rightarrow \quad g = 10/20 = 0.5$$

Now that we know g, we can substitute its value in equation (2):

$$\begin{aligned} 14 \times 0.5 + 7r &= 9.8 \\ \Rightarrow \quad 7 + 7r &= 9.8 \\ \Rightarrow \quad 7r &= 9.8 - 7 \\ \Rightarrow \quad 7r &= 2.8 \\ \Rightarrow \quad r &= 0.4 \end{aligned}$$

Grey slabs are 0.5 m long and red slabs are 0.4 m long.

Summing up:

If one equation has only one unknown, use that equation to find the value of the variable, then *substitute* this value in the other equation

Test yourself 20.1

Solve these pairs of equations by substituting.

1 $4x + 3y = 20$
$7y = 28$

2 $2x - 5y = 6$
$3x = 24$

3 $3b = -15$
$5a + 2b = 0$

4 $3m = 2n - 17$
$2m = -6$

5 $\dfrac{x}{2} = 2$
$\dfrac{3x - 4y}{4} = 4$

Eliminating

If *both* equations contain *both* unknowns, we need to eliminate one of the unknowns, so that we can find the value of the other.

Examples

Solve these equations:

$$4x + 5y = 30 \qquad (3)$$

$$2x + 5y = 20 \qquad (4)$$

Adding and subtracting equations

An equation remains balanced (and therefore true) if we add the same amount to each side (page 159).

Given equation A:	Left side of A = Right side of A
and equation B:	Left side of B = Right side of B
Adding the equations	Left side of A + Left side of B = Right side of A + Right side of B

In adding the two sides of equation B to equation A, we have added the *same amount* to both sides of A. The combined equation is balanced and true.

Example

Equation A is $2x + 3y = 13$
Equation B is $5x + 2y = 16$
Adding them $7x + 5y = 29$

Similar reasoning applies if we subtract B from equation A.

Inspecting the equations, we note that y has the same coefficient in both equations. This makes it easy to eliminate y just by subtracting one equation from the other. Subtract equation (4) from equation (3):

On the left side: $4x - 2x = 2x$
and $5y - 5y = 0$
On the right side: $30 - 20 = 10$

The result of subtracting is:

$$2x = 10$$
$$\Rightarrow \quad x = 5$$

Having found x, we find y by substituting $x = 5$ in equation (3) or (4). It does not really matter which one we substitute into. If we substitute in (3):

$$\Rightarrow \quad 4 \times 5 + 5y = 30$$
$$\Rightarrow \quad 20 + 5y = 30$$
$$\Rightarrow \quad 5y = 10$$
$$\Rightarrow \quad y = 2$$

The solution of the pair is $x = 5$ and $y = 2$.

In the next pair of equations, the coefficients of y are equal in size but *opposite in sign*.

$$2x + 3y = 21 \qquad (5)$$
$$7x - 3y = 6 \qquad (6)$$

With opposite signs we eliminate y by *adding* the equations:

On the left side: $2x + 7x = 9x$
and $3y + (-3y) = 0$
On the right side: $21 + 6 = 27$

The result of adding is:

$$\Rightarrow \quad 9x = 27$$
$$\Rightarrow \quad x = 3$$

Now substitute $x = 3$ in equation (5):

$$2 \times 3 + 3y = 21$$
$$\Rightarrow \quad 6 + 3y = 21$$
$$\Rightarrow \quad 3y = 15$$
$$\Rightarrow \quad y = 5$$

The solution of the pair is $x = 3$ and $y = 5$.

Summing up:

If the coefficient of one of the variables is the same size in both equations, add or subtract the equations to eliminate that variable

When coefficients are unalike

A theatre stage is lit by floodlamps, some of which are blue and others white. When 7 blue lamps and 2 white lamps are turned on, the total current needed is 20 A. When 4 blue lamps and 6 white lamps are turned on, the total current is 26 A. Find the amount of current taken by each type of lamp.

This information is expressed in two simultaneous equations, using b for blue lamps and w for white lamps:

$$7b + 2w = 20 \qquad (7)$$

$$4b + 6w = 26 \qquad (8)$$

Neither of the variables have coefficients of equal size, so simply adding or subtracting the equations will not eliminate anything. But we can *make* the coefficient of w equal to 6 in both equations if we multiply equation (7) by 3.

Equation (7) $\times$ 3 is $\quad 21b + 6w = 60 \quad$ call it (9)
Equation (8) is $\quad 4b + 6w = 26 \quad$ (8)
Subtract (8) from (9) $\quad 17b = 34$
$\Rightarrow \quad b = 2$

Substitute $b = 2$ in equation (7):

$$7 \times 2 + 2w = 20$$
$$\Rightarrow \quad 14 + 2w = 20$$
$$\Rightarrow \quad 2w = 6$$
$$\Rightarrow \quad w = 3$$

Blue lamps take 2 A, white lamps take 3 A.

Summing up:

If the coefficient in one equation is a multiple of the coefficient in the other equation, multiply one equation to make coefficients equal in size

After multiplication, if the coefficients have the *same* sign (as in the example above), *subtract* one equation from the other. If they have *opposite* signs, *add* the equations.

Here is another pair to solve:

$$4a - 5b = 32 \qquad (10)$$

$$10a + 7b = 2 \qquad (11)$$

Coefficients are not equal in size; none are multiples of one of the others. This time we have to multiply *both* equations.

There are several ways of tackling this:

- Multiply (10) by 7, multiply (11) by 5, so that the coefficients of b are -35 and $+35$. Then add equations.
- Multiply (10) by 10, multiply (11) by 4, so that the coefficients of a are both 40. Then subtract equations.
- Multiply (10) by 5, multiply (11) by 2, so that the coefficients of a are both 20. Then subtract equations.

Any of these three ways would give the same result, but it is slightly better to choose the third way as smaller numbers are involved, with less risk of errors. This way makes the coefficient of a equal to the LCM of the original two coefficients, the LCM of 4 and 10 being 20.

Equation (10) × 5 is	$20a - 25b = 160$	call it (12)
Equation (11) × 2 is	$20a + 14b = 4$	call it (13)
Subtract (13) from (12)	$-39b = 156$	
$\Rightarrow$	$b = -4$	

Substitute $b = -4$ in equation (10):

$$4a - 5 \times (-4) = 32$$

$$\Rightarrow \quad 4a + 20 = 32$$

$$\Rightarrow \quad 4a = 12$$

$$\Rightarrow \quad a = 3$$

The solution of the pair is $a = 3$ and $b = -4$.

There may often be equation pairs in which coefficients have no common factors, so there is no LCM. In such cases multiply both equations to make the coefficients of one variable the same size, as in this example:

$$3p - 4q = 9 \qquad (14)$$

$$5p + 3q = -14 \qquad (15)$$

There are no coefficients of equal size or with common factors. We could eliminate p or q, and decide to eliminate q:

Equation (14) × 3 is	$9p - 12q = 27$	(16)
Equations (15) × 4 is	$20p + 12q = -56$	(17)
Add (17) to (16)	$29p = -29$	
$\Rightarrow$	$p = -1$	

Substitute $p = -1$ in equation (14):

$$3 \times (-1) - 4q = 9$$
$$\Rightarrow \quad -3 - 4q = 9$$
$$\Rightarrow \quad -4q = 12$$
$$\Rightarrow \quad q = -3$$

The solution of the pair is $p = -1$ and $q = -3$.

Test yourself 20.2

Solve these pairs of equations, using substitution or elimination, as appropriate.

1 $3a - 7b = -15$
$4a + 7b = 29$

2 $7p + 3q = -15$
$5q = -10$

3 $9x + 5y = -1$
$3x - 2y = -26$

4 $7j + 3k = 4$
$-4j - 5k = 1$

5 $5c + 2d = -16$
$-5c - 7d = -44$

6 $5/x = 2$
$7x - 3y = 7$

7 $3a = 15b + 3$
$6b - 89 = 7a$

8 The digits of a 2-digit number add up to 16. Three times the first digit minus twice the second digit equals 3. Find the number.

9 Two steel rods 3 m long are cut into pieces of two sizes, referred to as size A and size B. One rod is cut into 15 size A and 10 size B. The other rod is cut into 24 size A and 4 size B. What lengths are sizes A and B?

10 The equations of two straight-line graphs are $y = 3x - 9$ and $y = 16 - 2x$. Without plotting the graphs, find the point at which the lines cross.

11 To assess your progress in this chapter, work the exercises in *Try these first*, page 293.

21 Equations with squares

Quadratic equations crop up often in maths. This chapter presents the techniques for solving them.

You need to know about factorizing algebraic expressions (Chapter 11), handling equations (Chapter 12), and square roots (page 180).

Try these first

Solve these equations.

1 $x^2 - 5x - 14 = 0$

2 $a(a - 6) - 3(a + 10) = 6$

3 $7 + 4p^2 = 13p$

Quadratic equations

A quadratic equation has:

- Only one variable
- A term in the square of the variable (but not the cube or higher power)
- Possibly a term in the variable
- Possibly a constant.

Here are some examples of quadratic equations:

$$x^2 - 5x + 6 = 0 \quad (1)$$

$$y^2 - 9 = 0 \quad (2)$$

$$p^2 = p + 12 \quad (3)$$

$$a(a - 12) = -35 \quad (4)$$

$$b^2 + 7b = 0 \quad (5)$$

They all conform to the conditions listed above. Equation (4) does not have a^2 in it, as the equation is written but, if we clear the brackets, the first term is a^2 and the equation is quadratic.

In this chapter we discuss two techniques for solving quadratic equations. One technique makes use of factorizing, and the other uses a formula.

Factorizing

Try this technique first, as it is the easiest. As an example, we will solve equation (1):

$$x^2 - 5x + 6 = 0$$

This equation is already written out in the standard format for quadratics:

$$\begin{bmatrix}\text{Term in} \\ x^2\end{bmatrix} \pm \begin{bmatrix}\text{Term in} \\ x\end{bmatrix} \pm \text{Constant} = 0$$

It is essential to have zero on the right. The first step is to factorize the left side of the equation:

$$\Rightarrow \quad (x - 2)(x - 3) = 0$$

We have two expressions, $(x - 2)$ and $(x - 3)$. The equation states that, when these two are multiplied together, the product is zero. If a product is zero, one or possibly both of the expressions must be equal to zero. These are the only ways in which we can get zero as the product.

This being so, then:

EITHER $\quad x - 2 = 0$

OR $\quad x - 3 = 0$

If $x - 2 = 0$, then: $\quad x = 2$

If $x - 3 = 0$, then: $\quad x = 3$

The equation has *two* solutions $x = 2$ or $x = 3$.

Either of the two solutions satisfy the equation. Check this by substituting $x = 2$ in equation (1):

$$2^2 - (5 \times 2) + 6 = 4 - 10 + 6 = 0 \qquad \text{True}$$

Now substitute $x = 3$ in the same equation:

$$3^2 - (5 \times 3) + 6 = 9 - 15 + 6 = 0 \qquad \text{True}$$

As a general rule, *all* quadratic equations have two solutions.

For another example, try equation (2). Factorizing this gives:

$$(x - 3)(x + 3) = 0$$

Either $\quad x - 3 = 0 \quad \Rightarrow \quad x = 3$

Or $\quad x + 3 = 0 \quad \Rightarrow \quad x = -3$

Often we accept either solution of a quadratic equation as correct, but, when solving practical problems, we may have reasons for saying that one of the solutions is impossible. For example, a negative value may not make sense. Here we accept both solutions.

The solutions are $x = 3$ and $x = -3$.

Equation (3) is not in the standard format so it needs rearranging before we try to solve it:

$$p^2 = p + 12$$
$$\Rightarrow \quad p^2 - p - 12 = 0$$

Now it is ready to factorize:

$$\Rightarrow \quad (p + 3)(p - 4) = 0$$

Either $\quad p + 3 = 0 \quad \Rightarrow \quad p = -3$

Or $\quad p - 4 = 0 \quad \Rightarrow \quad p = 4$

The solutions are $p = -3$ or $p = 4$.

The two solutions of a quadratic equation are often called the **roots** of the equation.

Equation (4) needs expanding before we can give it the standard format:

$$a(a - 12) = -35$$
$$\Rightarrow \quad a^2 - 12a = -35$$
$$\Rightarrow \quad a^2 - 12a + 35 = 0$$

Factorizing:

$$(a - 5)(a - 7) = 0$$

Either $\quad a - 5 = 0 \quad \Rightarrow \quad a = 5$

Or $\quad a - 7 = 0 \quad \Rightarrow \quad a = 7$

The solutions are $a = 5$ or $a = 7$.

Equation (5) has a common factor:

Factorizing: $\quad b(b + 7) = 0$

Either $\quad b = 0$

Or $\quad b + 7 = 0 \quad \Rightarrow \quad b = -7$

The solutions are $b = 0$ or $b = -7$.

Test yourself 21.1

Solve these quadratic equations. Some of them are already factorized.

1 $(a - 7)(a + 3) = 0$
2 $(x + 2)(x - 4) = 0$
3 $x(x - 5) = 0$
4 $a^2 - 3a + 2 = 0$
5 $y^2 - y - 6 = 0$
6 $r^2 - 15r + 44 = 0$

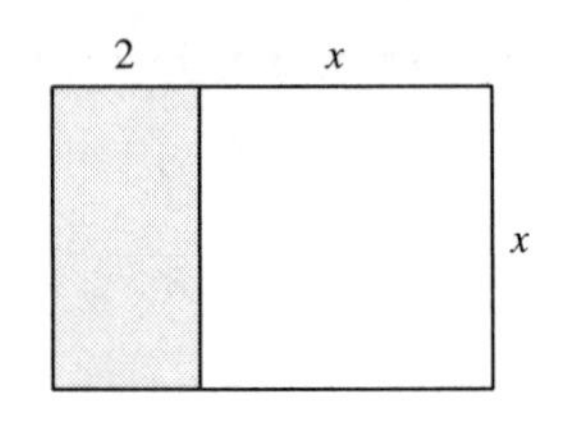

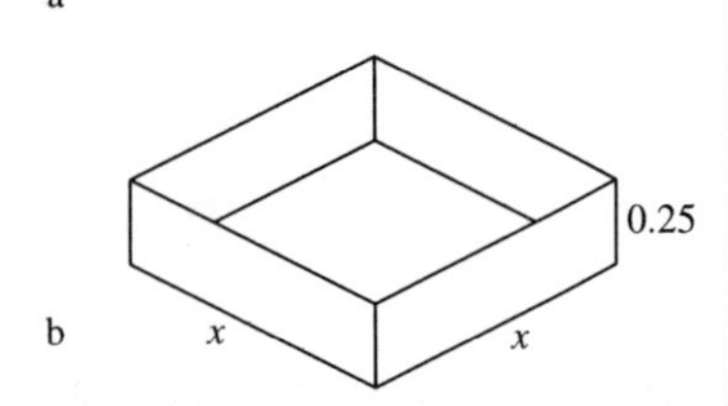

Figure 21.1

7 $s^2 = s - 6$

8 $z^2 = 7z$

9 $(d - 7)(d + 6) = 48$

10 $j(j - 8) = -15$

11 A business card has a coloured border 2 cm wide down the left side, with the company's logo printed on it (Figure 21.1a). The remainder of the card is square, and the total area of the card is 24 cm^2. Find x and from this find the length and width of the card.

12 A square metal tray is 0.25 m deep (Figure 21.1b). The area of sheet metal used in its construction is 24 m^2. Find x, the length of the sides of the tray.

When x^2 has a coefficient

All the examples so far have had x^2 without a coefficient. If x^2 has a coefficient, factorizing requires an extra step.

Examples

$$3x^2 + 18x - 21 = 0$$

All terms on the left have a common factor, 3:

$$\Rightarrow \quad 3(x^2 + 6x - 7) = 0$$

The expression in the brackets can be factorized in the usual way, so we have three factors:

$$\Rightarrow \quad 3(x - 1)(x + 7) = 0$$

It is evident that the first factor, 3, cannot be equal to zero.

But:

Either $\quad x - 1 = 0 \quad \Rightarrow \quad x = 1$

Or $\quad x + 7 = 0 \quad \Rightarrow \quad x = -7$

The solution is $x = 1$ or $x = -7$.

$$3x^2 - 11x + 6 = 0$$

This has no common factors. It has to be factorized as it stands. There *are* ways of doing this, which are extensions of the methods we have already used. But the calculations can become very involved, and there is always the possibility that, after a long struggle, the conclusion is reached that the expression has no factors. For equations of this type, the best technique is to apply the formula described in the next section.

Quadratic formula

To apply this formula, the equation must be written in the standard format. We use letters a and b to represent the coefficients of x^2 and x, and we use c to represent the constant:

$$ax^2 + bx + c = 0$$

The formula for the solutions or roots of the equation are:

$$x = \frac{-b \pm \sqrt{b^2 - 4ac}}{2a}$$

This formula applies to any quadratic equation, whether it can be factorized or not. Here is an example which does not factorize:

$$x^2 - 3x - 9 = 0$$

Identify the coefficients and constant:

$$a = 1 \qquad b = -3 \qquad c = -9$$

Substitute these in the formula:

$$x = \frac{-(-3) \pm \sqrt{(-3)^2 - (4 \times 1 \times -9)}}{2 \times 1}$$

$$= \frac{3 \pm \sqrt{9 - (-36)}}{2}$$

$$= \frac{3 \pm \sqrt{45}}{2} = \frac{3 \pm 6.708}{2}$$

The $\pm$ sign means that we obtain the two solutions (or roots) either by adding or subtracting the value after the sign.

Adding: $x = \frac{3 + 6.708}{2} = \frac{9.708}{2} = 4.854$

Subtracting: $x = \frac{3 - 6.708}{2} = \frac{-3.708}{2} = -1.854$

The solution is $x = 4.85$ or $x = -1.85$ (2 dp).

The coefficient b or the constant c can be zero. Obviously a cannot be zero, because the equation would not then be quadratic. Take this example, with $b = 0$:

Solve $5x^2 - 7 = 0$

Identify the coefficients and constant:

$$a = 5 \qquad b = 0 \qquad c = -7$$

Substitute these in the formula:

$$x = \frac{-0 \pm \sqrt{0^2 - (4 \times 5 \times -7)}}{2 \times 5}$$

$$= \frac{\pm\sqrt{140}}{10}$$

$$= \frac{\pm 11.832}{10}$$

The solution is $a = \pm 1.18$ (2 dp).

This example produces a slightly different kind of result:

Solve $9x^2 - 6x + 1 = 0$

Identify the coefficients and constant:

$a = 9 \qquad b = -6 \qquad c = 1$

Substitute these in the formula:

$$x = \frac{-(-6) \pm \sqrt{(-6)^2 - (4 \times 9 \times 1)}}{2 \times 9}$$

$$= \frac{6 \pm \surd\sqrt{36 - 36}}{18}$$

$$= \frac{6 \pm \sqrt{0}}{18}$$

The expression after the ± sign has zero value. It makes no difference whether we add or subtract it. The equation appears to have has only one solution.

The solution is $x = 1/3$.

Although it seems as if the equation has only one root, a quadratic equation *must* have two roots. It is just that the two roots are equal.

This example produces an awkward result:

Solve $2x^2 + 3x + 7 = 0$

Identify coefficients and constant:

$a = 2 \qquad b = 3 \qquad c = 7$

There is a problem here. Without writing out the formula in full, look at the value of the expression under the square-root sign:

$$\sqrt{b^2 - 4ac} = \sqrt{3^2 - (4 \times 2 \times 7)}$$

$$= \sqrt{9 - 56}$$

$$= \sqrt{-47}$$

We have to find the square root of –47. In other words, we need a number which, when multiplied by itself (is squared), comes to –47. But we have already stated (page 60) that *all* squares positive. There is no way of obtaining a negative square. The answer to this difficulty is discussed in Chapter 23.

Test yourself 21.2

Use the formula to solve these equations (2 dp).

1 $4x^2 - 3x - 6 = 0$

2 $5a^2 + 6a - 2 = 0$

3 $16p^2 + 1 = 8p$

4 $7n + 2 = -2n^2$

5 One of the equations of motion is $s = ut + \frac{1}{2}at^2$, in which s is the distance travelled in time t, with an initial speed of u and an acceleration a. A car leaves a roundabout travelling at 8 m/s (about 20 mph). It accelerates at $2\,\text{m/s}^2$. How long does it take to travel 100 m from the roundabout?

6 A cylindrical can, 10 cm tall, is made from sheet metal. It is enclosed at the top and bottom. The area of sheet metal (ignoring overlaps) is $300\,\text{cm}^2$. Find the radius of the can.

7 To assess if you have covered this chapter fully, answer the questions in *Try these first*, page 299.

22 Directed quantities

Technologists often deal with force and motion, quantities which have direction as well as size. This chapter shows how such quantities are represented mathematically and how they are added and subtracted. The chapter also describes and relates the two major systems of coordinates.

You need to know about triangles (Chapter 14) and Pythagoras' theorem (Chapter 16).

Try these first

1 Using the scale drawn on the diagram and a protractor, specify the size and direction of the vectors in Figure 22.1.

2 Draw vectors to these specifications:

a size = 7.5, direction = 65°

b size = 4.2, direction = 210°

3 This data specifies 4 vectors, **a**, **b**, **c**, and **d**.

a size = 4, direction = 50°

b size = 5, direction = 160°

c size = 4, direction = 280°

d size = 2, direction = 210°

By drawing diagrams or by calculation find:

$\mathbf{e} = \mathbf{a} + \mathbf{b}$ $\quad\quad$ $\mathbf{f} = \mathbf{c} + \mathbf{d}$

$\mathbf{g} = \mathbf{a} + \mathbf{b} + \mathbf{c}$ $\quad\quad$ $\mathbf{h} = \mathbf{b} - \mathbf{d}$

4 A vector has size = 5, direction = 53°. Resolve it into its components with directions 0° and 90°.

5 Convert the rectangular coordinates (7, –3) into polar form (2 dp).

6 Convert the polar coordinates (5, 130°) into rectangular form.

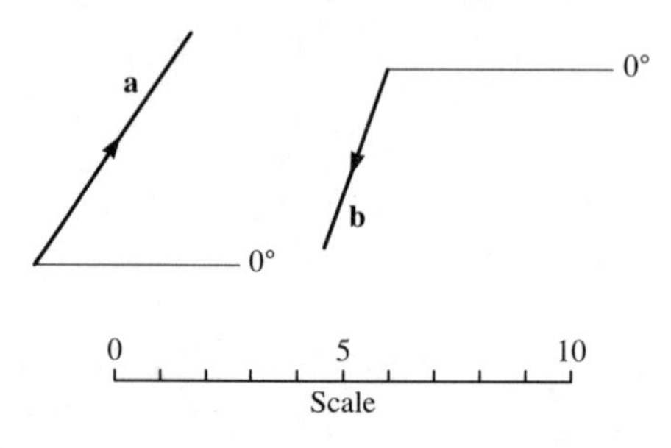

Figure 22.1

Size and direction

Quantities such as area, volume, temperature and electric charge are specified by their **size** alone. We say, for example, that a rectangle of card has an area 450 mm^2. We say that the temperature of a greenhouse is 27°C. The charge on a capacitor may be 2.5 coulombs. In each case we need *only one number* to specify the exact size of the quantity. Quantities such as these, which need only one number to specify them completely, are known as **scalar** quantities, or **scalars**, for short. In previous chapters in this book, we have always had scalars to work with, and have learned how to add them, subtract them, multiply them and carry out various other maths operations on them.

Now we come to quantities which are specified by both **size** and **direction**. We call these **vector** quantities, or just **vectors**. For example, an aeroplane flies due east at 300 kilometres per hour. We specify its velocity vector by:

Size: 300 kilometres per hour
Direction: due East

Another example: a slug crawls up a bean stem at 20 mm per minute. Its velocity vector is:

Size: 20 mm per minute
Direction: upward

Force is another vector. For example, a person presses their thumb down on a drawing-pin with a force of 100 N:

Size: 100 N
Direction: downward

Drawing vectors

Working with vectors is made easier if we draw a map or plan of them, to scale. Each vector is shown as a straight line. The **size** of the vector is represented by the **length** of the line. The **direction** of the vector is represented by the **direction** of the line. The line has an arrowhead on it to show which way it points.

Symbols for vectors

The symbol for a vector is a small letter printed in bold type.

Examples

a **b** **n** **z**

In your notes, write the letter with a wavy line underneath it.

Examples

$\utilde{a}$ $\utilde{b}$ $\utilde{n}$ $\utilde{z}$

The **size** of a vector is printed or written using the same letter with vertical lines on either side.

Example

The size of vector **a** is printed or written as $|a|$.

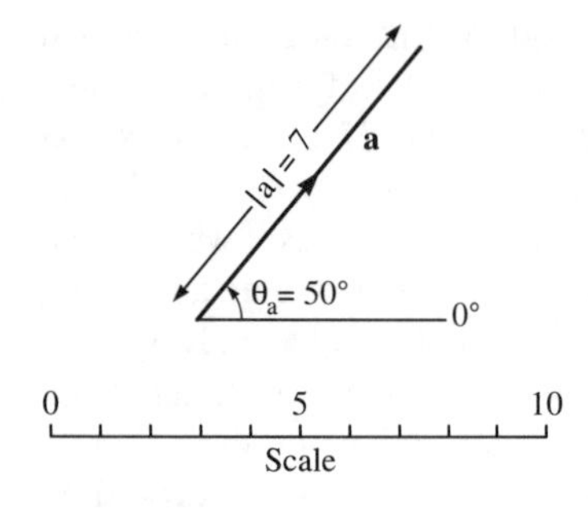

Figure 22.2

In Figure 22.2 the arrowed line represents a force vector of 7 N, directed at an angle of 50° with reference to the 0° line. The 0° direction is towards the right edge of the page, as is usual in plans of vectors.

The symbol for this vector is **a**. Mathematically, this single symbol includes both the size *and* the direction of the vector. The symbol for the size only of vector **a** is $|a|$. We have said that the size of **a** is 7 N, so we write:

$$|a| = 7\,\text{N}$$

In Figure 22.2, the line is drawn 7 units long, to scale. For example, if we decide on the scale 1 cm = 1 N, it is drawn 7 cm long.

The symbol for the direction of vector **a** is θ_a, so we write:

$$\theta_a = 50°$$

Directions may also be expressed in radians.

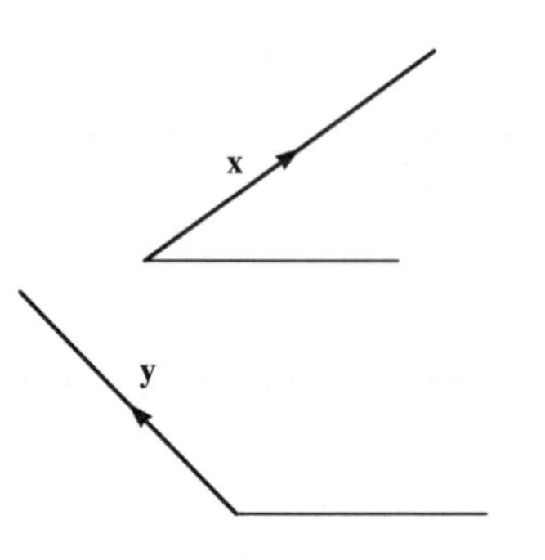

> **A symbol for angles**
>
> Ordinary letters such as x, $\hat{A}$, $\hat{b}$, and M are often used as symbols for variables which are angles, but a more distinctive symbol for angles is the Greek letter, *theta*.
>
> This is printed and written θ.
>
> We use it in many of the equations in this and later chapters.

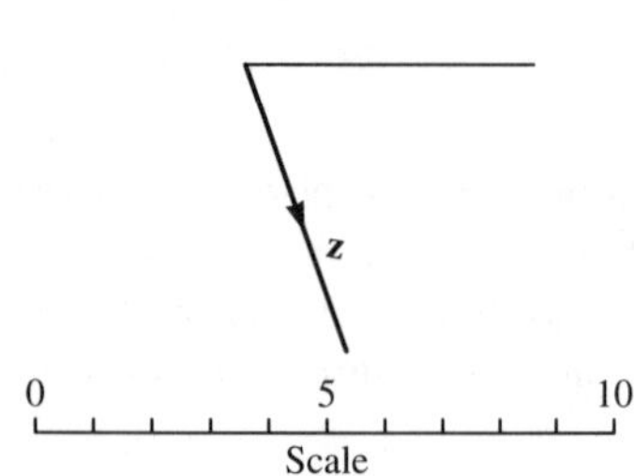

Figure 22.3

Test yourself 22.1

1 Measure the size and direction of the vectors in Figure 22.3.

2 Draw these vectors.

a $|a| = 5$, $\theta_a = 67°$ **b** $|b| = 6.5$, $\theta_b = 112°$
c $|c| = 2.1$, $\theta_c = 325°$

Adding vectors

A vector is a 'double quantity' (size and direction, $|a|$ and θ_a) represented by a single symbol, such as **a**. If we have two vectors **a** and **b**, we can add them together, to obtain a third vector **c**. Here is the vector equation:

$$\mathbf{a} + \mathbf{b} = \mathbf{c}$$

It looks like ordinary algebra, but this addition has to be done in a special way. We cannot just add their sizes and directions separately. With 'double' quantities, size and direction have to be added in one operation.

The simplest way to add two vectors is to draw them end-to-end.

Example

Given **a**, for which $|a| = 8$, $\theta_a = 40°$, and given **b** for which $|b| = 10$ and $\theta_b = 140°$, find their sum **c**.

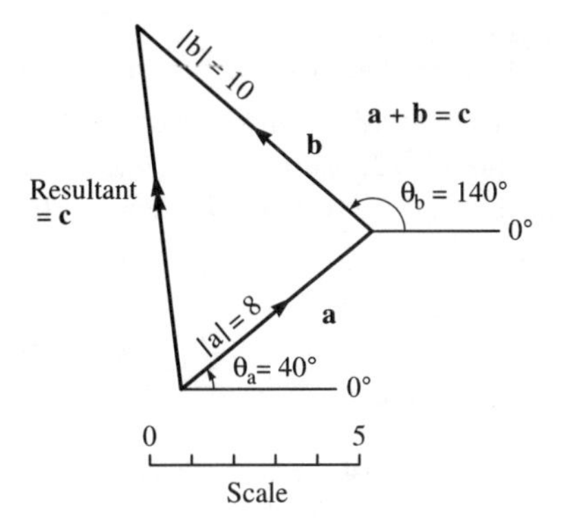

Figure 22.4

Figure 22.4a shows the two vectors drawn end-to-end. The drawing has a scale on it for measuring lengths, and the 0° direction is toward the right edge of the page. Their sum is the vector which runs from the beginning of **a** to the end of **b**. This is called **c**. We use the scale (or a ruler) to measure its length. We use a protractor to measure its angle.

Size (length): $|c| = 11.5$
Direction: $\theta_c = 98°$

When two or more vectors are added, their sum is called the **resultant vector**. A resultant vector is indicated by marking it with *two* arrowheads.

A practical example of this is a boat steering a course across a flowing river (Figure 22.5). Here we are adding two velocity vectors. Vector **a** is the velocity of the boat through the water, as steered by the pilot. Vector **b** is the velocity of flow of the river. The resultant **c** is the actual velocity of the boat, relative to the banks of the river. Note that both vectors act *at the same time*, not one after the other.

Any number of vectors may be added, simply by drawing them end-to-end, and measuring the size and direction of their resultant.

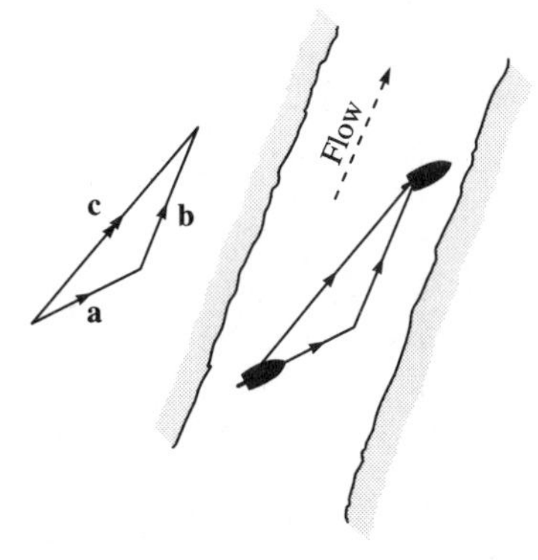

Figure 22.5

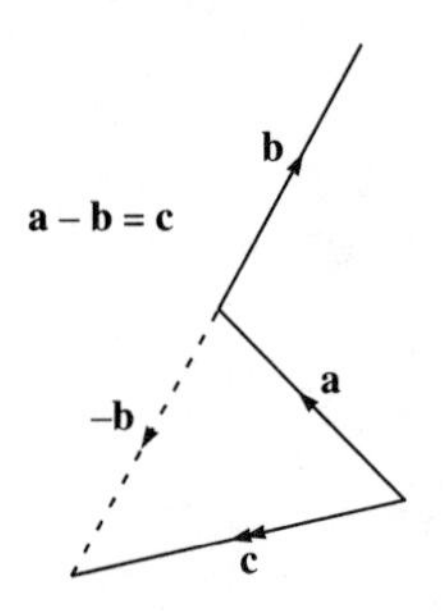

Figure 22.6

Subtracting vectors

A vector is subtracted from another vector by first reversing its direction, then adding it. Figure 22.6 shows how to subtract **b** from **a**. Reversing the direction of the arrowhead on **b**, gives –**b** (dashed line). Then we add **a** and –**b**, to get **a** – **b** = **c**.

Vectors in line

If two vectors that are to be added have the same direction, we add the vectors simply by adding their sizes. The direction of the resultant is the same as that of the two vectors being added. There is no need for a scale drawing.

Examples

Find **r** = **p** + **q** (Figure 22.7)

$|p| = 4, |q| = 5.5, \theta_p = \theta_q = 56°$

Resultant **r** is: $|r| = 4 + 5.5 = 9.5, \theta_r = 56°$

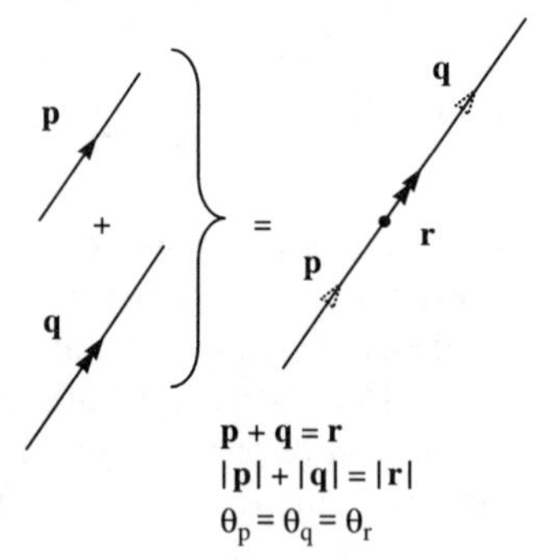

$\mathbf{p} + \mathbf{q} = \mathbf{r}$
$|\mathbf{p}| + |\mathbf{q}| = |\mathbf{r}|$
$\theta_p = \theta_q = \theta_r$

Figure 22.7

The same applies to subtraction.

Find **u** = **s** – **t** = **s** + (–**t**)

$|s| = 6, |t| = 4, \theta_s = \theta_t = 59°$

Resultant **u** is $|u| = 6 - 4 = 2, \theta_u = 59°$

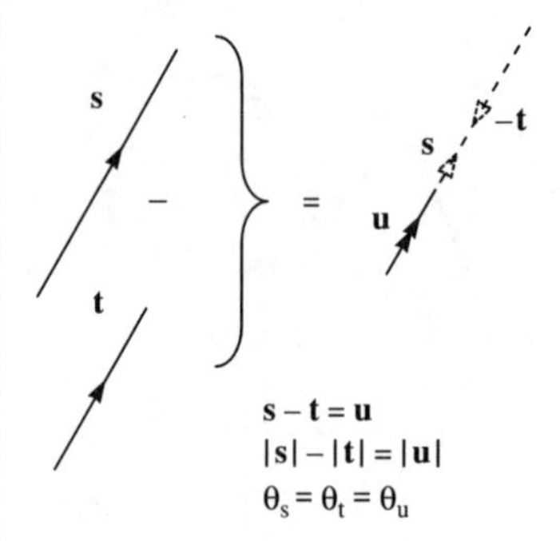

$\mathbf{s} - \mathbf{t} = \mathbf{u}$
$|\mathbf{s}| - |\mathbf{t}| = |\mathbf{u}|$
$\theta_s = \theta_t = \theta_u$

Figure 22.8

Test yourself 22.2

Four vectors are specified as below:

Vector	*Size*	*Direction*
a	5	140°
b	6	270°
c	3	50°
d	4	330°

Make scale drawings to find these resultant vectors (size to 2 dp, direction to nearest degree).

e = a + b   **f = c + d**

g = b + c   **h = a + b + c**

i = b − d   **j = a − b**

k = c − d + b   **l = a + a**

A different diagram

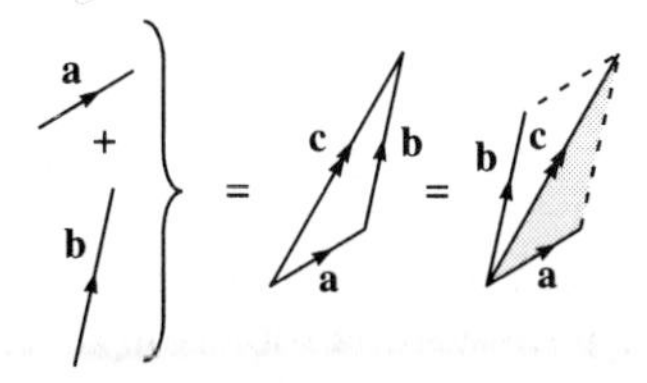

Figure 22.9

Figure 22.9 shows two vectors being added in the way described above, by drawing them end-to-end. On the right we see a different way of obtaining the same result. Instead of drawing the vectors end-to-end, we draw them both beginning at the *same* point. Then we draw dashed lines parallel with the vectors to complete a *parallelogram*.

The triangle in the ordinary vector diagram and the shaded triangle in the parallelogram both have sides equal to $|a|$ and $|b|$. The directions of **a** and **b** are the same in both triangles. The two triangles are congruent. This means that the diagonal of the parallelogram has the same length and direction as the resultant **c** of the two vectors.

Summarizing this way of adding vectors:

Draw the vectors from the same point, then complete the parallelogram. The diagonal is the resultant.

This other way of finding the resultant is often an easier way of thinking about vectors and their resultants. The end-to-end method is clear with velocity vectors, but the parallelogram method makes more sense with force vectors. In Figure 22.10a, we see an object being supported from two points by thin wires. Figure 22.10b is the vector diagram. The tensions in the wires

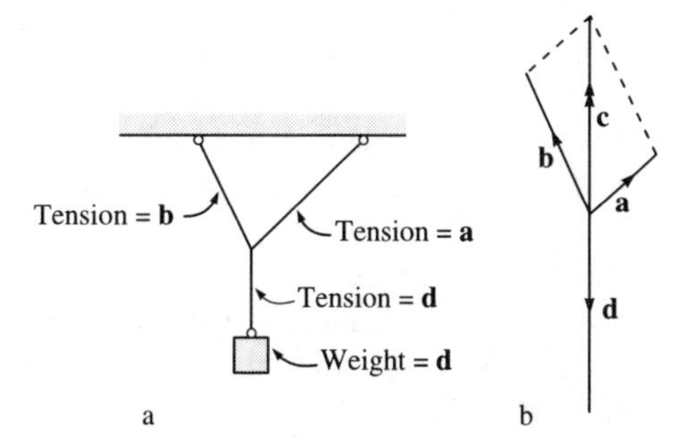

Figure 22.10

More vectors

Weight is a vector. Weight is the force by which the Earth pulls on an object, so weight is a force vector. Its size is proportional to the mass of the object. Its direction is down toward the Earth's centre.

Example

The weight of an apple, mass 100 g, is approximately 1 N. This is a vector, size = 1 N, direction = down.

Displacement is distance moved in a given direction.

Example

A person starts at A and walks to B which is 3 km to the north-east of A. Assuming that due east is taken as the 0° direction, the displacement of the person is a vector, size = 3 km, direction = 45°.

Acceleration is increase in velocity per second, in a given direction. Acceleration is caused by a force acting on an object. If the object is free to move in any direction, it accelerates in the same direction as the force.

These vectors can be added and subtracted, or resolved into components just like the other vectors in our examples. One point to remember is that, as with addition or subtraction of scalars, we can only add or subtract things that are of the *same kind* (page 8). We cannot, for example, add a force vector to a velocity vector.

are the force vectors **a** and **b**. Their resultant **c** is in the vertically upward direction. This force is balanced by the vertically downward tension **d** in the lower wire, caused by the weight of the object. **c** and **d** are equal in size but opposite in direction. Their resultant is zero. The object stays in the same place, moving neither up nor down.

The situation in Figure 22.10 is one in which we already know the force **d**, the weight of the object, and want to find the sizes of forces **a** and **b**. It might be important, for example, to know that the forces would not over-strain the wires. In other words, we would want to be able to **resolve** the resultant **c** into its two **components a** and **b**. This is done by using trig ratios, as explained in the next section.

Resolving vectors

Resolving a vector is the reverse of adding two vectors to find their resultant. In Figure 22.11 we have a vector **c** to be resolved into two components, **a** and **b**. We have specially chosen the directions of the components so as to make it easy to find their sizes. Component **a** is horizontal on the page. Its direction is 0° (from which other directions are measured). Component **b** is vertical on the page. Its direction is 90°.

The dotted lines complete the parallelogram which, since **a** and **b** are at right angles, is a rectangle. **c**, the vector to be resolved, is the diagonal of the rectangle. The dotted sides have the same lengths as the vectors **a** and **b**. They have lengths $|a|$ and $|b|$.

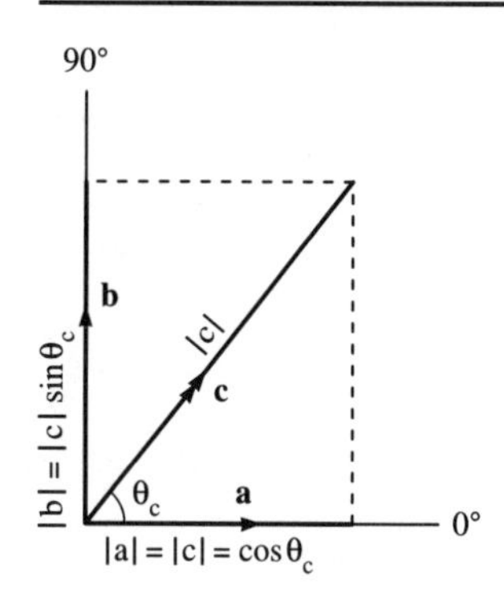

Figure 22.11

$$\cos\theta_c = \frac{\text{adjacent}}{\text{hypotenuse}} = \frac{|a|}{|c|}$$

$$\Rightarrow \quad |a| = |c|\cos\theta_c \qquad (1)$$

Equation (1) gives the size of **a** in terms of $|c|$ and θ_c, which are values we already know.

Similarly,

$$\sin\theta_c = \frac{\text{opposite}}{\text{hypotenuse}}$$

But the opposite side is equal in length to **b**:

$$\Rightarrow \quad \sin\theta_c = \frac{|b|}{|c|}$$

$$\Rightarrow \quad |b| = |c|\sin\theta_c \qquad (2)$$

Equation (2) gives us the other component of **a**.

Example

Resolve the vector **c**, when $|c| = 3.4$ and $\theta_c = 55°$.

Substituting in equation (1):

$|a| = 3.4\cos 55° = 3.4 \times 0.5736 = 1.95$ (2 dp)

Substituting in equation (2):

$|b| = 3.4\sin 55° = 3.4 \times 0.8192 = 2.79$ (2 dp)

The components of **c** are **a**, size 1.95, direction 0° and **b**, size 2.79, direction 90°.

Adding components

In Figure 22.4 we added two vectors by drawing scale diagrams, then measuring the size and direction of the resultant. Careless drawings give inaccurate results; careful drawing takes a long time. Here we show how to add two vectors without drawing diagrams and obtain accurate results.

The stages of this method are:

1 Resolve the vectors into their components
2 Add the components in the 0° direction
3 Add the components in the 90° direction
4 Find the resultant of the sums.

At stages 2 and 3 we are allowed to add the lengths of the sets of components because they are in the same directions. (See box page 310.)

Examples

Figure 22.12a shows two vectors **a** and **b**, to be added.

1 Resolve into components:

The components of **a** (Figure 22.12b) are:

Direction = 0° Size = 6 cos 25° = 5.4378
Direction = 90° Size = 6 sin 25° = 2.5357

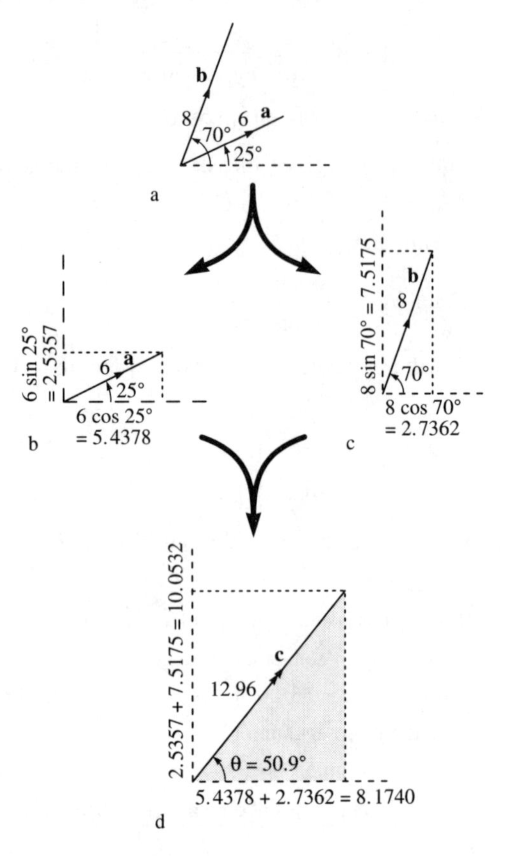

Figure 22.12

The components of **b** (Figure 22.12c) are:

Direction = 0° Size = 8 cos 70° = 2.7362
Direction = 90° Size = 8 sin 70° = 7.5175

2 Add the components in the 0° direction:

$$5.4378 + 2.7362 = 8.1740$$

3 Add the components in the 90° direction:

$$2.5357 + 7.5175 = 10.0532$$

4 Find the resultant of their sums (Figure 22.12d):

Finding the size of the resultant

The shaded triangle has base = 8.1740 and height = 10.0532. The size of **c** is the length of the hypotenuse. By Pythagoras:

$$|\mathbf{c}|^2 = 8.1740^2 + 10.0532^2$$
$$= 167.8811$$

$$\Rightarrow \quad |\mathbf{c}| = \sqrt{167.8811} = 12.96$$

Tangents in the four quadrants

Calculators usually do not evaluate inverse tangents correctly unless they are in the first quadrant. A simple sketch will show which quadrant an angle is in. You can also tell this by looking at the signs of the components in the 0° and 90° directions, as in the table below. The last column shows what to do to the result to obtain θ for each quadrant.

Quadrant	*Signs* 0°	*Signs* 90°	θ *sign*	*To obtain correct value of* θ	
1st	+	+	+	θ	(Value is correct)
2nd	−	+	−	$-\theta + 180°$	(Add 180°)
3rd	−	−	+	$\theta + 180°$	(Add 180°)
4th	+	−	−	$-\theta + 360°$	(Add 360°)

Finding the direction of the resultant

$$\tan \theta_c = \frac{\text{height}}{\text{base}} = \frac{10.0532}{8.1740} = 1.2299$$

$$\Rightarrow \quad \theta_c = \tan^{-1} 1.2299 = 50.9°$$

The resultant **c** has size = 12.96 and direction = 50.9°.

Figure 22.13a shows two vectors **a** and **b**, to be added.

1 Resolve into components:

The components of **a** are:

Direction = 0° Size = 5 cos 40° = 3.8302
Direction = 90° Size = 5 sin 40° = 3.2139

The components of **b** are:

Direction = 0° Size = 7 cos 150° = −6.0622
Direction = 90° Size = 7 sin 150° = 3.5000

One of the components of **b** has negative size; it is directed towards the left instead of towards the right.

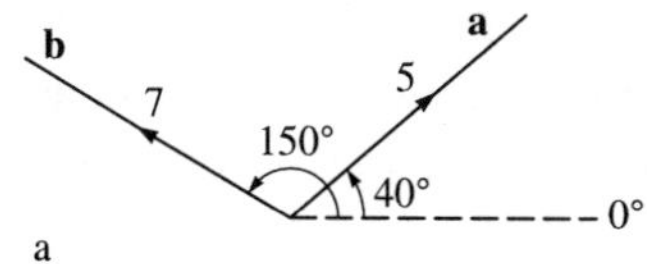

2 Add the components in the 0° direction:

3.8302 − 6.0622 = −2.2320

3 Add the components in the 90° direction:

3.2139 + 3.5000 = 6.7139

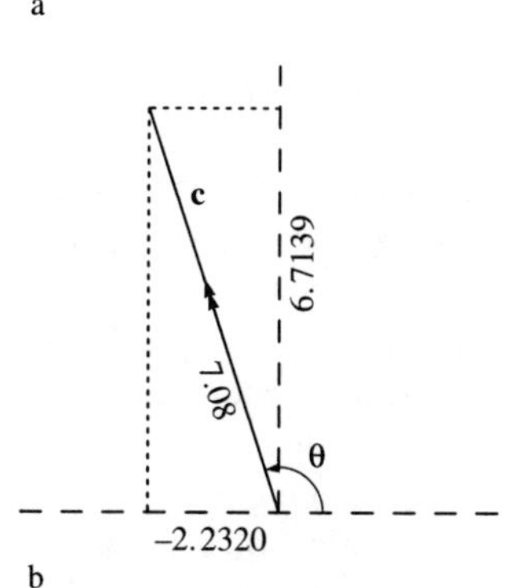

Figure 22.13

4 Find the resultant of their sums as in Figure 22.12:

Size

$$|c|^2 = (-2.2320)^2 + 6.7139^2 = 50.0583$$

$$\Rightarrow \quad |c| = \sqrt{50.0583} = 7.0752$$

Direction

$$\tan \theta_c = \frac{6.7139}{-2.2320} = -3.0080$$

$$\Rightarrow \quad \theta_c = \tan^{-1} -3.0080 = -71.60°$$

The sketch (Figure 22.13b) shows that θ is in the second quadrant (see box), so the direction of the resultant is 180° – 71.61° = 108.39°.

The resultant **c** has size = 7.08 and direction = 108.4°.

Test yourself 22.3

1 Resolve these vectors into their components, directions 0° and 90° (2 dp).

a Size = 4, direction = 33°

b Size = 9, direction = 85°

c Size = 6, direction = 130°

d Size = 5, direction = 230°

2 Add these pairs of vectors from question 1 (2 dp). [*Hint*: Sketches will help to put angles in the correct quadrants.]

e = **a** + **b**

f = **c** + **d**

3 A vehicle is pushed 34 m up a 27° slope, as measured along the slope. What distances is it moved

a horizontally **b** vertically?

4 A hot air balloon rises at the rate of 2.5 ms^{-1}. It is carried along horizontally by the breeze at 1.3 ms^{-1}. What is its resultant velocity and its angle of climb?

5 A picture is suspended as in Figure 22.14. The tension in each wire is 12 N. What is the weight of the picture, in kg, given that 1 kg ≈ 10 N?

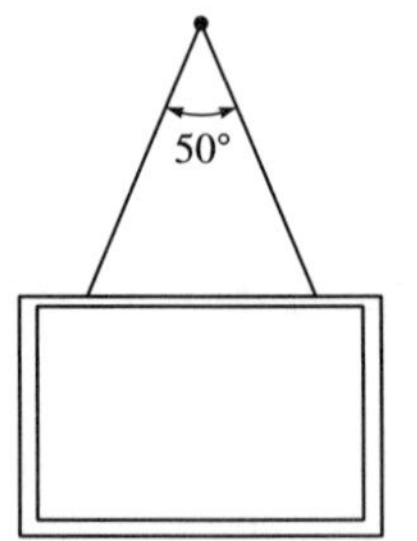

Figure 22.14

Coordinate systems

A person is given directions for finding treasure buried in a field: 'Start at the tree, facing east. Walk 30 paces forward, turn left, walk 40 paces forward. Dig there.'

Figure 22.15a shows the route to the treasure.

Another person is given directions for finding the treasure: 'Start at the tree, facing 53.1° north of east. Walk 50 paces forward. Dig there.'

Figure 22.15b shows this route to the treasure. Both routes lead to the same spot. They illustrate two different systems for locating the position of a point in a two-dimensional area such as a field or a sheet of paper.

The first system (Figure 22.15a) is the one we use when plotting graphs. A point on the graph is specified by quoting two numbers, the x-coordinate and the y-coordinate. The starting point is the origin. This system, which uses **rectangular coordinates**, is also used in geographical mapping.

The second system (Figure 22.15b) specifies the location of a point by quoting two numbers, the distance and direction of the point from the origin. Distance is usually referred to as r, and the direction as θ, where 0° is the 'due

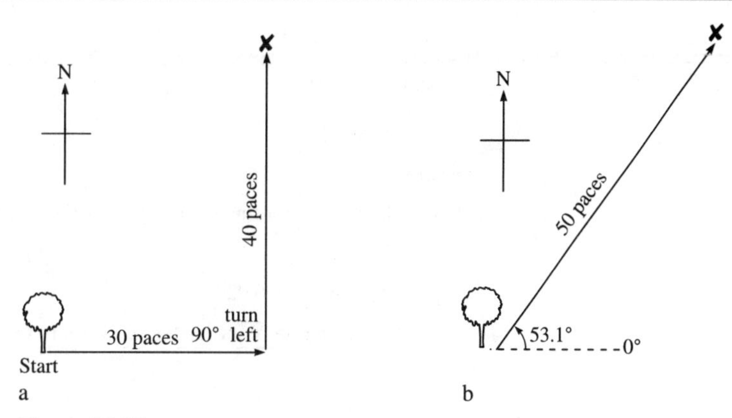

Figure 22.15

east' direction. The coordinates in this system are known as **polar coordinates** and the starting point or origin is sometimes known as the **pole**. Polar coordinates may be used for plotting special types of graphs (polar graphs). Some types of industrial equipment record temperature and other quantities as polar graphs.

Converting coordinates

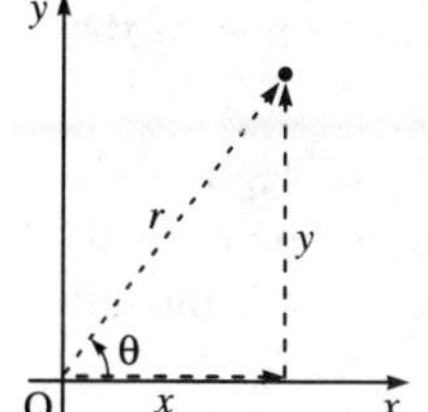

Figure 22.16

Rectangular to polar

In Figure 22.16 we are given x and y, the rectangular coordinates of point X, and want to find r and θ:

By Pythagoras:

$$r^2 = x^2 + y^2$$

From trig:

$$\theta = \tan^{-1} y/x$$

Example

Given the rectangular coordinates of a point P(12,17), find its polar coordinates:

$$r^2 = 12^2 + 17^2 = 433$$

$$\Rightarrow \quad r = \sqrt{433} \doteq 20.81 \text{ (2 dp)}$$

$$\theta = \tan^{-1} 17/12 = 54.78° \text{ (2 dp)}$$

The polar coordinates of P are (20.81, 54.78°).

A sketch is useful for making sure that you have evaluated θ in the correct quadrant (see box, page 315).

Polar to rectangular

If we are told r and θ, we can calculate x and y by a method similar to that used for resolving a vector into two components.

In Figure 22.16:

$x = r \cos \theta$

$y = r \sin \theta$

Example

Given the polar coordinates of a point Q(9,42°) find its rectangular coordinates:

$x = 9 \cos 42° = 6.69$

$y = 9 \sin 42° = 6.02$

The rectangular coordinates of Q are (6.69, 6.02).

Test yourself 22.4

1 Convert these rectangular coordinates into polar coordinates (2 dp).

A(4.5, 7.6) B(12.4, 45.8)
C(–7.7, 0.2) D(3.6, –10.4)
E(–34, –88) F(–9.01, 5.33)

2 Convert these polar coordinates into rectangular coordinates (2 dp).

G(23, 56°) H(7.45, 78°)
I(6.6, 128°) J(12, 200°)
K(77, 302°) L(2.5, 1.2 rad)

3 On an open flat plain, a driver is told to drive 5 km east, then 7 km north, to reach a given point. What is the most direct route to the point?

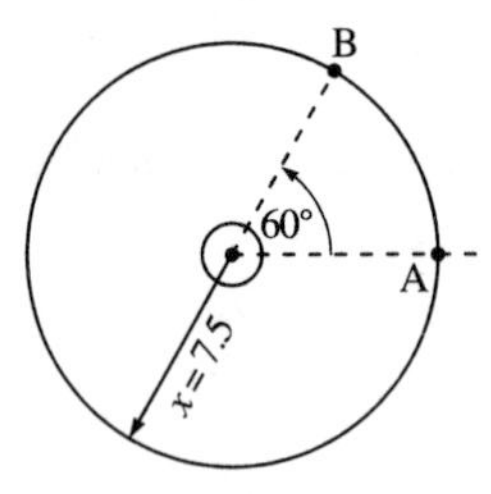

Figure 22.17

4 A point is marked on the rim of a pulley, radius 7.5. To begin with, the point is at A. Then the wheel turns 60° to bring the point to B. How far does the point rise and how far does it move towards the left as the wheel turns?

5 To assess your progress in this chapter, work the exercises in *Try these first*, page 306.

23 Using j

The symbol j often appears in engineering books. This chapter explains how to work with j and how to use it for solving problems connected with electrical and electronic circuits.

You need to know about the number line (page 172), trig ratios (Chapters 14 and 19) and vectors (Chapter 22).

Try these first

1 Simplify.

a $(4 + j7) + (6 + j2)$ **b** $(5 - j3) - (7 + j3)$

2 Solve $2x^2 - 3x + 12 = 0$

3 Convert to polar form (2 dp)

a $7 + j3$ **b** $-4 - j6$

4 Convert to rectangular form (1 dp)

a $16\underline{/57^\circ}$ **b** $256\underline{/355^\circ}$

5 Find the resultant of the vectors $3\ \underline{/25^\circ}$ and $2\ \underline{/75^\circ}$.

Coded instructions

In Chapter 22 we directed a person to the site of buried treasure, using rectangular coordinates. The instructions were:

> 'Start at the tree, facing east. Walk 30 paces forward, turn left, walk 40 paces forward. Dig there.'

The instructions are illustrated in Figure 22.15 and repeated in a shortened form in Figure 23.1. We simplify the instructions by assuming that the person must always start at the tree facing east. We assume that the person is to walk forward and that all numbers refer to paces. We assume that the last instruction is to dig. Making these assumptions, the instructions can be simplified to:

> '30 turn left 40'

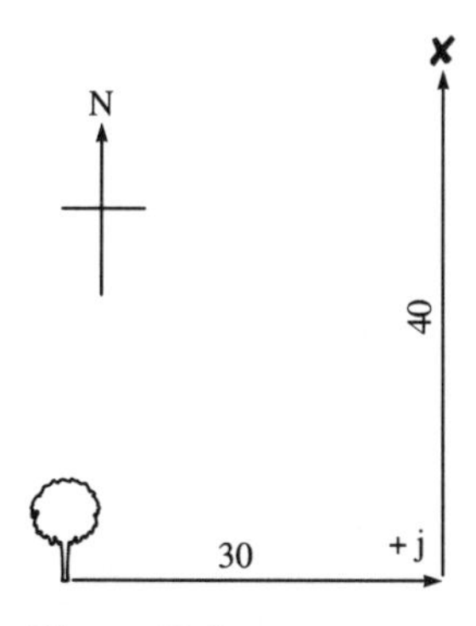

Figure 23.1

Given these coded instructions, we can direct the person to any point in the north-east quarter of the field. If we allow negative numbers to mean 'walk backward', we can reach *any* part of the field.

The code can be shortened even further by using a symbol instead of the words 'turn left'. That symbol is j. The instructions are reduced to a maths expression:

> $30 + j40$

It is essential to remember that j is not a variable. It is an instruction. The term j40 is not j multiplied by 40. It means a left turn followed by moving 40

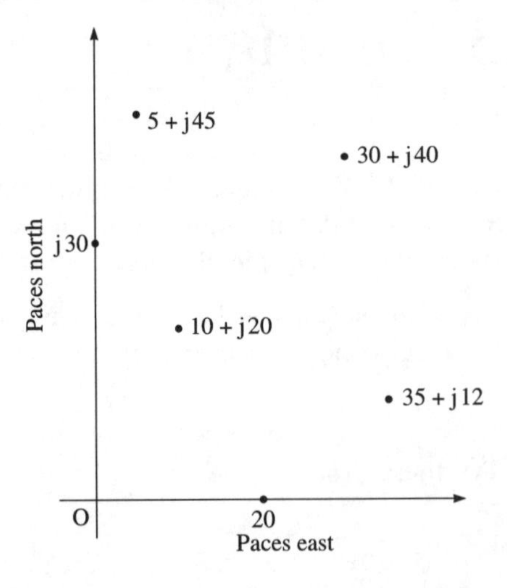

Figure 23.2

units. The instructions may be varied by changing the numbers. Figure 23.2 shows more points specified by using j. We have put in a pair of axes to help locate the origin (the tree) and indicate directions.

Negative numbers

If negative numbers mean 'walk backward', points in any of the four quadrants can be reached. For example, the instruction 10 + j(–20) means:

'Walk 10 paces forward, turn left, walk 20 paces backward'

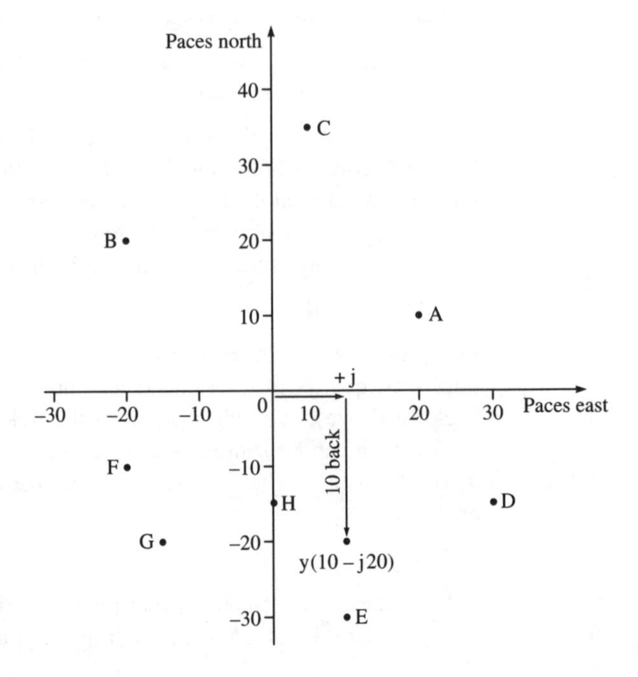

Figure 23.3

This brings us to point Y in Figure 23.3. We could also get to this position by walking 10 paces forward, turning *right* and walking 20 paces *forward.* If j means 'turn left' then let –j mean 'turn right', and:

$$10 + j(-20) = 10 - j20$$

It begins to look as if, in some ways at least, j may be *treated as* a variable, even though we have said that it is not a variable.

Test yourself 23.1

1 Write the coded form of the instructions for reaching points A to H. Figure 23.3.

2 On squared paper, plot the points specified by these coded instructions.

a $6 + j2$ **b** $2 + j8$

c $7 - j3$ **d** $-6 + j4$

e $j6$ **f** $-3 - j5$

g $5 - j4$ **h** $-j2$

More instructions

Figure 23.4 shows what happens when the first set of instructions have another set added to them.

The first instruction is	$3 + j5$
The second instruction is	$4 + j3$
The sum is	$7 + j8$

Using this system, it is easy to add two instructions together and find their sum. Addition follows the usual rules, but we add the two parts of the instructions separately. This makes sense because the number before the j refers to walking east, and the number after the j refers to walking north. These are in different directions and cannot be added.

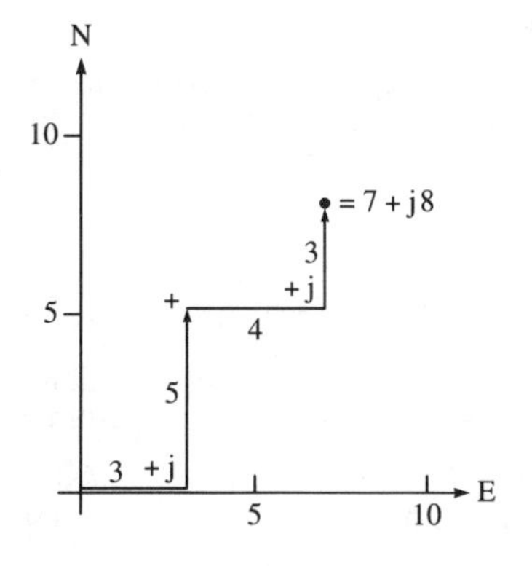

Figure 23.4

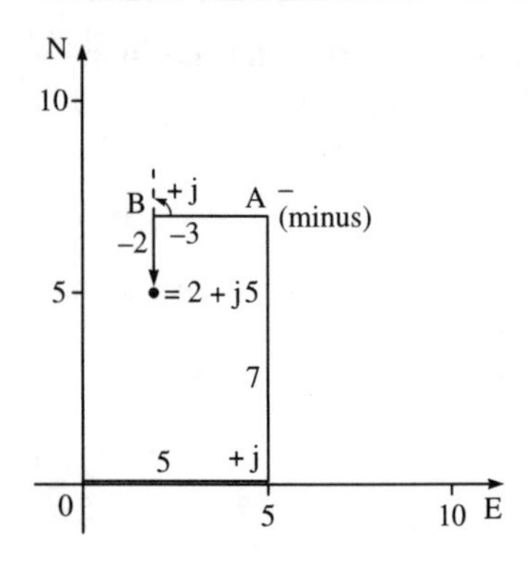

Figure 23.5

Instructions may also be subtracted (Figure 23.5):

The first instruction is	$5 + j7$
The second instruction is	$3 + j2$
The difference is	$2 + j5$

From point A in Figure 23.5, facing east, walk *backward* 3. Turn left to face north and walk *backward* 2.

Another example: From $(6 - j4)$ subtract $(2 + j3)$

$$\begin{aligned}(6 - j4) - (2 + j3) &= 6 - j4 - 2 - j3 \\ &= 4 - j7\end{aligned}$$

Follow the usual rule for signs when clearing the brackets.

Test yourself 23.2

1 Evaluate these additions.

a $(2 + j5) + (7 + j4)$ **b** $(4 + j3) + (4 + j2)$
c $(4 - j3) + (2 + j8)$ **d** $(-5 + j6) + (3 - j4)$
e $(5 + j) + (-5 + j2)$ **f** $(6 + j2) + (5 - j2)$
g $(-7 - j2) + (8 + j3)$ **h** $(10 - j) + j6$

2 Evaluate these subtractions.

a $(6 + j5) - 4$ **b** $(5 + j3) - (2 + j2)$
c $(2 + j7) - (4 + j4)$ **d** $(8 - j) - (3 + j)$
e $(4 - j3) - (4 + j3)$ **f** $(12 + j) - (10 - j7)$
g $(3 + j6) - (2 + j3)$ **h** $(-2 - j5) - (4 - j4)$

Vectors and j

The instruction format is ideal for specifying a vector.
If the horizontal (x-axis) component is a
and the vertical (y-axis) component is b
The vector is:

$a + jb$

Expressed in this format, vectors may be added and subtracted using the techniques described on this page.

Why use j?

We now have three ways of specifying a point:

	Example
Rectangular coordinates	P(3,4)
Polar coordinates	P(5,53°)
Instruction with j	$3 + j4$

Polar coordinates have a different basis and give different numbers, but the other two ways have the same pair of numbers written in different formats. We must justify inventing one more way of writing out what is almost the same thing. Let us extend the idea of j a little.

Here is a variation on the instructions:

$$20 + jj15$$

Or, we might write this as:

$$20 + j^2 15$$

j followed by another j means 'turn left, turn left again'. After turning left twice, the person is facing west. The effect of these instructions are to make the person walk 20 east, then 15 west. The point reached is 5 paces to the east of the origin. This point is also reached by the single instruction:

$$5$$

The instructions $(20 + j^2 15)$ and (5) have the same effect. They are equal:

$$20 + j^2 15 = 5$$

Although we have said that j is not a variable, let us see what happens if we treat it as a variable.

$$\begin{aligned} 20 + j^2 15 &= 5 \\ \Rightarrow \quad j^2 15 &= 5 - 20 \\ \Rightarrow \quad j^2 15 &= -15 \end{aligned}$$

Divide both sides by 15:

$$\Rightarrow \quad j^2 = -1$$

Taking the square root of both sides:

$$\underline{j = \sqrt{-1}}$$

Although j^2 has an easily understandable value, –1, the value of j is more difficult to understand. All squares are positive, so it is not possible to have the square root of a negative number. But, we can *imagine* such a number, even if it is not real. We represent it by the symbol j, and say that it is an **imaginary number**. Numbers such as 3 + j4 have a real part (3) and an imaginary part (j4). As these numbers have two different parts, we call them **complex numbers**.

Using i

You may find books that use the symbol i instead of the symbol j. It means exactly the same thing. Electrical and electronics books usually prefer to use j, because i may be confused with *i*, the symbol often used for electric current.

Although the numbers with j may be imaginary in whole or in part, this does not prevent them from being a useful tool for solving practical engineering problems.

Charting complex numbers

We already have a way of representing complex numbers in a diagram. All the diagrams in this chapter are examples. A diagram which represents complex numbers in this way is called an **Argand diagram**. It has two axes:

The horizontal axis represents real numbers, and we call it the **real axis** (Figure 23.6). All real numbers have a place on this axis, which extends infinitely far in both directions.

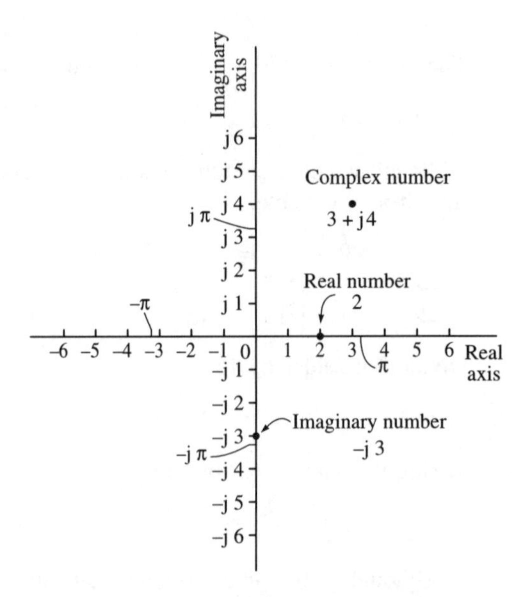

Figure 23.6

The vertical axis represents imaginary numbers (start at the origin, facing in a positive direction, turn left and move up or down). This too extends far in both directions. Numbers on this axis begin with a j to indicate the initial left turn. Imaginary numbers can include imaginary integers (examples: j2, j7, –j4), imaginary fractions (examples: j4.5, j0.39, –j3.72), and imaginary versions of numbers, such as jπ.

The area between the axes is where we find the complex numbers, such as 3 + j4.

A diagram such as this is used in electrical and electronic engineering for representing a type of vector known as a **phasor**. Figure 23.7 shows two

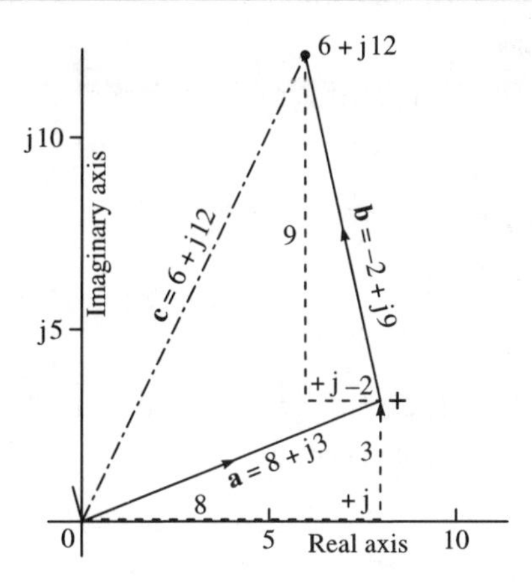

Figure 23.7

phasors being added to find their resultant. We add them by adding the complex numbers which represent them:

$$\mathbf{a} + \mathbf{b} = (8 + \mathrm{j}3) + (-2 + \mathrm{j}9)$$
$$= 8 + \mathrm{j}3 - 2 + \mathrm{j}9$$
$$= 6 + \mathrm{j}12$$

Polar form

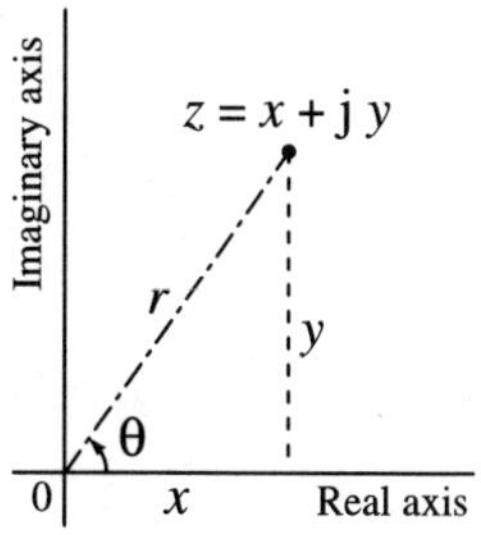

When applied to an Argand diagram, complex numbers are a form of rectangular coordinate. They can be converted to an equivalent in polar form, using the same routine as on page 317. In Figure 23.8, in which the rectangular coordinates are (x,y):

$$r^2 = x^2 + y^2$$

$$\Rightarrow \quad r = \sqrt{x^2 + y^2} \qquad (1)$$

$$\text{and} \quad \theta = \tan^{-1} y/x \qquad (2)$$

Figure 23.8

We also know by trig that:

$$x = r \cos \theta$$
$$y = r \sin \theta$$

Given a complex number $z = x + \mathrm{j}y$:

$$z = r \cos \theta + \mathrm{j}r \sin \theta$$

$$\Rightarrow \quad z = r(\cos \theta + \mathrm{j} \sin \theta) \qquad (3)$$

To convert (x, y) into polar form, we first find r and θ, using equations (1) and (2). Then we write them in the complex form, as in equation (3).

Examples

Express (7 + j6) in polar form.

$$x = 7 \qquad y = 6$$

$$r = \sqrt{x^2 + y^2} = \sqrt{7^2 + 6^2}$$

$$= \sqrt{85} = 9.22$$

$$\theta = \tan^{-1} 6/7 = 40.60°$$

In polar form, z = 9.22(cos 40.60° + j sin 40.60°)

Express (3 – j5) in polar form.

$$x = 3 \qquad y = -5$$

$$r = \sqrt{3^2 + (-5)^2} = \sqrt{34} = 5.83$$

$$\theta = \tan^{-1}(-5/3) = -59.04°$$

This is in the 4th quadrant, so θ= –59.04 + 360 = 300.96°

In polar form, z = 5.83(cos 300.96° + j sin 300.96°)

The polar form of all complex numbers is in the form $r(\cos\theta + \text{j}\sin\theta)$. Only r and θ vary. For this reason, we usually write polar form in a shorter format:

$$r\ /\underline{\theta}$$

In the two examples above:

$$z = 9.22\ /\underline{49.40°}$$

$$z = 5.83\ /\underline{300.96°}$$

Polar form to rectangular form

This follows the routine from page 317.

Example

Convert 4.6 $/\underline{58°}$ into rectangular form.

$$r = 4.6 \qquad \theta = 58°$$
$$x = r\cos\theta = 4.6\cos 58° = 2.44$$
$$y = r\sin\theta = 4.6\sin 58° = 3.90$$

$$z = 2.44 + \text{j}3.9$$

Reasons for converting

Addition and subtraction of complex numbers is much simpler if they are in rectangular form. If we are given two vectors or phasors in polar form and have to find their resultant:

1 Convert them to rectangular form
2 Add them
3 Convert the resultant back to polar form.

Example

Find the resultant of **a** = 4.5 $\angle 18°$ and **b** = 3.6 $\angle 87°$.

1 Converting **a** into rectangular form:

$r = 4.5 \quad \theta = 18°$
$x = r \cos \theta = 4.5 \cos 18° = 4.2798$
$y = r \sin \theta = 4.5 \sin 18° = 1.3906$
$\mathbf{a} = 4.2798 + j1.3906$

Converting **b** into rectangular form:

$r = 3.6 \quad \theta = 87°$
$x = 3.6 \cos 87° = 0.1884$
$y = 3.6 \sin 87° = 3.5951$
$\mathbf{b} = 0.1884 + j3.5951$

2 Adding:

$$\mathbf{c} = \mathbf{a} + \mathbf{b} = 4.2798 + j1.3906 + 0.1884 + j3.5951$$
$$= 4.4682 + j4.9857$$

3 Converting **c** into polar form:

$x = 4.4682 \qquad y = 4.9857$

$$r = \sqrt{4.4682^2 + 4.9857^2} = \sqrt{44.8220} = 6.6949$$
$$\theta = \tan^{-1}(4.9857/4.4682) = 48.13°$$

The resultant in polar form is **c** = 6.69 $\angle 48.13°$

When we need to multiply two complex numbers together, or divide one by another, the calculations are much simpler if the numbers are in polar form. Multiplication and division are not included in this book, so we shall not illustrate this.

Test yourself 23.3

1 Convert from rectangular form to polar form.

a 4 + j6 **b** 7.2 + j5.1
c 23 + j17 **d** 8.3 – j4
e –8.1 + j6 **f** –12 – j10

2 Convert from polar form to rectangular form.

a 67 $\angle 55°$ **b** 1.24 $\angle 10°$
c 3.5 $\angle 0°$ **d** 4 $\angle 167°$
e 100 $\angle 61°$ **f** 2.11 $\angle 101°$

3 Calculate the resultant of these two vectors, in polar form.

a = 5.3 $\angle 50°$ **b** = 2.9 $\angle 15°$

4 If you have a scientific calculator or graphic calculator with built-in routines for conversions, use it to check your answers to the questions above.

Quadratic roots

We now have an answer to the difficulty met with on page 304. We were trying to solve the equation:

$$2x^2 + 3x + 7 = 0$$

The difficulty was that we needed to evaluate $\sqrt{-47}$. Now we can do so:

$$\sqrt{-47} = \sqrt{-1 \times 47} = \sqrt{-1} \times \sqrt{47}$$

$$= \mathrm{j} \times 6.856 = \mathrm{j}6.856$$

Using this in the quadratic formula:

$$x = \frac{-3 \pm \mathrm{j}6.856}{4}$$

There are two solutions:

$$x = \frac{-3 + \mathrm{j}6.856}{4} = -0.75 + \mathrm{j}1.71$$

$$x = \frac{-3 - \mathrm{j}6.856}{4} = -0.75 - \mathrm{j}1.71$$

The solutions are complex numbers: $x = -0.75 + \mathrm{j}1.71$ or $x = -0.75 - \mathrm{j}1.71$.

If you are solving a quadratic equation using the formula and find that $(b^2 - 4ac)$ is negative, the equation has two imaginary roots. Perhaps your practical problem does not allow imaginary roots, in which event there is no solution. If imaginary roots are allowable, proceed to solve the equation as above.

Test yourself 23.4

1 Solve $5x^2 + 2x + 7 = 0$, to 3 sf.

2 Solve $4x^2 - 5x + 2 = 0$, to 3 sf.

3 To assess your progress in this chapter, work the exercises in *Try these first*, page 319.

24 Rates of change

Concrete hardens, furnaces cool, springs stretch, vehicles accelerate – *change* is an essential feature in most branches of technology. Knowing or predicting the rates of change can be vitally important. The branch of maths dealing with rates of change is known as *differential calculus*. Here we explain how it works and how to use it.

You need to know about powers (Chapter 13), trig ratios (Chapters 14 and 19), graphs (Chapter 17), and quadratic equations (Chapter 21).

Try these first

1 Find the derivatives of:

a $2e^{3x} - 3e^{-2x}$

b $3x^2 - 5x + 2$

c $4 \sin 2x - 8$

2 The voltage across a capacitor is rising according to the function $v = 5(1 - e^{-2t})$, where t is the time elapsed in seconds since charging began. Find v and the rate of increase of v after 1.5 s (3 sf).

3 A metal sphere is released and allowed to fall freely. Ignoring air resistance, the vertical distance fallen by a metal sphere is $s = 5t^2$, where s is in metres and t in seconds. Find the vertical velocity of the sphere after 3.5 s and its acceleration.

4 Given that $y = x^3 - 7x^2 - 24x + 3$, find the stationary points and state if each is a minimum or a maximum.

A matter of timing

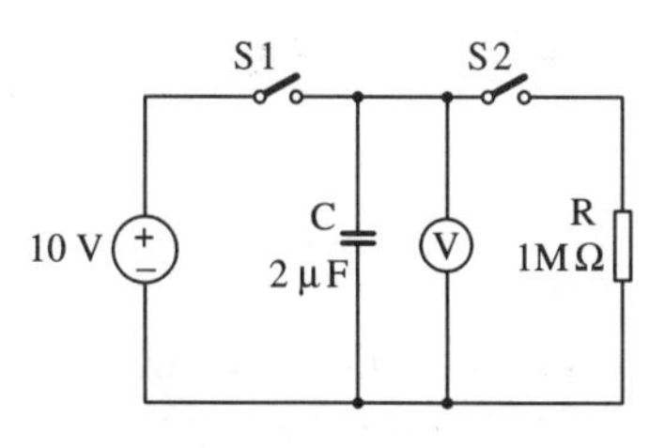

Figure 24.1

A discharging capacitor is at the heart of many types of electronic timing circuit. Figure 24.1 shows a simple circuit for measuring the rate of discharge. Switch S1 is closed and current flows from the voltage source on the left, charging the capacitor until the voltmeter V reads 10 V. Then S1 is opened and S2 closed. Now the current flows from the capacitor, through the resistor R. The voltage falls. The rate at which it falls depends on the capacitance of C and the resistance of R. If we read the voltage once a second and plot a graph of the results, we obtain the upper curve of Figure 24.2.

The voltage falls rapidly at first because the voltage is high, forcing a relatively large current through the resistor. As the capacitor becomes more and more discharged, the voltage falls, and less current flows. The voltage

Functions

Given the equation:

$$y = x + 4$$

We can substitute any value of x, and obtain a corresponding value of y.

Examples

If $x = 3$, then $y = 7$
If $x = -1$, then $y = 3$
If $x = 0$, then $y = 4$
If $x = 5.621$ then $y = 9.621$

. . . and so on.

Because each value of x produces one *and only one* value of y, then

(x + 4) is a *function* of x

In the example above, x could have *any* value. Or there might be reasons why the values of x are limited in some way. They may have to be integers, for example, or be restricted in size.

The set of values that x can take is called the **domain** of the function.
The corresponding set of values of y is called the **range** of the function.

falls more and more slowly as time goes on. With the values shown in Figure 24.1, the **function** for the voltage v at any given time t is:

$$v = 10e^{-0.5t}$$

where e is the exponential constant (page 187). As time passes, each new value of t produces a new value of v.

This is an example of exponential decay (page 248). Figure 24.2 shows what the voltage is at any given time. What we would also like to know is the **rate** of fall of voltage at any given time. Among other things, this allows us to calculate the current at this time.

Note for electronics students

A more specific equation is:

$$v = v_0 e^{-t/RC}$$

in which v_0 is the initial voltage (10 V in this example)
R is the resistance in ohms (1 MΩ)
C is the capacitance in farads (2 μF)

RC is the **time constant** of the circuit. $-t/RC = -0.5t$, the value given in the text.

The rate of fall

The rate of fall of voltage is the gradient of the upper curve of Figure 24.2, at any given point. In theory, it is possible to find the gradient at any point by drawing a **tangent** (see box, page 203). In practice, it is difficult to draw a

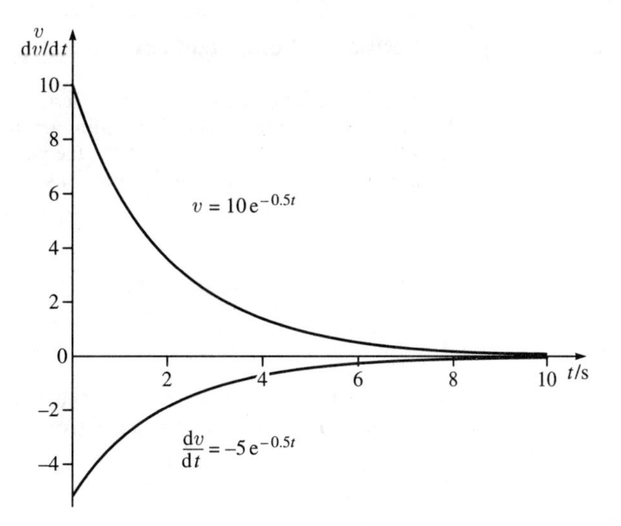

Figure 24.2

tangent which passes through exactly the right point with exactly the right slope. We need a mathematical way of finding gradients precisely.

Finding the gradient is made easy by using a technique known as **differentiating**. Given a function, we **differentiate** this and obtain another function, its **differential**, which tells us its rate of change. This is also known as the **derivative** because it is **derived from** the original equation.

Differentiating is a matter of applying a few simple rules. We shall not attempt to explain *how* these rules work, but will demonstrate that they *do* work and will show you how to make use of them for solving practical problems.

Differentiating an exponential function is easy. If the function is:

$$e^{ax}$$

where a is a constant, the derivative is:

$$ae^{ax}$$

We simply take the coefficient of x and place it in front of the function, as shown in these examples:

Function	*Derivative*
e^{2x}	$2e^{2x}$
e^{5x}	$5e^{5x}$
e^{-3x}	$-3e^{-3x}$
$4e^{2x} + 5$	$8e^{2x}$

The last example shows that, if the function already has a coefficient, we multiply this by the coefficient of x. It also shows that constants (for example, +5) are ignored when differentiating (see box).

Derivative of a constant term

The derivative of a constant term (a term that does not contain the variable) is zero. As we saw on page 237 adding or subtracting a constant to a function simply shifts the curve up or down the page. It has no effect on its shape and therefore has no effect on the gradients of the curve.

In the decay equation of Figure 24.2:

$$v = 10e^{-0.5t}$$

Here the variable is t instead of x, but the rule applies just the same. The coefficient of t is –0.5, so the derivative is:

$$\frac{dv}{dt} = -5e^{-0.5t}$$

The symbol dv/dt means the derivative of v with respect to t. In other words the **rate of change** of v, with respect to t. This symbol is pronounced 'dee-vee by dee-tee'. Note that it is the **symbol** for a derivative; it does *not* mean d times v, divided by d times t.

The graph of dv/dt is plotted as the lower curve in Figure 24.2. It has its most *negative* value when $t = 0$. This is when the capacitor is discharging most rapidly and the voltage is *falling* at the greatest rate. As time passes, dv/dt becomes less and less negative. This corresponds to the voltage falling less and less rapidly. Toward the right of the graph, it *almost* reaches zero. In the same way, the voltage curve flattens out until the voltage *almost* stops falling.

We can do more than show the general way in which the voltage falls. We can find the exact rate of fall at any given time. For example, to find the rate of fall after 3 seconds, substitute $t = 3$ in the equation of the derivative:

$$dv/dt = -5e^{-0.5t} = -5e^{-0.5\times3}$$
$$= -5e^{-1.5} = -1.116$$

The negative result indicates falling voltage. The calculation is quickly done, using the e^x key of a calculator.

The rate of fall after 3 s is $1.116\,Vs^{-1}$.

Similarly, the rate of fall after 5 seconds is:

$$dv/dt = -5e^{-2.5} = -0.410\,Vs^{-1}$$

We can also find when the rate of fall reaches any given value. For example, find when the rate of fall is $1\,Vs^{-1}$.

$$dv/dt = -5e^{-0.5t} = -1$$
$$\Rightarrow \quad = e^{-0.5t} = 0.2$$

Take natural logs of both sides:

$$-0.5t = \ln 0.2 = -1.6094$$

$$\Rightarrow \quad t = 3.2188$$

The rate of fall is 1 Vs^{-1} after 3.22 s.

Test yourself 24.1

1 Find the derivatives of these functions.

a $2e^{3t}$ **b** $7e^{2x}$

c $3.5e^{4x}$ **d** $2e^{1258n}$

e $2e^{-4t}$ **f** $6e^{-0.25x} + 3$

g $-7e^{-2t}$ **h** $8e^{0.125t} - 7$

2 In another circuit similar to Figure 24.1, the equation for decay of the voltage is $v = 4e^{-0.2t}$. Find the derivative of this and use it to calculate

a the rate of fall of voltage after 1.5 s

b the time taken for the rate of voltage fall to reach 0.5 Vs^{-1} (3 sf)

3 A lagged tank of water is heated by an immersion heater. When timing begins, $t = 0$, and the temperature of the water is $h = 40°C$. The power is switched off and the water gradually cools, external temperature remaining constant at 0°C. The temperature h of the water at time t hours is given by:

$$h = 40e^{-0.15t}$$

Find

a the rate of cooling after 3 hours

b how long the water takes to cool to 20°C (3 sf).

4 A 220 nF capacitor is charged to 10 V, then discharged through a 1 MΩ resistor. The instantaneous voltage across the capacitor is:

$$v = V_o . e^{-t/RC}$$

The constants are: $V_o = 10$, $R = 1 \times 10^6$ and $C = 220 \times 10^{-9}$. Calculate the rate of change of v:

a as discharge begins ($t = 0$)

b 0.25 s later (3 sf).

Derivatives of trig ratios

The trig ratios are especially important because they feature so much in the maths of alternating currents. They also have applications in the operation of machinery, if there are rotating parts or levers.

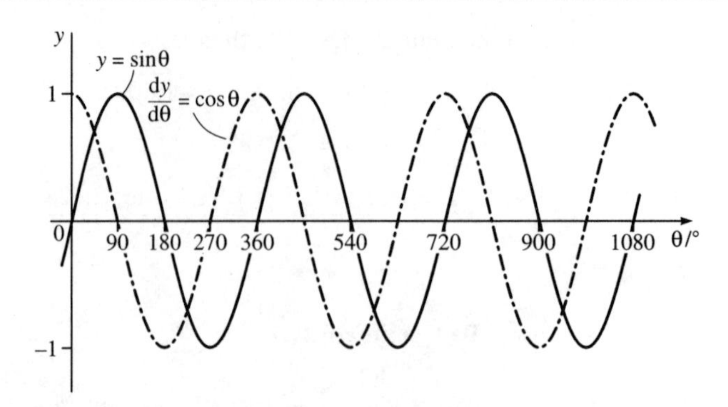

Figure 24.3

Two useful derivatives are:

Function	*Derivative*
$\sin a\theta$	$a \cos a\theta$
$\cos a\theta$	$-a \sin a\theta$

In these terms, a is a constant. In Figure 24.3 we make $a = 1$ and have plotted the $y = \sin \theta$ and its derivative $dy/d\theta = \cos \theta$. The cosine curve shows how the gradient of the sine curve varies with θ:

- When $\theta = 0°$, the sine curve slopes up to the right at its steepest. At this value of θ, $\cos \theta$ has its maximum positive value, +1.
- When $\theta = 90°$, the sine curve reaches a maximum and its gradient is zero. At this point, $\cos \theta = 0$.
- When $\theta = 180°$, the sine curve slopes most steeply down to the right. This is its greatest negative gradient. This is when $\cos \theta$ has its minimum value, –1. From $\theta = 180°$ onward, the value of $dy/d\theta$ continues to match the slope of $y = \sin \theta$.

Example

A crane has an arm 12 m long. The load is on the ground when the arm is horizontal and $\theta = 0°$. The arm is raised at the rate of 0.002 radians per second. What is the vertical velocity of the load when it is 4 m above ground?

The height h of the load above ground is given by the function:

$$h = 12 \sin \theta$$

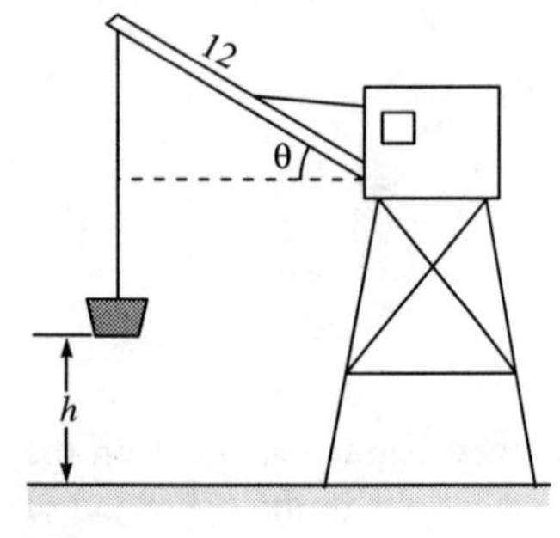

Figure 24.4

When the load is 4 m above ground, $h = 4$, and

$$4 = 12 \sin \theta$$

$$\Rightarrow \quad \theta = \sin^{-1} (1/3) = 0.340 \text{ rad}$$

The derivative is found by using the rule above:

$$dh/d\theta = 12 \cos \theta$$

When $\theta = 0.340$ rad

$$dh/d\theta = 12 \cos 0.340 = 12 \times 0.9428 = 11.31$$

Changing the independent variable

In this example we differentiate with respect to θ but, really, we want to know the derivative with respect to t. Use this equation to convert from θ to t:

$$\frac{dh}{dt} = \frac{dh}{d\theta} \times \frac{d\theta}{dt}$$

Although we have said that these are *symbols* for gradients, we are allowed, under certain conditions, to 'cancel' parts of the symbols. On the right, cancelling the $d\theta$'s leaves dh/dt, so the equation is valid.

In this example, we calculate $dh/d\theta$ and are told $d\theta/dt$. Multiplying them together gives dh/dt, which is the derivative we require. This technique applies to other examples in which we need to change the independent variable.

The rate of motion of the load is 11.31 m per *radian*. This result gives the rate at which the load is lifted with respect to the increase in the *angle* or the arm. To find the lifting of the load with respect to *time*, we multiply this by $d\theta/dt$, the rate of increase of θ. The problem quotes this as 0.002 rad s^{-1}. The box explains that:

$$\frac{dh}{dt} = \frac{dh}{d\theta} \times \frac{d\theta}{dt} = 11.31 \times 0.002 = 0.02262\,\text{ms}^{-1}$$

The load is lifted at 22.6 mm s^{-1} when it is 4 m above the ground.

Derivatives of logs

The derivative of ln ax is $1/x$. The value of the constant a makes no difference to the gradient of the curve, as can be seen from Figure 24.5. At any value of x, all three curves have the same gradient. The gradient is high for low values of x ($1/x$ is large) but approaches zero at very high values of x.

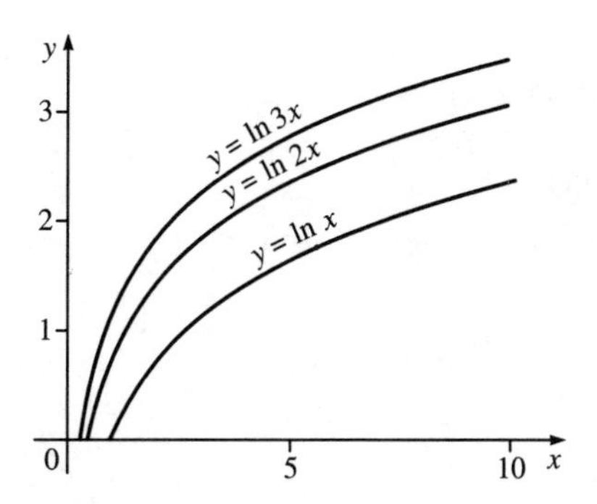

Figure 24.5

Example

Logarithmic amplifiers are used in certain electronic measuring instruments because they allow a wide range of values to be measured on a single scale. In a certain logarithmic amplifier:

$$u = 5 \ln 4v$$

where v is the input voltage and u is the output voltage. The amplification of the amplifier is, by definition, the rate of change of output voltage with respect to input voltage: In symbols:

amplification, $A = \mathrm{d}u/\mathrm{d}v$

Find u and A when: **a** $v = 2\,\mathrm{V}$ **b** $v = 200\,\mathrm{V}$

a When $v = 2\,\mathrm{V}$, $u = 5 \ln 8 = 10.397$

$A = \mathrm{d}u/\mathrm{d}v = 5/v = 5/2 = 2.5$

When $v = 2\,\mathrm{V}$, $u = 10.4\,\mathrm{V}$ and $A = 2.5$

b When $v = 200\,\mathrm{V}$, $u = 5 \ln 800 = 33.423$

$A = 5/200 = 0.025$

When $v = 200\,\mathrm{V}$, $u = 33.4\,\mathrm{V}$ and $A = 0.025$

Although the input is increased by 100 times, the output increases by only 3.3 times. This is due to the gradual reduction in amplification as voltage increases.

Derivative of functions with more than one term

Given a function with two or more terms, connected by + and – signs (no ×, no ÷, no brackets), each term is differentiated separately.

Example

$$y = 4 \ln 5x + e^{2x} - \sin 2x - e^{-3x}$$
$$\mathrm{d}y/\mathrm{d}x = 4/x + 2e^{2x} - 2 \cos 2x + 3e^{3x}$$

Keep the signs the same, except when the term itself changes sign, as in the last term above.

Test yourself 24.2

1 Find the derivatives of these functions.

a $\sin \theta$, where $\theta = 36°$ **b** $\sin 2\theta$, where $\theta = 18°$

c $3 \cos 1.5x$ **d** $4 \ln 2x$

e $4 \sin 2\theta + 4 \cos 2\theta$ **f** $\ln 0.5x - e^{0.5x}$

g $4 \ln 7x$ **h** $3 \ln 7x + \sin x$

2 The instantaneous voltage across a capacitor is $v = 5 \sin 4t$. The instantaneous current entering the capacitor is:

$$i = C \times \frac{\mathrm{d}v}{\mathrm{d}t}$$

Calculate the current when $t = 0.1$ s and $C = 1 \times 10^{-6}$ F (3 sf). [*Hint*: angles in radians]

3 A piston is driven by a crank and lever mechanism so that, as the crank rotates, the piston slides to and fro in a cylinder. The distance of the piston from the end of the cylinder is y cm, the length of the crank r is 15 cm, and the crank rotates at 2 rad per second. The geometry of the mechanism is such that:

$$y = r(1 - \cos\theta) + \frac{r^2}{360}(1 - \cos 2\theta)$$

where θ is the angle turned by the crank from the horizontal. If $d\theta/dt = 0.5$ rad s^{-1}, find the position y of the piston and its velocity, dy/dt, 1.5 s after the crank has passed through the $\theta = 0°$ position.

4 The output u of a logarithmic amplifier, with input v is:

$$u = 1.2 - 4.5 \ln 3.5v$$

Find the output and the amplification (du/dv) when $v = 0.2$ V.

Derivatives of polynomials

Polynomials are functions in which the terms are all powers of a single variable, possibly with one or more constant terms. The terms are added to or subtracted from each other, but not multiplied or divided. Terms are differentiated separately. There are just three simple rules:

- Ignore constants (as usual)
- Reduce the power of x by 1 in each term
- Multiply each term by the power of x.

Example

Find the derivative of:

$$y = 2x^3 - 10x^2 + 7x + 60$$

Term	*Its power is*	*Reduce the power*	*Multiply by the power*	*Derivative*
$2x^3$	3	$2x^2$	$3 \times 2x^2$	$6x^2$
$-10x^2$	2	$-10x$	$2 \times -10x$	$-20x$
$+7x$	1	$+7$	$1 \times +7$	$+7$
$+60$	Ignore constant			0

Assembling the derivatives into a new function:

$$\frac{dy}{dx} = 6x^2 - 20x + 7$$

Figure 24.6 shows y and dy/dx on the same pair of axes. The equation for y yields a typical cubic curve (page 250). The derivative is a parabola (page 237). Compare these two curves, to see how the value of the derivative matches the slope of the cubic curve for every value of x. In particular, note how the parabola cuts the x-axis ($dy/dx = 0$), at two values of x. These are the values of x when the cubic curve has zero gradient.

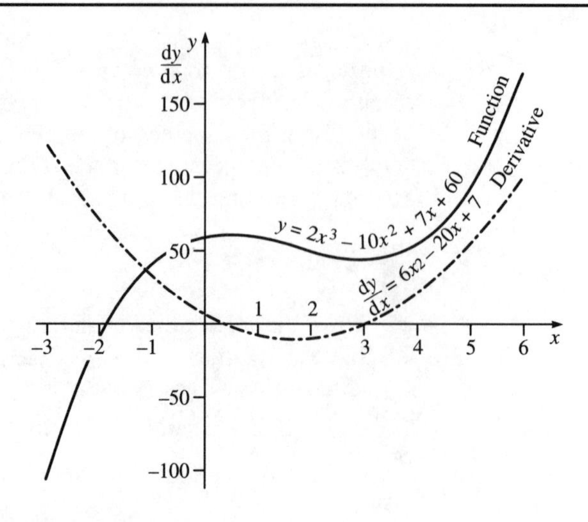

Figure 24.6

Equations of motion

A vehicle is travelling along a straight road. Timing begins as it passes a bus shelter. Its distance s, in metres, from the shelter is given by:

$$s = 4t + 3t^2$$

The derivative of this is:

$$ds/dt = 4 + 6t \qquad (1)$$

ds/dt is the rate of change of distance with time. In other words, ds/dt is the velocity. The velocity of the vehicle after 5 seconds is:

$$v = ds/dt = 4 + 6 \times 5 = 30$$

The velocity is $30\,\text{ms}^{-1}$.

Second derivatives

A derivative of a derivative is a **second derivative**. The symbol for a second derivative of y with respect to x is:

$$\frac{d^2y}{dx^2}$$

read as 'dee-two-why by dee-ex-squared'.

In the example on this page:

Original function	s is the distance, or displacement
1st derivative	ds/dt is the velocity (= v)
2nd derivative	$d^2s/dt^2 = dv/dt$ is the acceleration

More uses for 2nd derivatives on page 345.

We can go one stage further than this by finding the derivative of the velocity. Equation (1) gives:

$$v = 4 + 6t$$

Differentiate this:

$$dv/dt = 6$$

The rate of change of velocity with time is its accleration.

The acceleration is $6\,ms^{-2}$.

Test yourself 24.3

1 Find the derivatives of these functions:

a $4x$ **b** $5x^2$ **c** 12

d $6x^2$ **d** $2x + 3$ **e** $x^3 - x^2$

f $4x^2 + 3x$ **g** $x^4 - 6$ **h** $5x^3 + x$

2 Given the function $y = 3x^2 - 5x + 9$, find its derivative, dy/dx. Then find dy/dx when $x = 12$.

3 Given the function $y = 4x^3 + 3x^2 - 2x + 9$, find its derivative, dy/dx. Then find dy/dx when $x = 12$.

4 The volume of a sphere is given by $v = \dfrac{4\pi r^3}{3}$, find dv/dr the rate of change of volume with respect to changes in radius, when $r = 6$.

5 A railway wagon is set in motion and then runs up a sloping track. The distance s travelled from the lower end of the track, when timing begins is:

$$s = 36t - 2t^2$$

where t is in seconds. The velocity of the truck is ds/dt. Find this derivative. Then find how long the wagon takes to come to a halt. From this, find the total distance travelled. Find its acceleration up the slope.

6 The trough in Figure 18.17 is supplied with water from a tap which delivers $4\,m^3$ per hour. If x is the depth of water in the trough, show that the volume $v = 40x^2/3$. The volume of water delivered by the tap in t hours is $4t$. Show that $x = 0.5477t$. Calculate the depth after 4 hours. Find the derivative dx/dt and find the rate of rise of the water at this time.

7 A ball is thrown vertically upward. The vertical distance s metres travelled in t seconds is:

$$s = 20t - 5t^2$$

Write an equation for its velocity, ds/dt. Use this to find when the ball reaches its maximum height. Find the maximum height. Find the acceleration of the ball.

Explore this

Here are two spreadsheet ideas to help you understand derivatives of polynomial functions.

The first one calculates dy/dx:

```
   A.... B.... C.... D.... E.... F.... G.... H....
 1 DERIVATIVES
 2 Functions
 3   y =         x3   +        x2   +        x    +
 4
 5 dy/dx  =
 6               x2   +        x    +
 7
 8 Gradient at a point
 9   x =
10   y =                  m =
```

These are the formulae:

```
B6=B3*3
D6=D3*2
F6=F3
B10=B3*B9*B9*B9+D3*B9*B9+F3*B9+H3
E10=B6*B9*B9+D6*B9+F6
```

Key in the coefficients of the functions and also a constant, if required. When the spreadsheet is updated, the coefficients of dy/dx are calculated and appear in the second function. Also, if you key in a value for x, the spreadsheet calculates the value of the function and the gradient of the graph at that point.

The second spreadsheet calculates values for plotting a graph of a function and its derivative:

```
   A.... B.... C.... D.... E.... F.... G.... H.... I.... J.... K.... L....
 1 GRAPHING DERIVATIVES
 2 Functions
 3   y =         x3   +        x2   +        x    +
 4
 5 dy/dx  =
 6               x2   +        x    +
 7 Plot graph from:
 8   x =
 9   y =
10 dy/dx
```

There are the formulae:

```
B6=B3*3
D6=D3*2
F6=F3
C8=B8+1
D8=C8+1
E8=D8+1
F8=E8+1
G8=F8+1
H8=G8+1
I8=H8+1
J8=I8+1
K8=J8+1
L8=K8+1
B9=B3*B8*B8*B8+D3*B8*B8+F3*B8+H3
C9=B3*C8*C8*C8+D3*C8*C8+F3*C8+H3
D9=B3*D8*D8*D8+D3*D8*D8+F3*D8+H3
E9=B3*E8*E8*E8+D3*E8*E8+F3*E8+H3
F9=B3*F8*F8*F8+D3*F8*F8+F3*F8+H3
G9=B3*G8*G8*G8+D3*G8*G8+F3*G8+H3
H9=B3*H8*H8*H8+D3*H8*H8+F3*H8+H3
I9=B3*I8*I8*I8+D3*I8*I8+F3*I8+H3
J9=B3*J8*J8*J8+D3*J8*J8+F3*J8+H3
K9=B3*K8*K8*K8+D3*K8*K8+F3*K8+H3
L9=B3*L8*L8*L8+D3*L8*L8+F3*L8+H3
B10=B6*B8*B8+D6*B8+F6
C10=B6*C8*C8+D6*C8+F6
D10=B6*D8*D8+D6*D8+F6
E10=B6*E8*E8+D6*E8+F6
F10=B6*F8*F8+D6*F8+F6
G10=B6*G8*G8+D6*G8+F6
H10=B6*H8*H8+D6*H8+F6
I10=B6*I8*I8+D6*I8+F6
J10=B6*J8*J8+D6*J8+F6
K10=B6*K8*K8+D6*K8+F6
L10=B6*L8*L8+D6*L8+F6
```

Key in coefficients and a constant for the expression for function y. As above, the spreadsheet calculates the derivatives. Key in the lowest value for x at cell B8. The spreadsheet then calculates a sequence of 10 more values of x, incrementing by 1 each time. For each value of x it calculates the corresponding value of y, and of dy/dx. The data in rows 8 to 10 can be saved and used in a graphic program. This spreadsheet could be modified to allow you to key in the increments of x, so that the graphs are plotted in steps of 0.5 or 0.2, for example.

Explore this too

Use a graphic calculator or graphic software to plot graphs of a function and its derivative and display both at the same time. This is a good way of getting to understand how the derivative varies with respect to the function. First key in the function, for example: $y = 5x^2 - 7x + 9$. Then, without clearing the screen, key in its derivative: $y = 10x - 7$. It really should be $dy/dx = \ldots$, but the calculator insists that the equation for plotting begins with $y = \ldots$). Note how the slope of the derivative graph corresponds to changes in the gradient of the graph of y. Note where the derivative graph cuts the x-axis; at the same value of x, the graph of y should have zero gradient.

Comparing graphs in this way is a check that you are learning to calculate derivatives correctly. Here are some functions to try, with values of x from about -5 to $+7$, and with the calculator set to operate in radians:

$y = 3 \sin x$ $\quad\quad$ $y = 2 \cos 3x$

$y = 2e^{0.5x}$ $\quad\quad$ $y = 3e^{-0.2x}$

$y = 2 \ln 8x$ (set x in range 0 to 10 for this)

$y = 4x^2 + x - 12$ $\quad\quad$ $y = -2x^2 + 3x - 8$

$y = 2x^3 - 3x^2 + 25$ $\quad\quad$ $y = x^4 - 7x^2 - 10$

We leave you to work out the derivatives for yourself. Write some functions of your own and try these too.

Negative and fractional powers

The rules for derivatives also apply to negative and fractional powers.

Example

Find the derivative of $3x^{-2}$

The power is -2

Reduce the power by 1 to give -3

Multiply by -2 to give $-2 \times 3x^{-3} = -6x^{-3}$

The derivative of $3x^{-2}$ is $-6x^{-3}$.

Another way of writing a negative power is in the form of a reciprocal (page 178), so the result above could also be written:

The derivative of $3/x^2$, is $-6/x^3$.

We find the derivative of a square root by using the fact that:

$\sqrt{x} = x^{0.5}$

Example

Find the derivative of $7x^{0.5}$

The power is 0.5

Reduce the power by 1 to give $7x^{-0.5}$

Multiply by 0.5 to give $3.5x^{-0.5}$

The derivative of $7x^{0.5}$ is $3.5x^{-0.5}$.

Because $x^{-0.5}$ is the same as $1/x^{0.5}$, which is the same as $1/\sqrt{x}$, we can also write this result as:

The derivative of $7\sqrt{x}$ is $3.5/\sqrt{x}$.

Stationary points

Stationary points are points on a curve at which the gradient is zero; a rising curve turns and begins to fall, or a falling curve turns and begins to rise. On the turns, the value of y is stationary for an instant. Figure 24.7 shows a cubic curve with two stationary points. At the point on the left, the gradient changes from positive to negative and the value of the function is a **maximum**. At the point of the right, the gradient changes from negative to positive and the value of the function is a **minimum**.

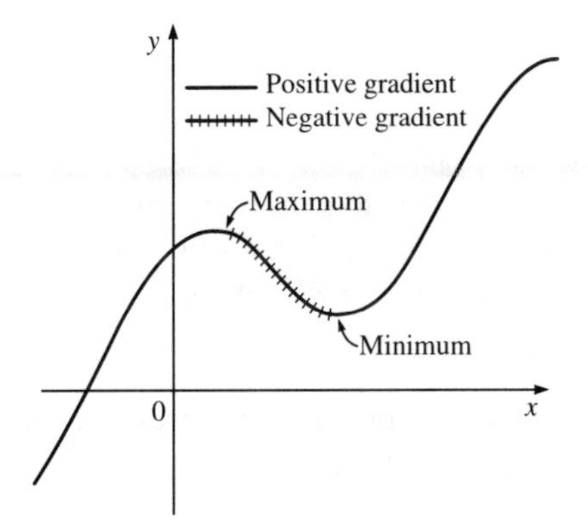

Figure 24.7

In Figure 24.8 the curve labelled 'Function' is a cubic curve. It is the same curve as in Figure 24.6. There is a maximum when x is approximately 0.5 and a minimum when x is approximately 3. The line curves so gradually that it is not possible to find the exact values of x simply by looking at the graph. We need to find a way of calculating these values.

The way to find the values is already shown in Figure 24.8. This is the curve of the derivative. This cuts the x-axis at the values of x for which the gradient of the function curve is zero. These points are easier to determine. It looks as if the maximum is when $x \approx 0.4$ and the minimum is when $x = 3$ (almost exactly?).

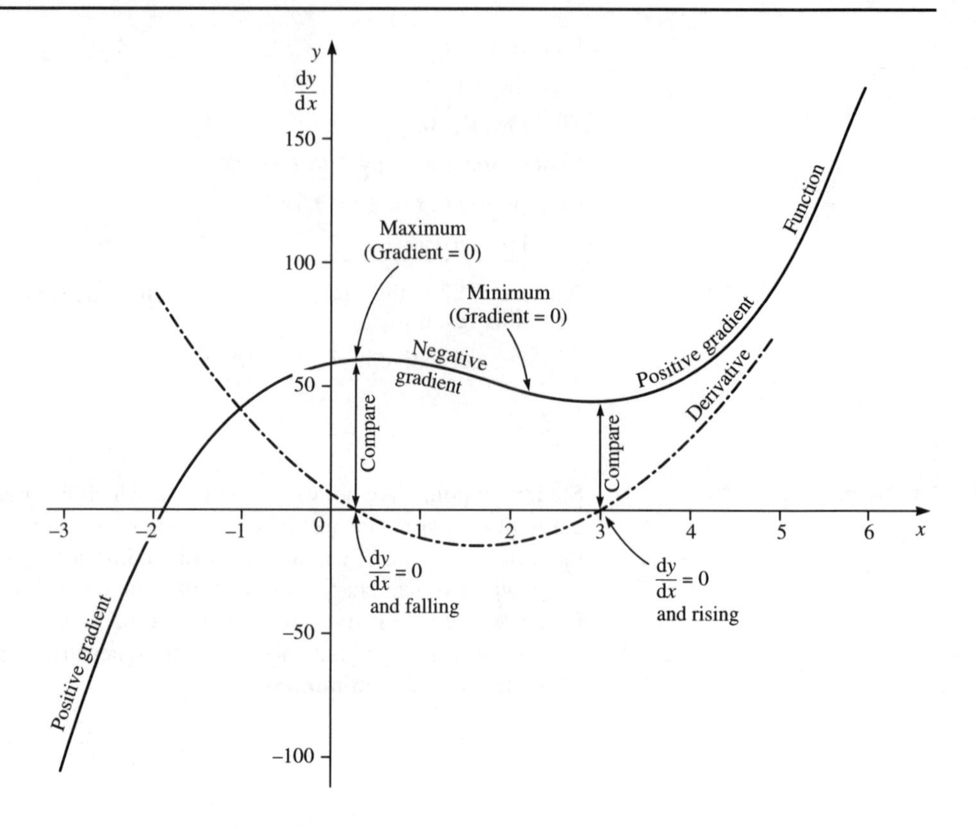

Figure 24.8

There is no need to draw the graph and measure points on the x-axis. All we need to do is to solve the equation of the derivative when $dy/dx = 0$. This is where it cuts the x-axis. Making the derivative equal to 0:

$$6x^2 - 20x + 7 = 0$$

The left side of the equation does not factorize, so we use the quadratic formula (page 303):

$$a = 6 \qquad b = -20 \qquad c = 7$$

$$x = \frac{-(-20) + \sqrt{(-20)^2 - 4 \cdot 6 \cdot 7}}{2.6}$$

$$= \frac{20 + \sqrt{232}}{12} = \frac{20 \pm 15.2315}{12}$$

$$x = 4.7685/12 = 0.397$$

or $\quad x = 35.2315/12 = 2.936$

Our estimates were reasonably correct. Next to find the coordinates of the stationary points.

If $x = 0.397$, substitute this value in the equation for y and find that $y = 61.33$.

If $x = 2.936$, we find that $y = 44.97$.

The stationary points are (0.397, 61.3) and (2.94, 45.0) (3 sf).

Having plotted the graphs we can see that the first point is a maximum and the second is a minimum.

Maximum or minimum?

It is more useful if we can identify a maximum or minimum without having to plot the graph. Let us look for a way of finding if a given stationary point is a maximum. In Figure 24.7 we see that, at a maximum, the gradient changes from positive to zero to negative. In Figure 24.8, the curve for dy/dx passes *down* through the x-axis. Without plotting this curve, we could tell if it was going *down* by the fact that it has a *negative* gradient. We need to know the gradient *of the derivative* when $x = 0.397$.

Derivatives – a summary

a and n are constants; x and θ are independent variables.

Function	*Derivative*	*Examples* *Function*	*Examples* *Derivative*
1 ax^0	anx^{n-1}	$2x^3$	$6x^2$
2 $\sin a\theta$	$a \cos a\theta$	$\sin 2\theta$	$2 \cos 2\theta$
3 $\cos a\theta$	$-a \sin a\theta$	$\cos 3\theta$	$-3 \sin 3\theta$
4 e^{ax}	ae^{ax}	e^{4x}	$4e^{4x}$
5 $\ln ax$	$1/x$	$\ln 7x$	$1/x$

The gradient of the derivative can be found by finding the derivative of the derivative (see box, page 338). The derivative is

$$dy/dx = 6x^2 - 20x + 7$$

The second derivative is

$$d^2y/dx^2 = 12x - 20$$

When $x = 0.397$:

$$d^2y/dx^2 = 12 \times 0.397 - 20 = -15.24$$

The second derivative is *negative*, showing that dy/dx slopes down. The gradient is becoming less, so (0.397, 61.3) is a **maximum**.

Now substitute $x = 2.94$:

$$d^2y/dx^2 = 12 \times 2.94 - 20 = 15.28$$

The second derivative is *positive*, showing that dy/dx slopes up and that (2.94, 45.0) is a **minimum**.

Maximums and minimums – a summary

Start with the function of x.

Find the derivative, dy/dx.

Put the derivative equal to 0, and solve it for one or more values of x.

For x (or for each value of x, if more than one) find y. This gives the stationary point(s).

Find the second derivative, d^2y/dx^2

Substitute the value(s) of x into this: if negative, a maximum; if positive, a minimum.

Differentiating – in brief

y = function of x
↓
differentiate
↓
dy/dx
(first derivative, rate of change of y)
↓
differentiate
↓
d^2y/dx^2
(second derivative, rate of change of dy/dx)

Applications

Finding rates of change
Finding maximum and minimum values

Test yourself 24.4

1 Find the derivatives of these functions.

a $2x^{0.5}$ **b** $5x^{-2}$

c $1/6x$ **d** $5x^{-3} + 3x^{-1}$

e $7x^{-0.5} + 3x^{0.5} + 3^{0.5}$ **f** $x^{-6}/6$

2 For each of these functions, without plotting their graphs, find the stationary point or points and state whether each is a maximum or a minimum.

a $y = 7x^2 + 3x - 9$ **b** $y = -6x^2 - 5x$

c $y = 8x^3 - 3x^2 + 12$ **d** $y = -2x^3 + 10x$

3 Two numbers add up to 16. Find the two numbers, if their product is to be as great as possible.

4 The sum of the base and height of a triangle is 40 mm. Find the maximum area of the triangle.

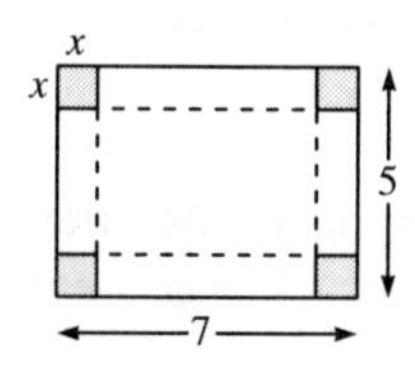

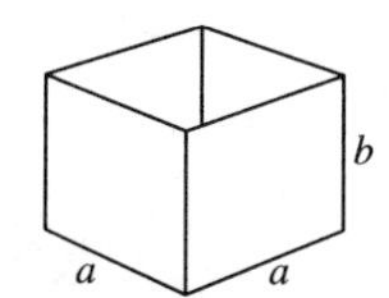

Figure 24.9

5 A person buys 100 m of fencing to fence off a rectangular area. One side is to be provided by an existing hedge. What is the largest area that can be fenced off, and what are its dimensions?

6 A rectangle of card 7 cm × 5 cm (Figure 24.9a) has squares (shaded) cut from the corners and is folded up along the dashed lines to make a tray. Write an equation which expresses the volume v of the tray as a function of x, the side of the squares. Find the derivative of this, then find the value of x for which v is a maximum. What is the maximum volume of the tray?

7 A box with no lid is to be made from 100 cm^2 of sheet metal (Figure 24.9b). The base of the box is to be square. What is the maximum volume of such a box and what are its dimensions?

8 To assess your progress in this chapter, work the exercises in *Try these first*, page 329.

25 Summing it up

Integration is a powerful method for finding areas, averages, and other quantities when we know the equations that define them. It has many applications in engineering.

You need to know about derivatives (Chapter 24).

Try these first

1 Find the integral of $3x^2 - 2x - 6$, given that $\int y = 3$, when $x = 2$.

2 Evaluate $\int_0^{\pi} (2 \sin \theta + 3 \cos 3\theta)\mathrm{d}\theta$

3 Find the average value of $2 + 3e^{-0.5t}$, between $t = 1$ and $t = 3$.

Reversing the derivative

In Chapter 24 we began with a function of x and derived another function from it, the derivative. Now we take this in reverse. Begin with a function of x and find the function of which this is the derivative.

Example

Begin with a function e^{ax}. The rule for finding the **derivative** of this is to multiply by the coefficient a of the index to give ae^{ax} (page 345). To reverse the process, in other words, to find the **anti-derivative**, all we need to do is to *divide* by the coefficient of the index, obtaining:

$$\frac{e^{ax}}{a}$$

We need a symbol to represent this operation. The anti-derivative is usually called the **integral** and the process of obtaining it is known as **integration**. The symbol for integration is a 'long s':

$$\int$$

Integrating e^{ax} to obtain e^{ax}/a (see box Rule 4) is written:

$$\int e^{ax}\mathrm{d}x = \frac{e^{ax}}{a}$$

The 'dx' indicates that we are integrating with respect to x, the independent variable.

Rules for finding integrals of other functions are given in the box. Integrating a constant is a special case of integrating a polynomial (Rule 1). In effect we are integrating a term in which the power of x is zero ($x^0 = 1$, page 177). Rule 1 reduces this to placing an x after the constant.

Useful integrals

a and n are constants

Function	Integral	Example Function	Example Integral
1 ax^n	$\frac{ax^{n+1}}{n+1}$	$2x^3$	$\frac{2x^4}{4}$
2 $\sin ax$	$\frac{-\cos ax}{a}$	$\sin 4x$	$\frac{-\cos 4x}{4}$
3 $\cos ax$	$\frac{\sin ax}{a}$	$\cos 2x$	$\frac{\sin 2x}{2}$
4 e^{ax}	$\frac{e^{ax}}{a}$	e^{3x}	$\frac{e^{3x}}{3}$
5 a/x	$a \ln x$	$3/x$	$3 \ln x$

Remember to add c, the constant of integration, to all integrals.

In Rule 5 we integrate $1/x$, in which the power of x is -1. The rule for polynomials does not work here, as $-1 + 1 = 0$, and it is not possible to divide by 0. The clue to overcoming this difficulty is in the box on page 345. There Rule 5 stages that the derivative of $\ln x$ is $1/x$. So the integral of $1/x$ is $\ln x$.

The disappearing constant

Consider this function:

$$y = 2x^2 + 3x + 6 \qquad (1)$$

According to the rules on page 345, the derivative of (1) is:

$$dy/dx = 4x + 3 \qquad (2)$$

According to the rules on page 345, the integral of (2) is:

$$\int (4x + 3)dx = 2x^2 + 3x \qquad (3)$$

In going from (1) to (2) we obey the rules and ignore the constant 6. But, in going from (2) to (3), we cannot replace the constant, for we have no way of working out what it might be. It has disappeared without trace.

There are millions of possible functions like equation (1), but with different constants, *and* all of them produce the same derivative. Figure 25.1 shows the graph of the derivative $4x + 3$ (dot-dashed line) and graphs of a few of the many functions which could produce it.

Conversely, Figure 25.2 shows the graph of a function $4x + 3$ and graphs of a few of the many possible integrals it can produce. The members of this 'family' of parabolas all have the same shape and are really the same parabola just shifted up and down the page because they have different values for the constant.

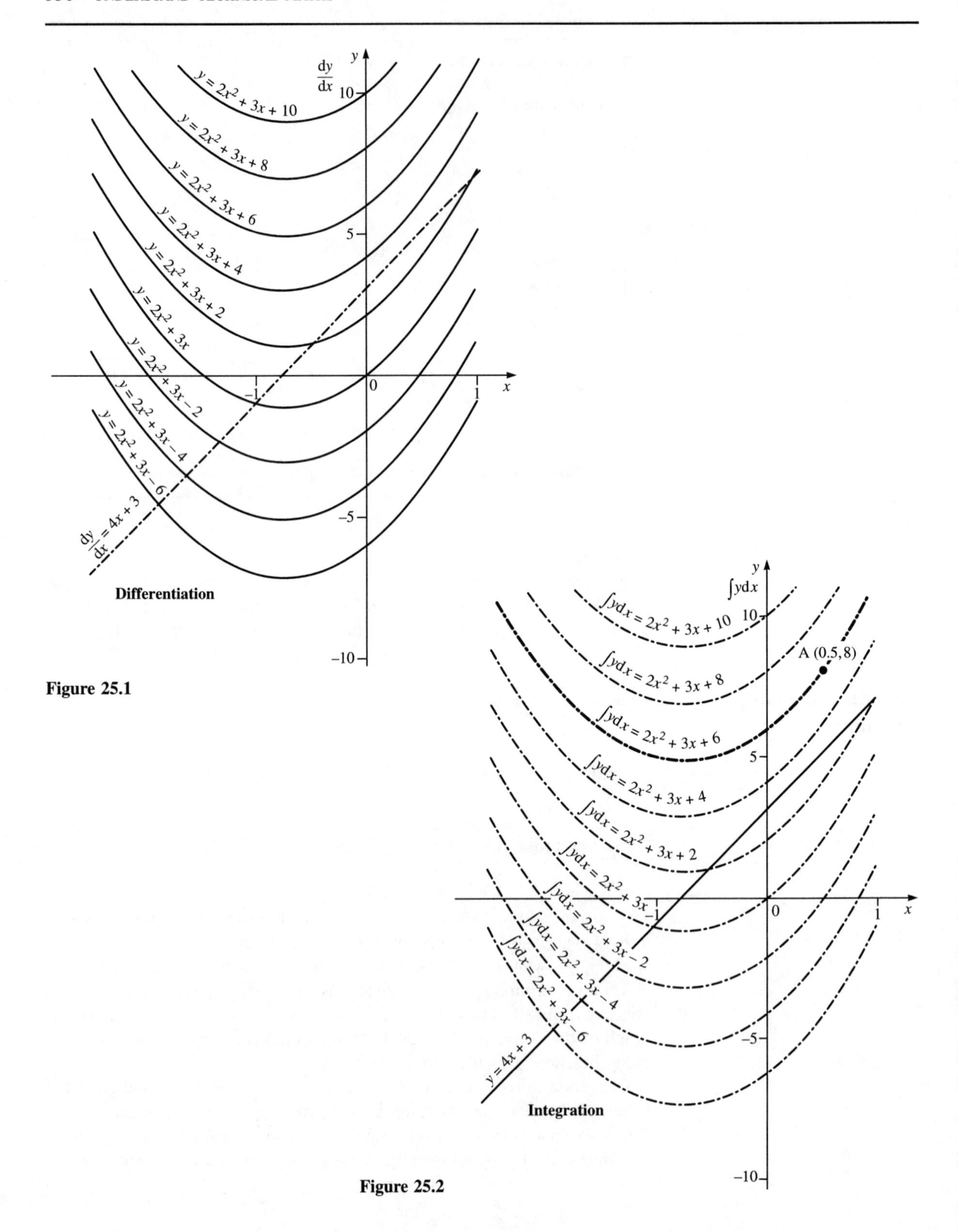

Figure 25.1

Figure 25.2

When we integrate a function, there is nothing to tell us the value of the constant. It could be anything. The best we can do is to put in an unknown constant, c and say:

$$\int(4x + 3)\mathrm{d}x = 2x^2 + 3x + c$$

This constant is known as the **constant of integration.**

The need to include the constant of integration applies to integration of any function.

Examples

$$\int \mathrm{e}^{ex}\,\mathrm{d}x = \mathrm{e}^{3x}/3 + c$$

$$\int 3\sin 4\theta\,\mathrm{d}\theta = 0.75\sin 4\theta + c$$

$$\int 3/x\,\mathrm{d}x = 3\ln x + c$$

Finding c

It is possible to fix a value for c if we are given extra information. For example, in Figure 25.2 we might be given the extra information that, when $\int y = 8$ then $x = 0.5$. Substituting the given values of y into the equation:

$$\begin{aligned} \int y &= 2x^2 + 3x + c \\ \Rightarrow \quad 8 &= 2 \times 0.5^2 + 3 \times 0.5 + c \\ &= 0.5 + 1.5 + c \\ \Rightarrow \quad c &= 8 - 2 = 6 \end{aligned}$$

The constant of integration has the value 6. There is only one equation which fits all the facts:

$$y = 2x^2 + 3x + 6$$

This is the equation of the parabola drawn with a thicker line in Figure 25.2, which passes through point A(0.5,8). We have singled out one curve by locating a point which lies on it. The parabola representing $\int y$, must be the one which passes through this point.

Short forms

Given $y = 5x^2 - 2x + 7$ (for example) we write its integral as:

$$\int(5x^2 - 2x + 7)\,\mathrm{d}x$$

We can also write it as:

$$\int y\,\mathrm{d}x$$

This means the same thing. If there is no risk of confusion, we can also write it as:

$$\int y$$

Test yourself 25.1

1 Find the integrals of these functions. Check your answers by working back to the derivatives.

a $6x$	**b** 5
c $3e^{2x}$	**d** $2 \sin 2\theta$
e $7x$	**f** $2x^2$
g $-\cos 5\theta$	**h** $6x^5 - 5x^4$
i $4/x$	**j** $x + 1$
k $5x^2 - 3x - 8$	**l** $1/x^2$

2 Find the integrals of these functions, using the extra information to evaluate the constant of integration.

a $y = 4x + 1$, given $\int y = 6$ when $x = 1$

b $y = 5$, given $\int y = 9$ when $x = 2$

c $y = 4 + e^{2x}$ given $\int y = 10$ when $x = 3$

d $y = 9x^2 — 4x + 1$, given $\int y = 16$ when $x = 2$

e $y = \sin \theta + 2 \cos 2\theta$, given $\int y = 0$ when $\theta = 2$. [*Hint*: θ in radians]

What is an integral?

In Figure 25.3, we show the graph of $y = x + 2$, and the graph of one of its integrals, $\int y \, dx = x^2/2 + 2x + 4$. The integral is found by using Rule 1 in the box on page 349. We have *chosen* 4 as the value of the constant simply to put

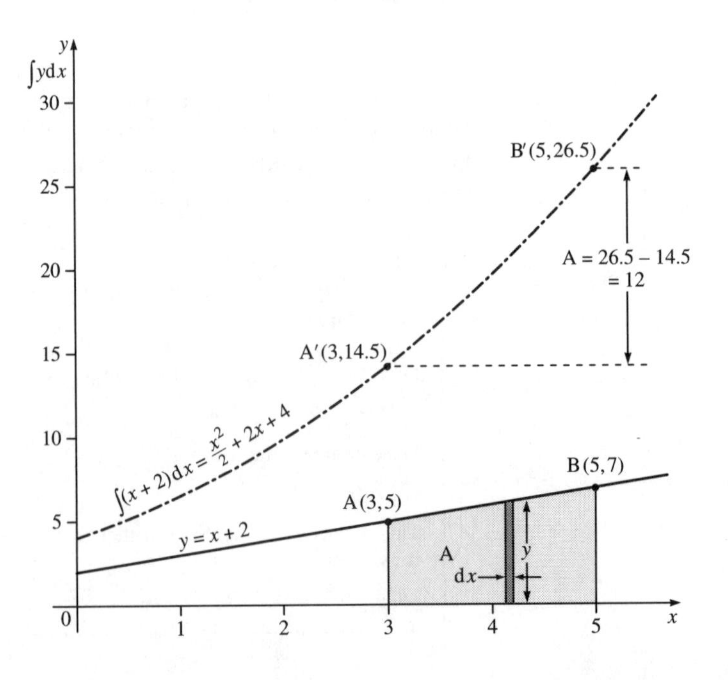

Figure 25.3

the curve on a convenient part of the page, a little higher up than the graph of $y = x + 2$. The value of c makes no difference to what follows.

An area of the graph below the graph of y is shaded in. This area extends from $x = 3$ to $x = 5$. Think of the area being divided into many narrow vertical strips. One of these is shaded in the figure. The width of the strip is very small; call it dx, where the 'd' means 'a small distance in the x direction'. The strip is so narrow that its top can be taken to be horizontal. The strip is approximately equal in area to a rectangle measuring dx by y. The length y varies according to the value of x. Its range (page 330) is from 5 when $x = 3$, to 7 when $x = 5$. The area of a strip is:

$$\text{length} \times \text{width} = y \times \mathrm{d}x$$

The *total* shaded area is the *sum of* the areas of *all* such strips. We write this as:

$$A = \int_3^5 y \, \mathrm{d}x$$

We have used the long 's', the integral symbol to mean 'the sum of'. This is what integration does. It sums the areas of the strips. The 3 and the 5 below and above the symbol mean that we start summing from $x = 3$ and take it as far as $x = 5$. An integral like this, which is summed over a definite domain (page 330), is known as a **definite integral**.

Integration

According to dictionaries, the word **integration** means adding parts together to make a whole. This is exactly what we do when we add the areas of the strips to obtain the whole area under the curve.

Now to find the size of the shaded area:

When $x = 3$: $\int y = 3^2/2 + 2 \times 3 + 4 = 14.5$

When $x = 5$: $\int y = 5^2/2 + 2 \times 5 + 4 = 26.5$

What we have done is to find the coordinates of points A' and B' on the integral curve. The area A is the difference between these two values of $\int y \, \mathrm{d}x$:

$$A = \int_3^5 y \, \mathrm{d}x = 26.5 - 14.5 = 12$$

On the graph, A is also the vertical distance between A' and B'.

Shaded area = 12

We can check this, to show that integration has given the correct result. Using the method of Figure 6.7 we find that the area of the shaded trapezium, is 12. The integration technique gives the right answer in this case. We shall not attempt to prove that it gives the right answer in every case, but in fact it does.

Note that the chosen constant 4 is included in *both* values of $\int y$. When we find the *difference* between the two values, the constant is subtracted out. We could have chosen any other value for the constant without having any effect on the difference. This is the same as saying that the value of the constant merely shifts the integral curve up or down the page, without altering its shape and slope. When working with definite integrals, we can usually forget about the constant of integration.

Another example

Having shown that the technique works, let us try it on a longer polynomial:

$$y = 3x^2 - 4x + 5$$

We will calculate the area under the curve from $x = 4$ to $x = 7$. Figure 25.4 shows the area, the upper edge of which is part of a parabola. The area is equal to the definite integral of y from $x = 4$ to $x = 7$:

$$A = \int_4^7 (3x^2 - 4x + 5)\,\mathrm{d}x$$

Applying Rule 1 the integral of y is:

$$\int y\,\mathrm{d}x = x^3 - 2x^2 + 5x$$

We should add the integration constant to this. In the figure we have chosen to add 7 to bring the curve to a convenient position. But, as we have said, the value of c makes no difference to the calculation of definite integrals, so we shall leave it out altogether.

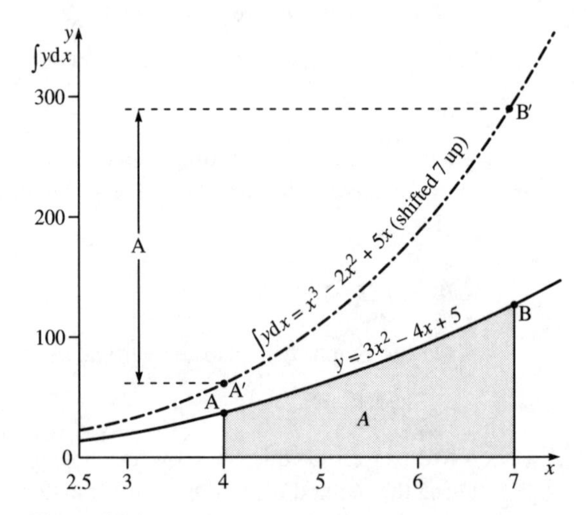

Figure 25.4

The definite integral is:

$$A = \int y\,dx = [x^3 - 2x^2 + 5x]_4^7$$

We have written the integral in square brackets. The 4 and 7 at the closing bracket indicate that we have to evaluate the integral with $x = 7$ and with $x = 4$, and then find their difference. Evaluating the integrals:

$$\int_4^7 y\,dx = [7^3 - 2 \times 7^2 + 5 \times 7] - [4^3 - 2 \times 4^2 + 5 \times 4]$$
$$= 280 - 52$$
$$= 228$$

The area of the shaded part of Figure 25.4 is 228.

A trig example

The function is:

$$y = 3 \sin 2\theta$$

This function is plotted in Figure 25.5 over the domain $\theta = 0$ to π radians (0 to 180°). We are working in radians in this example. We wish to find the area under the curve from $\theta = 0.3$ to $\theta = 1.1$. By Rule 2 the definite integral is:

$$\int_{0.3}^{1.1} 3 \sin 2\theta\, d\theta = [-1.5 \cos 2\theta]_{0.3}^{1.1}$$

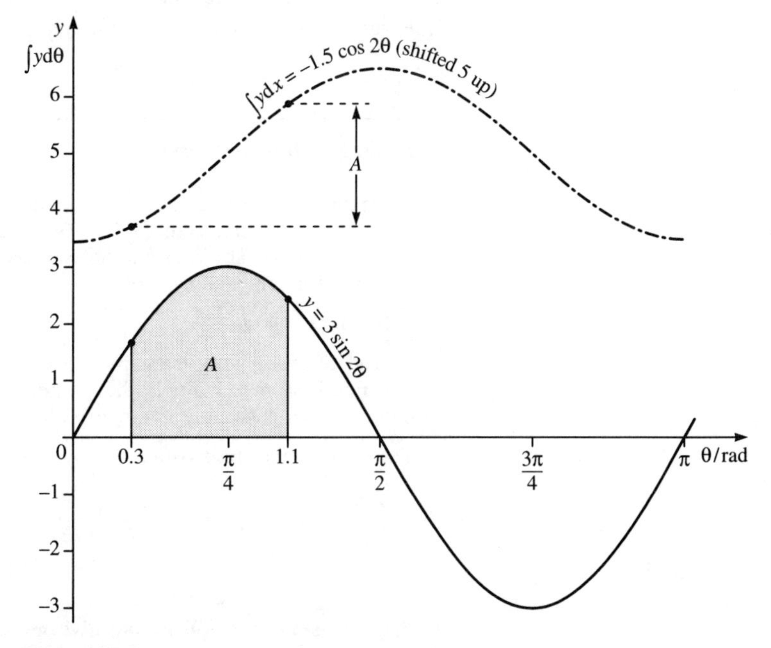

Figure 25.5

As in the previous example, we have written the integral in square brackets, with the lower and upper values of θ beside the closing bracket. Next we find the difference between the integrals for these two values:

$$A = [-1.5 \cos 2.2] - [-1.5 \cos 0.6]$$

$\cos 2.2 = -0.5885$ and $\cos 0.6 = 0.8253$. There are several minus signs in this equation, so we must be sure to take all of them into account.

$$\begin{aligned} A &= [-1.5 \times -0.5885] - [-1.5 \times 0.8253] \\ &= 0.88275 - (-1.23800) \\ &= 0.88275 + 1.23800 \\ &= 2.1208 \end{aligned}$$

The area of the shaded part of Figure 25.5 is 2.121 (4 sf).

It is interesting to evaluate this integral for another domain, for example, from $\theta = 0$ to $\theta = \pi$

$$A = [-1.5 \cos 2\pi] - [-1.5 \cos 0]$$

$\cos 2\pi = 1$, also $\cos 0 = 1$, so the difference between the two integrals is zero:

$$A = [-1.5] - [-1.5] = 0$$

This zero result is explained by saying that the area *below* the line (from $\pi/2$ to π) is *negative*. It is exactly equal to the area above the line, between 0 and $\pi/2$. The two areas cancel out.

Integrating from $\pi/4$ to $3\pi/4$ similarly gives zero area.

Dealing with multipliers

Integration is the summing of strips to make a whole. If all strips are made 4 times longer, the area is made 4 times bigger. In other words, if a function to be integrated has any constant **factor**, we can remove the factor and place it in front of the integral symbol:

$$\int 4x \, dx = 4 \int x \, dx$$

Instead of multiplying the x by 4 and *then* integrating, as on the left, we integrate first, *then* multiply the integral by 4, as on the right. It is usually better to do this because it simplifies the expression to be integrated.

The same applies to divisors:

$$\int \frac{x}{2\pi} \, dx = \frac{1}{2\pi} \int x \, dx$$

Rule:

Place constant multipliers and divisors in front of the integral symbol

Test yourself 25.2

Evaluate these definite integrals.

1 $\int_0^3 3x \, dx$ **2** $\int_1^5 2x \, dx$

3 $\int_2^8 (x + 4) \, dx$ **4** $\int_1^4 (5x^2 - 2x + 12) \, dx$

5 $\int_1^3 5e^{2x} dx$ **6** $\int_0^3 (4 \sin 3\theta) \, d\theta$

Using the area technique

In a number of practical problems, a quantity we want to evaluate can be thought of in terms of an area below the line of a curve. If so, integrating the function over a given domain produces a value for the quantity.

Example

A vehicle moves along a road with varying velocity:

$$v = 10t + 4$$

where v is in ms^{-1} and t is in s. Find the distance it travels in the period from $t = 2$ until $t = 6$. Imagine the area under the curve being divided into narrow vertical strips, as in Figure 25.3. If dt is small, v does not vary significantly during that time. The area of the strip is:

$$A = \text{velocity} \times \text{time}$$

But, for a moving object:

$$\text{velocity} \times \text{time} = \text{distance covered}$$

Summing the strips by integration tells us the total area beneath the curve, which equals the total distance covered.

Integrating v:

$$\text{Total distance} = \int_2^6 (10t + 4) \, dt$$

$$= [5t^2 + 4t]_2^6$$

$$= [180 + 24] - [20 + 8] = 204 - 28 = 176$$

Distance travelled is 176 m.

In a similar way, the area under a graph of acceleration plotted against time gives the increase in velocity during a given interval of time.

An electronic example is the charging of a capacitor. If a graph is plotted to show current i against time t, the area of a narrow strip under the graph is $i dt$. This is the amount of charge q accumulating during that time. The total area gives the total charge accumulating.

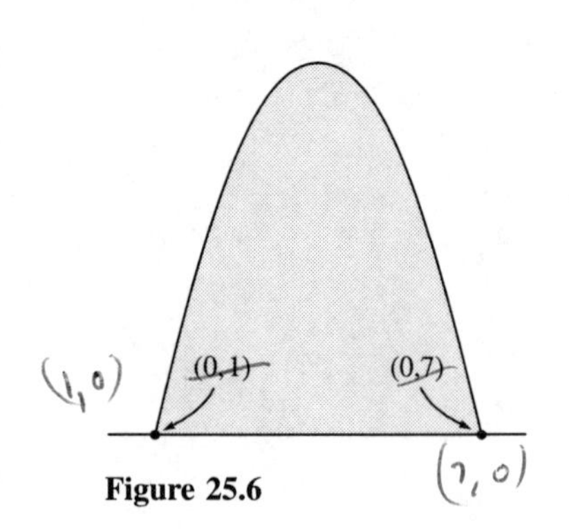

Figure 25.6

Test yourself 25.3

1 The acceleration of a rocket in ms^{-1} is given by:

$$a = 10e^{-0.5t} + 3$$

By how much does the velocity increase between 1 s and 3 s after launching?

2 A balloon rises at a velocity given by:

$$v = 2 - 0.05t$$

How far does it rise in 15 s?

3 A flower bed has the shape of a parabola (Figure 25.6) with the equation $y = -x^2 + 8x - 7$. Find the area of the bed, if y is in metres.

Average values

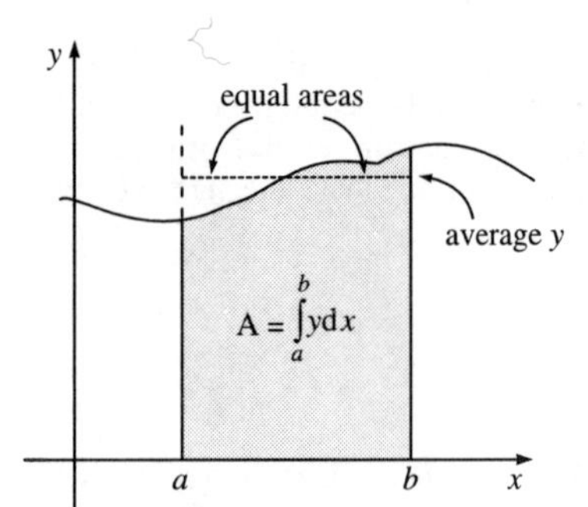

Figure 25.7

In Figure 25.7, the value of y varies according to a function of x which we do not need to define. The shaded area under the curve between a and b is:

$$A = \int_a^b y \, dx$$

If we think of this area as being roughly a rectangle, its area is:

$$A = \text{length} \times \text{height}$$

For the shaded area:

$$A = \text{length} \times \text{average height}$$

$$= (b - a) \times \text{average height}$$

Putting the area as an integral:

$$(b - a) \times \text{average height} = \int_a^b y \, dx$$

$$\Rightarrow \qquad \text{average height} = \frac{\int_a^b y dx}{b - a}$$

This provides us with a way of finding averages.

Example

A reflector concentrates a circular beam of light on to a surface so that the light intensity L at a point distant r from the centre of the circle is given by:

$$L = 4/r^2 \qquad r \geqslant 0.5$$

Find the average intensity in the region $r = 0.6$ to $r = 2.6$.

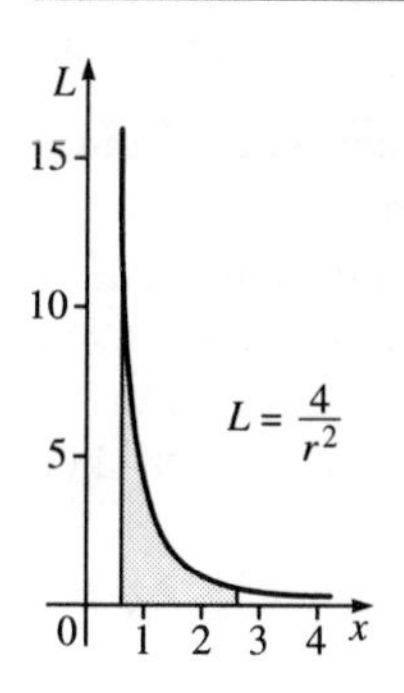

Figure 25.8

Figure 25.8 shows how the intensity falls off in the region from 0.5 outwards. The average is given by:

$$L_{av} = \frac{\int_{0.6}^{2.6} 4/r^2 dr}{2.6–0.6}$$

$4/r^2 = 4r^{-2}$, so its integral is $-4r^{-1} = -4/r$

$$L_{av} = 0.5 \times [-4/r]_{0.6}^{2.6}$$
$$= 0.5([-4/2.6] - [-4/0.6])$$
$$= 0.5([-1.5385] - [-6.6667])$$
$$= 0.5 \times 5.1282$$
$$= 2.5641$$

The average light intensity is 2.56 (3 sf).

Test yourself 25.4

1 The braking force F in a machine is given by:

$F = 5t^2 - 2t^3 \qquad 0 \leqslant t \leqslant 2.5$

When t is the time for which the brake is applied. Find the average force during the first 2 s after the brake is applied. (3 sf)

2 Find the average value of $\sin \theta$ over the domains

a 0 to π **b** 0 to 2π

3 The signal voltage from a tone generator is:

$v = 6 \sin \theta + 3 \sin 2\theta + 2 \sin 3\theta$

Find the average voltage from $\theta = 0$ to $t = \pi$. (3 sf)

4 The distance travelled by a car is:

$s = 1.8t + 0.2t^2$

Find its average *velocity*, over the domain $t = 10$ to $t = 40$ (3 sf).

Explore these

1 Use a graphic calculator or software package to view and evaluate the integrals in the problems of *Test yourself 25.2, 25.3 and 25.4.*

2 Design and program a spreadsheet for finding the integrals of polynomial functions, basing this on the spreadsheet for derivatives, as described on page 340. As an extension of this, program the spreadsheet to calculate $\int y \, dx$ for different values of x over a chosen domain. Use the data produced by the spreadsheet to plot graphs of the integral.

Volumes of solids

Figure 25.9a is the graph of $y = 0.2x$. Imagine the line of this graph from $x = 0$ to $x = 10$ spun round the x-axis. The area A under the graph sweeps out a volume of space, having the shape of a cone. This is shown as a perspective drawing in Figure 25.9b. Our aim is to find the volume of this cone.

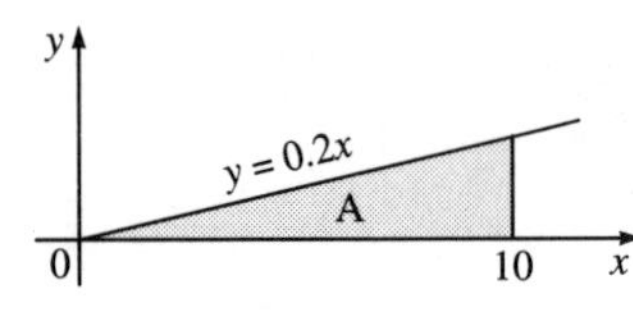

a

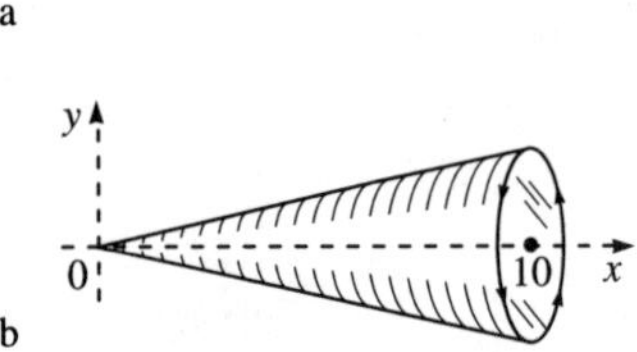

b

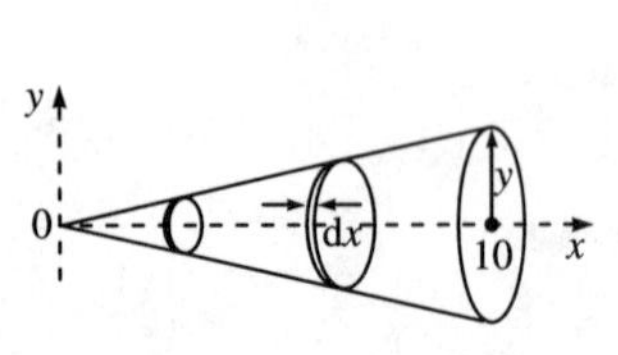

Figure 25.9

The cone can be thought of as a stack of thin circular slices (Figure 25.9c). The thickness of each slice is dx, where dx means a very small distance in the x-direction. The area of a slice is πy^2, where y is its radius. For any given slice $y = 0.2x$ we substitute this into the formula for the volume:

$$v = \text{area} \times \text{thickness}$$
$$= \pi(0.2x)^2 \times \mathrm{d}x$$
$$= 0.04\pi x^2 \mathrm{d}x$$

The total volume v_{tot} of the cone is the sum of the volumes of the slices from $x = 0$ to $x = 10$:

$$v_{tot} = \int_0^{10} 0.04\pi x^2 dx$$
$$= \left[\frac{0.04\pi x^3}{3}\right]_0^{10}$$

Bringing the constants outside the brackets:

$$v_{tot} = \frac{0.04\pi}{3}[x^3]_0^{10}$$
$$= \frac{0.04\pi}{3}([10^3] - [0^3])$$
$$= \frac{0.04\pi}{3} \times 1000$$
$$= 41.89$$

The volume of the cone is 41.89 (4 sf).

The effect of integrating is to make the slices exceedingly thin, so that they form a smooth-surfaced cone, not one with stepped surfaces. The calculated volume is that of a smooth cone.

On page 255 the formula for the volume of a cone is given as:

$$v_{tot} = \frac{1}{3}\pi r^2 h$$

Substitute the given values for the cone of Figure 25.9 into this formula. $h = 10$, $r = 0.2 \times 10 = 2$ and:

$$v_{tot} = \frac{\pi \times 2^2 \times 10}{3} = 41.89$$

This is the same result as obtained by integration, confirming that the technique works. There is not much point in using the integration technique for a solid of such simple shape as a cone. But it is useful for more complex solids, as some of the following examples show.

Example

Find the volume of a wine-glass 5 cm deep, shaped in section according to the function:

$$y = 4 - 4e^{-x}$$

Figure 25.10 shows the curve. The dashed lines complete the outline of the glass, as seen in section. The volume of a single disc is:

$$\begin{aligned} v &= \pi(4 - 4e^{-x})^2 \\ &= \pi(16 - 32e^{-x} + 16e^{-2x}) \\ &= 16\pi(1 - 2e^{-x} + e^{-2x}) \end{aligned}$$

Summing this volume for the whole glass (Rule 4):

$$\begin{aligned} v_{tot} &= 16\pi\int_0^5 (1 - 2e^{-x} + e^{-2x})dx \\ &= 16\pi\Big[x + 2e^{-x} - e^{-2x}/2\Big]_0^5 \\ &= 16\pi([5 + 2e^{-5} - e^{-10}/2] - [0 + 2e^0 - e^0/2]) \\ &= 16\pi(5.0135 - 1.5) \\ &= 16\pi \times 3.5135 = 176.6 \end{aligned}$$

The volume of the glass is $176.6\,cm^2$.

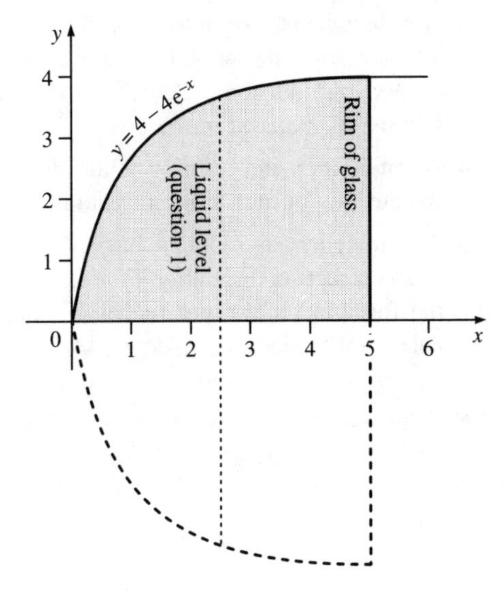

Figure 25.10

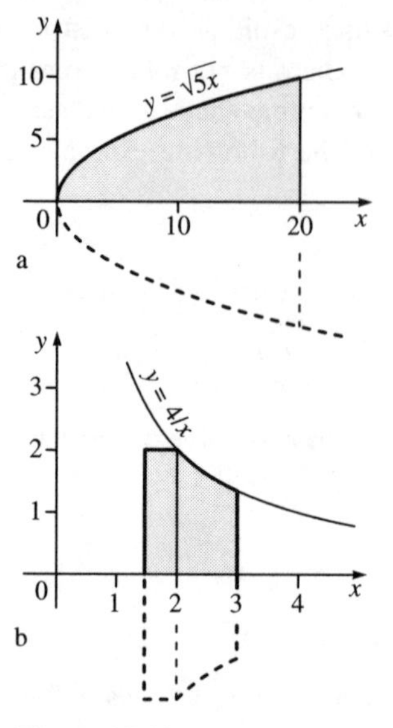

Figure 25.11

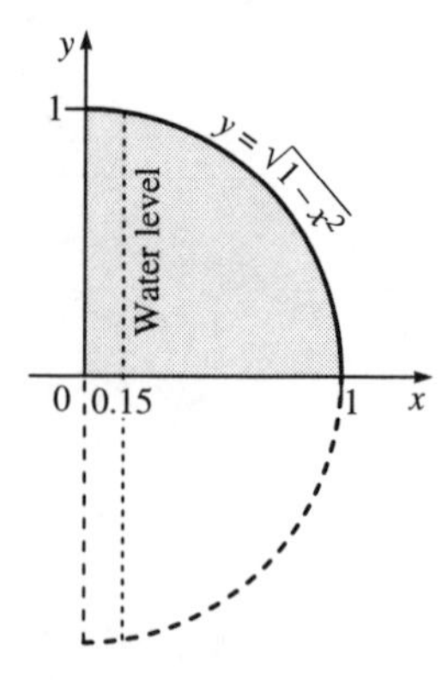

Figure 25.12

Test yourself 25.5

1 The wine glass with the shape shown in Figure 25.10 is filled to a depth of 2.5 cm in the centre. Find the volume of liquid it contains.

2 Use the method of integration to find the volume of a circular plastic bottle cap, radius 10 mm at the top, radius 15 mm at the bottom and 20 mm high.

3 A nose cone is 20 mm long, with a radius of 10 mm at its widest end. The equation defining its shape is $y = \sqrt{5x}$ (Figure 25.11a). The dashed outline completes the shape, as seen in section. Find the volume of material in the cone.

4 A circular metal turning knob is shaped as in Figure 25.11b, dimensions in mm. Find its volume.

5 A hemispherical fountain basin has a radius of 1 m. Figure 25.12 shows a section through it. Find its volume and confirm this by using the formula for the volume of a sphere on page 255. The basin is filled with water to a depth of 0.85 m. Find the volume of the water.

6 To assess if you have covered this chapter fully, work the exercises in *Try these first*, page 348.

26 Rules for areas

There are instances in which it is difficult, time-consuming, or even impossible to integrate a function. This is when the three techniques described in this chapter can be of help.

You need to know about integration (Chapter 25).

Try these first

Evaluate these integrals, using the Trapezium Rule, the Mid-ordinate Rule or Simpson's Rule (4 sf).

1 $\int_0^5 xe^{-x}\,dx$

2 $\int_0^{\pi} \sqrt{\sin\theta}\,d\theta$

3 $\int_1^2 e^{-x^2}\,dx$

Difficult integrals

The functions in *Try these first* are examples of integrals that cannot be found by using the methods of Chapter 25. Question 1 asks us to integrate a **product**. Unfortunately we cannot just integrate x and e^{-x} separately, then find the product of their integrals. This does not give the correct result. Integrating products requires special techniques which are not included in this book. Question 2 asks us to integrate a **function of a function**; the first function is $\sin\theta$, and we are asked to integrate the **square root** (second function) of this. There are methods of doing this, but these are not included here. The third question is impossible to integrate, for there is no function which differentiates to give e^{-x^2}. There are also those cases in which x and y are related in such a complex way that it is not possible to write out an exact function.

In all such difficult cases it is usually sufficient to employ an approximate method to obtain the integral. There are four techniques, one of which is counting squares (Chapter 6), which is especially good for very irregular areas. The three techniques described in this chapter are very similar to each other, but each has its advantages and disadvantages.

Mid-ordinate rule

The area to be evaluated is divided into strips of equal width h (Figure 26.1). A line (the **mid-ordinate**) is drawn along the middle of each strip and its length is measured. The rule is

Area = h × sum of mid-ordinates

The method is based on thinking of each strip as a rectangle, ignoring the fact that it usually has a sloping upper end. For each strip:

area = h × length of ordinate

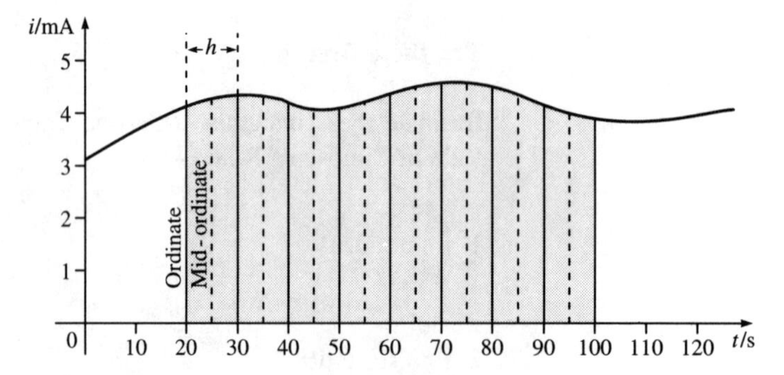

Figure 26.1

Summing all the strips to find the total area we sum the lengths and multiply that total by h. The method is an easy one to work and gives good results. As might be expected, it gives better results if we divide the area into many narrow strips. On the other hand, the more strips, the longer the calculation takes. Also rounding errors accumulate with many strips, so there is a limit to the accuracy which we may obtain.

The method may be used exactly as described above. Plot the graph of the function, rule it into strips, draw the mid-ordinates and measure their length against a scale. This is a suitable technique when working from experimental data, when the function might not be known exactly.

Example

A capacitor is charged by a varying current i. Figure 26.1 shows the readings obtained with a continuously recording milliammeter. Find how much charge accumulates in the capacitor between 20 s and 100 s.

The charge is equal to current multiplied by time, so is represented by the area under the curve. We need to evaluate:

$$q = \int_{20}^{100} i \, \mathrm{d}t$$

But there is no clear function to relate i to t. All we have is the data of Figure 26.1. The area is divided for convenience into 8 strips, width 10, by drawing vertical lines, the

ordinates. The dashed lines are the **mid-ordinates**. From left to right, their lengths arc:

4.21, 4.29, 4.02, 4.13, 4.45, 4.52, 4.28 and 4.00

Their total is 33.9.

Area = 10 × 3.39 = 33.9

Charge accumulated = 33.9 mC

If we are given a **function** to integrate by the mid-ordinate rule, there is no need to plot its graph. Everything can be done by calculation.

Example

Evaluate $\int_0^4 x \sin x \, dx$

The domain is 4 which, for convenience, we divide into 10 strips, each 0.4 wide. The first mid-ordinate is located half-way across the first strip, at $x = 0.2$. The remaining strips are at intervals of 0.4 after this. The table lists the mid-ordinates and the steps in the calculation:

Mid-ordinate location, x	*sin x*	*x sin x*
0.2	0.1987	0.0397
0.6	0.5646	0.3388
1.0	0.8415	0.8415
1.4	0.9854	1.3796
1.8	0.9738	1.7529
2.2	0.8085	1.7787
2.6	0.5155	1.3403
3.0	0.1411	0.4234
3.4	–0.2555	–0.8688
3.8	–0.6118	–2.3251
	Total strip length =	4.7010

Strip width = 0.4
Area = 0.4 × 4.7010 = 1.880 (4 sf)

$$\int_0^4 x \sin x \, dx = 1.880$$

Having listed the values of x, we use a calculator to find corresponding values of $\sin x$, and write these in the second column of the table. Then we find the product of each x and $\sin x$ pair. In short, we find the value of $x \sin x$, the length of each mid-ordinate. These products are written in the third column, then totalled. The last two strips have negative length, because $\sin x$ is negative when $x > 3.142$ radians (180°). These values are subtracted from the total of the other (positive) values.

The calculation is lengthy but repetitive. It is fairly easy to program a calculator, or a microcomputer to perform this technique automatically.

Mid-coordinate rule on a calculator

The integral of a **function** (but not a set of data values) may be evaluated quickly on a calculator with programming facilities.

Example

To integrate the function $x \sin x$ (see example on this page) key in a 3-stage program:

$$y = x \times \sin x\colon z = z + y\colon x = x + 0.4$$

The calculator asks for the first value of x to be keyed in; key in the value 0.2.

Press the EXEcute key for each stage of the program:

1st stage calculates y
2nd stage stores the running total of y as z.
3rd stage increments x by h, to give the next value of x

Press the key repeatedly (five times for each ordinate). x is automatically incremented to 0.6, 1.0, 1.4 . . . and the total strip length accumulates as z.
When the display shows that x has become 3.8, press the key four more times until the updated value of z is displayed.
Multiply z by 0.4 to obtain the area.

This technique can be adapted to summing the products (except the first and last) for the trapezium rule.

Trapezium rule

This technique considers each strip to be a trapezium, with the upper end straight, but approximating to the slope of the curve (Figure 26.2). The figure shows onlys three strips to make the discussion simpler, but usually we would work with more strips than this. The sides of the strips are ordinates.

The area of a trapezium is its width multiplied by its average length. For the first strip the area is:

$$h \times \frac{y_0 + y_1}{2}$$

where y_0 is the first ordinate and y_1 is the second ordinate. The total area for all 3 strips is:

$$h\left(\frac{y_0 + y_1}{2} + \frac{y_1 + y_2}{2} + \frac{y_2 + y_3}{2}\right)$$

$$= \frac{h}{2}(y_0 + y_1 + y_1 + y_2 + y_2 + y_3)$$

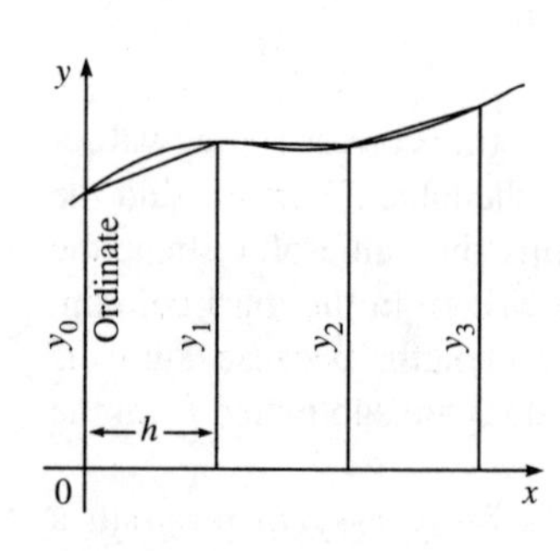

Figure 26.2

The first and last ordinates, y_0 and y_3, occur once each in the brackets; the other ordinates twice each. This means that we can generalize the expression to:

$$\textbf{area} = \frac{h}{2}\,(\textbf{first} + \textbf{last} + 2 \times \textbf{sum of others})$$

Example

Evaluate $\displaystyle\int_2^5 x^2 \ln x \, dx$

The domain is from 2 to 5, a difference of 3; divide this into 10 strips, width 0.3. In this technique, the first ordinate is the lowest limit of the domain, with other ordinates at intervals of 0.3.

Ordinate location x	x^2	$\ln x$	$x^2 \ln x$	
2.0	4.00	0.6931	2.7726	
2.3	5.29	1.6658		4.4061
2.6	6.76	1.9110		6.4593
2.9	8.41	1.0647		8.9542
3.2	10.24	1.1632		11.9107
3.5	12.25	1.2528		15.3463
3.8	15.21	1.3350		19.2774
4.1	16.81	1.4110		23.7187
4.4	19.36	1.4816		28.6839
4.7	22.09	1.5476		34.1857
5.0	25.00	1.6094	40.2359	
		Totals =	43.0085	152.9423
		Total doubled =		305.8846

The products $x^2 \ln x$ are summed in two columns. The first and last are totalled separately from the others. The total of the others is doubled.

$$\text{area} = \frac{0.3}{2}(43.0085 + 305.8846) = 52.33 \text{ (4 sf)}$$

$$\int_2^5 x^2 \ln x \, dx = 52.33$$

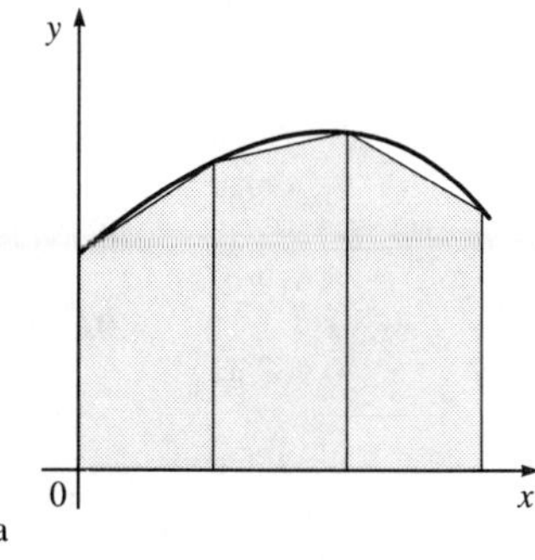

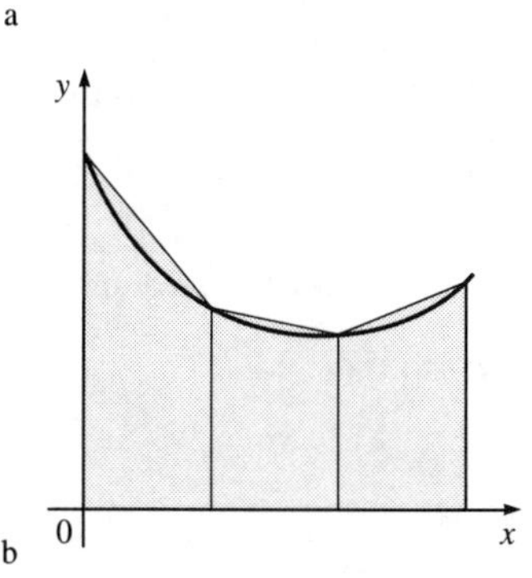

Figure 26.3

The advantage of the trapezium rule compared with the mid-ordinate rule is that the top end of the strip is closer to the line of the curve than if it is levelled at the half-way value (which is not necessarily the average length). The disadvantage of the trapezium rule is that it may show **systematic error** if the curve is wholly convex (a 'hump', see Figure 26.3a) or concave (a 'hollow', see Figure 26.3b). If the curve is convex the method under-estimates the area of every strip, so the total is smaller than it should be. Conversely the method over-estimates the area under a concave curve. These effects cancel out partly if the curve has both convex and concave sections. (Figure 26.2)

Simpson's Rule

This rule is based on the strips having *curved* upper ends (parts of a parabola) that fit closely to any given function. This gives a very accurate estimate of the area. Graphic calculators that are able to find integrals generally have Simpson's Rule programmed into them for this purpose.

As with the other methods, we divide the area into strips. The strips must be all the same width and there must be an *even* number of them. It follows from this that there will be an *odd* number of ordinates. The proof of the rule is complicated, but the rule is simple:

$$\textbf{area} = \frac{h}{3}\,(\textbf{first + last + 4[sum of odds] + 2[sum of even others]})$$

In this short formula:

- *first and last* refer to the first and last ordinates
- *odds* refers to the odd-numbered ordinates; 1, 3, 5, . . .
- *even others* refers to even-numbered ordinates, except the first and last; 2, 4, 6, . . .

Example

Evaluate $\int_2^3 (x - 2)\sqrt{x}\,\mathrm{d}x$

We will assume that the *positive* square root of x is to be used. Taking 10 strips, width 0.1 we have this table:

Ordinate location x	$x - 2$	$\sqrt{x}$		$(x - 2)\sqrt{x}$	
2.0	0.0	1.4142	0.00000		
2.1	0.1	1.4491			0.14491
2.2	0.2	1.4832		0.29664	
2.3	0.3	1.5166			0.45498
2.4	0.4	1.5492		0.61968	
2.5	0.5	1.5811			0.79055
2.6	0.6	1.6125		0.96750	
2.7	0.7	1.6432			1.15024
2.8	0.8	1.6733		1.33864	
2.9	0.9	1.7029			1.53261
3.0	1.0	1.7321	1.73210		
		Totals =	1.73210	3.22246	4.07329
			Total doubled =	6.44492	
				Total quadrupled =	16.29316

The products are summed in three columns, first + last, the even others, and the odds. The total of the even others is doubled. The total of the odds is quadrupled.

$$\text{area} = \frac{0.1}{3}(1.73210 + 6.44492 + 16.29316) = 0.8157 \text{ (4 sf)}$$

$$\int_2^3 (x - 2)\sqrt{x}\,\mathrm{d}x = 0.8157$$

The chief disadvantage of this method is that it is more complicated to use. Another possible difficulty is that, if it is to be used with data collected from experiments or observations, there must be an even number of readings to work on. If there is an odd number of readings, omit the first or last and apply the rule to the remainder.

Test yourself 26.1

Use one of more of the rules to answer these questions. This exercise is only for practice, so it is enough to work with only 5 or 6 strips, and to evaluate answers to 3 sf.

1 Evaluate these integrals.

a $\displaystyle\int_1^2 \frac{2}{1+x}\,dx$

b $\displaystyle\int_1^3 \sin\theta\cos\theta\,d\theta$

c $\displaystyle\int_{0.1}^{0.4} 4^x\,dx$

2 Estimate the area of a quadrant of a circle, radius 6. The function for y in Figure 26.4 is $y = \sqrt{36 - x^2}$. Compare your answer with the result obtained by using the formula $A = \pi r^2/4$.

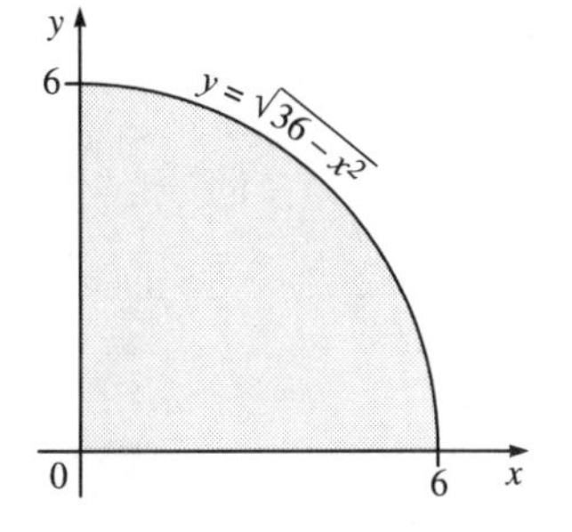

Figure 26.4

3 A bridge is built with two arches shaped as in Figure 26.5. The bridge is 17.5 m long, 8.5 m high and 5 m wide. Choose a suitable technique and use it to estimate the area of an arch. Find the volume of material used in the bridge.

4 The speedometer reading of a car is recorded at intervals of 15 min while the car is cruising along a motorway:

t (hours)	0	0.25	0.5	0.75	1.0
v (km h^{-1})	100	85	110	100	92

Choose a suitable technique and use it to find $\int v\,dt$, the distance travelled in 1 hour.

5 Use the techniques of this chapter to check on the results of some of the problems in Chapter 25.

6 Answer the questions in *Try these first*, page 363.

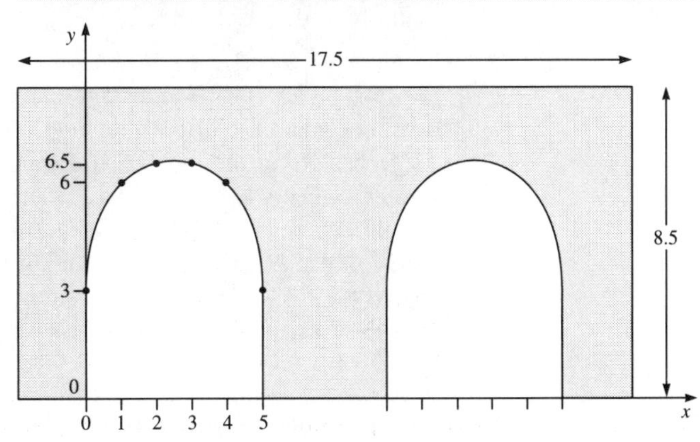

Figure 26.5

27 Charts for efficiency

Planning a project or organizing a workshop are made easier and more effective by the use of graphic techniques. Here we explain how to implement some of these methods.

You need to know about inequalities (page 171), and graphing inequalities (page 231).

Try these first

1 Two jobs are to be processed on three machines, both jobs require the machines to be used in the order M_1, M_2, M_3. The times in hours spent on each machine are:

Job	M_1	M_2	M_3
1	3	5	4
2	1	6	5

Draw Gantt charts to find which job should be processed first, so that the two jobs are completed in the minimum time.

2 The following data refers to renovating a kitchen:

Activity	*Description*	*Predecessor*	*Time (h)*
A	Renew electric wiring	–	4
B	Renew plumbing	–	3
C	Lay tiled floor	B	8
D	Tile walls	F, G	4
E	Paint ceiling	–	1
F	Install cooker	A, C	1
G	Install other units	C	6

Draw a network to show how the activities are linked. Find the minimum time for completion of the renovation. Identify the critical path. Find the slack time, if any, for each activity.

3 A block of 20 offices is to be provided with electric heaters of two powers, small (1 kW) and large (2.5 kW). Each office is to have at least one small heater for mild days. There are not to be more than 26 large heaters in the block. There must not be more than one small heater for every two large heaters. The maximum total power allowed is 100 kW. What is the largest number of heaters allowed, and how many of each sort? What numbers of heaters will consume the most power?

The best order

Doing things in the right order is often important if a task is to be completed as quickly as possible and with the least waste of time and resources. For example, suppose we are making two different parts for a caravan. The parts have to be machined on a lathe, after which they are to be spray painted. The time taken for processing the two parts on the two machines (lathe and paint spray) are:

	M_1	M_2
J_1	30	70
J_2	50	20

In the table, M_1 and M_2 refer to the two machines, or stages in manufacture. J_1 and J_2 refer to the two parts to be made, often referred to as jobs.

It is essential that machining is done first, followed by painting, so we have to begin by placing one of the jobs on the lathe, while the other job waits. We have to decide whether to start with J_1 or with J_2. A **Gantt Chart** helps us make the decision. Figure 27.1a shows the chart for the sequence J_1,J_2, that is to start with J_1 in M_1.

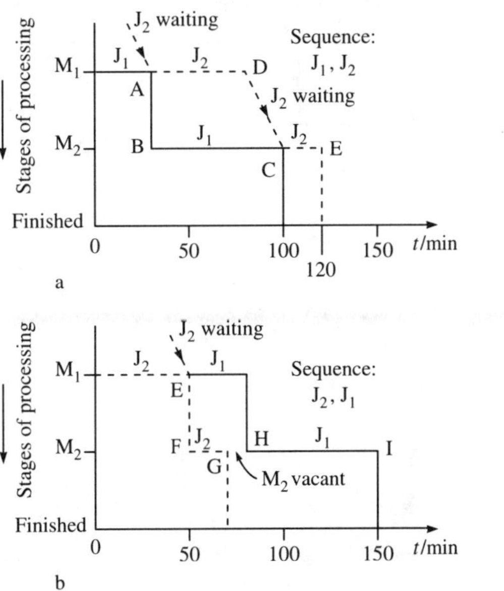

Figure 27.1

We begin by plotting the progress of J_1, representing this on the chart by a continuous line. We run *down* the chart from top to bottom and, at the same time, across the chart from left to right. The line starts on the level marked M_1, which means that this part is being processed on the lathe, while J_2 is kept waiting. J_1 requires 30 min on M_1, then goes for painting. At this stage the line drops down from point A to point B, on the M_2 level. Painting takes 70 min, so the line runs horizontally to point C. This completes all the work on J_1 and its line drops down from C to the base line. It has taken 100 min to complete J_1.

On the same chart, we plot the processing of J_2, representing it by a dashed line. It waits while J_1 is on M_1. At point A machining begins and lasts for

50 min. We have reached point D, but painting cannot begin immediately because J_1 is still using the paint spray (M_2). J_2 has to wait for 20 min and is then taken for painting (point C). Painting takes 20 min (E). J_2 is now complete and its line drops to the base line at 120 min. Completion of both jobs takes 120 min.

Figure 27.1b shows what happens if we process the parts in the reverse order, starting with J_2. First we draw the dashed line to show J_2 being machined for 50 min, then painted for 20 min. This takes a total of 70 min. Next we add J_1 to the chart. It cannot begin until M_1 is free, at 50 min (point F). After 80 min, J_1 goes to be painted. This can be done immediately, as the painting of J_2 is already completed. In fact there is a 10-minute period (G to H) when the spray is not in use. Painting takes a further 70 min (to point I), at which time the line for J_1 drops to the base line. The total time taken is 150 min.

The sequence J_1 followed by J_2 takes the shorter time.

Gantt charts can be applied to processes in which there are three or more stages. Figure 27.2 shows the chart for sequences based on these timings:

	M_1	M_2	M_3
J_1	2	4	5
J_2	5	3	4

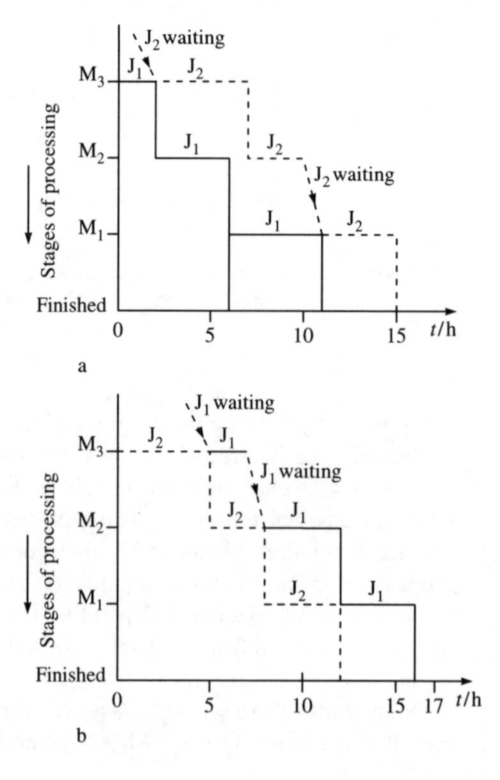

Figure 27.2

As in the previous example, there are two jobs and we have to decide which is the better sequence. The jobs take 15 hours if J_1 is done first, and 17 hours if J_2 is done first.

The sequence J_1 followed by J_2 takes the shorter time.

It is also possible to use Gantt charts when there are three or more jobs. With three jobs, we need to draw six charts to cover all possible sequences:

J_1, J_2, J_3	J_1, J_3, J_2
J_2, J_1, J_3	J_2, J_3, J_1
J_3, J_2, J_1	J_3, J_1, J_2

It may be possible to eliminate one or more of these sequences for various reasons. For example, J_1 may not be sequenced first because an essential part for it is not available early in the working day. With four jobs, the number of sequences is 24. Drawing 24 charts is *possible* but not really practicable. The Gantt chart is suitable only when there are two or three jobs in the sequence.

Longer sequences

Various routines have been worked out for finding the best sequence of jobs when there are more than two. We will look at one for sequencing any number of jobs through two processing stages or machines. The technique does not rely on a chart but we have presented the routine of the technique in chart form (Figure 27.3), as an illustration of another useful type of chart. This is a **flow chart** (see box).

Try the flow chart on this problem. There are two machines and seven jobs; the time for the jobs on each machine are:

Job no.	M_1	M_2
1	2	5
2	4	7
3	5	2
4	4	3
5	6	7
6	8	6
7	1	9

Start at the top of the chart. At the first box, the job with the shortest time (1 hour) is J_7. The decision box asks if there is only one job with this time, which there is, so leave the box by the 'Yes' arrow. The next decision box asks if this shortest time is on M_1, which it is, so follow the 'Yes' arrow. Obey the instruction in the next box; put J_7 first in the list. The sequence list at this stage is:

7 ? ? ? ? ? ?

Delete J_7 from the table. There are still some jobs left, so return to the box at the top of the chart. This time round, the shortest time is 2 hours. There are

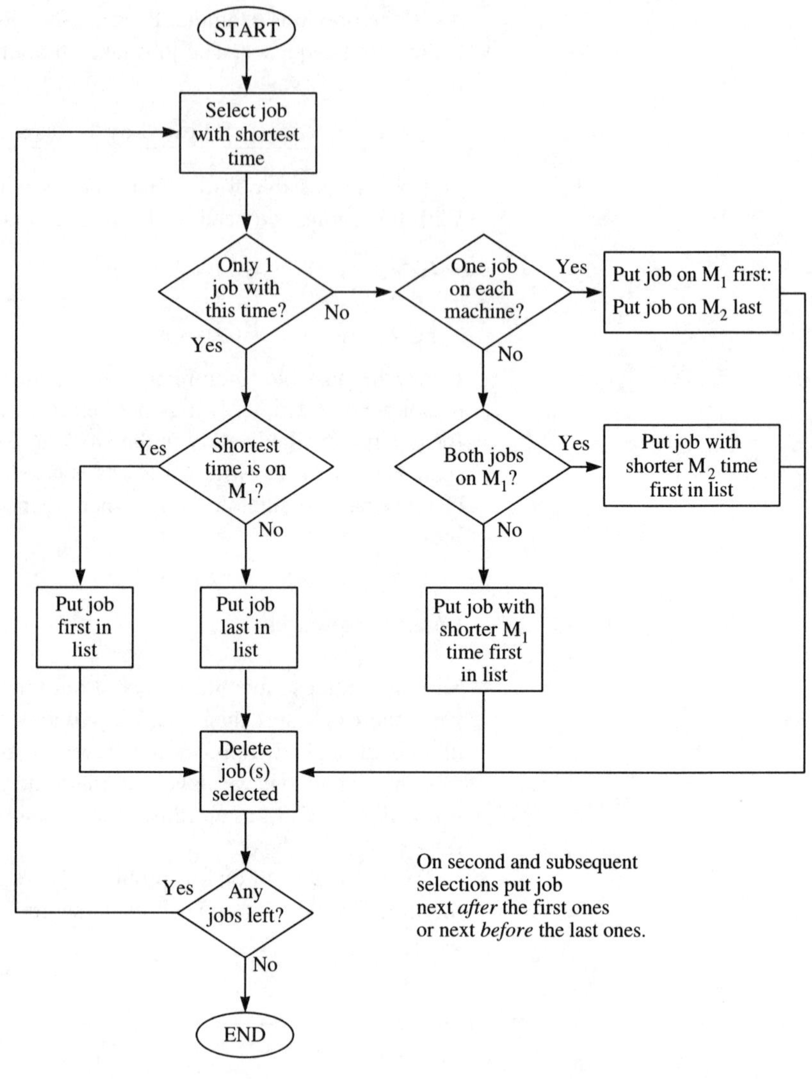

Figure 27.3

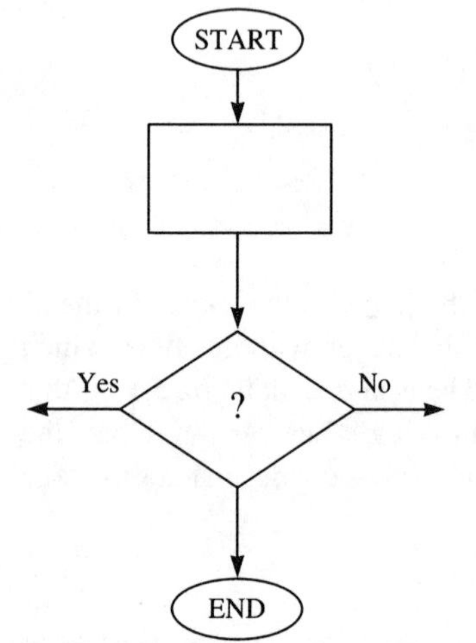

Figure 27.4

Following flow charts

Figure 27.4 illustrates the different types of boxes.

- Start at the **start box**, usually at the top of the chart.
- Follow the arrows.
- When you come to a **process box** (a rectangle), do what it says.
- When you come to a **decision box** (a rhombus), answer the question with 'Yes' or 'No'. Then follow the labelled arrow.
- Continue following the arrows and visiting the boxes until you come to the **end box**.

There is another flow chart for you to practise with on page 398.

two jobs with this time (J_1 and J_3) so follow the 'No' arrow from the first decision box. One of the jobs is on M_1 and one on M_2. Follow the next 'Yes' arrow to the top box on the right. Obeying these instructions, the list becomes:

7 1 ? ? ? ? 3

Delete J_1 and J_3 from the table and return to the top. Next time round, we select J_4 of 3 hours duration on M_2. The instructions are to put this last (or rather next-to-last):

7 1 ? ? ? 4 3

Going round again, we select J_2 on M_1. The list becomes:

7 1 2 ? ? 4 3

This leaves only J_5 and J_6 so we complete the list:

The shortest sequence is 7, 1, 2, 5, 6, 4, 3

Test yourself 27.1

1 Draw Gantt charts to find the sequence of jobs that takes the shorter time.

	M_1	M_2
J_1	4	4
J_2	3	6

2 Draw Gantt charts to find the sequence of jobs that takes the shortest time, given that, for other reasons, J_3 cannot follow immediately after J_2.

	M_1	M_2	M_3
J_1	3	2	4
J_2	2	1	6
J_3	4	4	3

3 Find the sequence of jobs that takes the shortest time.

Job no.	M_1	M_2
1	5	3
2	6	4
3	2	3
4	6	5
5	3	7
6	7	4

Organizing projects

A project (anything from making a curry to building a motorway) consists of a number of stages or **activities**. Some of these have to be done in a definite order. Others can be done at any time, and at the same time as other activities.

Each activity requires a certain amount of time. We may want to know how long the project will take to complete, or the latest date or time it should be started if it is to be completed by a given date or time. We may need to identify those activities which are **critical**. These are the ones which, if delayed, will delay the completion of the whole project. The **critical path method** (CPM) is a way of analysing a project to obtain information of this kind.

As an example, take the organizing of a display for an Open Day at a college. Preparing for this has two main stages. In the first stage, we plan and build the exhibits, design and draw charts, and write leaflets and other documentation. These are long-term activities, requiring many days or weeks to complete. The second stage is setting out the display on the day. This consists of short-term activities, taking only a few minutes or hours. Both stages can be analysed by CPM, but we shall take the setting-out stage as our example.

The first step is to list all activities, not necessarily in order. We assume that all the exhibits, charts and document are already prepared and are available in the same building as the room for the display. We also assume that there are enough people available for several activities to occur simultaneously. The list is shown in Table 27.1.

The table also lists **predecessors**. These are activities which must be completed before the activity can begin. For example, the benches must be cleaned before we can set up the exhibits and computer. The video display is free-standing, so this can be installed as soon as the room is tidy and need not

Table 27.1 Data

Activity	*Description*	*Predecessors*	*Time (min)*
A	Tidy the room; put away items not needed for the display	–	30
B	Clean bench-tops	A	5
C	Collect, set up and test exhibits	B	60
D	Collect computer, load and run program	B	15
E	Collect and set up video display	A	10
F	Place chairs for video display	E	10
G	Pin up charts	A	20
H	Place piles of leaflets beside exhibits, computer and video display	C, D, E	5
I	Last-minute check on whole display	F, G, H	5

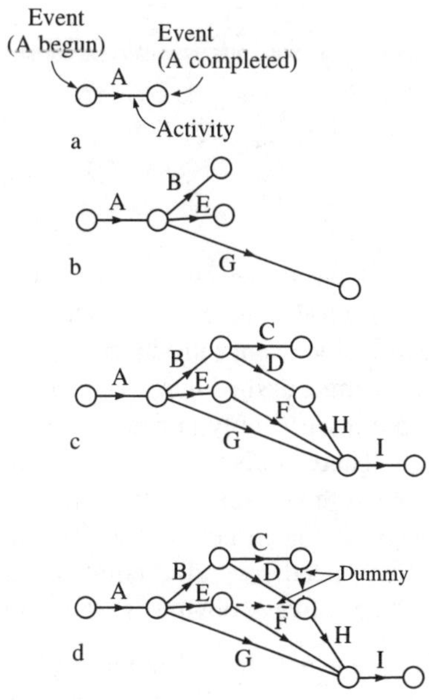

Figure 27.5

wait for bench-tops to be cleaned. The last column of the table lists the estimated time to complete each activity.

The second step is to draw a network to show the dependence of each activity on its predecessors. Figure 27.5 shows the stages in drawing the network. Each **activity** is represented by a line with an arrow-head. The activities link **events** drawn as small circles. These events mark the beginning or the completion of the activities.

The sequence of drawing the chart is as follows:

1 On the left, draw an event to represent the start of the project.

2 From this event draw lines to represent the activities which have no predecessor. In the example, A is the only such activity, so there is one line, labelled A (Figure 27.5a). End the line with another event, the completion of A.

3 From this new event, which also represents the *starting* of activities *following* A, draw lines to represent activities which have A as their predecessor. The table shows that these are B, E and G (Figure 27.5b). End these lines with events to show the completion of these activities.

4 From the events representing the completion of B, E and G draw lines to represent activities which have B, E, or G as their predecessors. In Figure 27.5c, C and D have B as predecessor, F has E, and I has G. We have also drawn in activities F and H, which follow from E and D, and lead to I. The circle at the far end of I is the last event, the completion of the whole project.

5 There are some activities which need to be synchronized. The data table shows that C, D and E must *all* be completed before H (putting out leaflets).

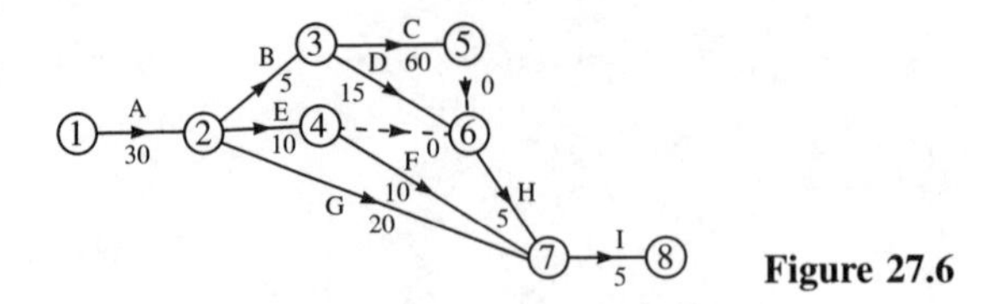

Figure 27.6

We could have drawn two lines C and D running side by side and both converging on the same event. But having two lines which both start and end at the same event introduces confusion later. So we draw in a **dummy activity**, using a dashed line, to connect C to H (Figure 27.5d). This activity takes no time: it simply indicates that C *and* D must be complete before H can start. There is also a dummy activity from E to H, because E must be finished before H can start. But putting out chairs (F) can go on at the same time as H, and need not wait until C and D are finished.

You may need several attempts to draw a network of this kind, as its layout is not always obvious at the beginning.

6 Number the events from 1 at the start, with the final event, the completion of the project, having the largest number (Figure 27.6).

7 Beside each arrow write its activity time, taken from the data table.

8 Using the chart and the data in the table, find the *earliest times* for reaching each event. Think of the network as a number of towns (the events) connected by roads (the activities). You travel from 1 to 8 by various routes. We want to know the earliest times that we could expect to reach each town.

For each event, its earliest time is the time to reach its predecessor *plus* the time taken by the activity. For example, it takes 30 min to tidy the room. In other words, it takes 30 min to reach event 2. After that, it takes 5 min to clean the bench-tops. The time to reach event 3 is $30 + 5 = 35$ min. We proceed like that across the chart adding up the time as we go along. But there are some events (examples, 6 and 7) that are reached by more than one route. Different routes usually take different times. In such cases we take the *longest* of the route times (marked * in Table 27.2). This is because if we have reached an event by a shorter route we have to wait there for others to catch up.

For each event, the network tells us the preceding event. In column *a* we enter the earliest time for reaching the preceding event, taken from column *c* *for that event*. In column *b* we enter the activity time, taken from the data table, to get from the preceding event to this event. The sum of these two times goes in column *c*.

Example

Event 3 is preceded by event 2; the earliest time to reach event 2 is 30 min (see column *c*, event 2). Going from 2 to 3 by way of activity B (see network) takes 5 min. $30 + 5 = 35$. Enter 35 in column *c* for event 3.

If an event is preceded by more than one event, work out the timings for each route separately. Then select the *longest* of these to put in column *c*.

Table 27.2 Earliest arrival times

Event	*Preceding event*	*Earliest arrival time at preceding event* (*a*)		*Activity time* (*b*)	*Earliest arrival time* (*c*)
1	–	–		–	0
2	1	0	A	30	30
3	2	30	B	5	35
4	2	30	E	10	40
5	3	35	C	60	95
6	3	35	D	15	
	4	40	d	0	
	5	95	d	0*	95
7	2	30	G	20	
	4	40	F	10	
	6	95	H	5*	100
8	7	100	I	5	105

d = dummy activity

Example

Event 7 is reached via 2, 4 or 6. It takes 95 min to reach 6, then 5 min to get to 7 by way of activity H: total 100 min. Other routes via 2 and 4 take less time. Enter 100 in column *c* for event 7.

When the table is complete, the time in column *c* for the final event is the shortest time in which the project can be completed. Completion time is 105 min.

If this is all we want to know, the analysis can stop here. But there is more that can be found out, as shown in the next section.

Critical path

There is obviously plenty of time for some of the activities. But certain **critical** activities must be completed in their allotted times, or the project will be delayed. We have found the **earliest** times we reach events. Now we will find the **latest** times at which we can leave each event to reach the last event in 105 min. The method is the same as for earliest times, but worked in reverse (Table 27.3).

For each event, the network tells us the event which follows it. In column *a* we enter the latest time to leave the following event, taken from column *c* for that event. In column *b* we enter the activity time to get from the event to the following event. The difference *c* – *b* goes in column *c*.

Table 27.3 Latest departure times

Event	*Following event*	*Latest departure time from following event* (*a*)	*Activity time*	(*b*)	*Latest departure time* (*c*)
8	–	–		–	105
7	8	105	I	5	100
6	7	100	H	5	95
5	6	95	d	0	95
4	6	95	d	0	
	7	100	F	10*	90
3	5	95	C	60*	35
	6	95	D	15	
2	3	35	B	5*	30
	4	90	E	10	
	7	100	G	20	
1	2	30	A	30	0

d = dummy event

Example

Event 6 is followed by event 7; the latest time to leave 7 is 100 min (see column *c*, event 7). Going from 6 to 7 by way of H takes 5 min. 100 – 5 = 95. Enter 95 in column *c* for event 6.

If an event is followed by more than one event, work out the timings for each route separately. Then select the *earliest* of these to put in column *c*. Whatever route we take from that event, we must be sure to allow enough time to go the longest way.

Example

The latest departure time from event 4 by way of event 6 is the same as that from event 6, because of the dummy event, and is 95 min. Event 7 must be reached in no more than 100 min. Going from 4 via activity F, which takes 10 min, we must leave event 4 no later than 90 min. This is the earlier time; enter 90 in column *c* for event 4.

When the table is complete, the time in column *c* for event 1 is 0. This fact is a good check on the calculation of both tables.

We know the earliest times to reach an event and the latest time to leave an event. For convenience, we list these in a table (Table 27.4).

For many events the two times are the same. There is no leeway between arriving at a node and leaving it. Any delay in starting or finishing an activity adds to the total time of the project. In this example, the events concerned are:

$$1 \rightarrow 2 \rightarrow 3 \rightarrow 5 \rightarrow 6 \rightarrow 7 \rightarrow 8$$

Table 27.4 Arrivals and departures

Event	*Earliest arrival*	*Latest departure*
1	0	0
2	30	30
3	35	35
4	40	90
5	95	95
6	95	95
7	100	100
8	105	105

We have joined the events by arrows to indicate that this is the **critical path** through the network. Any delays in these events delay the completion of the project.

The critical path can also be listed in terms of activities:

$$A \rightarrow B \rightarrow C \rightarrow H \rightarrow I$$

These are the activities where timing is important. Having identified them, we should reconsider their estimated times and, if there is doubt, increase the estimates a little.

The other activities, D. E. F, and G are non-critical. If we are content just to find which activities are critical and which are not, we can leave the analysis at this point. If we want to know more about the non-critical activities, we proceed to the next and final stage of the analysis.

Slack time

This is the amount by which an activity may be delayed, either by starting or finishing late, or by interruptions, without delaying the the completion of the project. If, for example, it happens that getting the computer program to run takes twice as long as estimated, it would be good to know that this will not delay the project.

To find slack times, we need to return to activities (Table 27.5).

The events column lists the events which mark the starting and completion of each activity. The entries in column *a* are taken from the earliest arrival table. For example, the earliest time that activity B can begin is the earliest time we arrive at event 2. The entries in column *b* are taken from the latest departure table. The latest time by which this activity must be completed is the latest time of departure from event 3.

For activity B, these times allow 5 min (column *c* above) for completion of the activity. This is the same as the estimated time of the activity, so there is

Table 27.5 Slack times

Activity	*Events*	*Earliest arrival*	*Latest departure*	*Available time*	*Time for activity*	*Slack time*
		(a)	(b)	(c)	(d)	(e)
A	1 → 2	0	30	30	30	0
B	2 → 3	30	35	5	5	0
C	3 → 5	35	95	60	60	0
D	3 → 6	35	95	60	15	45
E	2 → 4	30	90	60	10	50
F	4 → 7	40	100	60	10	50
G	2 → 7	30	100	70	20	50
H	6 → 7	95	100	5	5	0
I	7 → 8	100	105	5	5	0

no slack time. Putting it another way, we have 5 min to get from 2 to 3 and the whole of this time is needed for activity B. The same applies to activities A, C, H, and I, which are the other activities on the critical path. But for activity D, the available time ($c = b - a$) is 60 min. We arrive at event 3 with 60 min to get to event 6. Yet D takes only 15 min (column d). This means that the slack time is 45 min ($e = c - d$). There is time to spare to repair faulty computer connections or even obtain a replacement machine if necessary.

Test yourself 27.2

1 The data in Table 27.6 refer to preparing a curry.

Draw a network to relate these activities. Find the total time for preparing the curry. Identify the critical path. Find the slack times, if any, for each activity.

2 The data in Table 27.7 refer to building a motorway.

Draw a network to relate these activities. Find the total time for completing the motorway. Identify the critical path. Find the slack times, if any, for each activity.

The best mix

This type of analysis, known as **linear programming**, is used when, under certain given conditions, we have to decide what mix of two types of object will give the best results.

Table 27.6

Activity	*Description*	*Predecessor*	*Time (min)*
A	Measure out rice and water	–	5
B	Soak rice	A	30
C	Cook rice in rice cooker	B	40
D	Grind spices	–	10
E	Prepare other ingredients	–	20
F	Mix spices and other ingredients	D,E	5
G	Cook curry in microwave oven	F	30
H	Cook poppadums	–	10

Table 27.7

Activity	*Description*	*Predecessor*	*Time (months)*
A	Earth moving	–	9
B	Road levelling and foundation	A	14
C	Road surfacing	B,E	5
D	Laying electric mains	–	3
E	Drains, culverts and bridges	A	5
F	Lamps and road signs	B,D	3
G	Road markings	C	2
H	Building service station	D	20

Example

A machine shop is to be equipped with two types of lathe, type A and type B, and we have to decide what mix of A and B will require the least servicing time per lathe. The facts (or conditions) are:

1 Type A need 30 min servicing per week.
2 Type B need 20 min servicing per week.
3 The maximum number of lathes is to be 36.
4 There must be at least 6 type A.
5 There must be at least 10 type B.
6 The number of type A is not to be more than the number of type B.

The first step is to turn these sentences into equations. Let a and b represent the numbers of lathes of types A and B, respectively. Using the information from 1 and 2 above, we find that the total servicing time per week is $30a + 20b$. Dividing this by the number of lathes, the time per machine is:

$$t = \frac{30a + 20b}{a + b}$$

We have to discover what values of a and b, consistent with the other conditions, may be substituted into this equation to give the least possible value of t.

The other conditions are expressed in these equations:

Condition 3: $a + b \leqslant 36$
Condition 4: $a \geqslant 6$
Condition 5: $b \geqslant 10$
Condition 6: $a \leqslant b$

These are inequalities. The way to discover the possible values of a and b subject to these conditions is to draw them on the same graph (Figure 27.7). Having shaded in the areas which do *not* satisfy the inequalities, we are left with a clear area. This area, which is a polygon, contains all the points (a,b) which satisfy all the inequalities. This is known as the **feasible region**.

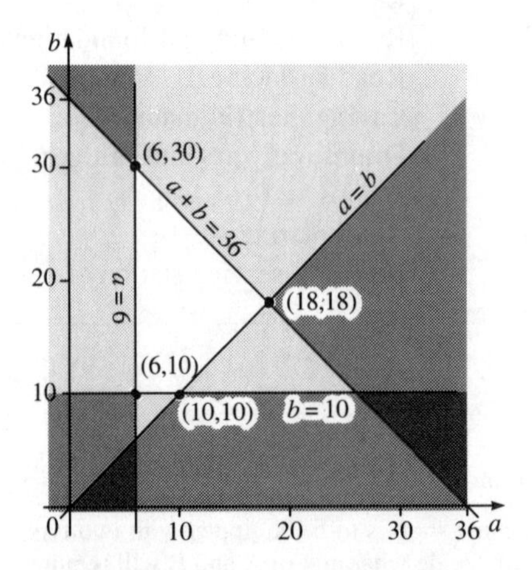

Figure 27.7

To find the least servicing time, we could take all the points in the feasible region, and use their values of a and b to calculate the servicing time t. From these times, we could pick out the shortest. There are 159 points with integer values of a and b, so such calculations would take a long time. Fortunately, it can be proved that the points which give minimum and maximum times for t are at the vertices of the feasible region. We need not consider any other points.

The feasible region has only four vertices, and these are the coordinates of these points:

Point	t
(6,10)	23.75
(10,10)	25.00
(18,18)	25.00
(6,30)	21.67

The table shows values of t calculated from a and b using the equation given earlier. The point with minimum value is (6,30).

The lowest service time per machine is 21.67 min per week, and occurs when there are 6 type A and 30 type B machines.

In this and many other examples, only integer values make sense. In this example, the vertices are all at points expressible in integers, so the maximum and minimum points are at the vertices. In other examples it may happen that some of the inequality lines cross at points which are not integers. In such cases, take the integer points inside the feasible area that are nearest to the vertices.

In most problems it is not possible for the numbers of objects to be less than zero, though this may be taken for granted when stating the conditions of the problem. If the feasible region extends to the axes, the axes act as boundaries.

Test yourself 27.3

1 An orchard is to be planted with apple trees and pear trees. The area of the orchard is 400 m^2. Each apple tree will eventually require 10 m^2 and each pear tree will require 8 m^2. The annual yield from an apple tree is 20 kg. The annual yield from a pear tree is 15 kg. The trees are to be purchased from a local nursery who can supply no more than 20 apple trees of the required variety. There must be at least 5 apple trees and 5 pear trees to ensure reliable pollination. Local market preferences suggest that no more than one quarter of the trees should be pear trees. What is the maximum yield from the orchard and what mix of apple and pear trees will provide it?

2 A stylist uses two sizes of roller, medium and large, for setting hair. He uses 24 rollers to give a full set, never using more than 6 of the large size and always at least 8 of the medium size. The time taken to wind the rollers is 30 s for combing the hair and applying setting lotion, plus 10 s for each medium roller or 15 s for each large roller. What are the minimum and maximum times required for putting the hair on rollers and how many of each size of roller are wound in each case?

3 A librarian is authorized to order up to 30 new books, both fiction and technical. She decides to order at least 5 fiction books. The library rules require that the number of technical books ordered may be no more than half the number of fiction books. The average cost of a fiction book is £10 and the average cost of a technical book is £25. The total amount available for the order is £400. What is the maximum number of books she can order and how many of each sort? If she wants to spend as much of the allowance as possible, how many books of each sort may she order?

4 To assess your progress in this chapter, work the exercises in *Try these first*, page 370.

28 Using spreadsheets

Spreadsheet software is ideal for handling certain types of calculation and has many applications in technology. Here we demonstrate some of its possibilities.

You need to know proportions (page 143), and how to handle formulae (Chapter 12).

The basic idea of a spreadsheet is extremely straightforward. It is a table, set out in columns (lettered A, B, C, . . .) and rows (numbered 1, 2, 3, . . .). A place in the table is called a **cell** and is referred to by its column letter followed by its row number. For example, cell A1 is at the top left corner of the table. Cell B5 is in the second column, fifth row.

Into any cell, we can put *one* of the following:

- Text – used for headings, instructions and other things to help or inform the user
- A number
- A formula for calculating a number.

When the spreadsheet is displayed on the computer screen or printed out on paper, text and numbers are shown in their cells. But formulae are not shown; instead, we see a number calculated by the formula.

With many spreadsheets the formula can include many kinds of mathematical operation, such as adding, subtracting, multiplying, dividing, square roots, trig ratios, logs and more. For the spreadsheets we describe in this book, we need only the first four operations and the INT function (see later).

Once you have set up a spreadsheet complete with formulae, all you have to do is key in numbers and get results.

You key the numbers into one or more of the cells. The computer then takes these numbers, substitutes them into the formulae, and displays the results. If, after this, you enter some different numbers, the computer takes these and gives you a different set of results. You can change the numbers as often as you like; the computer does all the hard work of recalculating the results every time.

If you obtain a set of results that you wish to keep for reference, you can save these on to disk, or print them on paper. With many spreadsheets, there are facilities for displaying and printing graphs of the results.

We cannot describe how to handle any particular spreadsheet here. Consult the instruction manual of the spreadsheet you are using. One problem you could meet is that the manual contains a lot of information that you may never need to use. If you have difficulty, ask an experienced person to write out a few short instructions for getting the program running and for using the spreadsheet for the simple operations described below.

The best way to understand spreadsheets is to use one. Below we demonstrate how a spreadsheet is used to perform a few simple tasks.

Multiplying

This is how to set up a spreadsheet to calculate a multiplying table. Figure 28.1 shows the table as it appears on the screen. Your spreadsheet display may have additional features, but Figure 28.1 has all the essentials. The column headings are at the top of the screen and the row numbers are on the left. In Figure 28.2 the same spreadsheet is printed out in the more usual way, without column headings and row numbers.

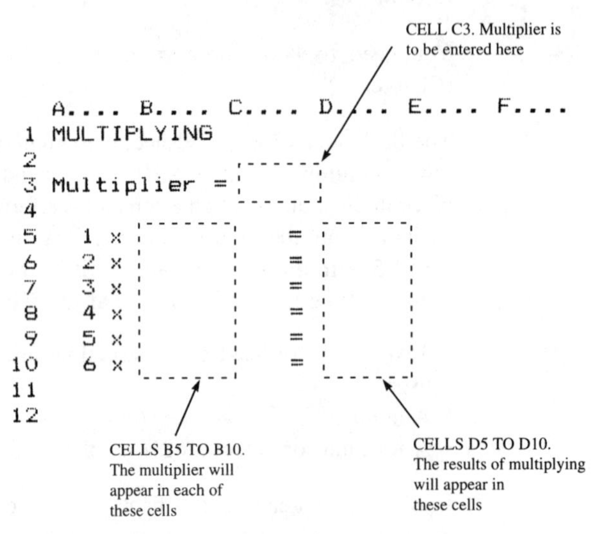

Figure 28.1

```
MULTIPLYING

Multiplier =

  1 x          =
  2 x          =
  3 x          =
  4 x          =
  5 x          =
  6 x          =
```

Figure 28.2

```
B5=C3
D5=1*C3
B6=C3
D6=2*C3
B7=C3
D7=3*C3
B8=C3
D8=4*C3
B9=C3
D9=5*C3
B10=C3
D10=6*C3
```

Figure 28.3

The first step is to plan the layout on paper. Then type in the text, cell by cell, as in Figure 28.1.

Next key in the formulae. These are not shown in the display or printout of the spreadsheet, but can be listed separately (Figure 28.3). The formula for cell B5 is:

B5 = C3

This means that when a number has been typed into C3 and the table is updated (the computer is instructed to calculate results), the number in C3 also appears in B5. Suppose, for example, that we type 4 into cell C3, and then update the table, this will make row 5 of the table read:

$1 \times 4 =$

But this will not happen yet; we have to finish typing in the table before we start to use it. The formula for cell D5 is:

D5 = 1 * B5

The * symbol means **multiply**. The number in B5 is to be multiplied by 1 and displayed in D5. Later, when the table is updated, row 5 will read:

$1 \times 4 = 4$

This will be the first line of the multiplying table.

```
MULTIPLYING

Multiplier =     4

   1 x       4       =       4
   2 x       4       =       8
   3 x       4       =      12
   4 x       4       =      16
   5 x       4       =      20
   6 x       4       =      24
```

Figure 28.4

```
MULTIPLYING

Multiplier =    12

   1 x      12       =      12
   2 x      12       =      24
   3 x      12       =      36
   4 x      12       =      48
   5 x      12       =      60
   6 x      12       =      72
```

Figure 28.5

The list shows that cells B5 through to B10 all contain the equivalent formula, so that cells B5 through to B10 will all display the same number, the multiplier. Cells D5 through to D10 multiply this number by 1, 2, 3, . . . , 6 to produce the first 6 lines of the multiplying table. We could extend the table by adding more rows, to calculate the table up to 12 times or even beyond.

When the table is complete, we key a multiplying number into cell C3. For example we may key ‘4’. When we update the table, we obtain the result shown in Figure 28.4. This is the multiplying table for 4. To obtain the table for 12, all we have to do is to key 12 into C3 and update. In a few seconds, the updated display appears (Figure 28.5). We are not limited to multiplying by 12. We can key in any number of reasonable size and obtain a table of its multiples.

Curtain calculator

Given the dimensions of a window and the width of the fabric used for curtaining, this spreadsheet calculates the amounts of material required for the curtains and for hanging them.

Figure 28.6 shows the spreadsheet, complete with column headings and row numbers, and with a set of results already calculated. You key in only three numbers, all in metres:

C3: Window width
C4: Window length
C5: Fabric width

```
    A.......  B.......  C.......  D....
 1  CURTAIN CALCULATOR
 2
 3  Window width =          2.65
 4  Window length =         1.75
 5  Fabric width =          1.50
 6
 7  Number of drops =                    3
 8  Curtain length =        1.87
 9  Fabric length =         5.61
10  Tape length =           4.50
11
12  Rail length =           3.05
13  Number of brackets =                10
14  Number of runners =                 60
```

Figure 28.6

```
D7=INT(C3*1.5/C5+0.9)
C8=C4+0.12
C9=D7*C8
C10=C5*D7
C12=C3+0.4
D13=INT(C12/0.3+1.5)
D14=INT(C5/0.08+1.5)*D7
```

Figure 28.7

When the table is updated, all the other numbers are calculated and displayed. The formulae are listed in Figure 28.7. The explanations of these are:

D7: Number of drops. This is the number of curtains required. The basic formula for this is:

$$\text{No. of drops} = \frac{\text{window width} \times 1.5}{\text{fabric width}}$$

This formula allows for the total width of the curtains to be 50% wider than the window. This is divided by fabric width to give the number of drops. But it is usual to work with whole widths of fabric, not fractions. The way to do this is to add almost 1 width, then cut off the fraction. For example, if the number of drops found by division is 2.0, or only slightly over, then 2 drops are sufficient. But, if the number is greater than 2.1 but less than 3, we need 3 drops. For example, if the number is 2.15, add 0.9 to make 3.05. Cut off the 0.05, leaving 3 drops.

The INT function

The INT function used in spreadsheets removes the decimal places from a number.

Examples

INT (3.45) = 3
INT (6.83) = 6

Note that it simply *removes* the decimal places. It does not round the number to the nearest integer.

C8: Curtain length. The curtain length is the window length plus 0.12 m to allow for hems at the top and bottoms of the curtains.

C9: Fabric length. This is the length of one curtain (C8) multiplied by the number of drops (D7).

Rounding up with INT

If you spreadsheet does not have a ROUND function, you can round numbers to integers using INT. The routine is to add 0.5, then use INT.

Examples

Round 3.24

INT (3.24 + 0.5) = INT (3.84) = 3 The number is rounded *down* to 3.

Round 3.86

INT (3.86 + 0.5) = INT (4.36) = 4 The number is rounded *up* to 4.

Round 3.50

INT (3.50 + 0.5) = INT (4.00) = 4 The number is rounded *up* to 4.

C10: Header tape length. This is the width of the fabric (C5) multiplied by the number of drops (D7).

C12: Rail length. The width of the window plus 0.4 m to allow the rail to extend beyond the window. If a curtain is being fitted into a recessed window, the rail length equals the width of the window; this result is not then used.

D13: Number of rail brackets. This allows for the brackets to be spaced 30 cm apart. Having calculated the number in this way, we need to add 1 bracket at the starting end. This is then rounded up to the next integer, by adding an extra 0.5 before finding the integer.

D14: Number of runners. The runners are spaced 8 cm apart on the tape. The tape is in separate lengths, each as long as the fabric width. For each drop:

Number of runners per drop = INT(fabric width/0.8 + 1.5)

Total number of runners = INT(fabric width/0.8 + 1.5)*number of drops

This spreadsheet tells us what materials to obtain. It could be extended to calculate the total cost of dressing the window, given the cost of the fabric per metre and the prices of the other items.

Accounting

Probably the most common use for spreadsheets is in preparing invoices, financial accounts, stock sheets, inventories and similar tables. We shall not go into these aspects here, but most spreadsheet manuals describe examples.

Explore these

Design and use spreadsheets for the following tasks.

1 To find fractions as decimals. You key in a number n, and the spreadsheet calculates, $1/n$, $2/n$, $3/n$ . . . , $12/n$.

2 To calculate the volume of a rectangular box, given its three dimensions. Also to calculate the surface area and, given the weight of $1\ m^3$ of the sheet material used to make it, the weight of the box.

3 To calculate the materials needed for building a wooden or wire fence, given height and length.

4 To find the energy content of a person's daily diet, given the quantities of carbohydrate, protein and fat consumed.

5 There are many suggestions for spreadsheet work under the heading *Explore these* in other chapters.

29 Hunting for solutions

There are many equations, often quite simple to look at, but which cannot be solved by any of the analytical methods we have described in earlier chapters. For these, we use a numerical method, relying on the power of a calculator or computer to help.

You need to know about equations and formulae (Chapter 12), graphs (Chapter 17), differentiating (Chapter 24), and to be familiar with using a calculator.

Try these first

1 Use the bisection method to find (to 2 dp) the positive root of the equation: $x^2 + x - 4$.

2 Use the Newton–Raphson method to find (to 4 dp) the root of the equation $2x^5 - 3 = 0$.

3 Use the secant method to find (to 4 dp) the positive non-zero root of the equation $\sin x - x^2 = 0$.

Hunting for the solution

The methods described in this chapter are for use with equations in which there is only one unknown, and which can be expressed in the form:

$$f(x) = 0$$

Function notation

A short way to refer to 'a function of x' is to write:

$$f(x)$$

The f is a **symbol** for 'function'. It does not mean 'multiply the x in brackets by f'.

The function of x when x has a particular value may be written with the value in the brackets. For example,

$$f(3.5)$$

is a short way of writing 'the value of the function when x equals 3.5'.

If we differentiate a function of x, the derivative is represented by the symbol:

$$f'(x)$$

Example

If $\quad f(x) = 3x^2 - 3x + 5$

then $\quad f'(x) = 6x - 3$

That is to say, there are terms in x and possibly some constants on the left side and zero on the right side. An example of such an equation is:

$$x^3 - x - 5 = 0 \tag{1}$$

There is no way of solving this equation by ordinary algebra, and it cannot be factorized. It is not quadratic, so we cannot use the quadratic formula. We use one of the **numerical methods** (also called **iterative methods**). The essentials of an iterative method are:

- Make an intelligent guess at the solution
- Go through a repeated sequence of calculations ('iterate' means 'to repeat') gradually getting nearer to the true solution
- Stop when the value is as precise as it needs to be.

Starting with graphs

Graphs help us make an intelligent guess at the solution. Equation (1) in the previous section is re-written:

$$x^3 = x + 5 \tag{2}$$

x appears on both sides of this equation. We split the equation into its two sides and write:

$$y = x^3 \tag{3}$$

$$y = x + 5 \tag{4}$$

Plot graphs of these two equations, as in Figure 29.1. Where the lines cross, x and y have the same value in both equations (page 229). If y has the same value, then the right side of (3) equals the right side of (4), and:

$$x^3 = x + 5$$

This brings us back to equation (2). The value of x when the lines cross is the solution to this equation. Estimating the value from the graph:

$$\underline{x = 1.9}$$

With some equations there may be more than one solution (or **root**, page 301). But for this example, Figure 29.1 shows the cubic curve to begin to slope steeply down for negative values, so there is only one point at which the

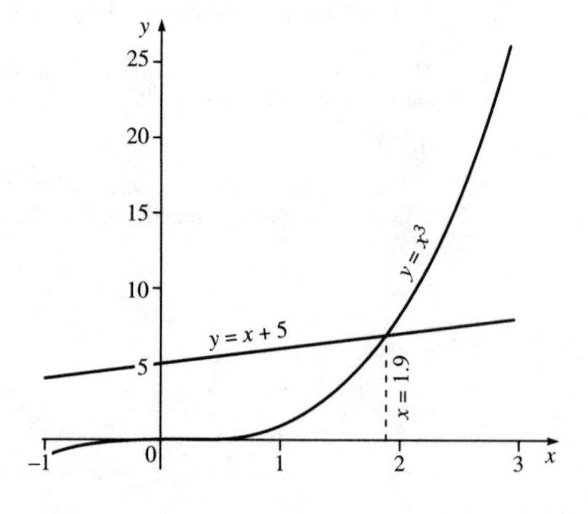

Figure 29.1

lines cross and only one root. If the lines cross at two or more places, we have two or more roots. It may be necessary to ignore some of these; for example, a negative root is not usually acceptable in practical problems.

Substituting $x = 1.9$ into equation (1) gives:

$$x^3 - x - 5 = 6.859 - 1.9 - 5 = -0.041$$

The result is not *exactly* zero as equation (1) requires, so $x = 1.9$ is not the exact solution. But the result is reasonably close to zero and we may decide that this solution is close enough to the true root. If so, we need go no further. For greater precision, we proceed to the next stage.

Bisection method

Figure 29.2 shows how the value of $x^3 - x - 5$ varies with x. When the curve cuts the x-axis:

$$x^3 - x - 5 = 0$$

The value of x at this point is the solution of the equation. We can see that this is when x is approximately 1.9. Now to find this point more precisely. One clear fact about it is that, when the curve cuts the x-axis:

The value of $x^3 - x - 5$ changes from negative to positive

With other functions, the curve may slope the other way but there is still a **change**, in this case from positive to negative. Our hunt for the solution or root of the equation is to:

Look for the change in sign

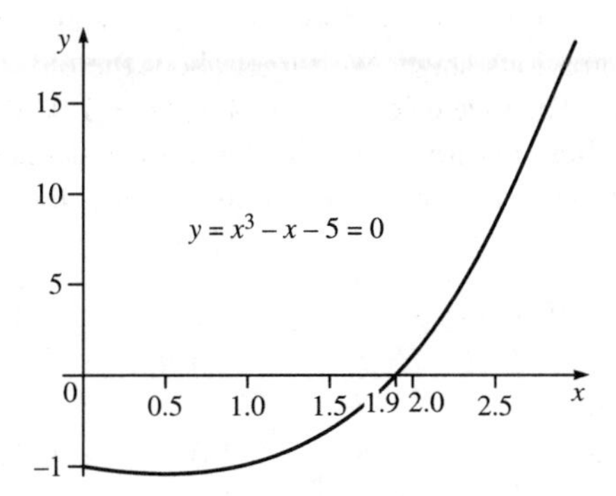

Figure 29.2

> **Graphs and roots**
>
> Splitting the equation, plotting graphs and finding where they cross is one way of using graphs to find roots (Figure 29.1, Figure 29.5).
>
> Another way is to plot a graph of the whole equation and finding where it cuts the x-axis (Figure 29.2).
>
> Generally the first method is preferred because the equations to be plotted are simpler. This makes it easier to see in advance where they are likely to cross, and so to plot or sketch the curve over the appropriate domain.
>
> Remember that not every root found by these methods makes sense when applied to a practical problem.

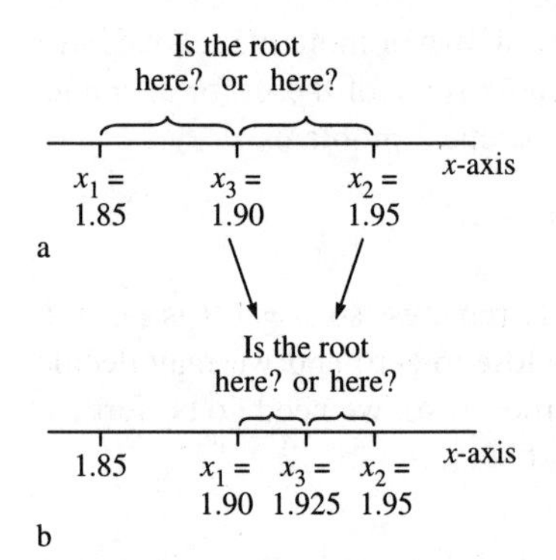

Figure 29.3

We search systematically for this:

1 Decide on two values of x, called x_1 and x_2, which lie on either side of the estimated root. In this example, suitable values are $x_1 = 1.85$ and $x_2 = 1.95$. We know that the value of the root lies somewhere between x_1 and x_2 (Figure 29.3a). The next step is to narrow the search.

2 Find the value which comes half-way between x_1 and x_2. Call this mid-point x_3. We have **bisected** the interval between x_1 and x_2. The root must lie between:

$$x_1 \text{ and } x_3$$

or between x_3 and x_2

where $$x_3 = \frac{x_1 + x_2}{2} = \frac{1.85 + 1.95}{2} = 1.90$$

We find the values of the function $f(x) = x^3 - x - 5$ for the three values of x:

For $x_1 = 1.85, f(1.85) = -0.518$

For $x_3 = 1.90, f(1.90) = -0.041$

For $x_2 = 1.95, f(1.95) = 0.465$

The change of sign comes between x_3 and x_2. This indicates that the value of the root lies between x_3 and x_2.

If we wish to continue to hunt, we examine the interval from x_3 to x_2, bisecting it again, and looking for a change of sign. This is really a repeat of the previous stage. We forget about the old value of x_1 and give it the value of x_3. Then we calculate a *new* x_3 ($x_3 = 1.925$) and find on which side of it the root lies (Figure 29.3b). To make the calculation clearer and easier to check, it is best to set it out as a table. This is the table for what we have done so far:

Values of x				*Values of function*			
Low	*Mid*	*High*	*Interval*	*Low*	*Mid*	*High*	*Reject*
x_1	x_3	x_2					
1.85		1.95	1.0	–0.518		0.465	
					change ↗↙		
	1.90				–0.041		x_1

We worked out the value of the function for the low, mid and high values of x. The **change** of sign comes between mid and high (between x_3 and x_2) so we reject x_1.

In the next line of the table, fill in the new value of x_1 and carry down the old value of x_2. Do the same for the values of the function. The interval is now $1.95 - 1.90 = 0.5$. The mid-value of x is $x_3 = 1.925$. The new mid-value of the function is 0.208. The next entries in the table are those shown between the dashed lines below:

Numerical methods on a calculator

An ordinary calculator is a great help, but a formula calculator is even better. Many types of scientific and graphic calculator allow you to key in a formula, for example:

$$Y = X^3 - X - 5$$

This the function used in the example on this page. We are using capital letters because some calculators display only capitals, not lower-case letters.

Press EXEcute and the display asks you to key in a value of X. Press EXEcute again and the calculator displays the value of Y, calculated by using the formula. Press EXEcute again and the next value of X is requested.

In the Newton–Raphson method and the Secant method, the program can include a second phase, to make the result of one calculation become the input to the next. For example:

$$Y = (2X^3 + 4)/(3X^2 + 2) : X = Y$$

This the formula for the example on page 401. The final $X = Y$ takes the result (Y) and gives its value to X, ready for the next iteration. All you have to do after entering the initial value of x_1 is to press the EXEcute key repeatedly. Successive values of x_2 are displayed. Stop when two successive values are identical, to the required number of decimal places.

Your calculator may have a different method of programming formulae; check with the user's manual. If you do not have access to a calculator, but are good at programming a computer, it is an interesting project to write programs for one or more of these numerical methods.

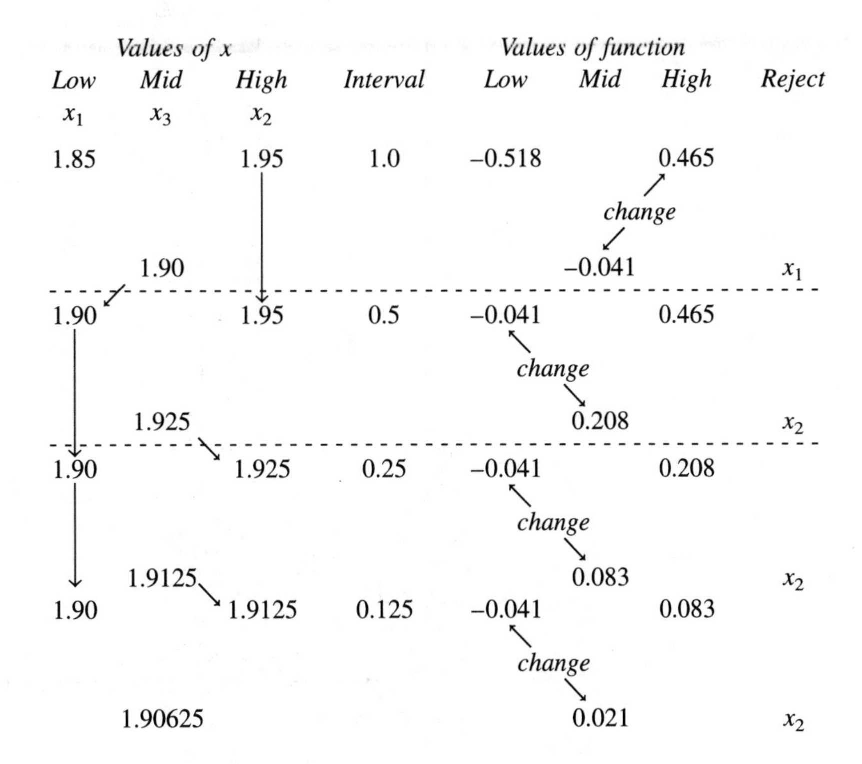

Values of x				*Values of function*			
Low	*Mid*	*High*	*Interval*	*Low*	*Mid*	*High*	*Reject*
x_1	x_3	x_2					
1.85		1.95	1.0	–0.518		0.465	
					change		
	1.90				–0.041		x_1
1.90		1.95	0.5	–0.041		0.465	
					change		
	1.925				0.208		x_2
1.90		1.925	0.25	–0.041		0.208	
					change		
	1.9125				0.083		x_2
1.90		1.9125	0.125	–0.041		0.083	
					change		
	1.90625				0.021		x_2

During the next 3 repetitions (or iterations), the search moves closer toward $x_1 = 1.9$. As we get closer to the root, the values of the function get closer to zero. At the next stage beyond the end of the table, we would retain $x_1 = 1.9$, and make x_2 equal to 1.90625. This is as far as we need to go. Take the average of x_1 and x_2 as our best estimate of the root:

Solution: $x = 1.903$ (3 dp)

Flow chart

Figure 29.4 is a flow chart which summarizes the stages of the bisection method. See the box on page 374 for the way to use the chart. Work through

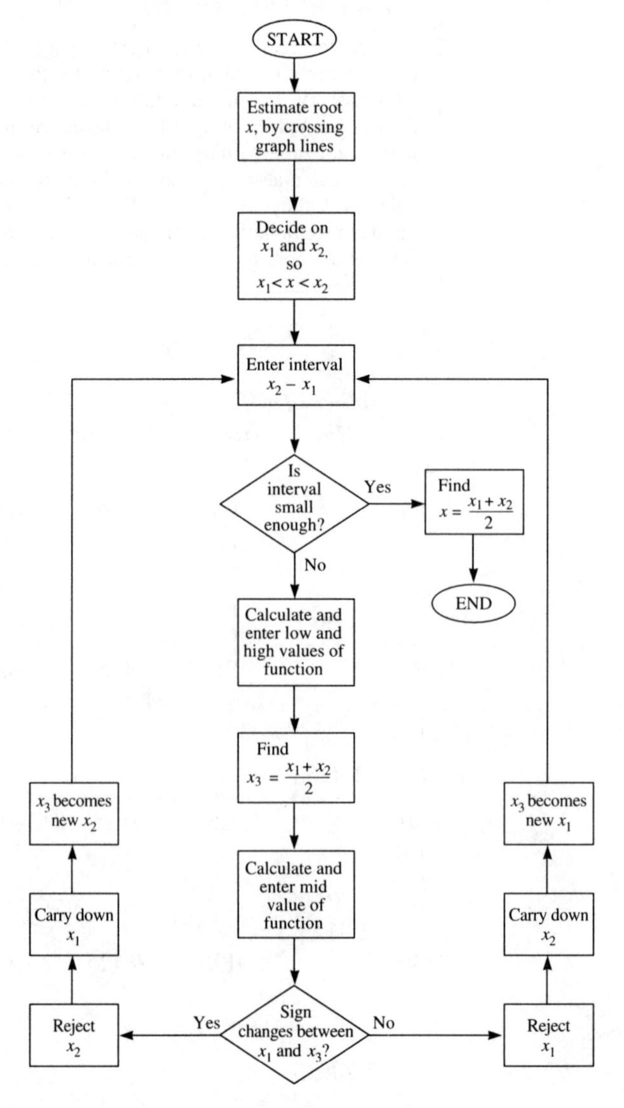

Figure 29.4

the table of the previous example, following the chart at each stage. Then see how the chart works with this example. The equation is:

$$\sin x - 2x + 1 = 0$$

Re-writing gives:

$$\sin x = 2x - 1$$

Splitting the equation:

$$y = \sin x$$
$$y = 2x - 1$$

The curves of these equations cross when $x = 0.9$ (Figure 29.5). This is a rough estimate, and we use the bisection method to improve upon it. As starting point, we take $x_1 = 0.85$ and $x_2 = 0.95$. Remember to put the calculator into radian mode.

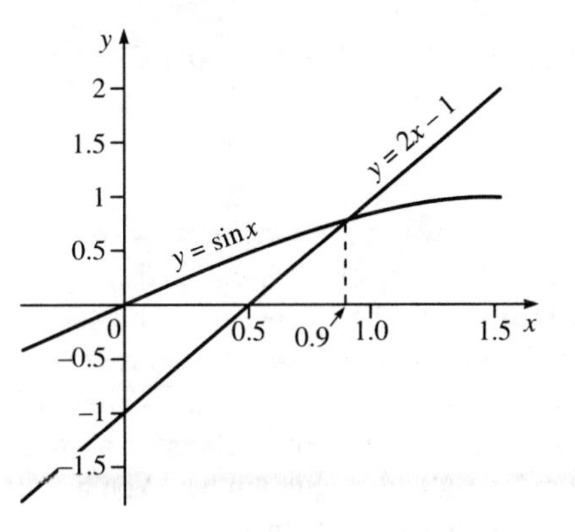

Figure 29.5

Values of x				*Values of function*			
Low	*Mid*	*High*	*Interval*	*Low*	*Mid*	*High*	*Reject*
0.85		0.95	0.1	0.0513		−0.0865	
↓					*change*		
	0.90				−0.0167		x_2
0.85		0.90	0.05	0.0513		−0.0167	
		↓			*change*		
	0.875				0.0175		x_1
0.875		0.90	0.025	0.0175		−0.0167	
		↓			*change*		
	0.8875				0.0005		x_1
0.8875		0.90	0.0125	0.0005		−0.0167	
					change		x_2
	0.89375				−0.0081		

The root lies between 0.8875 and 0.89375. Taking the average of these:

Solution: $x = 0.891$ (3 dp)

The bisection method always works (some of the other methods do not *always* work). The calculations are rather lengthy to do on paper, but are quick with a calculator that has a formula facility.

Test yourself 29.1

Use the graphical method to find approximate solutions to these equations. Then use the bisection method to obtain solutions to 2 dp. Equations 1 and 2 have only one root.

1 $x + \cos x = 0$

2 $e^x + x = 0$

3 $\sin x - 2x^2 = 0$ (has a root at $x = 0$; find the other root)

Newton–Raphson method

This is a fast way of finding roots with great precision, but is usable only with functions that can be differentiated. It also has the slight disadvantage that it may lead you to the wrong root. This happens if your original estimate is too far from the value of the required root.

It is best to begin by plotting a graph or in some other way finding a good first estimate of the root or roots. If a sketch graph is plotted, this will show how many roots the equation has. You may then decide to find only one of the roots, or perhaps all of them.

When an approximate value of a root has been found, one or two stages of the bisection method are used to get closer to the required root. From then on, the Newton–Raphson method relies on an equation that we state without proving. Given an estimate of the root x_1, we find a new (and closer) estimate, x_2 by using the formula:

$$x_2 = x_1 - \frac{f(x_1)}{f'(x_1)}$$

In this formula, $f(x_1)$ is the function with the values of x_1 substituted in it. $f'(x_1)$ is the **derivative** of the function, with values of x_1 substituted in it.

At each stage, the routine produces values $x_2, x_3, x_4 \ldots$ These gradually get nearer to the true root as shown in Figure 29.6. The routine is repeated until the latest value of x is exactly the same as the one before it, to the required number of decimal places. This is the value of the root.

Example

To 5 dp, find the root of:

$$x^3 + 2x - 4 = 0$$

This equation has only one root. We begin as before, re-writing the equation:

$$x^3 = 4 - 2x$$

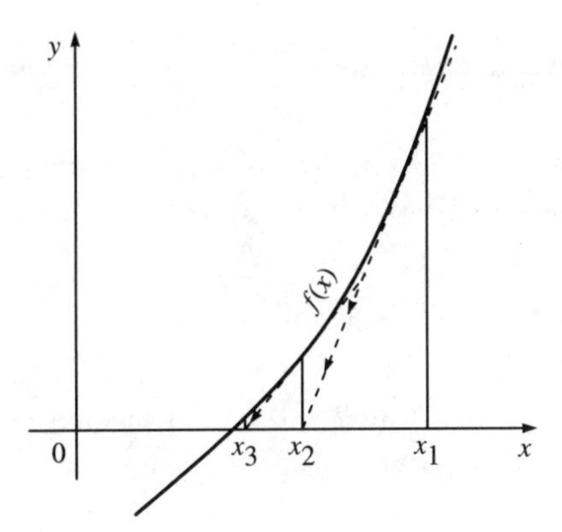

Figure 29.6

Plotting graphs of the two sides of the equation, and finding where the lines cross provides an estimate of the root, which lies somewhere between 1.1 and 1.2. Taking these as x_1 and x_2 for the bisection method, we find that the root is between 1.15 and 1.20. We select any *one* of these values, say 1.15, to be the first value, x_1, for the Newton–Raphson method.

The function is $x^3 + 2x - 4$

Its derivative; is $3x^2 + 2$

Putting these expressions into the formula:

$$x_2 = x_1 - \frac{x_1^3 + 2x_1 - 4}{3x_1^2 + 2}$$

Combining the terms on the right into a single term makes the calculations easier:

$$x_2 = \frac{x_1(3x_1^2 + 2) - x_1^3 - 2x_1 + 4}{3x_1^2 + 2}$$

$$= \frac{3x_1^3 + 2x_1 - x_1^3 - 2x_1 + 4}{3x_1^2 + 2}$$

$$\Rightarrow \quad x_2 = \frac{2x_1^3 + 4}{3x_1^2 + 2}$$

This is a fairly complicated formula, but it is easy to evaluate with a formula calculator.

Putting $x_1 = 1.15$ into this, we obtain the result $x_2 = 1.180\,02$

Make this the new x_1.

Putting $x_1 = 1.180\,02$ into the formula, we find $x_2 = 1.179\,51$

Make this the new x_1.

Putting $x_1 = 1.179\,51$ into the formula, we find $x_2 = 1.179\,51$

This is the same as the previous value, so this is the value of the root. We have reached a very precise result with only three iterations.

Solutions: the root is 1.179 51 (5 dp)

Test yourself 29.2

Use the Newton–Raphson method to solve these equations to 5 dp. The equations have only one root.

1 $2x^3 - 10x - 12$

2 $\sin x - 2x + 4 = 0$

3 $2 \ln x - 3x^2 + 5 = 0$

4 Use the Newton–Raphson method to find the square root of 7 (to 5 dp). [*Hint*: Solve the equation $x^2 - 7 = 0$]

Secant method

This is another quick method, not as fast as the Newton–Raphson method but having the advantage that differentiation is not required. This makes it suitable for functions that are difficult or impossible to differentiate. This method does not work with all functions. With some, instead of getting closer and closer to the root, it wanders wildly, getting further and further from the root. If this is happening it soon becomes obvious and one of the other methods can be tried.

We *explain* the method by using a diagram (Figure 29.7), though it is not necessary to draw this when using the method. We begin by locating two points, (x_1, y_1) and (x_2, y_2) on the curve of the function. On the diagram, a line is drawn through the two points. Such a line, which cuts a curve at two or more points, is called a **secant**, and this gives the name to this method. Contrast this with a **tangent** which just **touches** the line at **one** point (page 203).

The secant is continued to cut the x-axis at x_3, which is closer to the true root than x_1 or x_2. We then use x_3 and x_2 to find x_4, which is even closer to the root. The operation is repeated until we get as close to the root as necessary, depending on the number of decimal places required.

Figure 29.7 shows that the initial values x_1 and x_2 can be on the same side of the root.

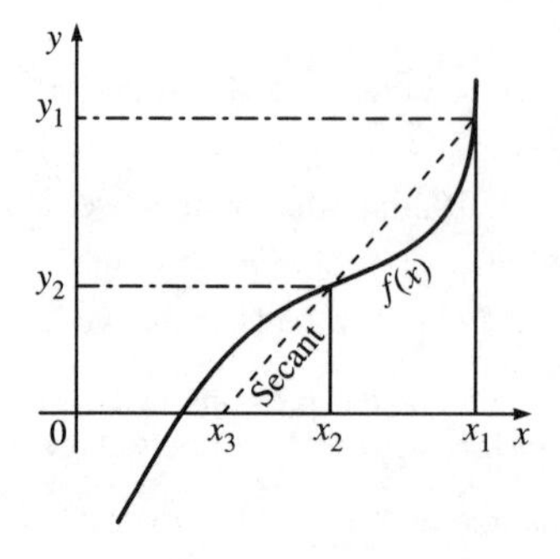

Figure 29.7

The gradient of the secant between the two points on the curve is:

$$\frac{y_1 - y_2}{x_1 - x_2}$$

The gradient of the secant between the points (x_2,y_2) and $(x_3, 0)$ is:

$$\frac{y_2 - 0}{x_2 - x_3}$$

These gradients refer to different parts of the same straight line, so they are equal:

$$\Rightarrow \quad \frac{y_1 - y_2}{x_1 - x_2} = \frac{y_2}{x_2 - x_3}$$

$$\Rightarrow \quad x_2 - x_3 = y_2 \times \frac{x_1 - x_2}{y_1 - y_2}$$

$$= \frac{x_1y_2 - x_2y_2}{y_1 - y_2}$$

$$\Rightarrow \quad -x_3 = \frac{x_1y_2 - x_2y_2}{y_1 - y_2} - x_2$$

Changing signs:

$$\Rightarrow \quad x_3 = x_2 - \frac{x_1y_2 - x_2y_2}{y_1 - y_2}$$

Simplifying the right side:

$$x_3 = \frac{x_2(y_1 - y_2) - x_1y_2 + x_2y_2}{y_1 - y_2}$$

$$= \frac{x_2y_1 - x_2y_2 - x_1y_2 + x_2y_2}{y_1 - y_2}$$

$$\Rightarrow \quad x_3 = \frac{x_2y_1 - x_1y_2}{y_1 - y_2}$$

This is the formula for calculating x_3 given x_1, x_2 and the corresponding values of y_1 and y_2.

Having found x_3, we approach the root even more closely by calculating x_4. To find this, we discard x_1. Then we use x_2 and x_3, with the corresponding values of y_2 and y_3 in the formula:

$$x_4 = \frac{x_3y_2 - x_2y_3}{y_2 - y_3}$$

If this is not precise enough, we can go on to find x_5, discarding x_2 and using x_3, x_4, y_3 and y_4. The secant method converges on the root very quickly, so it is often not necessary to go further than this.

Using the secant formula

As an example, we take the equation:

$$x \sin x + 2x - 4 = 0$$

A sketch of the curve shows that the root occurs when $x \approx 1.4$. Let us take $x_1 = 1.5$ and $x_2 = 1.4$.

If $x_1 = 1.5$, $y_1 = 0.496\,24$

If $x_2 = 1.4$, $y_2 = 0.179\,63$

Using the formula:

$$x_3 = \frac{1.4 \times 0.496\,24 - 1.5 \times 0.179\,63}{0.496\,24 - 0.179\,63}$$

$$= 1.343\,26$$

If $x_3 = 1.343\,26$, $y_3 = -0.004\,84$

At the next step we discard x_1. We use x_2 and x_3 to calculate x_4:

$$x_4 = \frac{1.343\,26 \times 0.179\,63 - 1.4 \times (-0.004\,84)}{0.179\,63 - (-0.004\,84)}$$

$$= 1.344\,75$$

If $x_4 = 1.344\,75$, $y_4 = 3.97 \times 10^{-5}$

Since y_4 is so small, we are obviously very close to the root. Try one more iteration, as a check. Discard x_2:

$$x_5 = \frac{1.344\,75 \times (-0.004\,84) - 1.343\,26 \times 3.97 \times 10^{-5}}{-0.004\,84 - 3.97 \times 10^{5}}$$

$$= 1.344\,75$$

$x_5 = x_4$ to 5 dp.

<u>Solution: $x = 1.344\,75$</u>

Although the calculations may seem complicated when compared with the bisection method, note that this result is correct to 5 dp while that on page 396 is correct to only 3 dp.

Test yourself 29.3

Use the secant method to solve these equations to 4 dp.

1 $x^2 - 3x + 1 = 0$

2 $xe^x - 2 = 0$

3 $2 \tan x - 1 =$ (first positive root only)

4 Find the fifth root of 5, by solving the equation $x^5 - 5 = 0$

5 To assess your progress in this chapter, work the exercises in *Try these first*, page 393.

30 Numbers for counting

We are so used to counting and working maths problems in the decimal system that we tend to forget that other counting systems exist, with special applications. This chapter describes the more commonly used number systems and explains some of their uses.

Try these first

Work these exercises without using a calculator.

1 Convert 28_{10} into its binary equivalent.

2 Convert 100110_2 into its decimal equivalent.

3 A computer instruction is written as the hexadecimal number A25B. Write this as a list of 16 bits, with the most significant bit on the left.

4 Find the sum of these binary numbers: 101101 + 110101.

5 Find the difference of these binary numbers: 111010 – 10011.

Counting the miles

The odometer (often called the **mileometer**) of a car is an instrument for counting the miles the car has travelled. It is mounted on the dashboard, connected to a road wheel by a flexible coupling. There is gearing to reduce the rate of rotation, and a mechanism to advance the reading every mile. In this way, it counts the number of miles travelled by the car since it came off the assembly line.

The six discs of the odometer have the numbers 0 to 9 on their rims (Figure 30.1). As each mile is completed, the disc on the right is advanced by a tenth of a turn. The figure shows the odometer after the car has travelled 2 miles. After 9 miles, the indicator shows 000009.

At the tenth mile, the mechanism of the mile disc completes one turn and shows 0 again. But, as it does so, the disc next to it is advanced a tenth of a turn, from 0 to 1. The indicator changes from 000009 to 000010. There is a **carry-over** from one digit to the digit on its left. The unit disc continues to count from 0 to 9 again. The 'tens' disc makes one complete turn every hundred miles. As it turns from 9 to 0, the disc to its left (the hundreds disc) is advanced a tenth of a turn. The indicator changes from 000099 to 000100. There have been two carry-overs, from tens to hundreds, as well as from units to tens.

There is a thousands disc which is advanced each time the hundreds disc goes from 9 to 0, as the indicator changes from 000999 to 001000. Altogether, the six discs are able to register up to 999999 miles.

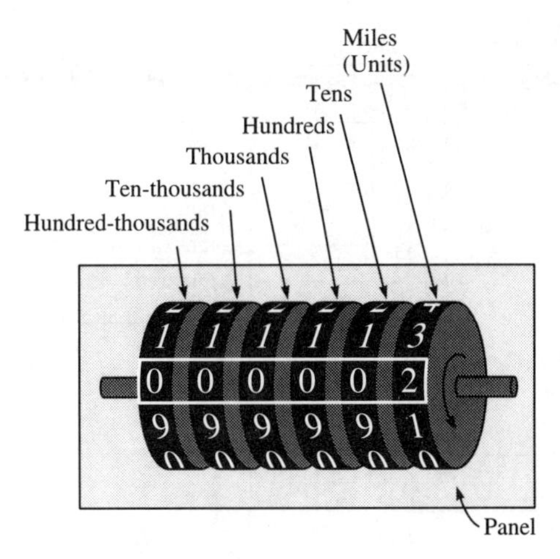

Figure 30.1

In this mechanism, the discs are **identical**, all bearing numbers from 0 to 9. But the **position** of a disc in the row decides whether the numbers on it represents hundreds of thousands, tens of thousands, thousands, hundreds, tens or units of miles. Thus, each disc has a **position value**, which is ten times the position value of the disc on its right.

It would be easy to make a similar odometer in which the discs had numbers around their rims not from 0 to 9 but from 0 to 7. The right-hand disc would be advanced an *eighth* of a turn at each mile. It would read miles from 0 to 7. Then, at the eighth mile, there would be a carry-over and it would turn from 7 to 0. The disc to its left would turn from 0 to 1. The indicator would change from 000007 to 000010, representing 8 miles.

Octal is the name given to a system of counting based on eights. In the octal system, the disc on the right registers miles, the disc to the left of it registers eights of miles. The disc to the left of that registers eight eights (or sixty-fours) of miles. The disc to the left of that registers eight sixty-fours (or five-

Specify the system

A number written as 345 is usually taken to be three hundred and forty-five. But, if it is a number in the octal system, its value (in decimal) is 229. Or it might be a number in a system based on fives, in which case its decimal value is 100. If there is any doubt we must make it clear which number system is being used.

One way of doing this is to write the base of the number system after the number, as a subscript.

Examples

$345_8 = 229_{10}$ $\quad$ $345_5 = 100_{10}$

On this page, $10_8 = 8_{10}$

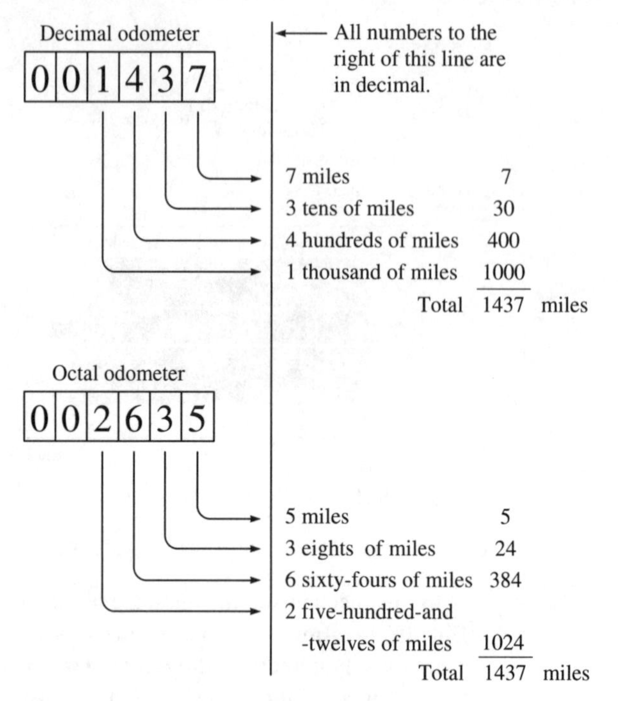

Figure 30.2

hundred-and-twelves) of miles. Each step to the left is an increase in position value of eight times the disc on its right. In Figure 30.2 we see the readings that would be shown on the odometers of two cars, one working on our usual decimal system, the other working on the octal system. Both cars have travelled the same distance but they produce different figures because of the two systems. However, when calculated they have registered the same number of miles.

Hex

The octal system is not widely used, but the **hexadecimal system** (**hex** for short) has many applications in the working of computers. It is a system based on sixteens. A hexadecimal odometer, if such an instrument were made, would have discs that advanced one-sixteenth of a turn each time. Their rims would be marked with the numbers 0 to 9, but beyond 9 there are no more single-digit numbers that we could use. We could invent some new symbols for the hex equivalents of 10, 11, 12, 13, 14, and 15. But this would mean six new symbols to learn and it would make the keyboards of calculators and computers even more crowded with keys than ever. Instead, we use the first six letters of the alphabet as symbols for numbers. We always use capital letters.

Counting in hex goes like this:

Hex	*Decimal equivalent*	
0	0	
1	1	
2	2	
3	3	
4	4	
5	5	
6	6	
7	7	
8	8	
9	9	
- - - - - - - - - - - - - -	- - - - - - - - - - - - - -	No difference so far
A	10	
B	11	
C	12	
D	13	
E	14	
F	15	
- - - - - - - - - - - - - -	- - - - - - - - - - - - - -	Now the carry over in hex
10	16	
11	17	
12	18	

and so on.

The next carry-over in hex comes at 1F, which carries to 20. In decimal, this is 31 becoming 32.

In the hex system, the digit on the right registers units (miles or whatever else is being counted). The digit to its left registers sixteens. The digit to its left registers sixteen sixteens (or two-hundred-and-fifty-sixes). The digit to its left registers sixteen 256's (or 4096's). At each shift to the left, a digit represents sixteen times more than the previous digit.

Names for digits

Consider the decimal number 427.

If we increase the 7 to its next value, 8, the number becomes 428. We have increased the number by 1.

If we increase the 2 to 3, the number becomes 437. We have increased it by 10.

If we increase the 4 to 5, the number becomes 527. We have increased it by 100.

Increasing (or decreasing) the 7 makes the **least** difference to the number. We say that the 7 is the **least significant digit** (or **LSD**).

Increasing (or decreasing) the 4 makes the **most** difference to the number. We say that the 4 is the **most significant digit** (or **MSD**).

Converting hex to decimal

To convert hex to decimal we multiply each hex digit by its position value, then sum the products.

Example

Convert 183 hex to decimal.

Take each digit in turn, starting from the LSD.

The digit is 3, the position value is 1:	3 × 1 =	3
The digit is 8, the position value is 16:	8 × 16 =	128
The digit is 1, the position value is 256:	1 × 256 =	256
	Total =	387

$186_{16} = 387_{10}$

If the hex number contains alphabetic digits, we have to remember their decimal equivalents, as listed in the table on page 409.

Example

Convert B6D hex to decimal.

Start with the LSD.

The digit is D, the position value is 1:	13 × 1 =	13
The digit is 6, the position value is 16:	6 × 16 =	96
The digit is B, the position value is 256:	11 × 256 =	2816
	Total =	2925

$B6D_{16} = 2925_{10}$

Converting decimal to hex

Examples

Convert 374 decimal to hex.

Start by finding the LSD. Divide the number by 16:

$$\frac{374}{16} = 23 \text{ remainder } 6$$

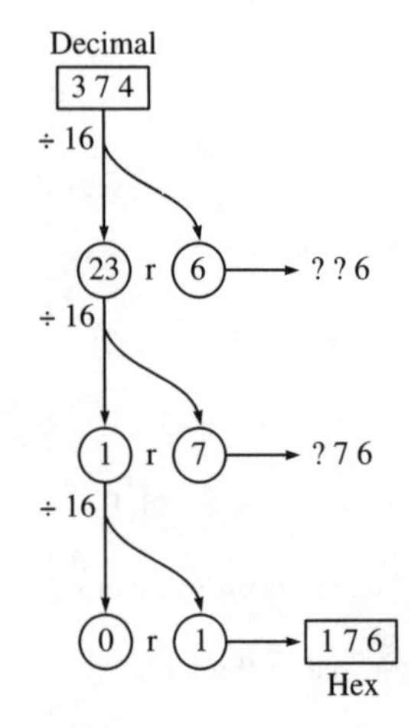

Figure 30.3

This shows us that 374 is split into two parts (Figure 30.3), 23 and 6. The 6 is the LSD. The 23 is to be represented by more digits to the left of the LSD. So far, the converted number is:

??6

where the question marks represent digits yet to be found. Now to find the next digit to the left. Divide 23 by 16:

$$\frac{23}{16} = 1 \text{ remainder } 7$$

We have now split the first part of the number into two parts, 1 and 7. The 7 is the next digit. The converted number is now

?76

The next step is to divide 1 by 16:

$$\frac{1}{16} = 0 \text{ remainder } 1$$

There is nothing left in the first part. There is only the remainder, which becomes the next digit:

176

With nothing left to divide by 16, there are no more digits to find:

$\underline{375_{10} = 176_{16}}$

Convert 15196 decimal to hex.

Divisions	*Hex equivalents*	*Converted no.*
$\frac{15196}{16} = 949$ remainder 12	$12_{10} = C_{16}$	???C
$\frac{949}{16} = 59$ remainder 5		??5C
$\frac{59}{16} = 3$ remainder 11	$11_{10} = B_{16}$	?B5C
$\frac{3}{16} = 0$ remainder 3		3B5C

$\underline{15196_{10} = 3B5C_{16}}$

Test yourself 30.1

1 Convert these hex numbers into decimal.

a 13 **b** 48
c A2 **d** 6E
e 523 **f** 17C
g 56F0 **h** ACD8

2 Convert these decimal numbers into hex.

a 35 **b** 63
c 64 **d** 43
e 500 **f** 128
g 12632 **h** 4095

3 Draw a flow chart (page 374) for converting hex numbers to decimal.

4 Draw a flow chart for converting decimal numbers to hex.

Binary numbers

The binary system, as its name suggests, is based on two's. A binary odometer, if such a thing existed, would have only two numbers on its rim, 0 and 1. Every time a disc is turned from 1 to 0, the disc to its left is turned half-way round, from 0 to 1, or from 1 to 0. Each step to the left is an increase in position value of twice the disc on its right. Counting goes like this:

Binary	*Decimal equivalent*	
0	0	
1	1	
- - - -	- - - -	1 carry over
10	2	
11	3	
- - - -	- - - -	2 carry overs
100	4	
101	5	
- - - -	- - - -	1 carry over
110	6	
111	7	
- - - -	- - - -	3 carry overs
1000	8	
continuing to		
1111	15	
- - - -	- - - -	4 carry overs
10000	16	

It is clear that a binary number has many more digits than a decimal number of the same value. This is a disadvantage to working with binary numbers on paper, but the box explains the very big advantage of using the binary system to represent numbers in electronic circuits, such as those in calculators and computers.

Binary numbers in electronics

In the binary system, a digit is either 0 or 1. These are the only two values it can have. By contrast the digits in the decimal system have 10 possible values, 0 to 9.

In a calculator or a computer it is possible to have circuits in which some quantity, such as a voltage, can take 10 different values. We might represent the numbers 0 to 9 by different voltages from 0 V to 9 V. But, in practice, it is much easier and cheaper to make fast, reliable circuits if only two voltages are used. In many such circuits we use 0 V to represent the digit 0, and 5 V to represent the digit 1. If we want to represent a number with more than one digit, we have a separate circuit for each, with its voltage at 0 V or 5 V, depending on whether the digit represented is 0 or 1.

Converting binary to decimal

The method follows the same sequence as for converting hex to decimal. The position values for binary numbers double at each step:

LSD	1
	2
	4
	8
	16
	32
	64
MSD	128

These position values are expressed in decimal. To convert binary to decimal, we multiply each binary digit by its position value.

Example

Convert binary 1101 to decimal.

Take each digit in turn starting from the LSD.

The digit is 1, the position value is 1: $1 \times 1 = 1$
The digit is 0, the position value is 2: $0 \times 2 = 0$
The digit is 1, the position value is 4: $1 \times 4 = 4$
The digit is 1, the position value is 8: $1 \times 8 = 8$
Total 13

$\underline{1101_2 = 13_{10}}$

Convert binary 111001 into decimal.

Starting from the LSD:

$1 \times 1 = 1$
$0 \times 2 = 0$
$0 \times 4 = 0$
$1 \times 8 = 8$
$1 \times 16 = 16$
$1 \times 32 = 32$
Total $= 57$

$\underline{111001_2 = 57_{10}}$

Converting decimal to binary

This follows the same sequence as converting decimal to hex, except that we divide by 2 at each stage.

Example

Convert 23 to binary.

Divisions	*Converted no.*
$\frac{23}{2} = 11$ remainder 1	????1
$\frac{11}{2} = 5$ remainder 1	???11
$\frac{5}{2} = 2$ remainder 1	??111
$\frac{2}{2} = 1$ remainder 0	?0111
$\frac{1}{2} = 0$ remainder 1	10111

$\underline{23_{10} = 10111_2}$

Converting between binary and hex

The hex equivalents of the 4-digit binary numbers are found by straightforward counting in the two systems:

Binary	*Hex*	*Decimal*
0000	0	0
0001	1	1
0010	2	2
0011	3	3
0100	4	4
0101	5	5
0110	6	6
0111	7	7
1000	8	8
1001	9	9
1010	A	10
1011	B	11
1100	C	12
1101	D	13
1110	E	14
1111	F	15

Binary and hex in computers

Computers work with numbers. For the reasons given in the previous box, these are usually represented in binary form. This applies whether the numbers are actually being used in calculations or if they are stored in memory for future use. They are also stored on disk in binary form. In many computers the numbers have 16 digits, in others they may have 32 digits. These numbers are also used to give coded instructions to the processing unit at the heart of the computer.

One of the problems with programming a computer or testing whether it is working correctly is writing out the binary digits. For example, a number such as:

1001 1101 1100 0010

is a long one to write out and it is easy to make mistakes. It simplifies things greatly if we convert the number into hex. In hex, the number above is 9DC2, which is much easier to deal with than its longer binary equivalent. We can program the computer to accept hex numbers and to convert them into binary for use in programs or for storage in its memory.

At the count of 16, we start a second group of 4 binary digits, and a second hex digit:

0001 0000	10	16
0001 0001	11	17
0001 0010	12	18
0001 0011	13	19
and so on to		
0010 0000	20	32
0010 0001	21	33
and beyond.		

When divided into groups of 4 digits (starting from the LSD), a binary number can be converted into hex simply by writing the hex equivalent of each group.

Example

Convert 1011101001_2 into hex.

In groups of 4:	10	1110	1001
Write hex equivalents (see table above):	2	E	9

$1011101001_2 = 2E9_{16}$

The reverse process is just as easy.

Example

Convert $4A2_{16}$ into binary.

Write the hex number:	4	A	2
Write binary equivalents:	0100	1010	0010

$4A2_{16} = 010010100010_2$

A number code

In electronic circuits, particularly in those driving numeric displays of clocks, watches, calculators and cash registers, it is usually convenient to code decimal numbers in binary form.

Take the decimal number, example 382:	3	8	2
Convert its individual digits *separately* into their binary equivalents:	0011	1000	0010
Run these together as a **binary coded decimal** (or **BCD**)	001110000010		

This is a **code**, a set of 0's and 1's produced by obeying a set of rules that are not completely mathematical. As an ordinary binary number, 001110000010 has the decimal value 898, but as BCD its value is 382.

Test yourself 30.2

1 Convert these binary numbers into decimal.

a 1101 **b** 110111

c 1001001 **d** 101010

2 Convert these decimal numbers into binary.

a 18 **b** 63

c 65 **d** 421

3 Convert these binary numbers into hex.

a 100110010100010 **b** 1101100111110001101

4 Convert these hex numbers into binary.

a 5C3A **b** 8F06

5 Draw a flow chart for converting decimal numbers into binary.

Binary maths

The addition table for binary numbers is:

$$\begin{aligned} 0+0 &= 0 \\ 0+1 &= 1 \\ 1+0 &= 1 \\ 1+1 &= 0, \text{ carry } 1 \end{aligned}$$

We use this table for adding pairs of binary numbers.

Example

Find 101110 + 100101.

```
    1 0 1 1 1 0
+   1 0 0 1 0 1
  1     1 1
  1 0 1 0 0 1 1
```

Follow this example through, working from right to left, noting where the carries occur.

101110 + 100101 = 1010011

Subtraction is based on the subtraction table:

0 – 0 = 0
1 – 0 = 1
1 – 1 = 0
10 – 1 = 1

Example

Find 110101 – 101001

```
  1⁰1¹0 1 0 1
– 1 0 1 0 0 1
  -----------
      1 1 0 0
```

Note the instance of borrowing when subtracting 1 from 0.

110101 – 101001 = 1100

Borrowing tends to occur fairly often in binary subtractions, leading to confusion. An easier routine for subtracting is as follows.

Form the **one's complement** of the number to be subtracted. This is done by re-writing the number with 1 for 0, and 0 for 1. Add 1 to the 1's complement; this forms the **2's complement**. **Add** the 2's complement to the other number, ignoring any carry beyond the MSD.

Try this on the example above.

The number to be subracted is: 101001
Its 1's complement is: 010110
Add 1 to this: 010110 + 1 = 010111

Now add this to the other number:

```
   110101
 + 010111
  -------
  1001100
```

The final carry is ignored. This gives the same result as before.

Binary and hex on a calculator

Scientific calculators are usually able to work in binary, hex and octal modes, as well as in decimal. If you key in a number when the calculator is in one mode, changing mode then displays the same number in the new mode. This makes conversions very easy to do.

Test yourself 30.3

Work these on paper, not on a calculator.

1 Find these binary sums.

a 101001 + 1101 **b** 100100 + 101101 **c** 111001 + 110110

2 Find these binary differences.

a 10011 – 101 **b** 101101 – 100011 **c** 1111110 – 111111

3 To assess your progress in this chapter work the exercises in *Try these first*, page 406.

31 Reasoning by numbers

Using equations to express facts and the way they are related logically is a branch of maths on which most aspects of control technology depend.

You need to know about the binary number system (page 412).

Try these first

1 Simplify: $\overline{A}\,\overline{B}\,\overline{C} + A\overline{B}\,\overline{C}$.

2 Simplify: $\overline{AB}(\overline{A} + B)$.

3 Use the Karnaugh map technique to simplify: $\overline{A}\,\overline{B}\,\overline{C} + \overline{A}B\overline{C} + \overline{A}BC + A\overline{B}C$.

Statements and symbols

This branch of maths deals with statements in terms of whether they are true or false (= not true).

Example

The lamp is on.

This statement refers to a ceiling lamp illuminating a corridor in an office block. A statement such a this is readily verified by looking at the lamp. The statement is either true or false (= not true). The lamp is either on or it is not on.

Truth and falseness are opposite states. There must be no half-truths, and a statement cannot be both true and false at the same time. We find the same sort of thing with the digits of a binary number. A digit is either 0 or 1. There is no half-way value and it cannot be both 0 and 1 at the same time. This is why we use binary numbers to represent truth and falseness.

Suppose that we represent the statement ‘The lamp is on’ by a symbol Z. Suppose also that 1 means ‘true’ and 0 means ‘not-true’ or ‘false’. When the lamp is on we write:

$$Z = 1$$

In full, this reads as ‘It is true that the lamp is on’. When the lamp is not on, we write:

$$Z = 0$$

Read this as ‘It is not true that the lamp is on’. We have represented the possible states of the lamp by a variable Z, which can take one of two binary values.

Logical statements

Logical statements usually have more to them than the one above. Often they state some kind of **condition**.

Example

If it is dark or the button has been pressed, the lamp is on.

This tells us more about what makes the lamp come on. This lamp is controlled by a light sensor, which automatically switches on the lamp whenever the light level falls, either at night or on a very dull day. There is also a manually operated push-button switch so that people can turn the lamp on at other times. The switch goes off after a few minutes. The circuit is shown in simple form in Figure 31.1.

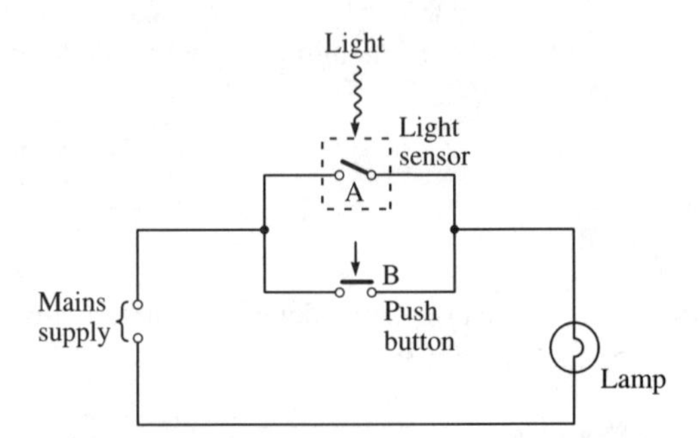

Figure 31.1

Let us analyse this statement, which really consists of three statements:

A = it is dark
B = the button has been pressed
Z = the lamp is on

A and B are statements of the conditions, and Z is a statement of what happens as a result of these conditions. These statements are linked by a logical connection:

IF . . . OR . . . THEN . . .

Expanding this to its fullest form:

IF *it is true that it is dark* OR *it is true that the button has been pressed* THEN *it is true that the lamp is on.* Or we could say:

If $A = 1$ OR $B = 1$ THEN $Z = 1$

Usually there is no need to write the '= 1'. When we write 'If A . . .' without stating a value, we take it to mean 'If $A = 1$ . . .' The statement reduces to:

IF A OR B THEN Z

To complete the analysis, we have to discover what happens if A or B are not true. We set this out systematically in a **truth table**:

A	B	Z
0	0	0
0	1	1
1	0	1
1	1	1

In the first two columns we have all four possible ways of fulfilling or not fulfilling the conditions. In the first row, it is daylight and nobody has pressed the button; the lamp is off. In the second row it is daylight and someone has pressed the button so the lamp is on. In the third row it is dark and no-one has pressed the button; the lamp has been turned on by the sensor. In fourth row it is dark, so the light is on; someone has pressed the button (perhaps absent-mindedly) but this makes no difference to the lamp. A and B are independent, and both may be true at the same time.

The way this works out in practice can be seen in the circuit diagram. Current can flow to the lamp either through switch A or through switch B:

IF A is closed OR B is closed, THEN current flows

The practical circuit is a model of the logical statement.

Logical equations

Statements can be represented by variables, and variables can have values. It makes sense that we can write **equations** when any of these values are equal. For example, the statements about the lamp are linked by this equation:

$$A + B = Z$$

The '+' sign means 'OR', not 'add', though it *almost* seems to mean 'add' too. Let us check this against the truth table.

In row 1: $A = 0,\ B = 0,\ Z = 0 \quad \Rightarrow \quad 0 + 0 = 0$
In row 2: $0 + 1 = 1$
In row 3: $1 + 0 = 1$

The values seem to be obeying the rules for adding. But in row 4 we have:

$$1 + 1 = 1$$

There is something a little different here. We have a kind of algebra which has many similarities to ordinary algebra, but at least one important difference. Obviously we could *not* have:

$1 + 1 = 2$ (decimal)
or $1 + 1 = 10$ (binary)

2 and 10 are values that Z can never have; it can only be 0 or 1, true or false.

This type of algebra, the algebra of logic, is known as **Boolean Algebra**. It has special rules of its own. For example, suppose we hold B constant at 0 (never press the button), but vary A. This corresponds to the first and third lines of the truth table:

A	B	Z
0	0	0
1	0	1

In both these lines, Z has the same value as A. In equation form:

$$A + 0 = A$$

This is another rule of Boolean Algebra. Suppose we hold B constant at 1 (keep pressing the button, night and day). The second and fourth lines of the truth table tell us:

A	B	Z
0	1	1
1	1	1

In these lines $Z = 1$, whatever the value of A. In equation form:

$$A + 1 = 1$$

Try ORing A with A to see what we get:

$$A + A = ?$$

Putting this into the words of our example: 'It is dark or it is dark'. This simplifies to 'It is dark':

$$A + A = A$$

Test yourself 31.1

1 A shop closes on Wednesday afternoons and Sundays. Given these statements:

A = It is Wednesday afternoon
B = It is Sunday
Z = The shop is closed

Link the statements as a logical statment including OR. Write out the truth table. Use the truth table to find the value of Z on a Saturday morning.

2 The interior lamp of a microwave oven comes on when the door is opened. It also comes on when food is cooking. Express these facts as a logically connected set of statements, including OR. Write out the truth table. Which line in the truth table never occurs in practice in a correctly operating oven?

3 An electric mower overheats if it is used on very rough grass, and also if it is run continuously for more than 10 min. Express these facts as a logically connected set of statements, including OR. Write out the truth table. If the mower is being run continuously on short grass, what will be the changes in the statement values after 10 min?

Another connection

Figure 31.2 shows the switching circuit for a table lamp plugged into a wall socket. There are two switches, one at the socket (the wall switch), one on the base of the lamp (the base switch). The logic that governs this lamp is:

IF the wall switch is on AND the base switch is on,
THEN the lamp is on.

The statements are:

A = the wall switch is on
B = the base switch is on
Z = the lamp is on (same as in the previous example)

The logic differs:

IF A AND B THEN Z

The relationship is AND, instead of OR. It has a different truth table:

A	B	Z
0	0	0
0	1	0
1	0	0
1	1	1

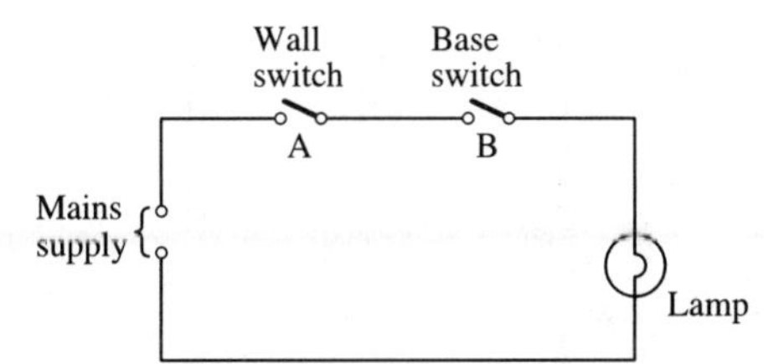

Figure 31.2

Comparing this table with the previous one, the first two columns are the same, setting out all possible values of A and B. The tables differ in the third column. The lamp is on only if A AND B are true.

In logic equations, we represent AND by the symbol $\bullet$, as here:

$$A \bullet B = Z$$

This 'heavy full-stop' is sometimes used in ordinary algebra to mean 'multiply'.

The AND operation is similar to multiplying in some (but not all) ways. As in ordinary algebra, we often leave out the multiply symbol. For example, we recognize that the expression pq mean 'p times q'. Similarly, AB means A AND B.

From the truth table, if B is held at 0 (base switch off), the lamp is off whatever we do to the wall switch:

$$A \bullet 0 = 0$$

If B is held at 1 (base switch on), the lamp is controlled by the wall switch:

$$A \bullet 1 = A$$

In words we may express $A \bullet A$ as 'The wall switch is on and the wall switch is on'. Obviously, the wall switch is on:

$$A \bullet A = A$$

More conditions

Statements are not limited to just three parts. We might have this longer statement:

> If the wall switch is on AND the base switch is on AND the main power switch is on, THEN the lamp is on

In logic form:

> If A AND B AND C THEN Z

where C = the main power switch is on. As an equation:

$$ABC = Z$$

As a truth table:

A	B	C	Z
0	0	0	0
0	0	1	0
0	1	0	0
0	1	1	0
1	0	0	0
1	0	1	0
1	1	0	0
1	1	1	1

Only in the bottom row, with all switches on, is the lamp on.

Have you noticed?

When we set out the logical values for a truth table, we need to list all possible permutations of 0's and 1's. This is most easily done by running through the binary numbers from 0 upward.

Example

For a table for two variables, A and B, there are four possible permutations:

A	B
0	0
0	1
1	0
1	1

This pattern of digits is the same as for the binary numbers 00 to 11 (0 to 3 in decimal). In a table with three variables (above) there are 8 permutations, corresponding to binary 000 to 111 (decimal 0 to 7).

There is no limit to the number of variables that can occur in Boolean equations.

Mixing the logic

The corridor lamp (page 420) is also under the control of the main power switch of the office block. We extend the statement to include this:

> IF it is dark OR the button has been pressed AND the main power switch is on, the lamp is on

Now we have OR and AND in the same statement. As a logic statement:

> IF (A OR B) AND C THEN Z

Note the brackets around A OR B. We use brackets for exactly the same reason as we do in an ordinary equation, to show which parts must be evaluated first. The wiring of the circuit (Figure 31.3) shows how the main power switch overrides the actions of the other switches. Its action has to be ANDed with the combined action of the other two ($A + B$). As an equation, the statement is:

$$(A + B)C = Z$$

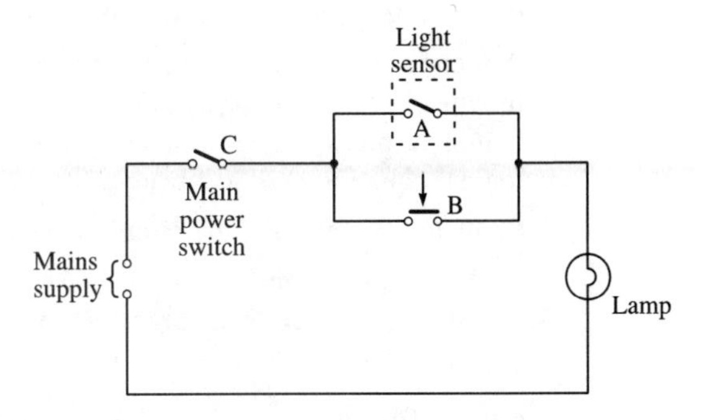

Figure 31.3

As a truth table, we develop it in stages:

1	2	3	4	5
A	B	C	$A+B$	Z
0	0	0	0	0
0	0	1	0	0
0	1	0	1	0
0	1	1	1	1
1	0	0	1	0
1	0	1	1	1
1	1	0	1	0
1	1	1	1	1

In columns 1 to 3, set out all possible combinations of values of A, B and C (see box, page 424). In column 4, write the values of $A + B$. To do this, look in columns 1 and 2; if either or both contain a 1, write 1 in column 4. Fill up the other spaces with 0's. In column 5 we have the values of $(A + B)C$, found by looking at columns 3 and 4; if both contain 1, write 1 in column 5. Fill up the other spaces with 0's. Check through column 5 of the table to see if it confirms what you would expect to happen with the circuit of Figure 31.3.

Test yourself 31.2

1 If the equation above was an ordinary algebraic equation, we would expand it by removing the brackets:

$$Z = (A + B)C = AC + BC$$

Find the truth table for $AC + BC$ and show that removing brackets in logic equations is a valid operation.

2 Figure 31.4 shows the circuit for the statement:

IF it is dark OR (the button has been pressed AND the power switch is on), the lamp is on

$$A + BC = Z$$

Find the truth table for this equation. Mark the rows of the table which show that this circuit is unsafe, since it allows the lamp to be on when the main power switch is off.

3 A security device is designed to sound an alarm if it detects a person moving, but to do this only at nights. It also sounds the alarm if its test button is pressed. Describe its action as a logical statement, as a logical equation and as a truth table.

4 Show the operation of the circuit of Figure 31.5 in the form of a truth table.

5 Use the rules of Boolean algebra to simplify these expressions.

a $A + A + 1$ **b** $A(A + 0)$

c $(A + 0) + A$ **d** $A + A(A + 1)$

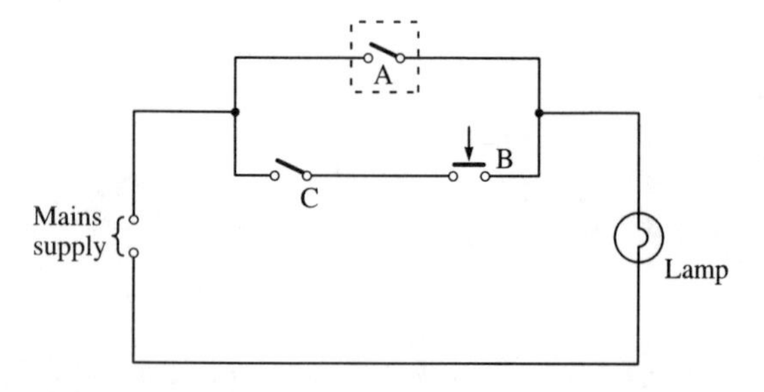

Figure 31.4

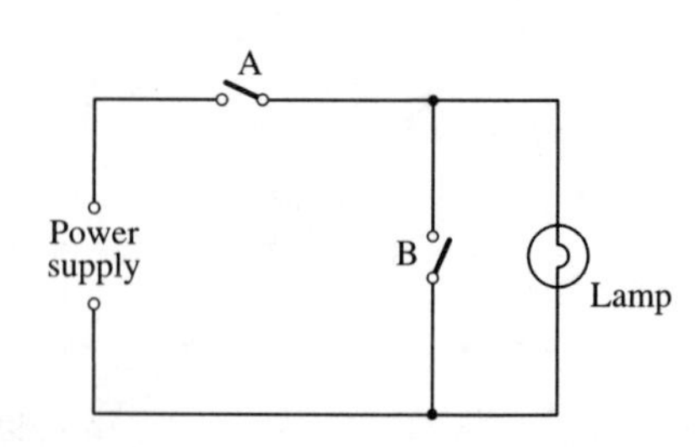

Figure 31.5

OR and AND summary

$Z = A + B$

A	B	Z
0	0	0
0	1	1
1	0	1
1	1	1

$Z = A \bullet B$

A	B	Z
0	0	0
0	1	0
1	0	0
1	1	1

Opposites

Given that A means 'It is dark', we may also make a statement which means the exact opposite:

It is not dark

The symbol for this is $\overline{A}$. Call this 'A-bar'. It follows that:

If $A = 1$ then $\overline{A} = 0$

If $A = 0$ then $\overline{A} = 1$

Here is the truth table:

A	$\overline{A}$
0	1
1	0

We may OR A with its opposite:

$$A + \overline{A} = 1$$

This equation simply states that it is true that it is dark or it is not dark. At any given time, it must be either dark or not-dark; no other possibilities exist. So this statement must *always* be true. Its value is 1, whether $A = 1$ or $A = 0$.

If we AND A with $\overline{A}$ we obtain:

$$A \bullet \overline{A} = 0$$

The statement says 'It is dark and it is not dark'. This is obviously false, so has the value 0.

We can have the opposite of whole statements. Take the statement about the table lamp (page 421) as an example:

$$A + B = Z$$

Opposites

$\overline{A}$, the opposite of A, is often called the INVERT of A. It is also known as NOT-A.

Rules for Boolean Algebra 1

Statements with OR	Statements with AND
$A + 0 = A$	$A \bullet 0 = 0$
$A + 1 = 1$	$A \bullet 1 = A$
$A + A = A$	$A \bullet A = A$
$A + \overline{A} = 1$	$A \bullet \overline{A} = 0$

Invert rules If $A = 0$, then $\overline{A} = 1$
If $A = 1$, then $\overline{A} = 0$

Double invert $\overline{\overline{A}} = A$

Taking the opposites of *both* sides maintains the equality:

$$\overline{A + B} = \overline{Z} \tag{1}$$

This is saying:

If NOT(A OR B) THEN NOT Z

In more conventional words:

If NEITHER the wall switch NOR the base switch B are closed, the lamp is NOT on.

There is another way of saying this which leads to the same result:

If the wall switch is NOT on AND the base switch is NOT on, the lamp is NOT on

Check the truth of this by referring to Figure 31.2. Contrasting this with the original statement, we have substituted the opposite for every statement and are now using AND instead of OR. This demonstrates that:

$$\overline{A} \bullet \overline{B} = \overline{Z} \tag{2}$$

INVERT bars

When reading equations, take care to note the exact extent of the inverting bar.

Example

$\overline{A}\,\overline{B}$ means NOT-A AND NOT-B Switch A is off and switch B is off – both switches definitely off.

but

$\overline{AB}$ means NOT (A AND B) Not true that A is on AND B is on – but A might be on without B, or B might be on without A.

Equations (1) and (2) have identical right sides. So their left sides must be equal:

$$\overline{A + B} = \overline{A}.\overline{B}$$

This is an important and useful rule, and is one of **De Morgan's Laws**.

The other law of De Morgan is:

$$\overline{\boldsymbol{A \bullet B}} = \overline{\boldsymbol{A}} + \overline{\boldsymbol{B}}$$

We could demonstrate this by reasoning it through from the corridor lamp circuit but let us prove it by using a truth table and our known rules.

1	2	3	4	5	6	7
A	B	$A.B$	$\overline{A \bullet B}$	$\overline{A}$	$\overline{B}$	$\overline{A} + \overline{B}$
0	0	0	1	1	1	1
0	1	0	1	1	0	1
1	0	0	1	0	1	1
1	1	1	0	0	0	0

Columns 1 to 3 are the usual truth table for $A \bullet B$. In column 4 we have written the invert of each value from column 3. The entries in this column are the values of the left side of the equation for each pair of values of A and B.

Columns 5 and 6 contain the opposites of the values in columns 1 and 2. Looking at these pairs of values, we have written 1 in column 7 when either or both of these are 1 (the OR operation). Thus column 7 shows the value of the right side of the equation.

The values in columns 4 and 7 are identical, proving that both sides of the equation are equal and that the law is true.

Rules for Boolean Algebra 2

De Morgan's laws	$\overline{A + B} = \overline{A}\overline{B}$
	$\overline{AB} = \overline{A} + \overline{B}$
Miscellaneous	$A(B + C) = AB + AC$
	$A + BC = (A + B)(A + C)$
	$A + AB = A$
	$A(A + B) = A$
	$A + \overline{A}B = A + B$

Using the rules

The boxes on this page and page 428 summarize the main rules of Boolean algebra. The prime use of these rules is to simplify expressions. There is a practical reason for wanting to do this. Control systems are built up using devices which behave according to the truth tables. Very often such devices are electronic. A logic **gate** may have two input terminals A and B. If the voltage at these terminals is made low (equivalent to 0, or false) or high (equivalent to 1 or true), the output from the gate changes accordingly.

Exactly what happens depends upon what type of gate it is. The output of an OR gate, for example, responds according to the truth table for OR. Its output (Z) is high (1) when either one or both of the inputs is made high. There are also gates for AND, NOT (opposites) and several other logical operations. Similar devices are available in pneumatic form, suitable for driving parts of machines, especially industrial robots.

Controlling a complex machine or electronic system requires complex logic. It may require many gates. In the case of a microcomputer, many tens of thousands of gates are involved. In designing such a system it is important to be able to keep the number of gates to a minimum. This is necessary for efficiency, reliability and speed of operation. We use Boolean algebra to simplify logical equations as much as possible, so that as few gates as possible are needed.

For example, the logic of a washing machine might require the motor to run according to this logic:

$$Z = A\overline{B} + B(A + \overline{B})$$

As it stands, this equation needs four gates (two OR, two AND). Let us simplify it, first by removing the brackets:

$$Z = A\overline{B} + AB + B\overline{B}$$

But $B\overline{B} = 0$ and $A + 0 = A$:

$$\Rightarrow \quad Z = A\overline{B} + AB$$

$$= A(\overline{B} + B)$$

But $\overline{B} + B = 1$

$$\Rightarrow \quad Z = A(1)$$

$$Z = A$$

In its simplified form we see that the value of Z follows the value of A. We do not need any logic gates. The motor can be controlled directly by the state of A. Logic reduction is not always as dramatic as this. Sometimes a few trials will show that there is no way of simplifying the system. In most cases a few savings of gates can be made.

Test yourself 31.3

1 By using truth tables, prove the final three rules in the box on page 429.

2 Use the Boolean rules to simplify these expressions.

a $AB + A\overline{B} + \overline{A}B$ **b** $(A + B)(A + C)$

c $ABC + AB\overline{C} + \overline{A}BC$ **d** $AB(A + \overline{B}) + (A + B)(\overline{A} + B)$

3 Use De Morgan's Laws and the Boolean rules to simplify these expressions.

a $\overline{AB}(\overline{A} + B)$ **b** $\overline{\overline{A} + \overline{B}} + \overline{\overline{A}\,\overline{B}}$

Karnaugh maps

These are another aid to simplifying logic expressions. They are really a kind of truth table but set out in a different way.

The circuit of Figure 31.5 produces this truth table:

A	B	Z
0	0	0
0	1	0
1	0	1
1	1	0

	$\overline{A}$	A
$\overline{B}$	0	1
B	0	0

Figure 31.6

This is not one of the standard tables such as OR or AND. We have to discover the minimum logic that will produce it. As a **Karnaugh** map, the table is set out as in Figure 31.6.

In the truth table we list the four possible states of A and B, by stating their values, 0 or 1. On the columns and rows of the map we do something slightly different. Consider the column headings to be telling us which is true, $\overline{A}$ or A. In the first column $\overline{A}$ is true. In other words $\overline{A}$ = 1 and, conversely, A = 0. So the first column of the map corresponds to the first two rows of the truth table, in which A = 0. The second column is headed by A, meaning A = 1, and corresponds to the last two rows of the truth table. Figure 31.7 shows the correspondences. It can be seen that there is a similar correspondence between values of B and the rows of the map.

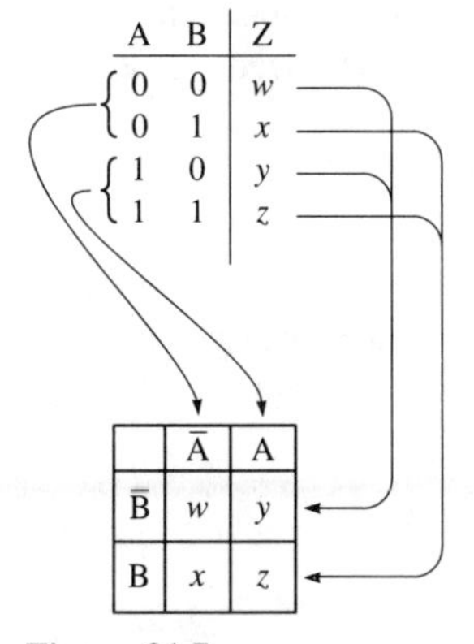

Figure 31.7

The four cells which form the body of the Karnaugh map correspond to the four cells of column Z of the truth table. The labels w to z indicate which cells correspond.

The 1 on the map is placed in cell y. It represents the term:

$$Z = A\overline{B}$$

For this term, A = 1. Also $\overline{B}$ = 1, so B = 0. This corresponds with the value in column Z of the truth table.

Because there is only a single 1 on the map, it is not possible to simplify the logic further. The lamp is on when switch A is on AND switch B is off. Note that the logic of a Karnaugh map is always AND. The column and row headings are ANDed together when we interpret what each 1 or 0 means.

Now to take a more complicated expression that might well lend itself to simplification:

$$Z = \overline{A}\,\overline{B} + \overline{A}B + A\overline{B}$$

	$\overline{A}$	A
$\overline{B}$	1^w	1^y
B	1^x	0^z

Figure 31.8

The expression consists of a set of AND terms, connected by OR logic. This is the type of logic statement for which the Karnaugh map is used. The map is shown in Figure 31.8. There are three terms and therefore the map has three 1's.

$\overline{A}\,\overline{B}$ gives a 1 in cell w
$\overline{A}B$ gives a 1 in cell x
$A\overline{B}$ gives a 1 in cell y

Fill the remaining cell(s) with 0's. In this case only cell z is left.

We now come to the whole point of the Karnauch map. Looking at cells w and y, we see a 1 in both of them. Draw a dashed loop to link these 1's. These cells have an edge in common. Crossing the edge from cell w to cell y corresponds to changing from $\overline{A}$ to A. But $\overline{B}$ *is true in both* cells. With these two 1's we are saying that $\overline{B}$ is true for either of them. But it makes no difference if A is true or $\overline{A}$ is true. In either case, $Z = 1$ and the outcome of the logic is unaffected. The value of A can be ignored in working out the logic. This tells us that:

$$\overline{A}\overline{B} + A\overline{B} = \overline{B}$$

Put this another way. The logic statement $\overline{A}\overline{B} + A\overline{B}$ places 1's in cells w and y. But so does the logic statement $\overline{B}$. We can replace the longer statement by the simpler one.

Another pair of cells sharing a common edge are cells w and x. Draw another dashed loop. Crossing their common edge changes from $\overline{B}$ to B, but $\overline{A}$ is common to both cells. Using reasoning similar to that above, we conclude that:

$$\overline{A}\overline{B} + \overline{A}B = \overline{A}$$

We have now accounted for all the 1's on the map, replacing the longer statements by simpler ones. To provide for all the 1's we need one OR the other of the simpler statements:

$$\underline{Z = \overline{A}\overline{B} + \overline{A}B + A\overline{B} = \overline{A} + \overline{B}}$$

Grouping the 1's into pairs is known as **coupling**. Two couples can share a 1 between them. If, at the end, there are isolated 1's on the map which cannot be coupled, the logic statement which produces them is left unsimplified and ORed with the other simpler statements.

Karnaugh map with 3 variables

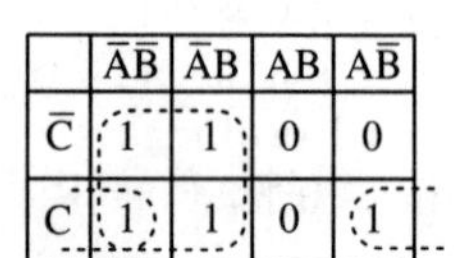

	$\overline{A}\overline{B}$	$\overline{A}B$	AB	$A\overline{B}$
$\overline{C}$	1	1	0	0
C	1	1	0	1

Figure 31.9

The principle is easily extended to three variables using an 8-cell map (Figure 31.9). Note that, in the column headings, only *one* variable (A or B) changes as we go from one column to the next. This is necessary so that there is only one change of variable as we cross a common edge between cells.

The 1's entered on this map are obtained as the result of the equation:

$$Z = \overline{A}\overline{B}\overline{C} + \overline{A}B\overline{C} + \overline{A}\overline{B}C + \overline{A}BC + A\overline{B}C$$

This is a complicated looking equation, but has the typical format of one solvable by the Karnaugh technique. There are five AND terms, each containing A, B and C or their inverts (opposites). The terms are connected by OR.

Again we look for couples. But this time we can identify a larger group, a couple of 4 adjacent squares arranged in a 2 by 2 block. This is circled in the diagram. Within this couple, B varies as we go from left to right, and C varies as we go up and down. But $\overline{A}$ is common to all four cells. This means that the expressions simplify like this:

$$\overline{A}\overline{B}\overline{C} + \overline{A}B\overline{C} + \overline{A}\overline{B}C + \overline{A}BC = \overline{A}$$

We have yet to account for the 1 at the bottom right of the map. This appears to be isolated but it is not. A Karnaugh map can be considered to wrap round from its left edge to its right edge, or from its top edge to its bottom edge. It behaves like a map of the World. This means that the 1 at bottom right can be coupled with the 1 at bottom left. C occurs in both cells, $\overline{B}$ occurs in both cells, so:

$$\overline{A}\overline{B}C + A\overline{B}C = \overline{B}C$$

Combining these two results:

$$\underline{Z = \overline{A}\overline{B}\overline{C} + \overline{A}B\overline{C} + \overline{A}\overline{B}C + \overline{A}BC + A\overline{B}C = \overline{A} + \overline{B}C}$$

Karnaugh map with 4 variables

	$\overline{A}\overline{B}$	$\overline{A}B$	AB	$A\overline{B}$
$\overline{C}\overline{D}$	1	0	1	1
$\overline{C}D$	0	0	1	0
CD	1	1	1	1
$C\overline{D}$	1	1	1	1

Figure 31.10

Four variables produce an array of 16 cells (Figure 31.10). This allows room for couplings of 8 cells, if any exist. The expression mapped in the figure is:

$$Z = \overline{A}\overline{B}\overline{C}\overline{D} + AB\overline{C}\overline{D} + A\overline{B}\overline{C}\overline{D} + AB\overline{C}D + \overline{A}\overline{B}CD + \overline{A}BCD + ABCD$$
$$+ A\overline{B}CD + \overline{A}\overline{B}C\overline{D} + \overline{A}BC\overline{D} + ABC\overline{D} + A\overline{B}C\overline{D}$$

A great simplification comes about because of the coupling of 8 cells in the bottom half of the map. C is true for all 8 cells, so:

$$\overline{A}\overline{B}CD + \overline{A}BCD + ABCD + A\overline{B}CD + \overline{A}\overline{B}C\overline{D} + \overline{A}BC\overline{D} + ABC\overline{D}$$
$$+ A\overline{B}C\overline{D} = C$$

There is also a coupling of four cells in a vertical row. Note how we use two of the cells already forming part of the 8-couple. For maximum simplification, couples should be made as large as possible. The common value for these four cells is AB, so:

$$AB\overline{C}\overline{D} + AB\overline{C}D + ABCD + ABC\overline{D} = AB$$

Finally, the two cells at top right and top left can be considered to be a wrap-round couple, with common term $\overline{B}\overline{C}\overline{D}$:

$$\overline{A}\overline{B}\overline{C}\overline{D} + A\overline{B}\overline{C}\overline{D} = \overline{B}\overline{C}\overline{D}$$

The simplified expression has three terms corresponding to the three couplings:

$$\underline{Z = C + AB + \overline{B}\overline{C}\overline{D}}$$

Test yourself 31.4

Use the Karnaugh map technique to simplify these expressions:

1 $AB + A\overline{B}$

2 $A\overline{B} + \overline{A}\overline{B}$

3 $AB + A\overline{B} + \overline{A}\overline{B}$

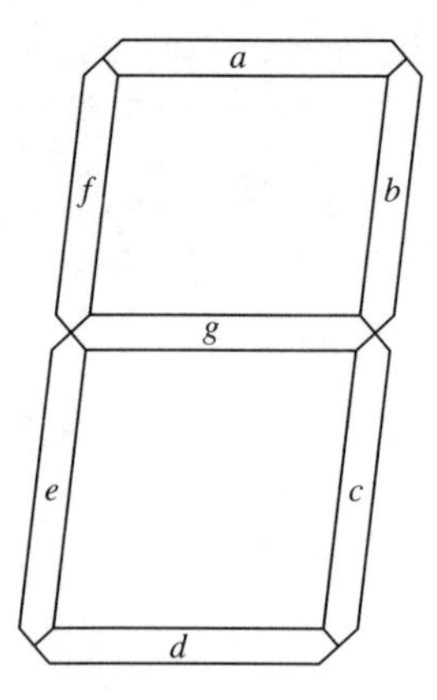

Figure 31.11

4 $\overline{A}\overline{B}\overline{C} + A\overline{B}\overline{C}$

5 $A\overline{B}\overline{C} + A\overline{B}C + AB\overline{C} + ABC$

6 $\overline{A}B\overline{C}\overline{D} + \overline{A}\overline{B}CD + \overline{A}BCD + \overline{A}\overline{B}C\overline{D} + \overline{A}B\overline{C}\overline{D}$

7 A 7-segment numeric display (Figure 31.11) is controlled by 4 inputs, A, B, C and D. These represent numeric values, coded as follows:

Decimal 0 → binary 0000 → $\overline{A}\overline{B}\overline{C}\overline{D}$
Decimal 1 → binary 0001 → $\overline{A}\overline{B}\overline{C}D$
Decimal 2 → binary 0010 → $\overline{A}\overline{B}C\overline{D}$

and so on. Segment *e* is to be lit when the number is 0 OR 2 OR 6 OR 8. Write a logic expression for the set of inputs that turn the lamp on. Use the Karnaugh map technique to simplify this. Repeat this for segment *d*, which is turned on by 0 OR 2 OR 3 OR 5 OR 6 OR 8.

8 To assess your progress in this chapter, work the exercises in *Try these first*, page 419.

Karnaugh maps, summary

1 Expression to be simplified is in the form:

$ABC + ABC + ABC$

May be fewer or more variables, fewer or more terms, some letters have bars above them.

2 Set out map grid, depending on number of variables.
3 Enter 1 for each term of expression. Fill remaining cells with 0's.
4 Look for couplings of 2, 4 or 8 cells (4 cells to form a line of 4 or a 2-by-2 square). Couplings may overlap and each should be as large as possible.
5 For each coupling, find the common variable(s).
6 List these, ORed together, as the simplified expression.
7 Isolated cells: include their terms in the ORed list.

Part 4 – Handling Data

How to collect, display and analyse data

32 Collecting and displaying data

A data analysis is only as reliable as the data on which it is based. This chapter explains different data types, how to collect data and how to present it most effectively.

Try these first

1 The annual production of glue by five factories is as follows:

Factory	*Tonnes*
A	12
B	14
C	5
D	21
E	8

Draw a pie chart to illustrate this data.

2 A sample of 39 tea-cups was taken from the production line at a pottery. They were weighed to the nearest gram, and the weights recorded below:

185	180	187	190	188	189	184	185	197	186
192	186	189	176	198	185	188	184	187	181
189	182	194	181	186	195	177	188	194	187
183	188	186	186	197	189	193	179	188	

Illustrate this data by drawing a histogram with class interval 5 g.

Data

Data is obtained in one of two ways, by **counting** or by **measurement**. An example of **counting data** (or **enumerative data**) is a count of the number of defective widgets produced daily at a factory. If this number rises above a given minimum level the methods of production must be checked to see what is wrong. Another example of counting data is a survey of the students of a college to discover what is the most popular newspaper read by them.

An example of obtaining measurement data is weighing the contents of packages filled by a packing machine, to check that every packet contains at least as large a quantity as is stated on the label, yet does not contain so much that there is a loss of profit. Another example is a series of measurements of the temperature of a cooling ingot.

Three essential qualities of data are that it must be:

- free of errors
- adequate in quantity
- unbiased.

The first quality is assured by careful counting and recording and by using suitable measuring instruments correctly. Counting must be done systematically, preferably using a ruled note-book or a form, so that there are no loose scraps of paper to be lost. A computer database program helps to organize data, but this is of no benefit if the person keying in the data is careless. Measuring instruments and the techniques employed for using them, should be checked for accuracy and precision and the operator must be properly trained. In short, freedom from error is a matter of careful planning and efficient supervision.

The more data we collect, the better informed we are. The results of a survey of newspaper preferences based on interviewing only 20 students usually would not be worth publishing. We would expect such a survey to be based on at least 100 students, preferably many more. It is not always easy to know how much data is needed to give a reliable result but, if in doubt, collect more data than you think you will need, rather than less. However, there are practical limits on the amount of data that can be collected. If a survey takes too long to collect the data, it may be out-of-date before the survey is complete. The cost of collecting data is another consideration. We may not be able to afford to collect as much data as we would like to have.

The quality of data which is often most difficult to achieve is absence of bias. For example, sets of records of the first spring arrivals of migrant birds have shown that there is a tendency for the birds to arrive on a Saturday or a Sunday. This does not show that birds know what day of the week it is. It is the result of biased observations. Amateur bird-watchers are able to follow their hobby only at the week-ends and so do not record arrivals which occur during the week.

In a survey of public opinion taken, for example, in a shopping precinct, there might be a tendency for the interviewers to select people who appear to be easy to approach, and to avoid others. To eliminate bias (which can be very difficult to detect) some definite procedure must be decided on, such as interviewing every tenth person to pass through a given doorway. Questions too must be carefully vetted, so that they do not produce biassed answers.

Bias also applies with measurement data. In selecting fruit for weighing we are likely to pick the juiciest, choicest-looking fruits and reject the discoloured, wrinkled and undersized fruits. The selection procedure must not allow this to happen. Fruits must be picked unseen from the container or, if they are on a conveyor belt, we must select at regular intervals, independently of appearance.

Sampling

One way to avoid bias is to interview *everybody*. If we really want to know the newspaper preferences of the students of a college we should obtain data from every student. This eliminates bias from the result. It also gives us as much information as possible. But, as remarked above, there may be practical reasons why we cannot interview the whole student population of the college. We have to be content with interviewing a **sample**. The word **population** is used deliberately here. It defines a group of people whose views of newspapers we wish to assess. The idea of population extends to other groups too, not necessarily to groups of people. If we are testing the quality of goods produced in a factory, the items produced in a given period, such as a day, or a week is regarded as a population. We cannot test every item produced (the whole population of items) but we can select a sample of items and test these. Results obtained from the sample are considered to be representative of the population.

Sampling is a critical aspect of data collection. A sample must represent the population as closely as possible. The sample must be big enough and unbiased. Sampling the student population of a college might be done in various ways. We might select just one of these groups to be interviewed:

- Senior students
- Female students
- Students taking technological (as opposed to scientific) subjects
- Students from country districts
- Students belonging to a particular political party.

It is clear that a survey based on such groups would be biased; none of these samples are likely to be representative of the student population as a *whole* (though surveys of particular groups and their differences from other groups might well be an interesting project in its own right). Other possible ways of picking out a sample are:

- Students whose surname begins with the letter A to F
- Students whose birthday falls on the 1st to 5th of the month
- Students whose telephone number is divisible by 4
- Or any other criterion which we suspect has no bearing on their preferences for newspapers.

Another way of selecting a sample is by a **random** method. At its simplest, we could work through the student roll and, for each student, toss a coin. If it falls 'tails' up, the student is not selected. This would produce a sample of about half the size of the population. To obtain a smaller sample, we could roll a die and select students for whom the die shows 6.

Flipping coins and rolling dice is time-consuming. A better way is to use a **random number generator**. Many calculators have a key for producing random numbers. One such calculator produces a different 3-digit number each time the key is pressed. An example of a random sequence generated in this way is:

0.486, 0.074, 0.787, 0.957, 0.425, 0.899, 0.407, 0.013, . . .

A sequence such as this can be used in many ways. For example, in a college with fewer than 1000 students, the numbers can be taken to be the students' roll numbers. Students 486, 74, 787, 957, . . . are selected for the sample. Or we could generate a random number for each student as we work through the roll; select the student if the random number ends in 2, 4 or 6. With the sequence above, we would select the first two students and reject the rest. Overall, we would expect to select 3 out of every 10 students, 30% of the population.

Random numbers can be used for selecting measurement data too. To monitor the growth rate of a flock of sheep, number each animal with a tag. For each weekly weighing, pick out a sample of, say 25 sheep, using random numbers. In a factory the product may be stacked in rows on shelves, which are on racks. Use random numbers to select each item for sampling, using its rack number, its shelf number and its numbered position on the shelf. In this way, there should be no bias in the selection, such as might be introduced by selecting the items nearest to the store-room door and which might tend to be items produced toward the end of the working day.

Test yourself 32.1

Prepare a detailed plan for collecting data of the types described below. Your plan should include the way the sample is to be selected, methods of obtaining the data (questionnaire, measuring instruments used), checks on accuracy and avoidance of error in the records, an assessment of how much data you will need, and precautions for avoiding bias.

1 A survey of the types of vehicle passing along a given section of road during a period of one week.

2 The means of transport used by students in coming daily to the college.

3 The preferences of college students for particular newspapers.

4 The weights of puddings served at a canteen.

5 The breaking strength of samples of string.

6 The wet strength of different brands of paper tissue.

7 The quantity of potato chips contained in packets marketed by different manufacturers.

8 The brand of toothpaste favoured by

a female students **b** male students

Displaying counting data

The passengers on a bus are recorded according to whether they are men, women, boys or girls:

Category	*Number*
Men	10
Women	12
Boys	6
Girls	8

Figure 32.1 shows this data presented as a **pie chart**. Although the table of data gives us all the facts, the pie chart demonstrates them more clearly.

> **Tallies**
>
> Counting the numbers of individuals in a group is known as making a **tally count**. A mechanical tally counter consists of register with usually four digits, rather like the odometer of a car (Figure 30.1). Pressing the button increases the reading by 1, so that we can quickly count people or objects, without losing count. There is a reset knob for returning the register to 0000.
>
> If we are keeping a tally count on paper we make a mark for each person or object. It makes totalling the count easier if the marks are made in groups of 5. The first 4 marks are made side-by-side: ||||
>
> The fifth mark is made diagonally across the group: ~~||||~~
>
> When totalling the tally, count the **groups**, multiply by 5, and then add on any odd marks.

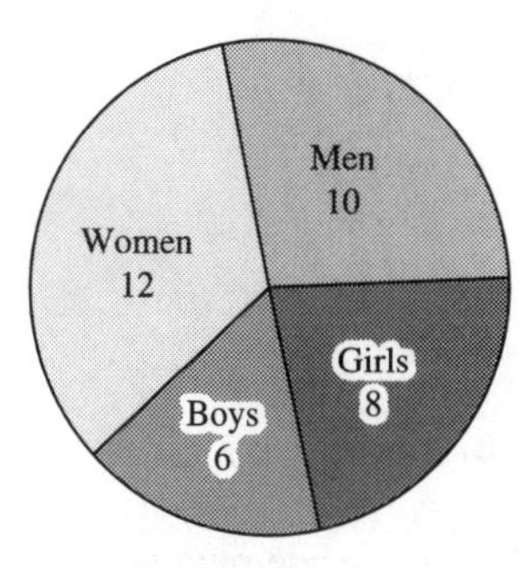

Passengers on a bus

Figure 32.1

Among other things, it shows that one third of the passengers are women, a fact not immediately apparent from the table. In general a pie chart is useful when we want to display the size of one or more groups in relation to the *whole*.

A pie chart shows the proportions between the various categories. This is how we calculate the proportions.

Find the total number of passengers: total = 36

The pie has to be divided into sectors with different angles. The angle of a sector is proportional to the number of passengers the sector represents. The total angle is 360°.

$$\text{Calculate a factor } \frac{360}{\text{passengers}} = 360/36 = 10$$

Each passenger is represented by 10°.

Calculate the angle for each category of passenger:

Category	*Number*	*Angle (°)*
Men	10	10 × 10 = 100
Women	12	12 × 10 = 120
Boys	6	6 × 10 = 60
Girls	8	8 × 10 = 80
	36	360

Check that the total angle is 360°.

Draw the pie, dividing it into sectors with the given angles. Label each sector with the name of the category and the number of passengers it represents. The numbers are not strictly necessary but they may be useful when interpreting the chart.

As an alternative to naming the sectors, shade each one differently and draw an explanatory key of the shading beside the chart. The chart is completed with a title to indicate its subject matter.

Bar chart

In Figure 32.2 the bus passenger data is presented as a **bar chart** in which the heights of the bars are proportional to the number of people in each group. Note that because each bar represents a distinct category of person, the bars are drawn separated, not touching. Each bar is labelled and there is a scale to show what the height means in terms of numbers of passengers.

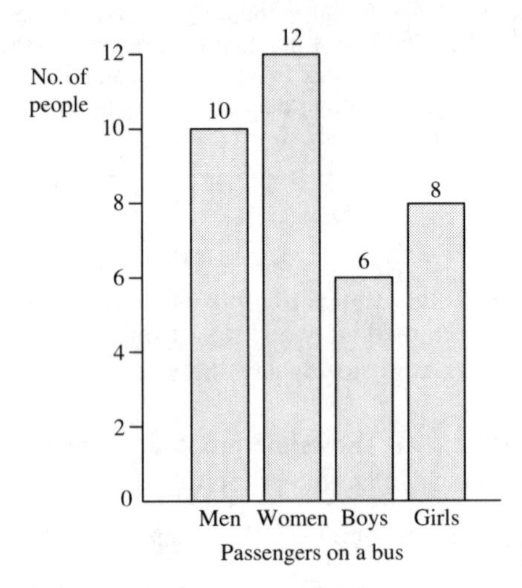

Figure 32.2

A bar chart is simple to draw; all we need to do is to decide on a suitable scale and draw the bars of the correct height. Preferably the tallest bar should fill the space available for the chart. From this chart it is clear that the biggest group of passengers is women and the smallest is boys. Bar charts are best when we want to pick out the largest and smallest groups. A bar chart is completed by adding an explanatory title.

The bar chart in Figure 32.3 shows fruit juice production at four factories. Here the bars are divided into sections corresponding to the three flavours of

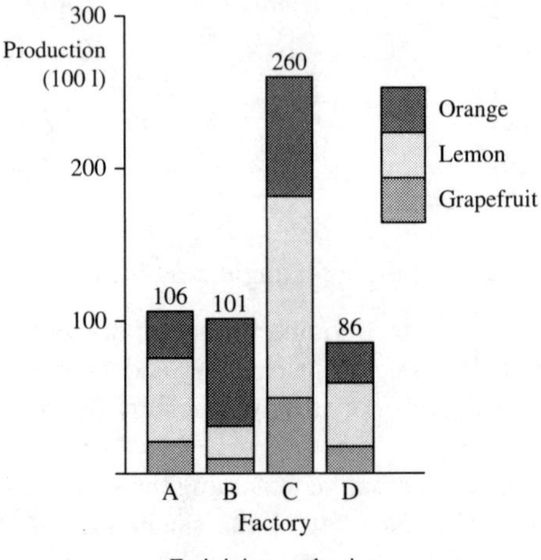

Figure 32.3

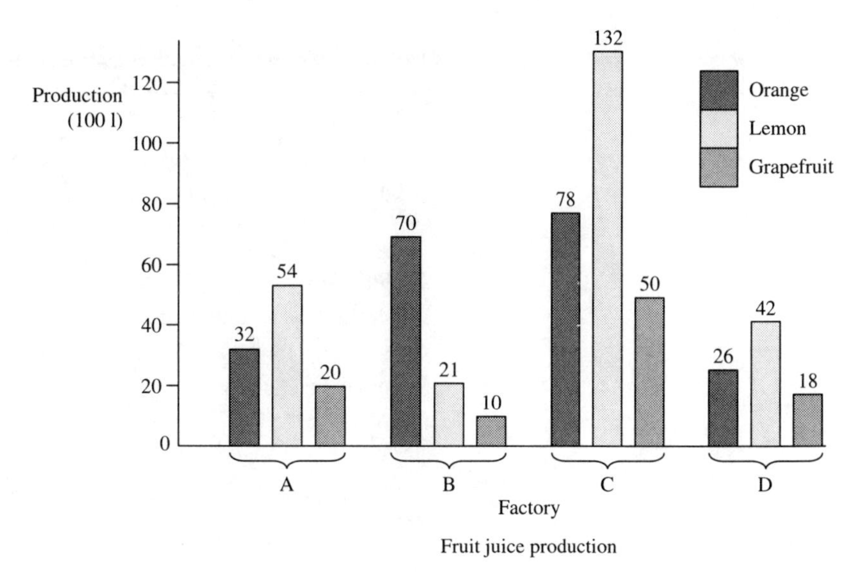

Figure 32.4

juice produced. This type of bar chart makes it easy to compare the *total* production at each factory. Although we can obtain a general idea of the relative amounts of the different flavours, it is not easy to compare production of these at the four factories. This is more easily done in Figure 32.4 where there is a bar for each flavour, and the bars are grouped according to factories. If preferred, the bars can be grouped by flavours, so that factories may be more readily compared.

Test yourself 32.2

1 A survey of students shows the following numbers taking courses in different subjects:

Subject	*Number of students*
Art	5
Building	8
Ceramics	2
Design	10
Electronics	7
Fashion	9

Illustrate this data by **a** a pie chart, **b** a bar chart

Approximately what proportion of the students are enrolled in design courses?

2 The composition by weight of 100 g of a brand of breakfast cereal is given as:

Protein	12
Carbohydrate	62
Fat	8
Soluble fibre	4
Insoluble fibre	3
Other solids	11

Illustrate this data by a pie chart. Approximately what proportion of the cereal is made up by protein and carbohydrate taken together?

3 Four fabric shops list their sales (in hundreds of metres) of fabrics according to the fibres.

Fabric	*Shop*			
	A	*B*	*C*	*D*
Cotton	102	135	178	117
Wool	35	62	42	38
Polyester	150	99	213	83

a Draw a bar chart to illustrate this data so that the *total* sales for each store can be compared.

b Draw a set of four pie charts, one for each shop, to show the proportions of the different fabrics sold in each shop. Which shop sells the greatest proportion of woollen fabrics? Which sells the smallest proportion of cotton? Which shops make almost half or over half of their sales from one type of fabric?

4 Why is it generally unreliable to conduct a survey of public opinion by telephone interviews?

5 Refer to Figure 32.3.

a Which factory has the greatest total production and how much?

b Which has the least total production and how much?

c What is different about the quantities of juice produced by Factory B?

d What is the total production of all four factories?

6 Refer to Figure 32.4.

a Which factory produces the most lemon juice and how much does it produce?

b Which factory produces the least orange juice and how much does it produce?

c What is the total production of grapefruit juice and how much is produced?

d What is the approximate ratio of orange juice produced to lemon juice produced in Factory B?

Which factory has the least total sales? Which has the greatest total sales?

Measurement data

Measurement data often consists of a single measurement carried out on each member of a sample. The aim is to use this data to find out the characteristics of the population from which the sample is drawn. As an example, consider a sample of concrete blocks tested for their compressive strength. The sample consists of 51 blocks and the compressive strength of each block is measured by crushing the block in a press until it collapses. The results obtained (in MNm^{-2}) are as follows:

30.2	27.3	24.1	27.2	32.2	28.6	22.1	27.4	27.0	29.5
31.9	25.9	28.3	28.3	32.1	28.0	29.7	25.0	26.3	30.0
23.7	25.3	33.8	28.7	28.3	29.3	24.6	27.8	31.4	34.3
26.1	27.9	30.2	25.4	28.4	30.7	32.9	26.9	26.3	24.0
33.9	31.7	29.3	29.6	28.9	26.4	26.7	29.2	29.7	30.3
31.3									

A mass of data such as this contains a lot of information but in a form that is very difficult to understand. Looking at the data table we can see that values are in the region of 28, that the lowest is 22.1 and the highest is 34.3. To get much further with analysing this data we need to be more systematic.

One way of sorting out the data is to arrange the items in numerical order. Then we can prepare a chart (Figure 32.5) to illustrate the data. Each small rectangle in the figure represents one concrete block. We can see that the values range from 22.1 to 34.3, but that the block for 22.1 has a much lower compressive strength than its nearest neighbours. The majority of the blocks have strengths between about 26 and 30. If two or more blocks have the same strength, we draw taller rectangles. The height of the rectangle represents the number of blocks (or the **frequency** of blocks) of that strength. There are two blocks each with strength 26.3, 29.3, 29.7 and 30.2. There are three blocks with strength 28.3. By arranging the data systematically and illustrating it by a chart, we have been able to discover far more about it. A chart of this type is known as a **histogram**. It is possible for blocks to have *any* strength in the range 22.1 to 34.3 (and maybe a little below or above these limits). This is why the rectangles in a histogram may touch adjacent rectangles. There are gaps only where blocks of a given strength are not present in the sample.

But the chart shows many irregularities; there are many gaps in it, especially at the lowest and highest values. Instead of plotting the individual blocks we could get a more overall view by grouping the blocks together. We

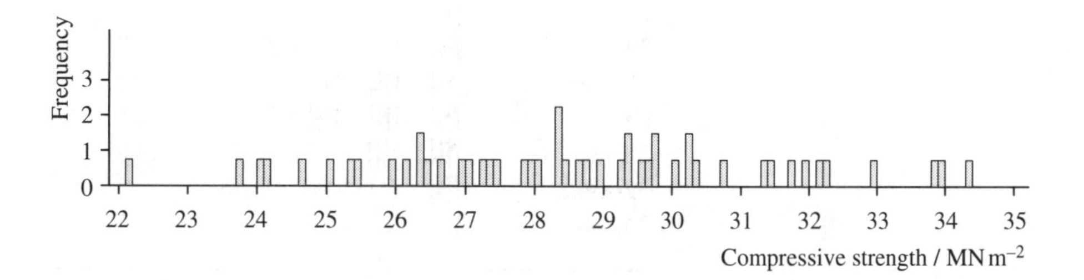

Figure 32.5

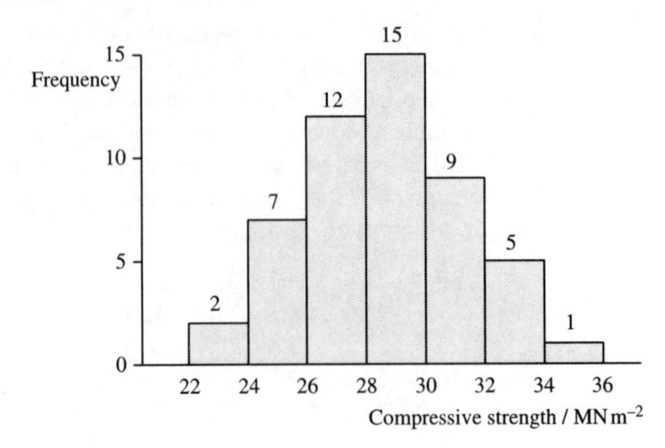

Figure 32.6

divide them into **classes**. In Figure 32.6 the blocks are divided into classes at intervals of 2 MNm^{-1}. It is as if we take the separate rectangles of Figure 32.5 and merge them into fewer but larger rectangles. As before, the rectangles are of standard width and their height varies according to the number of blocks in each class.

Preparing the figures for drawing a histogram is most easily done by setting out a table (Table 32.1).

The first column defines the classes. Although classes need not be the same size, it is more usual to have them all the same size. We begin the first class with a 'round' value a little lower than the smallest strength. In this case we begin with 22. We have also to decide on the **class interval**, the distance from the start of one class to the start of the next. It is usually best to aim to produce a histogram with between 5 and 10 columns (or classes). If we make the class interval 2, this gives 7 classes. Write the class start and finishing values in the first column. There is no overlap between classes and there are no gaps.

Next we run through the table of data and allocate each item to a class. As each item is allocated we record this as a tally (see box) in the second column.

Table 32.1

Class	*Tally*	*Class total*	*Cumulative total*
22–24	||	2	2
24–26	𝍸 ||	7	9
26–28	𝍸 𝍸 ||	12	21
28–30	𝍸 𝍸 𝍸	15	36
30–32	𝍸 ||||	9	45
32–34	𝍸	5	50
34–36	|	1	51

An item which is on the borderline between classes is allocated to the upper class. For example strength 26.0 goes into the class 26–28, not into 24–26. When all items have been allocated, count up the tallies and enter the class totals in the third column.

The fourth column contains the **cumulative** (or **running**) total. This consists of the total for a class together with the totals for the classes below it. For example, the class total for class 22–24 is 2, and so is the cumulative total. The class total for class 24–26 is 7, but the cumulative total is 2 + 7 = 9. Check that the cumulative total for the largest class equals the total number of data items.

The class totals are used for plotting the histogram. Choose a column width which makes the chart spread across the page. There must be no gaps between columns (compare with a bar chart, page 442) as the chart represents a range of possible values running continuously from 22 to 36. The fact that there is not actually a block with strength 23.2 does not mean that the histogram should have a gap at that point. The rectangle for class 22–24 is 2 units high, indicating that blocks with strength between 22 and 24 occur with a frequency of 2. It is not saying what the strengths of any particular blocks might be.

The histogram displays the features of the sample which we have already noted, but they are shown more clearly. Most blocks have strength between 26 and 30, with a slight tendency for more to be between 28 and 30. Above and below these peak values, the frequency drops away and there are few below 24 or above 34.

For comparison, we have drawn another histogram of the same data, with class interval 5 MNm^{-2} (Figure 32.7). Now there are only 3 classes and, although the histogram shows the same broad features, the picture lacks definition. When you are plotting a histogram, it may be necessary to try several different class intervals and select the one which displays the data most clearly. Occasionally a histogram has a very irregular outline and may even have a gap or two (Figure 32.8). Usually this happens when data is very variable. By chance, some classes receive too many items, some too few, and perhaps some receive none at all. It may be possible to cure this by selecting

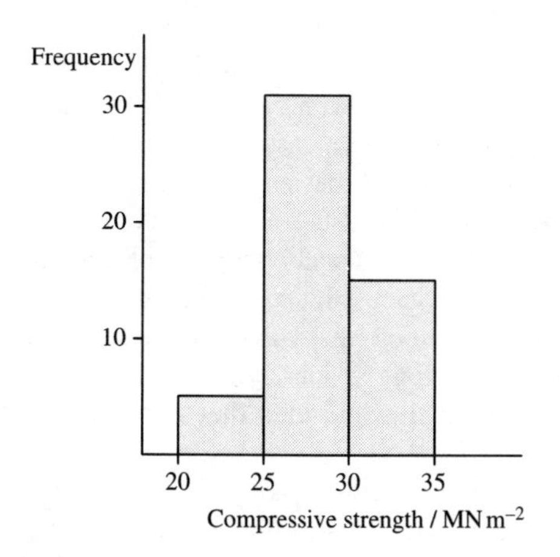

Figure 32.7

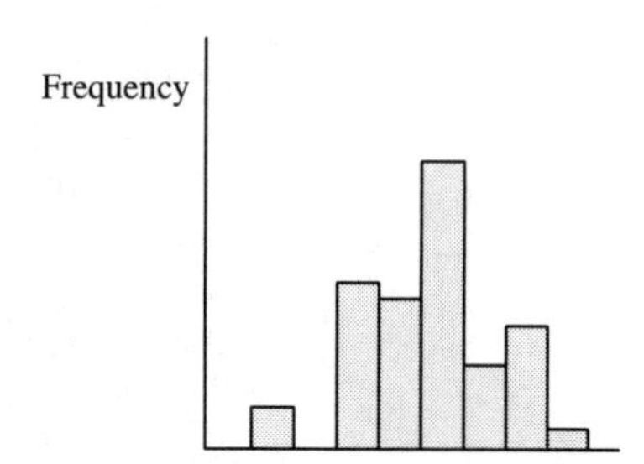

Figure 32.8

a larger class interval. But this may result in there being too few columns in the histogram. The solution then is to collect more data. This helps to smooth out the contours of the histogram and to close any gaps.

Test yourself 32.3

1 The lives of a sample of 27 electric cells are measured under standard conditions. The lives in hours are:

13.6	12.2	13.5	11.9	13.5	11.5	10.7	14.1	12.5
12.0	11.2	13.7	11.6	14.4	13.2	12.7	11.6	12.1
12.9	14.2	13.3	12.2	10.5	11.0	12.3	11.3	12.7

Plot a histogram to display this data. Comment on the shape of the histogram.

2 The weights (g) of a sample of 35 home-made biscuits are as follows:

20.3	19.2	21.5	22.4	18.9	18.3	20.9	21.0	22.4
18.5	19.6	20.1	21.9	20.4	19.0	19.7	22.5	21.4
20.8	21.5	21.7	18.2	20.3	22.0	18.7	20.7	20.2
22.2	21.5	18.8	22.4	19.2	22.6	21.8	22.7	

Plot a histogram to display this data. Comment on the shape of the histogram.

3 Collect some data of your own and draw a histogram of it. Suggestions are: lengths of nails, weights of nails, weights of bricks, heights of students, resistances of 100 Ω resistors (5% or 10% tolerance), reaction times of students, marks in class tests, weights of eggs, thickness of tiles.

Two-variable data

Data described in the previous section has had just one measurement or value associated with each unit in the sample. With some types of data we have two measurements or values. Experimental measurements often belong to this type of data. For example, we measure the length of a metal bar at various temperatures. Each observation consists of a measurement of temperature and a measurement of length. The usual way of displaying such data is to plot a line graph (Chaper 17), with the independent variable along the x-axis and the dependent variable along the y-axis. With reasonably accurate measurements, the points lie along a clearly defined line. It is usually possible to find the equation for this and thus the equation or law linking the two variables.

Another type of two-variable data is obtained by taking a sample of items and making two measurements on each item. This example concerns student projects. After the projects had been assessed and a percentage mark awarded for each, the students were asked to state how many hours they had spent working on the project. Here are the results for a sample of 12 projects:

Project no.		*Time (h)*	*Mark (%)*
1		34	80
2		33	56
3		20	68
4		45	90
5		38	66
6		20	52
7		27	60
8		43	72
9		15	32
10		36	82
11		19	46
12		27	64
	Totals	357	768
	Averages	30	64

The data is illustrated by plotting a **scattergram** (Figure 32.9). Each point on the scattergram represents a project. For example, Project 1 is represented by the point at (34,80). There is obviously no scientific law relating marks to time spent, so we do not find a single line of points running diagonally up the page. Instead, we have a cloud of points. The cloud is elongated and slopes up toward the right. This indicates a **tendency** for better marks to be obtained if more time is spent on the project. There is a **correlation** between marks and time spent. Because marks and time spent increase together, the correlation is positive. But the correlation is only loose; spending a long time on a project is no guarantee of scoring a high mark. Conversely, a clever (or lucky) student may obtain good marks for a project that is quickly completed.

Note that in a scattergram we do not define one variable as the dependent variable and one as the independent variable. Very often it is not possible to say which is which. In this example, we *might* say that getting a good mark depends on spending a long time on the project. But we might also say that an able student will think of more things to do to make the project a good one.

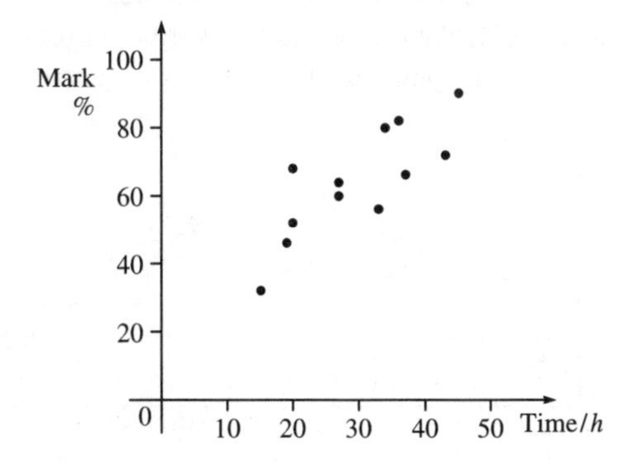

Figure 32.9

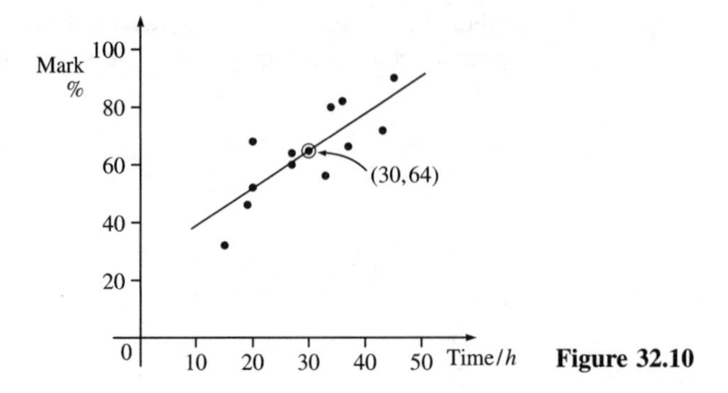

Figure 32.10

This will make the project take longer to complete. So the mark gained and the time taken might *both* depend on a third factor, the ability of the student. Interpreting scattergrams is not always as straightforward as it might appear.

It is often possible to draw a line through the cloud to establish an approximate relationship between the two quantities (Figure 32.10). Then we can begin to make estimates from the scattergram. With a clearly-defined cloud it is usually good enough to draw a straight line that passes centrally through the cloud. The line should pass through the point that corresponds to the averages of both sets of values. In this example, it should pass through the point (30,64). Having established this one point, it is usually easy to judge the correct slope of the line.

With rather more irregular clouds, a line can be drawn using routines for **regression lines**, described in a scientific calculator handbook or statistical textbook. The line in Figure 32.10 was drawn in this way, and minimizes the distances between the line and the points. We might deduce, for example, that to obtain a pass mark of 45% you should spend at least 15 hours on the project. Also that there is not much chance of obtaining a top grade (80% or more) unless you spend at least 42 hours. But all such deductions are subject to uncertainty because of the way the cloud is spread around the line instead of being lined up along it.

If the points are strung out in a very narrow cloud, it shows that the two variables are very strongly correlated (Figure 32.11a). With such strong correlation we may use the scattergram to make predictions with more certainty. We may even suspect that a natural law connects the two variables. Other sources of data may show negative correlation (Figure 32.11b), or the cloud may be so rounded that it shows no correlation at all (Figure 32.11c).

Figure 32.11

Test yourself 32.4

1 Ten broad bean seeds were weighed and their lengths measured:

Bean no.	*Weight (g)*	*Length (cm)*
1	0.7	1.7
2	1.2	2.2
3	0.9	2.0
4	1.4	2.3
5	1.2	2.4
6	1.1	2.2
7	1.0	2.0
8	0.9	1.9
9	1.0	2.1
10	0.8	1.6

Plot a scattergram of this data. State whether you think it shows correlation between weight and length and, if so, whether the correlation is positive or negative. Use the scattergram to predict the most likely length of a bean which weighs 1.2 g.

2 Twelve dried chillis were weighed. Then each was ground to make a paste which was assessed for 'hotness' by an expert taster. Their hotness was rated on a scale of 1 to 10. Here are the results:

Chilli no.	*Weight (g)*	*Hotness*
1	0.6	6
2	0.4	7
3	0.2	8
4	0.8	4
5	0.9	4
6	0.3	6
7	0.5	2
8	0.4	5
9	0.3	9
10	0.4	8
11	0.7	2
12	0.7	5

Plot a scattergram of this data and comment on what it shows. What is the largest weight for a chilli to rate at 5 or more on the hotness scale?

3 To assess your progress in this chapter, work the exercises in *Try theses first*, page 437.

33 Describing data

It is sometimes important to be able to summarize data, reducing a mass of figures to just a few that specify its essentials. This chapter introduces some techniques for this purpose.

You need to know about collecting data (Chapter 32).

Try these first

1 Use the data of question 1, page 437 to find the mean production of glue by the five factories.

2 Use the data of question 2, page 437 to find

a the median weight

b the mean weight

c the modal class

d the range

e the quartiles

f the interquartile range

Location

A histogram such as that of Figure 32.6 (repeated in Figure 33.1) gives us an overall picture of the **distribution** of the data. It shows that the majority of blocks have a compressive strength between 28 and 30 MNm^{-2}, but that values up to 4 MNm^{-2} lower or greater than this range are fairly common. The

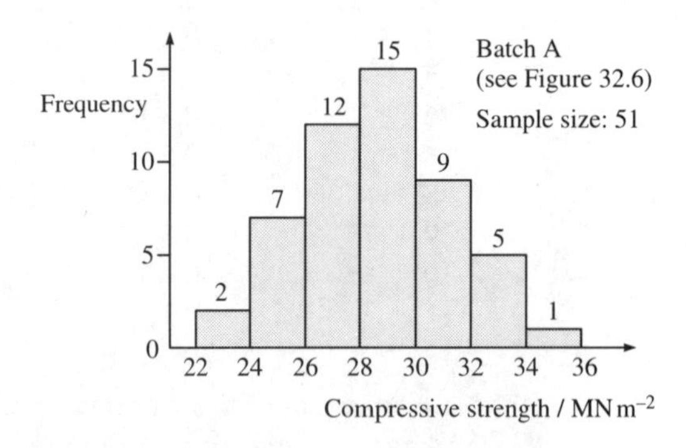

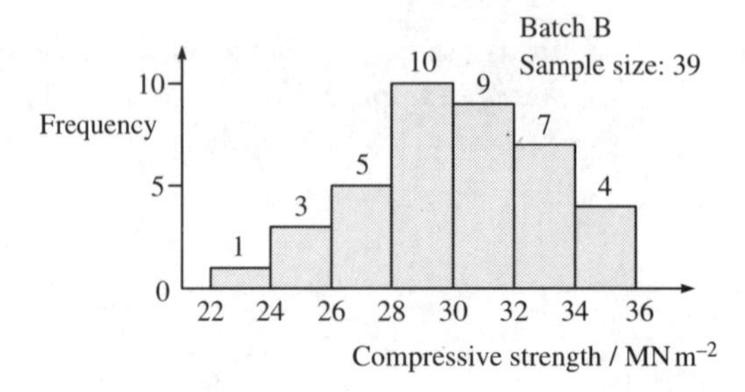

Figure 33.1

Batch B

This is the data for the compressive strength (MNm^{-2}) of a second sample (batch B) consisting of 39 concrete blocks. The data has been sorted into numerical order.

23.8 24.1 24.5 25.8 26.3 26.7 27.0 27.2 27.4 28.0 28.3 28.4
28.5 28.5 28.7 29.2 29.3 29.4 29.7 30.0 30.1 30.3 30.5 30.6
30.6 30.9 31.2 31.5 32.1 32.3 32.7 32.9 33.2 33.4 33.9 34.1
34.2 34.3 34.5

distribution falls away more-or-less equally on both sides. But suppose that we had another set of data based on blocks made from a different batch of concrete (Figure 33.1). We might want to compare these, to see if one batch has greater or lesser strength than the other. In order to be able to do this, we must specify certain key aspects of their distributions. One of these aspects is their **location**.

Location tells us the position of the distribution, in particular in relation to other distributions. One of the easiest way to locate a distribution is to state its peak. In other words, the strength of most samples of concrete. We call this the **mode** or, when data is classed, the **modal class**. In Figure 33.1, the modal class for batch A is 28–30 MNm^{-2}. The modal class for batch B is the same. Maybe the two batches are of equal compressive strength. But it looks as if batch B has relatively more samples in the classes above the mode, so perhaps there is a difference between the batches which we have not yet been able to quantify.

The mode is not an entirely reliable measure of location. Variations in sampling may make the distribution peak at a slightly different mode. A more accurate measure is to find the middle value, or **median**. Batch A consists of 51 blocks. One of these is the median block, leaving 50 blocks. There are 25 blocks less than the median, and 25 blocks greater than the median. To find the median we arrange the blocks in numerical order of strength, then identify the 26th block. This can be done on Figure 32.5. Counting along from the left, the 26th block is 28.4 MNm^{-2}. The median of batch A is 28.4 MNm^{-2}.

Batch B has 39 blocks. One is the median, leaving 38 blocks, of which 19 are below the median and 19 above it. The median is the 20th block. The strength of this is 30 MNm^{-2}. So the median of batch B is *not* in the modal class. This distribution has more values to the right of the modal class; we say this is a **skew** distribution. With a symmetrical distribution, the median is nearly always in the modal class but when a distribution is skew the median may be in one of the classes below or above the modal class.

The samples we have described each have odd numbers of values. There is no problem in taking one of these as the median and dividing the remainder into two sets of equal numbers. Some samples may have an even number of values. For example, this sample has 12 values:

34 36 37 38 38 41 42 43 47 48 50 53

There is no actual value that can be said to be the median. The median lies half way along the row, between the 6th and 7th values. Counting along 6 places we come to 41. The seventh value is 42. We say the median is between 41 and 42, or 41.5.

Mean

The most precise way of specifying location is to calculate the average, or **mean**. This involves more work than finding the mode or mean, especially if there are several hundred values in a sample, but it gives the best measure of location. To find the mean, total the values and divide by the total number of values.

For batch A, mean = (22.1 + 23.7 + . . . + 33.9 + 34.3)/51 = 1452.3/51

mean = 28.5

Give the mean to the same number of decimal places as the sample values.

Test yourself 33.1

1 Find the modal class, median and mean for the data on electric cells, page 448.

2 Find the modal class, median and mean for the data on biscuits, page 448.

3 Find the mean of the data of batch B, see box, page 453.

Dispersion

The other main characteristic of a distribution is how far it extends on either side of its middle point (mode, median or mean). The easiest method of specifying dispersion or **spread** is by finding the **range**:

range = highest value – lowest value

In the data for concrete blocks, batch A:

range = 34.3 – 22.1 = 12.2

For batch B:

range = 34.5 – 23.8 = 10.7

The range of batch A is greater, though Figure 32.5 shows that this is mainly due to a single block, strength 22.1, being so far from its nearest neighbours. Had this block not been selected in the sample, the range of batch A would have been only 34.3 – 23.7 = 10.6. This emphasizes the great disadvantage of range as a measure of spread. It is unduly affected by the odd values that appear rather haphazardly at the extremes of the range.

One way to avoid the extremes is to trim them off systematically. We establish two points known as the **quartiles**. These are two values that lie quarter-way from the two ends of the distribution. The readings are arranged in order, just as when we find the median. Then we divide it into 4 equal parts. The concrete data for batch A has 51 values which we divide as follows:

12 values	lower quartile (13th value)	12 values	median (26th value)	12 values	upper quartile (39th value)	12 values

Four sets of 12 values, plus the 2 quartiles, plus the median make up 51 items of data. Counting along on Figure 32.5 to find the 13th and 39th values we find:

lower quartile = 26.4
upper quartile = 30.2

The **inter-quartile range** = upper quartile – lower quartile
= 30.2 – 26.4 = 3.8

In batch B, there are 39 values, which we divide as follows:

9 values	lower quartile (10th value)	9 values	median (20th value)	9 values	upper quartile (30th value)	9 values

Lower quartile = 28.0
Upper quartile = 32.3

Inter-quartile range = 32.3 – 28.0 = 4.3

Batch B has a greater inter-quartile range than batch A. Its dispersion is greater.

If the sample size is such that the quartiles come *between* one value and the next, take the average of the values above and below each point. Take this sample with 20 data items:

6 7 7 8 8 9 10 12 14 15 17 19 20 21 23 23 25 26 27 29

The number of items is even; the median comes between the 10th and 11th item, between 15 and 17:

The median is 16.

The lower quartile comes between the 5th and 6th items, between 8 and 9, so it is 8.5. The upper quartile comes between the 15th and 16th items, which are both 23, so it is 23:

Inter-quartile range = 23 – 8.5 = 14.5

Test yourself 33.2

1 Find the range, the quartiles, and the inter-quartile range for the data on electric cells, page 448.

2 Find the range the quartiles and the inter-quartile range for the data on biscuits, page 448.

3 Find the quartiles and the inter-quartile range of this sample of data:

12 13 13 14 15 17 18 19 20 20 21 22 23 24 25 27

4 For data samples that you have collected for *Test yourself 32.3*, page 448, question 3, find the mode, median, mean, range and interquartile range.

5 To assess your progress in this chapter, work the exercises in *Try these first*, page 452.

34 Analysing data

In this chapter we discuss frequency distributions of different types and find out how to make a good estimate of the dispersion of a distribution. We also look at ways in which analysing data can help in decision-taking.

You need to know about collecting and displaying data (Chapter 32) and describing data (Chapter 33).

Try these first

1 Find the standard deviation of the data for glue production in question 1, page 437.

2 Find the standard deviation of the sample and the standard deviation of the population for the data on teacups, page 437, question 2. If another sample, containing 50 teacups, included one with weight 199 g and another with weight 201 g, what action should be taken?

Frequency distributions

When we are describing and analysing data, we need to work with the **raw data**, the original data obtained by measuring. For example, to find the mean, we total the items in the sample and divide by the number of items. But at the same time, if there is a large number of items in the sample, it is helpful to group the data into classes and plot a histogram (page 445) to show the frequency distribution. This may tell us little or it may tell a lot. Frequency distributions exist in a wide variety of shapes but there are certain distinctive shapes that indicate that the data is of a particular kind.

The rectangular distribution was seen in the biscuit data of *Test yourself 32.3*, page 448, question 2, where it seems that extra small and extra large biscuits were rejected. There is no distinctive mode and the frequency falls away sharply on both sides of the distribution. This sort of distribution can be obtained if we shake a die a number of times and record the scores. The scores 1 to 6 are all equally likely to occur, though some will occur slightly more often than others (Figure 34.1). But there is no possibility of scoring less than 1 or more than 6. There is not much point in using the mode to describe such a distribution, as the class which becomes the mode is mostly a matter of chance. Dispersion may be limited by the nature of the data, as in the case of die throws.

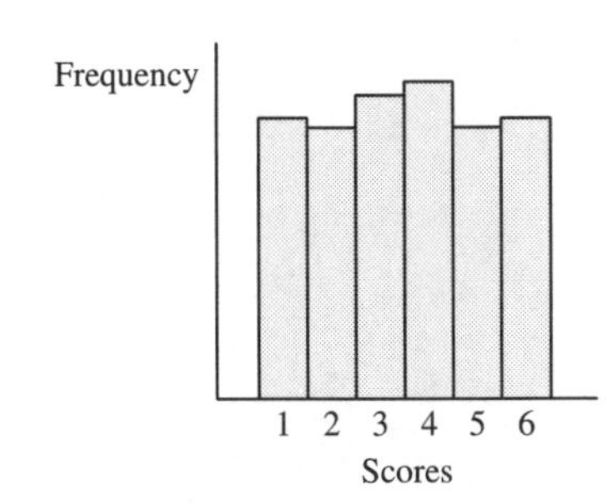

Figure 34.1

The most common shape is the humped distribution, illustrated by several of the sets of data we have analysed, for example, that of Figure 32.6. It has a distinctive mode and the distribution falls away more-or-less symmetrically on either side. The spread is the result of relatively small amounts of

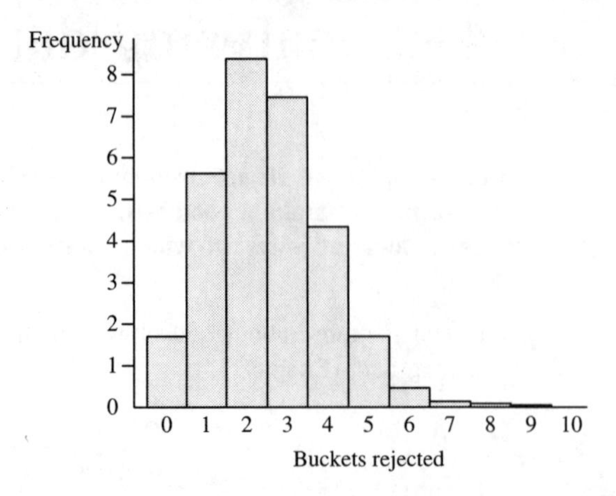

Figure 34.2

variability in the values of the individual measurements. Many kinds of data yield distributions of this type (see later).

Figure 34.2 shows a skew distribution, with the mode nearer one end of the distribution than the other. Distributions of this kind often result from what we call binary situations. For example, in a quality control scheme in a factory making plastic buckets, a sample of 10 buckets is tested every day. This histogram shows the frequency with which different numbers of defective buckets are found in a batch of 10. The frequency is given in days per month. For example, 2 buckets are rejected on just over 8 days each month. No buckets are rejected on 1 or 2 days each month. It is very rarely that we reject 7 or more buckets out of a batch. Given such a distribution as this, based on daily samples of 10, it is possible to calculate what percentage of faulty buckets occur in the total factory production (the population). A histogram shaped as in Figure 34.2 is obtained when 25% of the bucket population are defective.

If the percentage of defective buckets is lower than 25%, the distribution has its mode closer to the lower end of the scale. It is more strongly skewed. If the percentage is 50%, the distribution is symmetrical, with a mode at 5. If more than 50% of buckets are defective, the distribution is skewed in the opposite way, with its mode toward the higher end.

Sometimes a distribution may have two (or possibly more) modes. If it has two modes, we call it a **bimodal** distribution (Figure 34.3). A bimodal

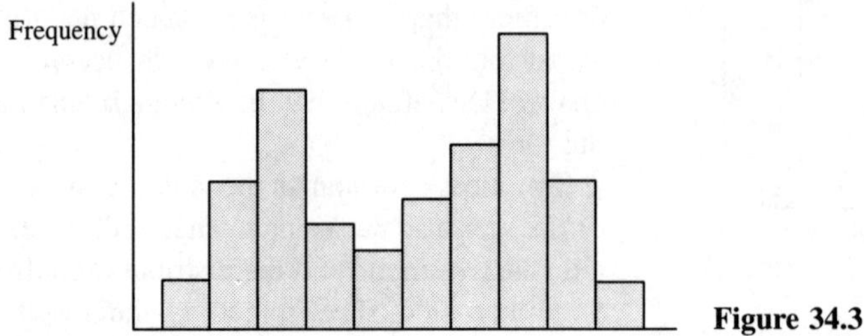

Figure 34.3

distribution may possibly be the result of collecting insufficient data. If we collect more data, the trough in the distribution may become filled in, resulting in a hump-backed distribution. But the two modes may be genuine, and may indicate that what we think is a single population is really two different populations mixed together. For example, if the histogram shows the weights of children, it may be that the lower peak represents mainly girls while the upper peak represents mainly boys. There are really two populations, girls and boys, with overlapping weight distributions. With a bimodal distribution, it is important to re-examine the data to see if there are any grounds for suspecting that it is taken from two differing populations. If so, each population should be sampled and analysed separately.

Standard deviation

This is another way of measuring the dispersion or spread of a sample. Take for example this sample of 10 values, which we have arranged in numerical order:

11 13 14 16 18 18 19 21 24 26

The mean of this data is $\frac{11 + 13 + \ldots + 24 + 26}{10} = \frac{180}{10} = 18$

Each value in the sample is a certain distance from the mean. For example, the difference between the mean and the value 21 is $21 - 18 = 3$. We call this its **deviation** from the mean. Here is a table of the deviations:

Value	*Deviation*
11	–7
13	–5
14	–4
16	–2
18	0
18	0
19	1
21	3
24	6
26	8

One way of measuring the spread of the data would be to find the **mean** deviation. We would add up all the deviations and divide by the number of deviations. This simple idea does not work. Deviations are negative as well as positive. If we add up the deviations in the table above, the total is zero. This happens with any set of data because, by definition, the mean is the value above and below which all other values are equally distributed.

The solution to obtaining a mean deviation is to *square* all the deviations, then find the mean of the squares, and finally take the square root of the result. Squaring any number, negative or positive, gives a positive result (page 60) so in this way we eliminate the negatives from the calculations. Let us try this on the data given above:

Value	*Deviation*	*(Deviation)*^{2}
11	–7	49
13	–5	25
14	–4	16
16	–2	4
18	0	0
18	0	0
19	1	1
21	3	9
24	6	36
26	8	64
	Sum of squares =	204

Dividing the sum of squares by n the number of values:

$$\frac{204}{10} = 20.4$$

This quantity is known as the **variance**. It is used directly in certain types of analysis. Its units are the squares of the units of the original measurements. To get back to the original units, we have to take the square root of the variance. This is the quantity known as the **standard deviation**, symbol s:

$$s = \sqrt{20.4} = 4.52$$

The standard deviation is a measure of the spread of the sample data about its mean. The standard deviation (or *s.d.* for short) is a better measure of dispersion than the range or inter-quartile range. As we have said (page 454), the range is influenced too much by the values that occur (rather irregularly) at the extremes of the distribution. On the other hand, the inter-quartile range ignores the first and last quarters of the data entirely. The s.d. takes all the data into account and gives equal importance to each.

The only possible disadvantage of the s.d. is the work involved in calculating it. But most scientific calculators have built-in routines for this. All that is necessary is to key in the items of data. When this is complete, you can read off n (to check that the correct number of items has been entered), the sum of squares, the s.d., and several other quantities. Refer to your instruction book for details.

Test yourself 34.1

1 Find the s.d. of the sample of electric cell data (page 448).

2 Find the s.d. of the sample of biscuit data (page 448).

Samples and populations

We usually work with a **sample** of data because it would take too long or cost too much to measure the whole population. Then we reduce the mass of data of the sample to two figures which define its location (mean) and dispersion (s.d.). The next step is to use these two pieces of information to *estimate* the location and dispersion of the population from which the sample is taken. For the sample, we *know* what its mean and s.d. deviation are, because we calculate these from the sample data. For the population, we can never know the exact mean and s.d. deviation. We can only make as good an estimate as is possible.

The best estimate of the mean of the population is the mean of the sample. We can not do better than this.

The best estimate of the s.d. of the population is obtained by a modification of the calculation for the s.d. of the sample. When we select a sample, it is unlikely that the sample will include many of the lowest and highest values from the population. It is likely that the sample will be less widely spread than the population it is drawn from. The s.d. of the population is likely to be a little greater than that of the sample. When calculating the s.d., we allow for this by dividing the sum of squares by $n - 1$ instead of by n.

In the example of the data on page 460:

$$\text{population variance} = \frac{\text{sum of squares}}{n-1} = \frac{204}{9} = 22.67$$

$$\text{population s.d.} = \sqrt{\text{population variance}} = \sqrt{22.67} = 4.76$$

The symbol for population s.d. is σ, so we estimate that:

$$\underline{\sigma = 4.76}$$

Compare this value of 4.76 with the value 4.52 for the s sample s.d., and note that it is slightly larger.

Test yourself 34.2

1 Estimate the population s.d. from the electric cell data (page 448).

2 Estimate the population s.d. from the biscuit data (page 448).

3 Find the sample and population s.d.'s of data that you have collected yourself.

The normal distribution

The **normal distribution** is a hump-shaped distribution with special properties. Imagine a histogram (Figure 34.4) made by taking an extremely large sample and measuring each individual with a high degree of precision. We could divide the sample into a very large number of classes. With a very large number of classes the outline of the histogram becomes very smooth. Imagine that the class intervals are so small that we can no longer see the steps

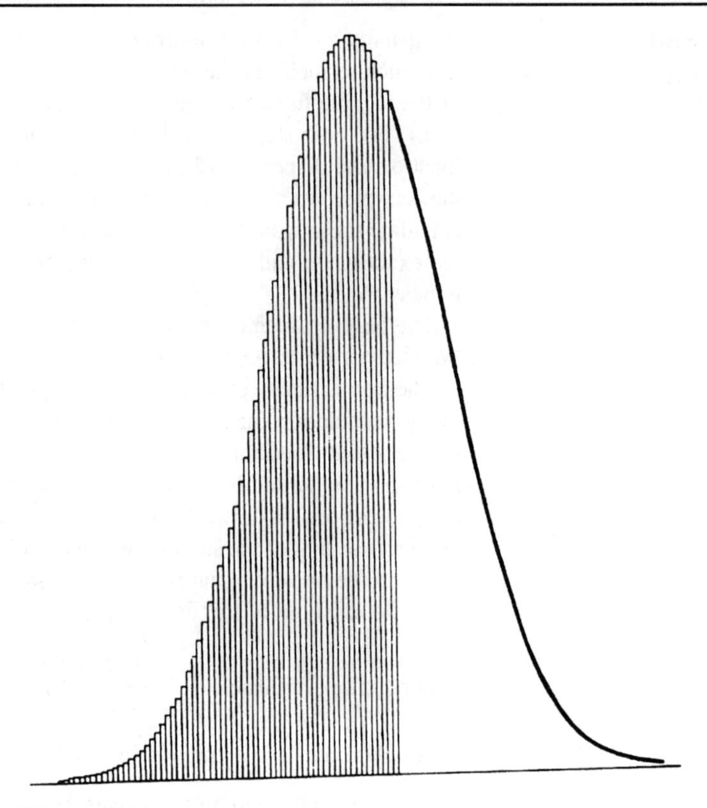

Figure 34.4

between one class and the next. The result is the bell-shaped curve partly shown on the right of the figure. It is symmetrical; the modal class is at the centre of the distribution but it is not prominent. At the extremes, the distribution flattens out like the rim of a bell.

Remember that when we use the word 'normal' in connection with data analysis, we do not mean 'ordinary' or 'usual'. The word has a special technical meaning. Many kinds of data are distributed normally, especially data obtained from living things, including humans. This is why the normal curve is so important. The bell-shaped curve in Figure 34.4 is typical of normal curves, but the bell can be wider or narrower than this. Its relative width depends on the scale on which it is plotted, and also on its s.d.

The s.d. is one of the important features of the normal curve. In Figure 34.5a we have drawn a normal curve and a central vertical line through it to indicate the location of the mean. We have then drawn two more lines spaced one s.d. above and below the mean. It is a property of the normal distribution that the shaded area between these lines represents 68% of the distribution. This applies whatever the value of the mean and of the s.d. deviation, because it is a result of the shape of the normal curve.

Let us see what this means in terms of an actual distribution. The data for batch A of the concrete blocks has a distribution that seems reasonably close

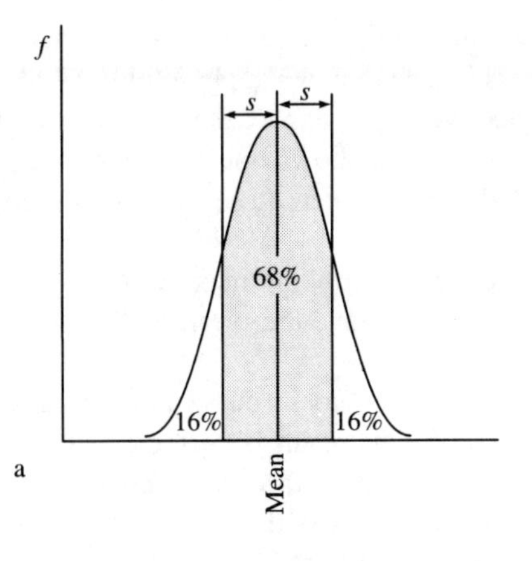

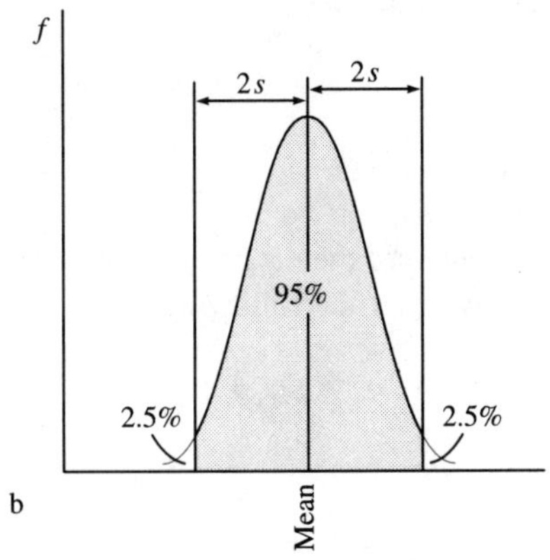

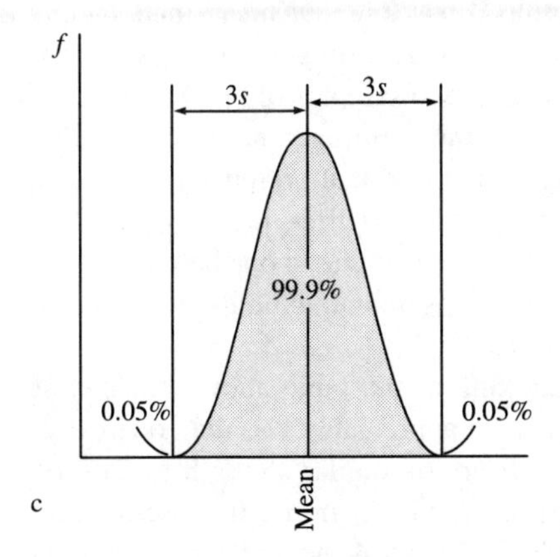

Figure 34.5

to a normal distribution (Figure 32.6). Calculations give the mean of the data as 28.5 and *s*, the s.d. of the sample, as 2.76.

One s.d. below the mean is 28.5 – 2.76 = 25.74
One s.d. above the mean is 28.5 + 2.76 = 31.26

If we check through the data, we find that 32 blocks fall within these limits. 32 blocks in 51 is equal to 63% of the blocks. We predicted that 68% would lie within the limits, but 63% is reasonably near to this, allowing for random effects in selecting a sample of only 51 blocks. With a larger sample, say 200 blocks, we might expect the number within the limits to be close to 68%.

Figure 34.5b shows lines drawn *two* s.d.'s below and above the mean. The shaded area between these lines represents 95% of the sample:

Two s.d.'s below the mean is 28.5 – 5.52 = 22.98
Two s.d.'s above the mean is 28.5 + 5.52 = 34.02

We find that 49 blocks out of the 51 fall within these limits, which is 96% of the blocks. This result is very close to the expected 95%.

If we set limits at three s.d.'s above and below the mean, we enclose 99.9% of the distribution. With only 51 blocks, this means that the whole sample falls between the limits at 20.22 and 36.78.

Same or different?

When we analyse the data for the two samples of concrete blocks (pages 445 and 453) we find the following:

Sample	*Mean*	*s.d.*
A	28.5	2.76
B	29.8	2.89

Sample B has a higher mean than sample A. Does this show that the blocks of sample B are stronger than those of sample A? One thing is almost certain; if we take two samples of blocks, even if they come from the same population (same manufacturing batch), we would be very unlikely to pick out two batches and find that both batches had *exactly the same* mean. The means are almost certain to differ, just by chance. The question is: is the difference due to chance, or is it due to one batch really being stronger than the other, perhaps because it is manufactured by a superior technique or from higher-grade materials?

Looking at the table above, we can see that the differences between the means are appreciably less than one s.d. In other words, most of the blocks in batch B are of similar strength to 68% of the blocks in batch A. Conversely, most of the blocks in batch A are of similar strength to 68% of the blocks in batch B. There is no evidence from these figures that the batches differ significantly. We can say that they could have been drawn from the *same population*. Even if batch B is manufactured in a different way from batch A, it is identical as far as strength is concerned.

Suppose that we tested a sample C taken from a new batch of blocks, made to a different formulation, and found:

Mean = 34.7
s.d. = 2.78

The mean of this batch differs from the mean of batch A by just over two s.d.'s. This puts it level with the top 2.5% of batch A. It is unlikely (though not impossible) that a sample would fall wholly at the top end of the distribution of batch A, if the sample was really from the same population. We prefer to believe that sample C is from a *different* population, a population of blocks made by a technique that gives greater strength.

In these two examples we have shown that there is reason to believe that samples A and B are the same, but A and C are different. We often aim to show that two samples are different, for example that a new technique gives a superior product. Or we may aim to show that two samples are the same, for example that it is possible to introduce less costly techniques without affecting the quality of the product. Another aim may be to investigate ways of reducing standard deviation so as to improve the reliability of the product.

Tests of this kind are important in quality control. For example, a factory making a vitamin concentrate will decide on the mean concentration of the vitamin to be contained in its concentrate, and also the acceptable standard deviation. Regular tests of samples of the concentrate should show a vitamin concentration that falls within two s.d.'s of the specified mean. If 95% of samples fall in this range, all is well. But if more than 5% are outside this limit, or worse, differ from the mean by three s.d.'s or more, this is a cause for concern and the production methods need to be checked urgently.

In the discussions above we have often used phrases such as 'it is likely that', 'it is almost certain that', and 'we believe that'. It would be more satisfying if we could replace such phrases by statements such as 'it is likely with a probability of 95% that', or 'with a probability of 1% that we are wrong, we believe that'. We need to be able to estimate how likely (expressed as a percentage) it is that two samples are the same or different. Later in your studies you may learn to make use of other tests based on the normal distribution and standard deviations, which have a greater degree of precision than the simple analyses described here.

Test yourself 34.3

1 Samples of the same transistor type, manufactured by two different companies are measured for gain. The data is:

Sample A: 51 52 54 54 55 57 57 58 59 60 61 62 62 63 64 65 67 69 71 73

Sample B: 67 71 73 75 77 77 80 81 81 82 83 83 83 85 86 88 89 90 91 96

Plot a histogram of each sample, based on a class interval of 5. Find the mean and standard deviation of each sample. Comment on the differences, if any, that exist between the two samples.

2 Glass tubing for fluorescent lamps is made by two different processes. The wall thicknesses (in mm) of samples from the two processes are:

Sample A: 1.57 1.58 1.61 1.63 1.69 1.70 1.71 1.73 1.73
1.78 1.80 1.82 1.82 1.84 1.84 1.89 1.93 1.93
1.94 1.98 2.01 2.02 2.07 2.13

Sample B: 1.70 1.72 1.73 1.73 1.74 1.78 1.79 1.80 1.81
1.81 1.82 1.84 1.84 1.84 1.86 1.87 1.89 1.92
1.94 1.98 2.01 2.03 2.09

Plot a histogram of each sample, based on a class interval of 0.1 mm. Find the mean and standard deviation of each sample. Comment on the differences, if any, that exist between the two samples.

3 A firm manufactures a heat-sensitive switch for fire alarm systems. Many samples of the switch are tested and measurements show that the mean temperature for triggering the switch is 57°, with a standard deviation of 0.55°C, which is taken to allow an acceptable amount of variability in the product. As part of a quality control scheme, a sample of 100 switches is tested each week. In one sample, it is found that one switch triggers at 58.7°C and another at 58.9°C. What does this indicate and what steps should be taken?

4 To assess your progress in this chapter, work the exercises in *Try these first*, page 457.

Appendix: Computer programs

Computer programs

The programs are intended to provide extended practice in mental arithmetic (see Chapters 1 and 2).

In TEN-PAIRS the screen displays a whole number between 0 and 10. Key in the number which makes this up to 10, and press ENTER or RETURN. You will be told 'Right' or 'Wrong' and your on-going score of correct answers is displayed. The computer then displays the next number for you to work on.

The routine is similar in ADDING, SUBTRACTING and MULTIPLYING, except that *two* numbers are displayed. In ADDING and SUBTRACTING, the numbers range from 0 to 10. In MULTIPLYING, the numbers range from 0 to 12. Depending on the program, you have to key in their sum, difference or product.

The programs are written in a standard form of BASIC and should run on almost any computer. The only variation that is likely to be met is the function for obtaining numbers at random (line 30). Consult the User Manual for your computer if in doubt. In BBC BASIC, line 30 should read:

	30 A = RND(10)	(in TEN-PAIRS)
or	30 A = RND(10):B = RND(10)	(in ADDING and SUBTRACTING)
or	30 A = RND(12):B = RND(12)	(in MULTIPLYING)

The number(s) in line 30 can be altered if you want to limit or extend the range of the test.

If you have an interest in programming, try adding a timing action to the program. Your program could measure the total time taken, or you could arrange for the numbers to be displayed for a limited time only (say 10 s), so that you have to answer quickly or lose the chance to score.

```
10 REM *** TEN-PAIRS ***
20 CLS: x = 0: y = 0
30 a = INT(RND*10)
40 PRINT
50 PRINT a
60 INPUT b
70 IF a + b = 10 THEN x = x + 1: PRINT "Right"
80 IF a + b <> 10 THEN PRINT "Wrong"
90 y = y + 1
100 PRINT "Your score is ";x;"/";y
110 GOTO 30
```

```
10 REM *** ADDING ***
20 CLS: x = 0: y = 0
30 a = INT(RND*10): b = INT(RND*10)
40 PRINT
50 PRINT a;"+";b;"=";
60 INPUT c
70 IF a + b = c THEN x = x + 1: PRINT "Right"
80 IF a + b <> c THEN PRINT "Wrong"
90 y = y + 1
100 PRINT "Your score is ";x;"/";y
110 GOTO 30
```

```
10 REM *** SUBTRACTING ***
20 CLS: x = 0: y = 0
30 a = INT(RND*10): b = INT(RND*10)
40 PRINT
50 PRINT a;"-";b;"=";
60 INPUT c
70 IF a - b = c THEN x = x + 1: PRINT "Right"
80 IF a - b <> c THEN PRINT "Wrong"
90 y = y + 1
100 PRINT "Your score is ";x;"/";y
110 GOTO 30
```

```
10 REM *** MULTIPLYING ***
20 CLS: x = 0: y = 0
30 a = INT(RND*12): b = INT(RND*12)
40 PRINT
50 PRINT a;"X";b;"=";
60 INPUT c
70 IF a*b = c THEN x = x + 1: PRINT "Right"
80 IF a*b <> c THEN PRINT "Wrong"
90 y = y + 1
100 PRINT "Your score is ";x;"/";y
110 GOTO 30
```

Answers

Chapter 1

Try these first (page 3)

1 6

2 3

3 a 10	**b** 11	**c** 14
d 57	**e** 163	**f** 821

4 44

5 a 4	**b** 5	**c** 34
d 15	**e** –5	**f** –44

Test yourself 1.1

1 8, 6, 9, 10, 3, 8, 5, 0, 4, 6, 7, 2, 1, 9, 3

2 4, 8, 0, 6, 3, 7, 3, 9, 4, 1, 7, 10, 8, 5, 2

3 a 11	**b** 11	**c** 12	**d** 9	**e** 8

Test yourself 1.2

1 a 10	**b** 10	**c** 11	**d** 9	
e 10	**f** 12	**g** 11	**h** 12	
2 a 13	**b** 14	**c** 13	**d** 12	
e 17	**f** 15	**g** 12	**h** 5	
3 a 37	**b** 51	**c** 59		
4 a 25	**b** 55	**c** 92	**d** 90	**e** 61
f 127	**g** 169	**h** 110	**i** 41	**j** 154
k 130	**l** 101	**m** 133	**n** 102	**o** 181
5 a 88	**b** 79	**c** 103		
d 93	**e** 91	**f** 128		
g 145	**h** 134	**i** 133		
j 219	**k** 222	**l** 231		
6 a 693	**b** 775	**c** 771		
d 621	**e** 1182	**f** 1000		

Test yourself 1.3

1 a 4	**b** 1	**c** –1	**d** 0
e 7	**f** –4	**g** –7	**h** 6
i –7	**j** 5	**k** 2	**l** –5
2 a 4	**b** –3	**c** 5	
d –2	**e** 0	**f** –11	
3 a 10	**b** 7	**c** 11	**d** 4
e 5	**f** –7	**g** 6	**h** 5
i –2	**j** –2	**k** 10	**l** 1
m 5	**n** –4	**o** –4	**p** –2
4 a 11	**b** 8	**c** –4	**d** 4
e 1	**f** 5	**g** 6	**h** 18
i –5	**j** 5	**k** –5	**l** –4
m –10	**n** 0	**o** –4	**p** –1

Test yourself 1.4

1 a 24	**b** 62	**c** 20	**d** 12
e 11	**f** 3	**g** 2	**h** 33
2 a 19	**b** 17	**c** 24	**d** 19
e 11	**f** 3	**g** 2	**h** 33
3 a 21	**b** –7	**c** 0	**d** –2057

Test yourself 1.5

1 a 28	**b** –28	**c** –22
d –21	**e** 43	**f** –43
2 a –9	**b** –34	**c** 21
d –31	**e** –2	**f** –94

Chapter 2

Try these first (page 17)

1 a 8	**b** 15	**c** 99	
2 a 204	**b** 231	**c** 2912	
3 a 5	**b** 12	**c** 11	
4 a 23	**b** 17	**c** 48	
5 a –28	**b** –185	**c** –19	**d** 27
6 a 80	**b** 72	**c** 2	

Test yourself 2.1

1 a 8	**b** 18	**c** 18	**d** 25
e 77	**f** 32	**g** 84	**h** 27
i 144	**j** 80	**k** 108	**l** 35
2 a 8	**b** 7	**c** 12	**d** 11
e 5	**f** 11	**g** 9	**h** 9
i 6	**j** 2	**k** 10	**l** 6

Test yourself 2.2

1 a 126	**b** 279	**c** 410	**d** 282
e 168	**f** 550	**g** 1408	**h** 996
i 672	**j** 1672	**k** 2628	**l** 2628
m 5518	**n** 10 396	**o** 9028	**p** 23 598
2 a 12 478	**b** 16 588	**c** 75 115	**d** 301 172
e 258 741	**f** 16 384	**g** 741 888	**h** 3 086 050

Test yourself 2.3

1 a –6	**b** –24	**c** –72	**d** 18
e –66	**f** 81	**g** 35	**h** –56
2 a –552	**b** –3293	**c** –4032	**d** 600
e –2223	**f** –3198	**g** 2277	**h** 1131
3 a 32	**b** 17	**c** 8	
d 31	**e** 32	**f** 0	
g 70	**h** 124	**i** 200	
4 a 36	**b** 83		
c 104	**d** –53		
e 286	**f** 0		
g 132	**h** –32		

Test yourself 2.4

1 a 4	**b** 11	**c** 12	**d** 6
e 9	**f** 7, r1	**g** 12	**h** 11, r1
i 12	**j** 60, r1	**k** 9, r2	**l** 7, r5
2 a 51	**b** 35	**c** 72, r1	**d** 127
e 13	**f** 14, r4	**g** 199, r4	**h** 187, r6

Test yourself 2.5

1 a 12 **b** –5 **c** –7
d 10 **e** 4 **f** –1
g –11 **h** 1 **i** –6
2 a 30 **b** 91 **c** 8
d 11 **e** –27 **f** 7
g 105 **h** –42 **i** 2
3 a 81 **b** 2 **c** 59

Test yourself 2.6

1 a 236 **b** 587 **c** 225 **d** 106
2 a 47, r2 **b** 712, r9 **c** –405 **d** –43, r1

Chapter 3

Try these first (page 28)

1 4.6 **2** $0.8\dot{3}$
3 $8\frac{1}{2}$ **4** 20/25
5 a 4.76 **b** $6\frac{3}{14}$ **c** 24.48
d $16\frac{1}{4}$ **e** 5.38 **f** $2\frac{3}{14}$
6 a 2 **b** 4 **c** 4
7 a 7.4 **b** 7.37
8 a 1730 mm **b** 0.007 57 m

Test yourself 3.1

1 a $3\frac{1}{2}$, 3.5 **b** $2\frac{2}{3}$, $2.\dot{6}$ **c** $3\frac{1}{7}$, $3.\dot{1}4285\dot{7}$
d $1\frac{4}{10}$ or $1\frac{2}{5}$, 1.4 **e** $1\frac{1}{5}$, 1.2 **f** $3\frac{1}{4}$, 3.25
g $\frac{5}{9}$ (already is a common fraction), $0.\dot{5}$
h $\frac{1}{8}$ (already is a common fraction), 0.125
i $2\frac{5}{6}$, $2.8\dot{3}$ **j** $\frac{2}{11}$, $0.\dot{1}\dot{8}$ **k** $\frac{3}{32}$, 0.09375
l $2\frac{1}{12}$, $2.08\dot{3}$
2 a 1/5 **b** 1/2 **c** 3/4 **d** 1/9
e 2/9 **f** 2/7 **g** 2/3 **h** 3/11
3 a 0.4 **b** 0.375 **c** $0.0\dot{2}$ **d** 0.75
e 0.7 **f** 0.0625 **g** $0.41\dot{6}$ **h** $0.\dot{6}$

Test yourself 3.2

1 a $\frac{1}{2}$ **b** $\frac{1}{5}$ **c** $\frac{1}{2}$ **d** $\frac{2}{3}$
e $\frac{1}{3}$ **f** $\frac{2}{3}$ **g** $\frac{1}{2}$ **h** $\frac{2}{7}$

2 a $4\frac{1}{11}$ **b** $8\frac{1}{3}$ **c** $3\frac{3}{17}$ **d** $3\frac{1}{5}$
e $1\frac{1}{32}$ **f** $1\frac{19}{20}$ **g** 3 **h** $3\frac{3}{8}$

3 a $\frac{7}{4}$ **b** $\frac{17}{5}$ **c** $\frac{39}{8}$ **d** $\frac{25}{11}$
e $\frac{33}{7}$ **f** $\frac{39}{20}$ **g** $\frac{53}{12}$ **h** $\frac{119}{5}$

4 a $\frac{2}{4}, \frac{3}{6}, \frac{4}{8}$ **b** $\frac{2}{8}, \frac{3}{12}, \frac{4}{16}$
c $\frac{2}{14}, \frac{3}{21}, \frac{4}{28}, \frac{5}{35}$ **d** $\frac{2}{10}, \frac{3}{15}, \frac{9}{45}, \frac{12}{60}$
e $\frac{4}{6}, \frac{6}{9}, \frac{8}{12}, \frac{10}{15}$ **f** $\frac{6}{8}, \frac{9}{12}, \frac{12}{16}, \frac{15}{20}$

5 a $\frac{2}{4}, \frac{3}{6}, \frac{4}{8}$ **b** $\frac{2}{12}, \frac{3}{18}, \frac{4}{24}, \frac{5}{30}$
c $\frac{2}{20}, \frac{4}{40}, \frac{5}{50}, \frac{9}{90}$ **d** $\frac{3}{15}, \frac{5}{25}, \frac{6}{30}, \frac{9}{45}$
e $\frac{6}{16}, \frac{9}{24}, \frac{30}{80}$ **f** $\frac{4}{14}, \frac{6}{21}, \frac{10}{35}$

Test yourself 3.3

1 a 4.82 **b** 1.990 **c** 5.0764
d 4.12 **e** 1.19 **f** −1.31

2 a $\frac{4}{5}$ **b** $\frac{5}{9}$ **c** 1 **d** $\frac{5}{6}$
e $\frac{9}{20}$ **f** $\frac{22}{35}$ **g** $1\frac{23}{60}$ **h** $1\frac{7}{12}$
i $\frac{1}{10}$ **j** $\frac{11}{60}$ **k** $\frac{88}{99}$ **l** $-\frac{4}{15}$

3 a $\frac{1}{2}$ **b** $8\frac{3}{20}$ **c** $1\frac{1}{2}$ **d** $1\frac{15}{16}$
e $8\frac{29}{60}$ **f** $-1\frac{1}{6}$ **g** $10\frac{3}{16}$ **h** $\frac{5}{16}$

Test yourself 3.4

1 a 14.4 **b** 2.85 **c** 4.32
d −0.912 **e** 0.96 **f** 0.078 32

2 a $\frac{2}{15}$ **b** $\frac{12}{35}$ **c** $\frac{14}{99}$
d $\frac{5}{12}$ **e** $\frac{1}{3}$ **f** $\frac{5}{44}$

3 a $1\frac{7}{8}$ **b** $-7\frac{19}{40}$ **c** 1
d $18\frac{9}{16}$ **e** $6\frac{1}{10}$ **f** $\frac{11}{24}$

Test yourself 3.5

1 a 7.8 **b** −4.8 **c** 59
d 0.04 **e** $4.\dot{0}\dot{9}$ **f** 78.63

2 a $1\frac{1}{20}$ **b** $\frac{35}{64}$ **c** $\frac{1}{3}$
d $\frac{1}{2}$ **e** $3\frac{17}{56}$ **f** $1\frac{1}{11}$

3 a $9\frac{1}{4}$ **b** 4 **c** 2 **d** $3\frac{1}{7}$

Test yourself 3.6

1 a 57 **b** 8 **c** 9 **d** 66
e 5 **f** 546 **g** 22 **h** 50
2 a 34.76 **b** 7.65 **c** 0.89 **d** 300.77
e 0.02 **f** 200.84 **g** 2.53 **h** 17.86
3 a 400 **b** 2400 **c** 12 500 **d** 50 000

Test yourself 3.7

1 a 35 000 **b** 57 000 **c** 54 800 **d** 6789
e 56 450 **f** 1 000 000 **g** 46 **h** 40 008
2 a 6.423 **b** 70.024 **c** 0.563 **d** 0.871
e 0.5629 **f** 0.083 **g** 0.009 **h** 0.000 67
3 a 5.890 **b** 2136 **c** 84 066
4 a 34 500 **b** 5000 **c** 0.632

Chapter 4

Try these first (page 49)

1 54 **2** 15.162
3 5 **4** 3
5 $3n + 9m + nm$ **6** $3r^2 - 28s - 6rs$

Test yourself 4.1

1 11 **2** 14
3 44 **4** 1
5 –2 **6** 8.65
7 3 **8** 12
9 9
10 $t = n + 5$ **a** £17 **b** £35
11 $b = p - 4$ **a** 8 **b** 46

Test yourself 4.2

1 20 **2** 32
3 7.65 **4** 44
5 5 **6** 36
7 5.625 **8** 11
9 32
10 $n = x(2 + 2y)$ **a** 60 **b** 70

Test yourself 4.3

1 4
2 5.75
3 12
4 14
5 6
6 3
7 8
8 12

Test yourself 4.4

1 $6m + 3n$
2 $8p + 4q$
3 $3x + 7$
4 $9g + 10$
5 $j - 3$
6 $6x + 5y$
7 $5d + de - 8$
8 $x - 8$
9 $y + 10$
10 $10x + 12$
11 $4a + 11ab$
12 $3y - x$
13 36
14 10
15 13

Test yourself 4.5

1 4
2 27
3 –6
4 –15
5 6
6 –19

Test yourself 4.6

1 $5a^2 + a + 4$
2 $f^3 - f^2 - f - 1$
3 $8x^2 - 5xy + 5y^3$
4 $a^2 + 2a + 8$
5 $3a^2 + 5ab - 2b^2$
6 $2j^2 - 4$
7 $r^3 + r^2 + r + 1$
8 $6 - 6m^2$
9 19
10 4
11 30
12 24
13 5.48
14 7
15 8
16 118

Chapter 5

Try these first (page 63)

1 $\hat{a} = 70°$, $\hat{b} = 110°$, $\hat{c} = 70°$, $\hat{d} = 40°$, $\hat{e} = 140°$
2 **a** equilateral triangle **b** rhombus **c** isosceles trapezium
d regular hexagon

Test yourself 5.1

1 $\hat{a}$ is acute, $\hat{b}$ is obtuse, $\hat{c}$ is a right angle, $\hat{d}$ is reflex, $\hat{e}$ is acute, $\hat{f}$ is obtuse

2 $\hat{a} = 40°$, $\hat{b} = 75°$, $\hat{c} = 300°$, $\hat{d} = 250°$, $\hat{e} = 170°$

3 All of them

4 [See Figure]

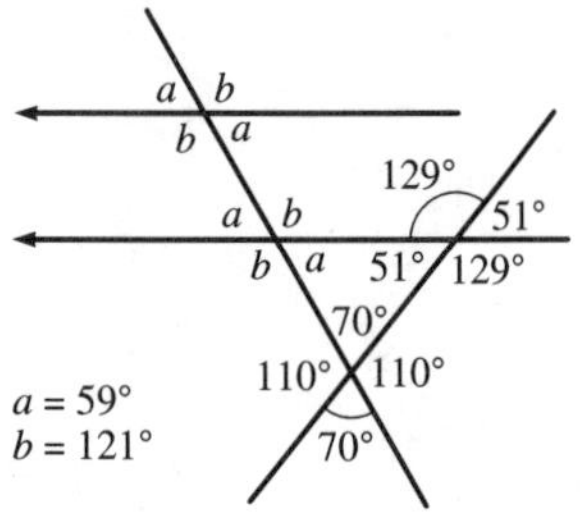

5 72°

6 60°

7 **a** 17° **b** 45° **c** 89°

8 **a** 130° **b** 20° **c** 90°

Test yourself 5.2

1 $\hat{a} = 60°$, $\hat{b} = 120°$, $\hat{c} = 60°$, $\hat{d} = 120°$, $\hat{e} = 60°$, $\hat{f} = 60°$, $\hat{g} = 30°$, $\hat{h} = 70°$, $\hat{i} = 70°$, $\hat{j} = 110°$, $\hat{k} = 30°$, $\hat{l} = 30°$, $\hat{m} = 120°$, $\hat{n} = 150°$, $\hat{o} = 60°$, $\hat{p} = 120°$, $\hat{q} = 30°$, $\hat{r} = 150°$, $\hat{s} = 20°$, $\hat{t} = 20°$, $\hat{u} = 20°$

2 At B turn 55° right, at C turn 90° left, at D turn 100° left, at E turn 135° right, at F turn 120° left

3 **a** c, d **b** a, b, f **c** f
d b, e **e** d, f **f** a, c

Test yourself 5.3

1 $\hat{a} = 100°$, $\hat{b} = 80°$, $\hat{c} = 90°$, $\hat{d} = 150°$, $\hat{e} = 55°$, $\hat{f} = 145°$, $\hat{g} = 120°$, $\hat{h} = 40°$, $\hat{i} = 50°$, $\hat{j} = 50°$, $\hat{k} = 75°$, $\hat{l} = 75°$, $\hat{m} = 30°$

2 360° (a complete turn, the same as for *all* polygons)

3 45°

4 **a** parallelogram **b** square **c** equilateral triangle
d rectangle **e** right-angled triangle **f** hexagon
g kite **h** obtuse-angled triangle **i** isosceles triangle

Chapter 6

Try these first (page 75)

1 11 250 m^2
2 23.37 m^2, 23 m^2 (2 sf)
3 297 000 mm^3
4 1750 m^3, 1.625 m
5 0.612 75 m^2
6 3 720 000 000 mm^3

Test yourself 6.1

1 125 990 mm^2, 126 000 mm^2 (2 sf)

2 1200 m^2, 1200 m^2 (2 sf)

3 6.24 m^2, 6.2 m^2 (2 sf)

4 49 200 mm^2, 49 000 mm^2 (2 sf)

5 17.857 m, 18 m (2 sf)

6 **a** 16 m^2 **b** 23.625 m^2 **c** 33.89 m^2

d 77.54 m^2 **e** 6.555 m^2

7 A = 9 m^2, B = 27 m^2, C = 45 m^2

Test yourself 6.2

1 48 000 mm^3

2 A = 6 000 000 mm^3, B = 8 064 000 mm^3, C = 5 670 000 mm^3. B holds the greatest volume, 0.005 67 m^3

3 0.0315 m^3

4 0.147 25 m^3, 0.15 m^3 (2 dp)

5 90 mm

Test yourself 6.3

1 11 850 mm^2, 5.9250 m^3

2 630 m^3

3 694 800 mm^3

4 1349.76 mm^3, 1300 mm^3

Chapter 7

Try these first (page 89)

1 $y = 6$

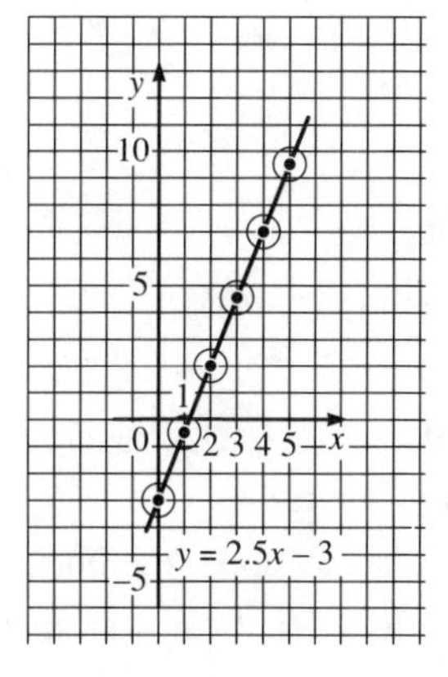

2 **a** $y = 2x + 3$

b $y = 12$

c $x = 2.5$

Test yourself 7.1

1

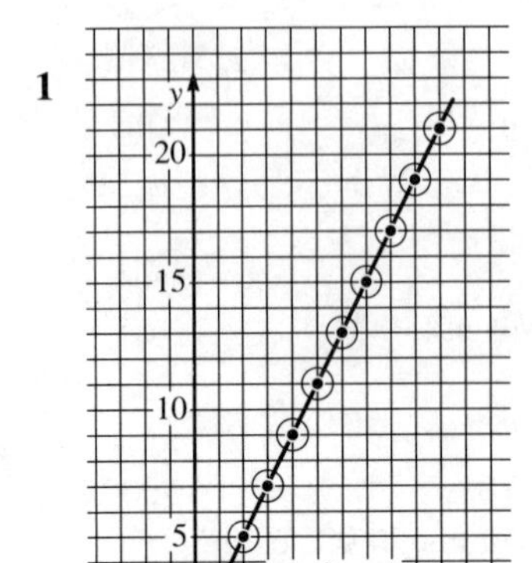

2

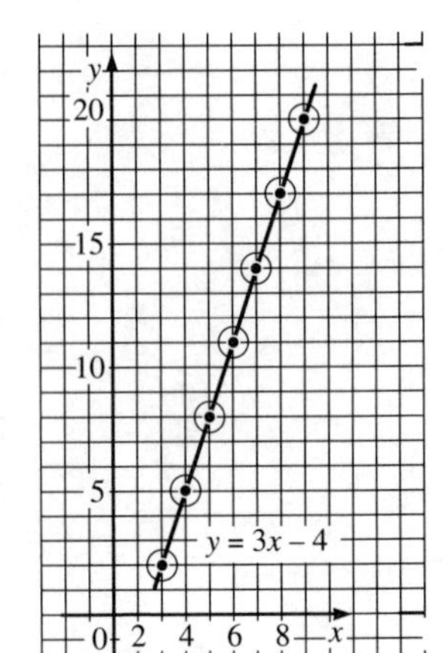

3

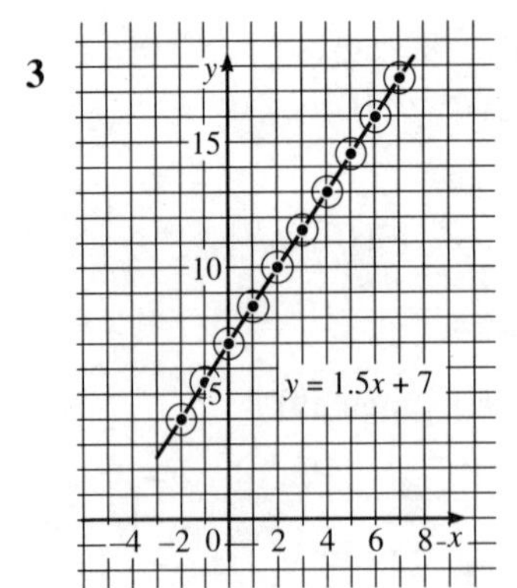

4

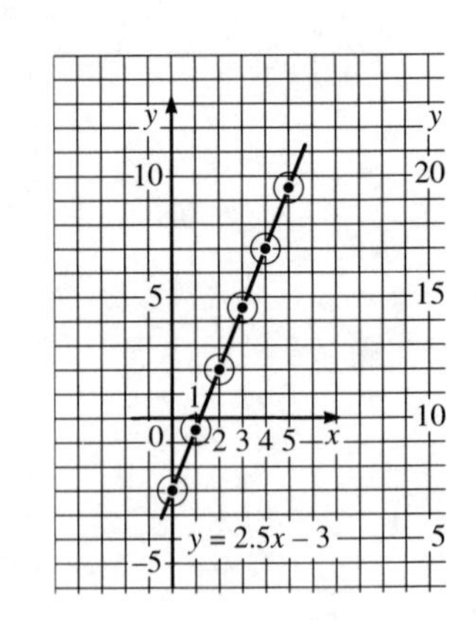

5 $y = 6$, $x = 6.5$

6 $y = -8$, $x = 3.5$

7 Steepest is **c**.

Least steep is **d**.

b slopes down to the right.

a and **e** have the same slope; the lines are parallel.

y-intercepts are: **a** 6 **b** 4 **c** –4 **d** 10 **e** –2 **f** 2

Test yourself 7.2

1–4

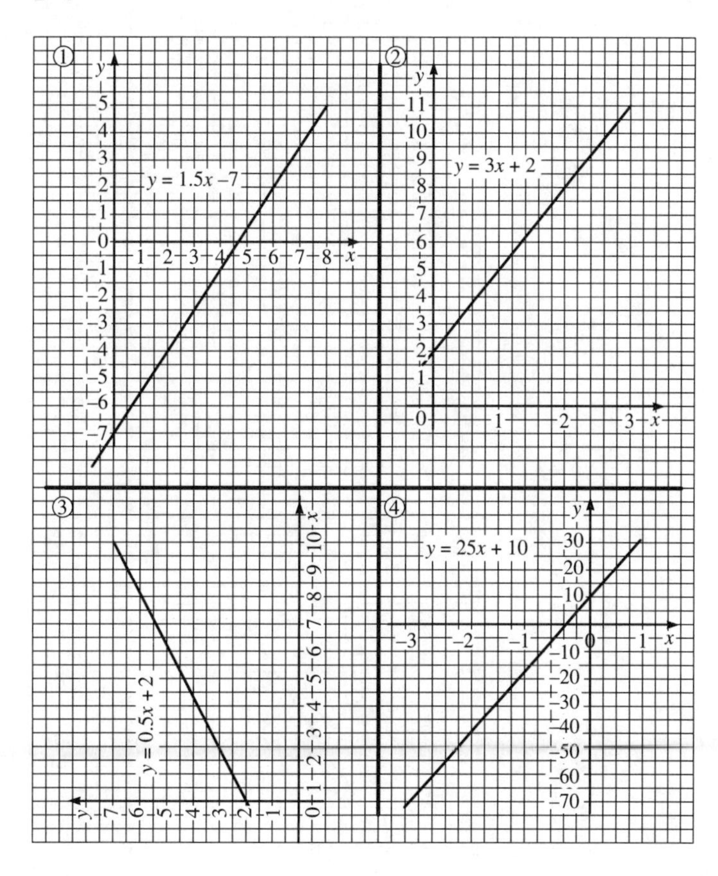

5–8

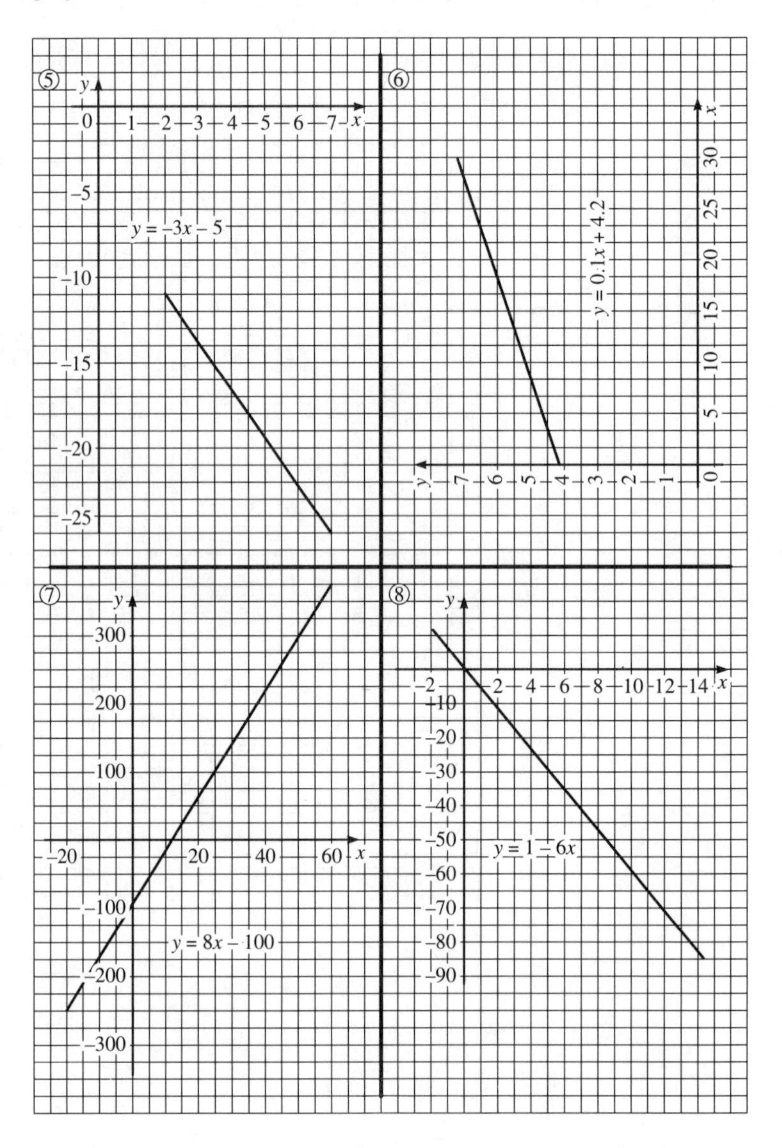

9 9.6 gallons, 12.5 h

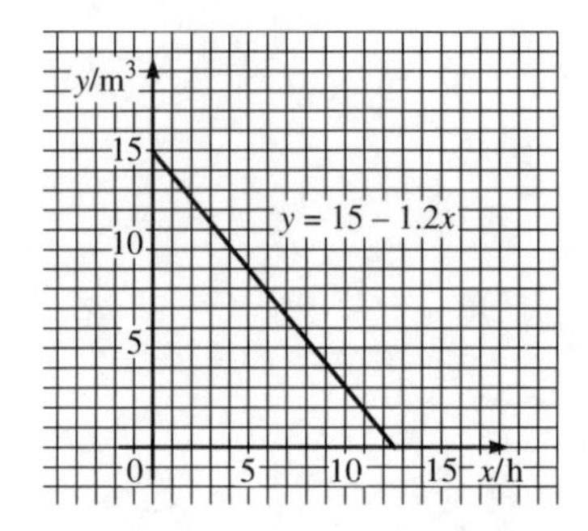

10 7 litres

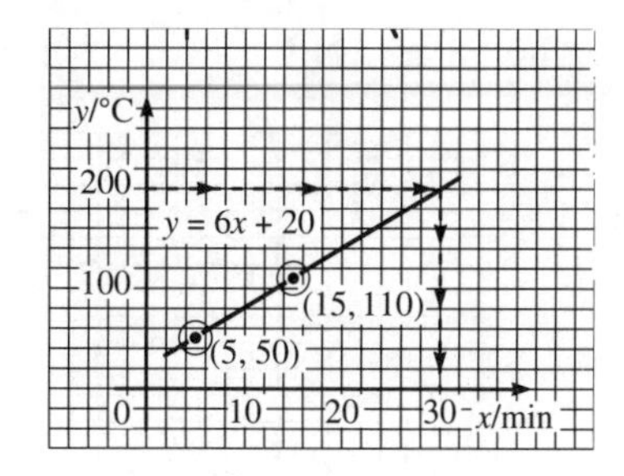

11 20°C, 30 min

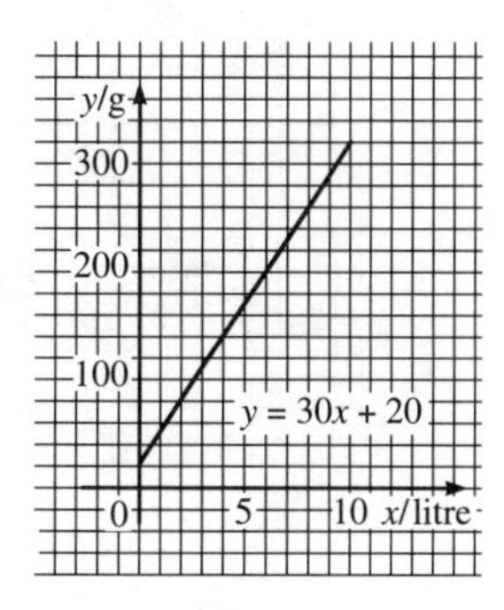

Chapter 8

On your own (page 113)

1 **a** 23.3 mm **b** 1.8 mm **c** $5300\,mm^2$

2

Margarine	1⅔ tablespoons
Water	6⅔ tablespoons
Eggs	5
Sugar	5/6 cups
Flour	2½ cups

3 **a** 775 bolts **b** £117.70 **c** M6, 9 p each

4 Arranged crossways, 27 boards

5 **a** $0.029\,m^3$ **b** 231 kg **c** 43 girders

6 13.5 m

7 $73\,500\,mm^2$

8 **a** 2.5 N **b** 9.5 N

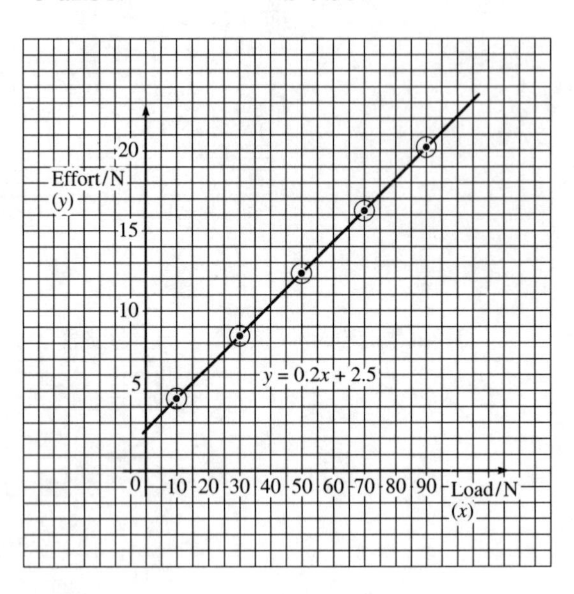

9 495 Ω at 0°C, 605 Ω at 75°C

10 **e** v = 0.009 35 m^3 (rounding the result to 3 sf means having 5 dp), w = 125 g

f h = 0.500 m, t = 3.00 m (zeros indicate 3 sf)

Chapter 9

Try these first (page 119)

1 354, 732, 5328, 73 281, 95 232
2 7, 17, 2, 47, 23
3 1, 2, 3, 5, 6, 9, 10, 15, 18, 30, 45, 90
4 35
5 1260

Test yourself 9.1

1 3, 23, 7, 59
2 6, 27, 123, 1110, 735
3 24, 624, 68, 112, 704
4 15, 100, 65, 4325, 25
5 56, 556, 60 016, 736, 72
6 108, 5301, 64 251, 5418

7 a 1, 3, 11, 33
b 1, 2, 5, 10, 25, 50
c 1, 3, 7, 9, 21, 63
d 1, 2, 3, 4, 6, 7, 12, 14, 21, 28, 42, 84
e 1, 2, 3, 4, 5, 6, 8, 10, 12, 20, 24, 40, 60, 120
f 1, 2, 4, 8, 16
g 1, 3, 5, 9, 15, 45
h 1, 2, 3, 4, 6, 12, 18, 36

Test yourself 9.2

1 a 2, 2, 3 **b** 3, 3, 5
c 7, 23 **d** 3, 7, 7
e 2, 2, 3, 7 **f** 3, 5, 5, 13
g 5, 11, 23 **h** 11, 11, 29
2 a 1, 2, 4, 8 **b** 1, 2, 3, 6, 9, 18
c 1, 2, 3, 4, 6, 9, 18, 36 **d** 1, 3, 5, 11, 15, 33, 55, 165

Test yourself 9.3

1 a 2 **b** 33 **c** 4
d 6 **e** 17 **f** 13
g No HCF **h** 14 **i** 33
2 a Cancel 14, giving 1/7
b Cancel 21, giving 2/5
c Cancel 6, giving 5/13
3 a 60 **b** 198 **c** 504
d 1104 **e** 210 **f** 7134
g 7560 **h** 5640 **i** 3420

Test yourself 9.4

1 a $1\frac{1}{2}$ **b** $\frac{1}{12}$ **c** $1\frac{2}{15}$
2 a 51/70 **b** 29/45 **c** 13/15

Chapter 10

Try these first (page 131)

1 4:3 **2** 1:25 000, 1/25 000, 925 m
3 135 mm **4** 100°, 120°, 140°
5 26 mm^3 **6** 75.6 g
7 1.7 A

Test yourself 10.1

1 12:3, 4:1 **2** 500:750, 2:3

3 a 1:5 **b** 10:1 **c** 3:1 **d** 2:3

e 6:1 **f** 1:7:2 **g** 3:10:1 **h** 2:7:5

4 175:75:2

Test yourself 10.2

1 187.5 mm **2** 56 g

3 Door: 26.3 mm high, 10.5 mm wide. House: 118 mm high

4 402, 950 **5** 1:87, 1/87, 23.0 mm, 103 mm

6 1:72, 1/72 **7** 1:50 000, 1/50 000, 2.9 km, 17.5 mm

8 15:1, 6 times **9** 240 V AC

10 0.030

Test yourself 10.3

1 12 rainy, 18 dry **2** A has 0.8 kg, B has 1.2 kg

3 0.84 m, 1.26 m, 1.40 m **4** 24, 16, 12, 12

5 8.3 kg

Test yourself 10.4

1 16.7% **2** 17%, 41%, 42%

3 8.75 kg **4** 15 ml

5 10.2 m **6** 2.38%

7 39.6 μF, 17.6 μF

8 a 1/2 **b** 3/4 **c** 1/3 **d** 1/50 **e** 3/20

9 a 20% **b** 27% **c** 0.083% **d** 98% **e** 37.21%

Test yourself 10.5

1 969 g **2** 96 mm, 110 mm

3 18 parts **4 a** 480 times per minute **b** 180 times per minute

5 50 m, 3 MHz

Chapter 11

Try these first (page 149)

1 $3ap - 6p - 9a - 9$
2 $x^2 - 2x - 15$
3 $\dfrac{n + 2}{n - 3}$
4 $2a(a - 2)(a - 1)$
5 $4(3 - y)(3 + y)$
6 $(x - 3)(x - 5)$

Test yourself 11.1

1 a $21xy$ **b** $10ab$
c $6a^2$ **d** $6x^2y$
e $4n^3$ **f** $20x^2y^2$
2 a $6 + 9x$ **b** $20 - 8a + 4b$
c $6n + 2n^2$ **d** $12pr - 8p^2 + 16p$
e $-10j^2 + 6j^3$ **f** $2st - 3t^3$
3 a $x^2 + 3x + 2$ **b** $a^2 + 6a + 9$
c $t^2 + 7t + 12$ **d** $n^2 + 7n + 10$

Test yourself 11.2

1 a $a^2 + 12a + 35$ **b** $n^2 + 8n + 16$
c $x^2 + 3.9x + 3.6$ **d** $p^2 - p - 6$
e $n^2 - 2n - 8$ **f** $q^2 + 4q - 12$
g $r^2 + 4r - 21$ **h** $d^2 + 8.5d - 15$
2 a $k^2 - 7k + 12$ **n** $n^2 - 3n + 2$
c $h^2 - 4$ **d** $a^2 - 14a + 49$
e $b^2 - 49$ **f** $s^2 - 4s + 3$
g $x^2 - 1$ **h** $r^2 - 1.9r + 0.7$
3 a $2x^2 + 16x + 24$ **b** $3a^2 + 24a + 21$
c $n^3 + 4n^2 + 3n$ **d** $4k^2 + 4k - 24$
e $5g^2 - 45$ **f** $2x^3 - 22x^2 + 48x$
g $ab^2 + 5ab + 4a$ **h** $x^3 + 6x^2 + 11x + 6$
4 a $a + ab + b + 1$ **b** $xy - 4x + 2y - 8$
c $mn - 2m - 7n + 14$ **d** $a^2 + 3ab + 2b^2$
e $c^2 - d^2$ **f** $2a^2 - a - 6$
g $12x^2 + 5x - 2$ **h** $6n^3 - 9n^2 + 3n$

Test yourself 11.3

1 a $2(1 + 2a)$ **b** $3(2 + n)$
c $a(1 - 2a)$ **d** No factors
e $2a(2a + 3)$ **f** $2y(x - 2 + 5y)$
2 a $3a(2b - c)$ **b** No factors
c $x(2x^2 + 3xy - 4y)$ **d** $3ab(3a + 2b + 1)$
e $2s(t + 2s)$ **f** $2p(3p + q)$

Test yourself 11.4

1 a $(x + 2)(x + 1)$ **b** $(x + 7)(x + 1)$
c $(a + 3)(a + 4)$ **d** $(n - 2)(n - 3)$
e $(r - 1)(r - 1)$ **f** $(b - 3)(b - 8)$
g $(q + 4)(q - 4)$ **h** $(x + 4)(x + 9)$
2 a $(k + 2)(k - 1)$ **b** $(n - 2)(n + 5)$
c $(a + 1)(a - 2)$ **d** $(q + 4)(q + 5)$
e $(b + 1)(b - 12)$ **f** $(j - 7)(j + 7)$
g $(t - 2)(t + 9)$ **h** $(p - 3)(p - 3)$
3 a $(y - 1)(y + 3)$ **b** $(6 - f)(6 + f)$
c No factors **d** $2(a - 2)(a + 5)$
e $d(d - 11)(d + 1)$ **f** $3(x - 3)(x + 3)$
g No factors **h** $2q(p + 2)(p - 1)$
i No factors **j** $4(n - 3)(n + 2)$

Test yourself 11.5

1 a $\frac{b}{2}$ **b** $\frac{2x}{y}$

c $\frac{x - 3}{2x - 1}$ **d** $\frac{x - 2}{x^2 - y}$

e $\frac{b}{b + 2}$ **f** $\frac{p + q}{4p + 3q}$

g $\frac{y}{y + 2}$ **h** $\frac{s + 2}{s - 1}$

i $\frac{t - 3}{t + 3}$ **j** $\frac{n - 5}{n + 5}$

Chapter 12

Try these first (page 159)

1 a 2 **b** 3
c 9

2 a $q = \frac{4}{3}(p-1)$ **b** $S = \frac{2}{3}R - \frac{T}{2}$

3 $x \leqslant 2$, 2

4 $n > -2$, -1

Test yourself 12.1

1 a 7 **b** 4
c 6 **d** 16
e −2 **f** 4
g −2 **h** 1

2 $x - 5 = 12$, $x = 17$

3 $x + 6 = 15$, $x = 9$

Test yourself 12.2

1 a 3 **b** 8
c 4 **d** 0
e −1 **f** 12
g 11 **h** −2
i 3 **j** 5

2 Working in minutes after 12, $x + 6 = 32$, $x = 26$. Cooking started at 12.26

3 $x - 3 = 7$, $x = 10$

Test yourself 12.3

1 a 3 **b** 2
c 4 **d** 5
e 3 **f** 4
g −2 **h** 3
i 1 **j** 2
k 2 **l** 0

2 $5x + 50 = 1300$, $x = 250$

3 $6x + 1.5 = 15$, $x = 2.25$

Test yourself 12.4

1 6
2 7
3 2
4 10
5 21
6 –8
7 9
8 –2
9 3
10 12
11 7
12 –3
13 5
14 5

Test yourself 12.5

1 a $u = v - at$
b $s = W + t$
c $x = a - y - z$
d $x = \dfrac{y-2}{3}$
e $b = \dfrac{3}{2a}$
f $B = \dfrac{AD - C}{2}$
g $z = \dfrac{y}{x} - w$
h $K = 4J - 1$

2 $A = 1.84\,\text{m}^2$, $W = A/L$, $W = 7\,\text{m}$

3 $z = 6(x + 3y)$, 102 lamps, $y = \dfrac{z}{18} - \dfrac{x}{3}$

or $y = \dfrac{z - 6x}{18}$, $y = 4$

Test yourself 12.6

1 $n \geqslant 4$, where n is an integer
2 $15 \leqslant t \leqslant 24$
3 $0 \leqslant v \leqslant 15$
4 $12.75 \leqslant n < 12.85$
5 $0 \leqslant n \leqslant 12$, where n is an integer
6 a $a \geqslant 1$
b $6 > j$
c $n \leqslant -2$
d $3 < x$

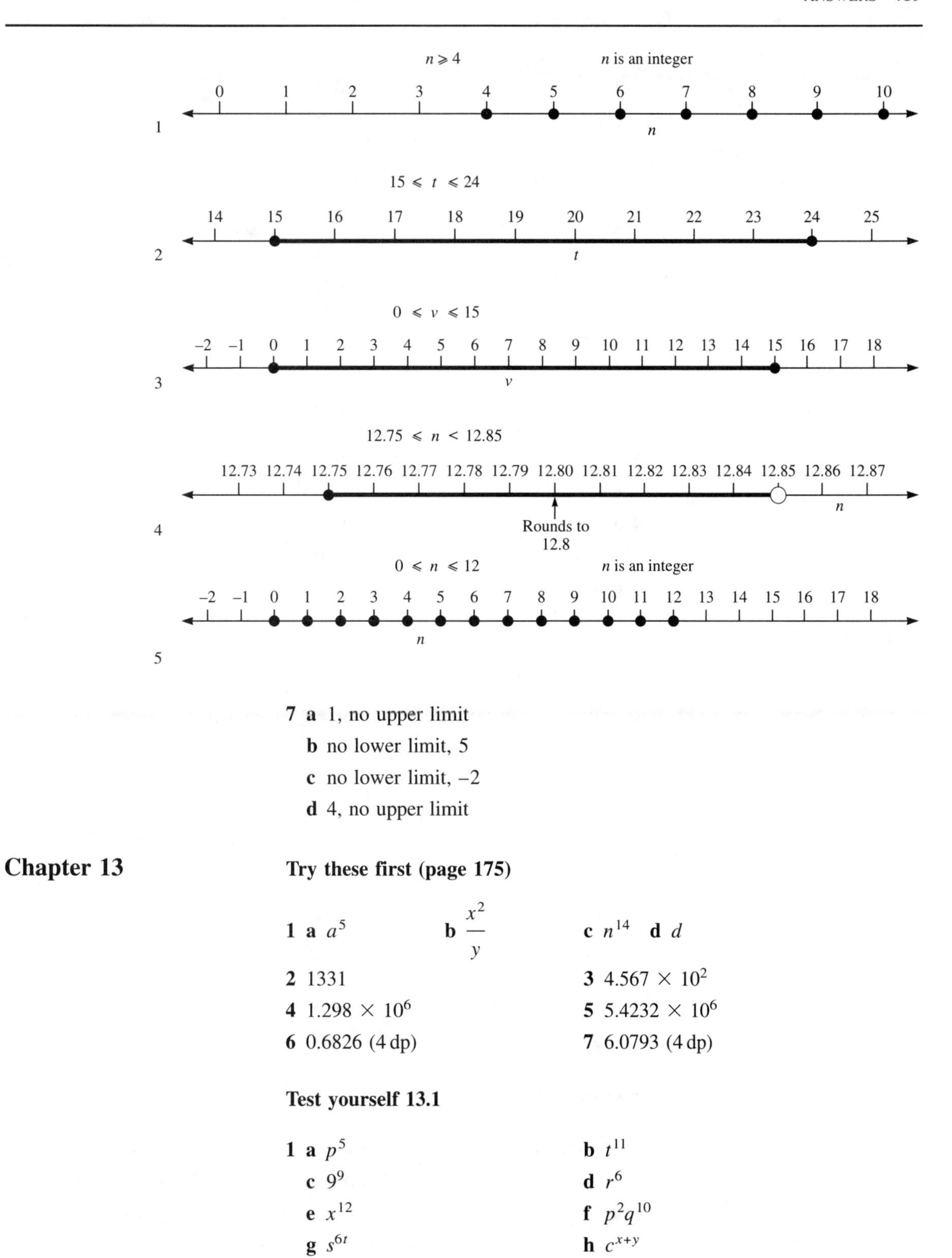

7 a 1, no upper limit
b no lower limit, 5
c no lower limit, −2
d 4, no upper limit

Chapter 13

Try these first (page 175)

1 a a^5 **b** $\dfrac{x^2}{y}$ **c** n^{14} **d** d

2 1331

3 4.567×10^2

4 1.298×10^6

5 5.4232×10^6

6 0.6826 (4 dp)

7 6.0793 (4 dp)

Test yourself 13.1

1 a p^5
b t^{11}
c 9^9
d r^6
e x^{12}
f p^2q^{10}
g s^{6t}
h c^{x+y}

2 a n^{10} **b** a^{14}
c n^{20} **d** k^{16}
e 8^8 **f** x^{3a}

Test yourself 13.2

1 a n^5 **b** $3d^4$
c q^9 **d** $5x$
e t^2 **f** r^4
2 a $\frac{1}{g^7}$ **b** $\frac{1}{a^5}$
c $\frac{1}{y}$ **d** $\frac{1}{n^9}$
3 a a^{-2} **b** b^{-7}
c p^{-14} **d** q^{-1}
4 a w^{-3} **b** g^{-4}
c f^{-2} **d** d^{-14}
e a^2 **f** s
g v **h** $6q^{-1}$

Test yourself 13.3

1 a $\sqrt[4]{x}$ **b** $\sqrt[4]{n^3}$ or $(\sqrt[4]{n})^3$
c $\sqrt{a^3}$ or $(\sqrt{a})^3$ **d** $1.4 = 7/5$, $\sqrt[5]{b^7}$ or $(\sqrt[5]{b})^7$
e $\frac{1}{\sqrt{f}}$ **f** $\frac{1}{\sqrt[3]{t^2}}$ or $\frac{1}{(\sqrt[3]{t})^2}$
2 a $n^{1/5}$ **b** $a^{2/3}$
c $t^{3/7}$ **d** $s^{4/9}$
e $x^{2/5}$ **f** $\frac{1}{k^{2/3}}$ or $k^{-2/3}$
3 a 11 **b** 5
c 8 **d** 9

Test yourself 13.4

1 a 1000 **b** 1 000 000 **c** 10
d 0.001 **e** 0.0001 **f** 0.000 001
2 a 10^2 **b** 10^7 **c** 10^4
d 10^{-1} **e** 10^{-4} **f** 10^{-6}

3 a 723 b 532 000 c 0.0072
d 0.6803 e 8000.03 f 0.000 000 071

4 a 1.215×10^7 b 1.089×10^8 c 73
d 3.4×10^{-4} e 7.8×10^2 f 5.7811×10^4

Test yourself 13.5

1 a 2 b 3 c 6

2 a 0.6021 b 1.8573 c 4.2767

3 a 1.5440 b 4.5203

Chapter 14

Try these first (page 190)

1 △ABC and △PQR are congruent. All three triangles are similar.

2 a 0.9205 b 87.68° c 0.6428
d 62.87° e 0.5317 f 18.66°

3 0.82 m

4 AB = 4.06 m, BC = 2.16 m

Test yourself 14.2

2 C = 56°, AB:BC is 2:3, tan C = 1.5

3 41 mm, 0.7536

4 a 3.0777 b 0.6009 c 0.2309
d 0.0000 e 1.0000 f 57.2900

5 a 52.43° b 26.57° c 42.00°
d 88.09° e 1.15° f 65.56°

6 53.75°

7 40.52 m

8 1.19 m

9 60.49°

10 The line is not straight. It curves up as x increases, becoming steeper with increasing values of x.

Test yourself 14.3

1 a 0.8988 b 0.4695 c 0.0872
d 0.0000 e 0.7071 f 1.0000

2 a 23.58° b 30.00° c 5.74°
d 60.00° e 14.48° f 64.16°

3 2.08 m

4 59.73°

5 15.10 m

6 The line is not straight. It starts at the origin ($y = 0$ when $x = 0°$). As x increases, the line slopes up, steeply at first, then gradually less steeply, until it reaches $y = 1$ when $x = 90°$.

Test yourself 14.4

1 a 0.9782 **b** 0.6820 **c** 0.1219
d 1.0000 **e** 0.7071 **f** 0.0000

2 a 45.00° **e** 75.52° **f** 28.36°

3 The line is not straight. It starts from $y = 1$ when $x = 0°$. As x increases, the line slopes down, gently at first, then gradually steeper, until it reaches the x-axis when $x = 90°$. Sin x = cos x where the lines cross, when $x = 45°$.

4 $\hat{x} = \sin^{-1}(0.1/2.5) = 2.29°$

5 a AE = 40/cos 35° = 48.83 m
b DE = 40 × tan 35° = 28.01 m

6 AC = 2.3/sin 26° = 5.25 m
BC = 2.3/tan 26° = 4.72 m

7 $\hat{X} = \sin^{-1}(10.4/15.7) = 41.48°$
$\hat{Z} = \cos^{-1}(10.4/15.7) = 48.52°$ (or $\hat{Z} = 180 - 90 - \hat{X} = 48.52°$)

8 $\hat{P} = \tan^{-1}(2/5) = 21.8°$
$\hat{R} = \tan^{-1}(5/2) = 68.20°$ (or $\hat{R} = 180 - 90 - \hat{P} = 68.20°$)

Chapter 15

Try these first (page 203)

1 141 mm

2 67.3 mm

3 radius = 39.79 mm, area = 4974 mm^2

4 916 mm^2, 36.65 mm

Test yourself 15.1

1 a 14.1 m, 14 m **b** 13.2 m, 13 m **c** 6.28 mm, 6 mm
d 4.08 km, 4 km **e** 21.05 m, 20 m **f** 911 mm, 900 mm

2 $c = 2\pi r,\ r = \dfrac{c}{2\pi}$

3 a 2.31 m **b** 0.51 mm
c 0.88 m **d** 8.69×10^2 km

4 918 times

5 13.4 m

Test yourself 15.2

1 **a** 20.9 mm **b** 1.26 m **c** 373.8 mm

2 0.52 m, 0.105 m^3

3 **a** 3848 mm^2 **b** 141 m^2 **c** 1963 mm^2
d 7.07 mm^2 **e** 7162 mm^2 **f** 6.36×10^{-5} mm^2

4 66.96 m^2

5 15, 15 773 mm^2, 225 mm, 3, 4205 mm^2, 5.72%

6 1.6 m

7 393 000 mm^2

Chapter 16

Try these first (page 214)

1 9.9 m

2 16 m

3 **a** $1/\sqrt{2} = 0.71$ **b** 0.5 **c** $1/\sqrt{3} = 0.58$

Test yourself 16.1

1 10 m
2 17.0 m
3 5.83 m
4 7.14 m
5 13.6 mm
6 AD = 26.46 CD = 14.14

Test yourself 16.2

1 BC = AB/sin 60° = 289 mm
AC = AB/tan 60° = 144 mm

2 height = 0.5/sin 60° = 0.433 m
area = 0.108 m^2

3 AB = 56.6 mm

Chapter 17

Try these first (page 218)

1 $m = 3.5$, x-intercept = 2, y-intercept = $c = -7$.

2 25 m^2, after 83 min. A finished first (100 min), B next (104 min).

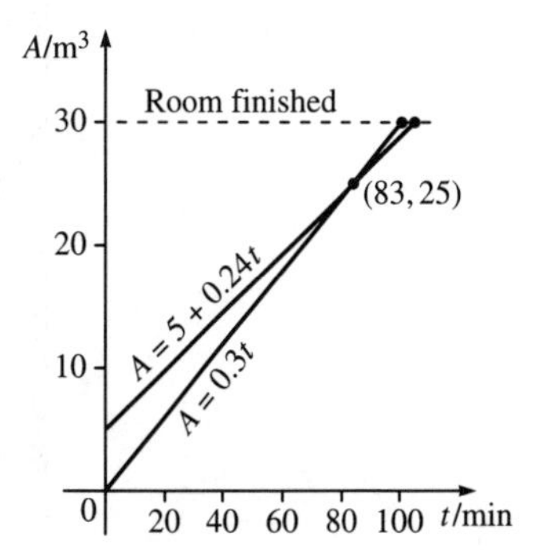

3 $x = 4,\ y = 13$
$x = -2,\ y = 1$

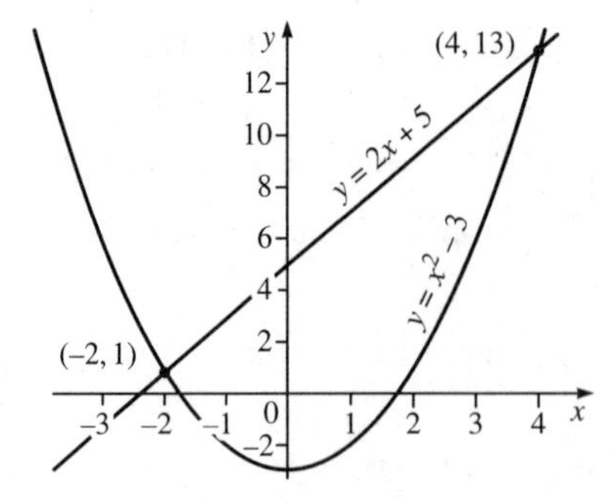

4 Greatest possible integer value of y is 3 at point A(1,3). At point (1,4) the line for the second equation is dashed, so y cannot equal 4.

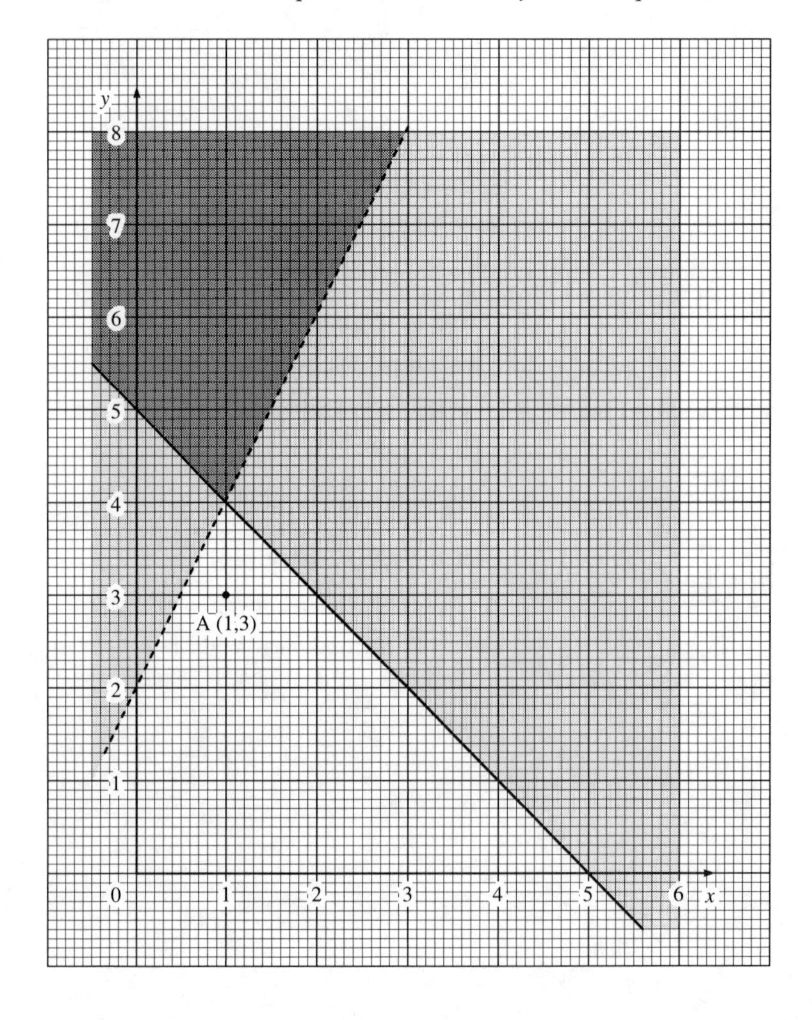

5 This is an exponential relationship, as shown by the straight line obtained when we plot ln y against x. a = 3, b = 0.3, $y = 3e^{0.3x}$.

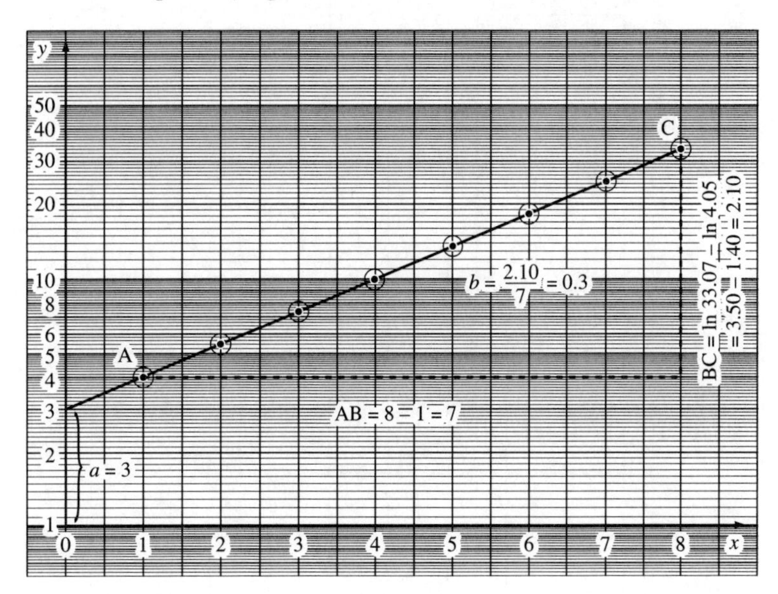

Test yourself 17.1

Answers give c followed by m.

1 a 1, 2 **b** –3, 3 **c** 4, 0.5

d 6, 1 **e** 5, 0 **f** –3, 1/3

2 a 2, 8, up **b** 8, –7, down **c** 7, –5, down

d 10, –6, down **e** –2.5, 3, up **f** 3, –1.2, down

3 See Figure A24

4 See Figure A25

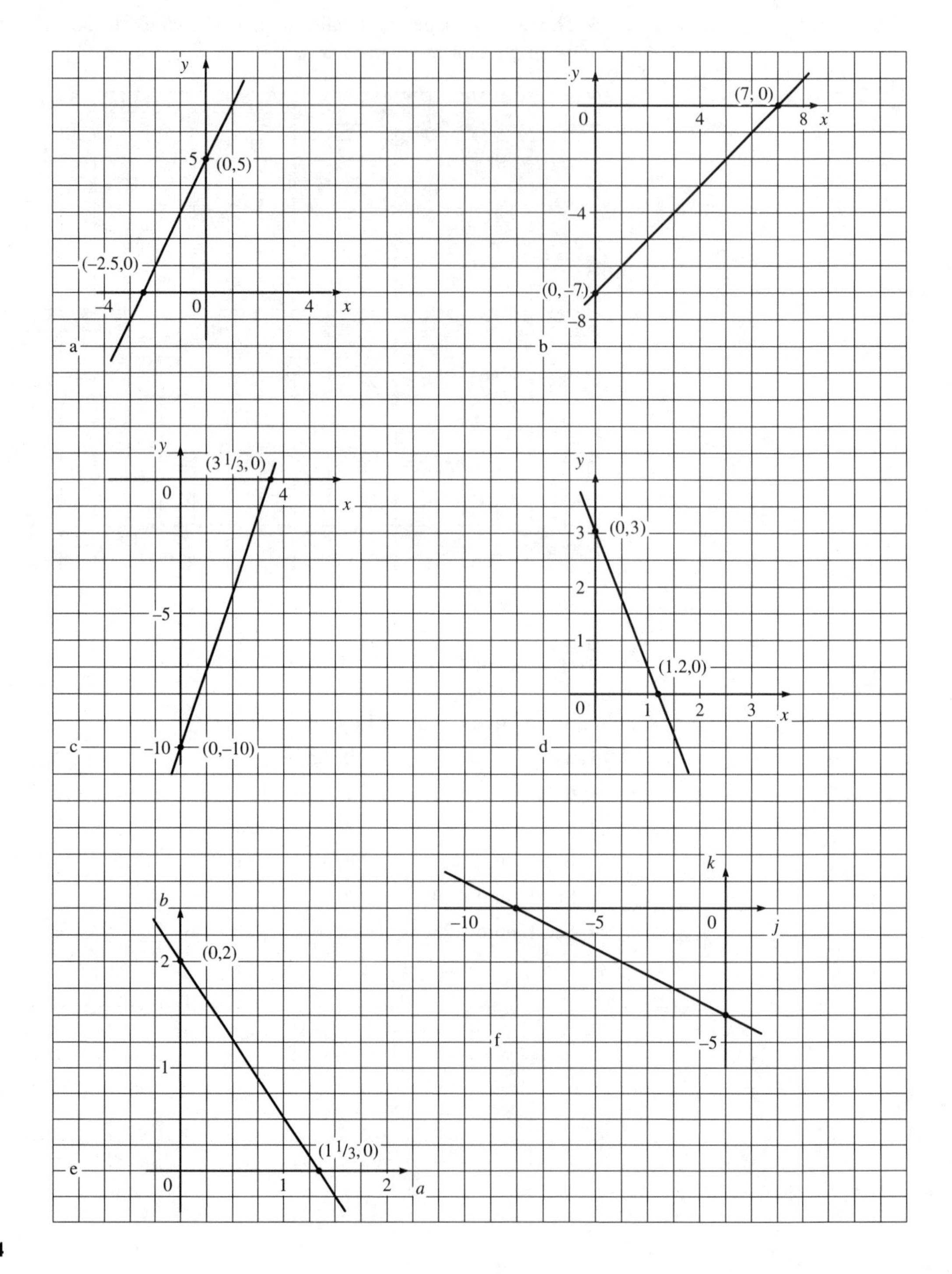

a
y
x
(0,5)
5
(−2.5,0)
−4
0
4
b
y
x
(7, 0)
0
4
8
−4
(0, −7)
−8
c
y
x
(3 1/3, 0)
0
4
−5
−10
(0,−10)
d
y
x
(0,3)
3
2
1
(1.2,0)
0
1
2
3
e
b
a
(0,2)
2
1
(1 1/3, 0)
0
1
2
f
k
j
−10
−5
0
−5

Figure A.24

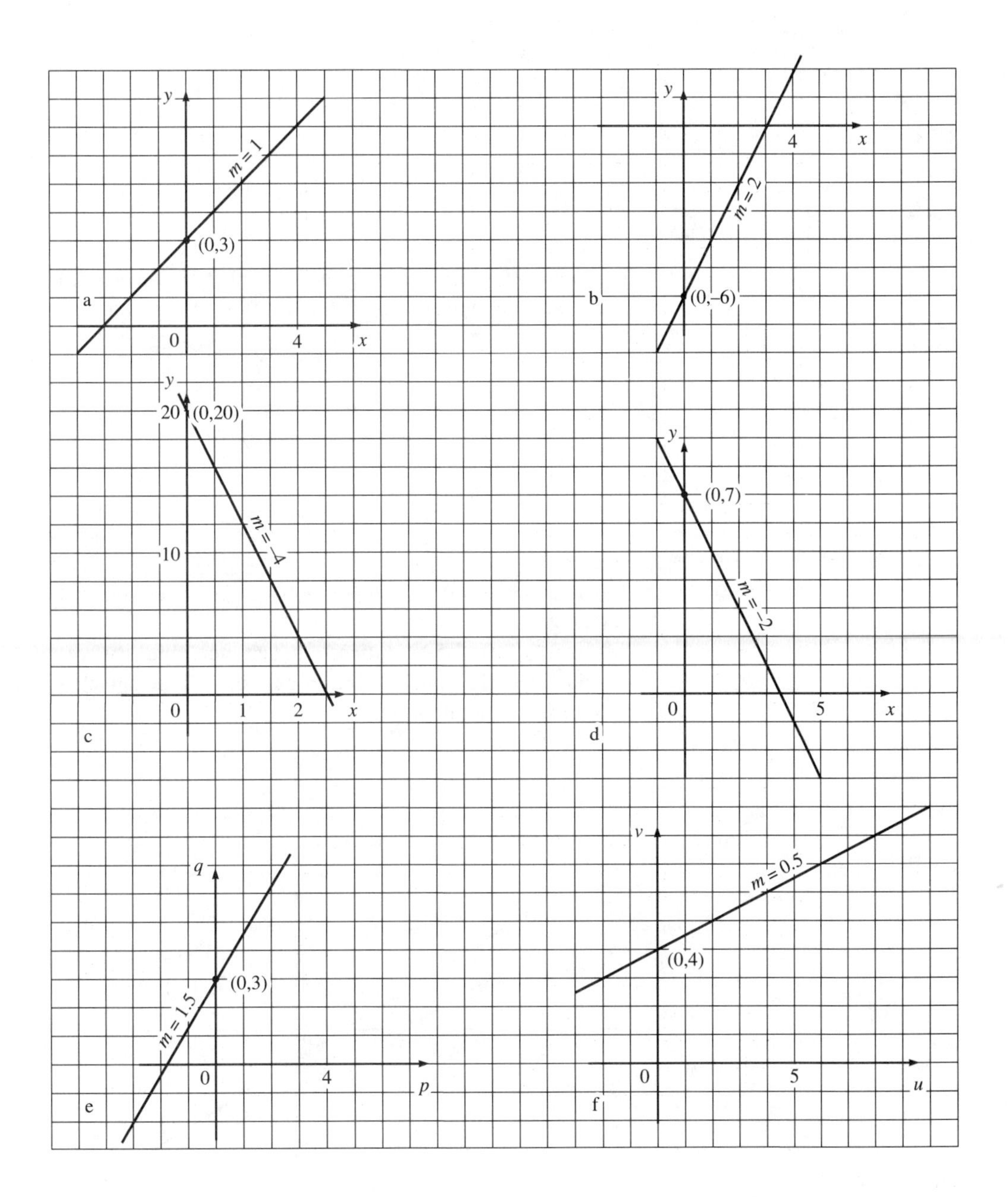
a
y
x
0
4
(0,3)
m = 1
b
y
x
4
(0,–6)
m = 2
c
y
x
20
10
0
1
2
(0,20)
m = –4
d
y
x
0
5
(0,7)
m = –2
e
q
p
0
4
(0,3)
m = 1.5
f
v
u
0
5
(0,4)
m = 0.5

Figure A.25

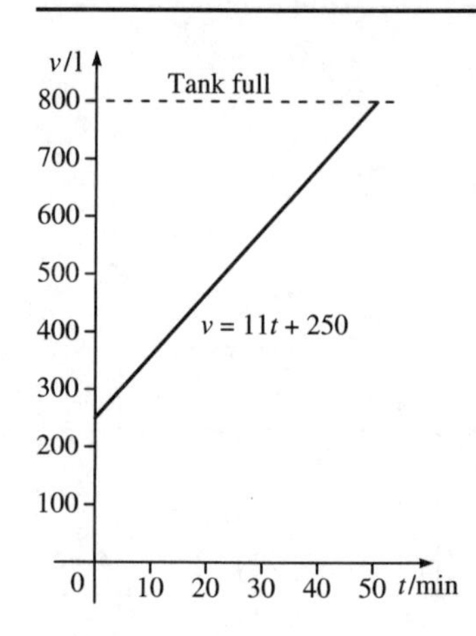

Test yourself 17.2

1 y-intercept = 250 (the initial volume of water in the tank, in litres). Gradient = 11 (the rate of supply of water, in litres per minute). When $t = 50$, the tank is full.

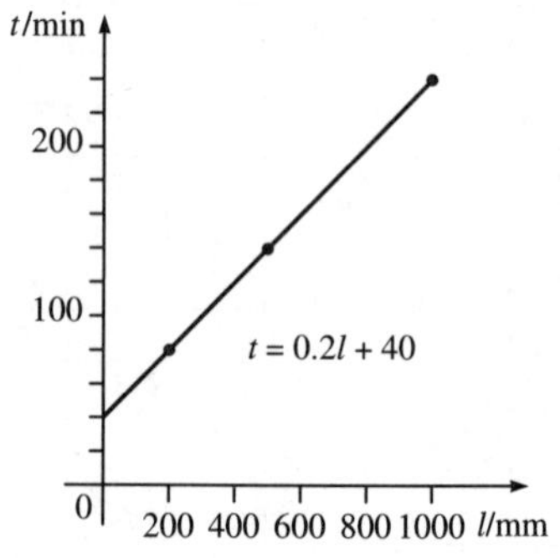

2 y-intercept is the setting-up time (40 min). Gradient = 0.2 (the rate at which the fabric is woven). It takes 240 min to weave 1 m.

3 x-intercept = −0.125 (input voltage for zero volts output, input offset voltage)
y-intercept = −1.5 V (output voltage when input is 0 V)
gradient = −12 (amplifier has a gain of −12; negative sign means that this is an inverting amplifier)

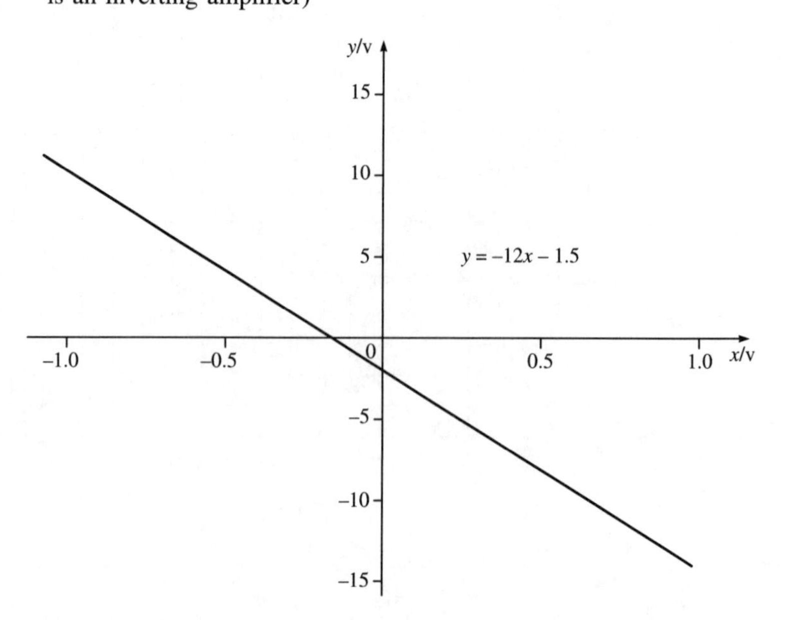

Test yourself 17.3

1 $x = 2, y = 5$

2 $x = 8, y = 8$

3 $x = -2, y = 4$

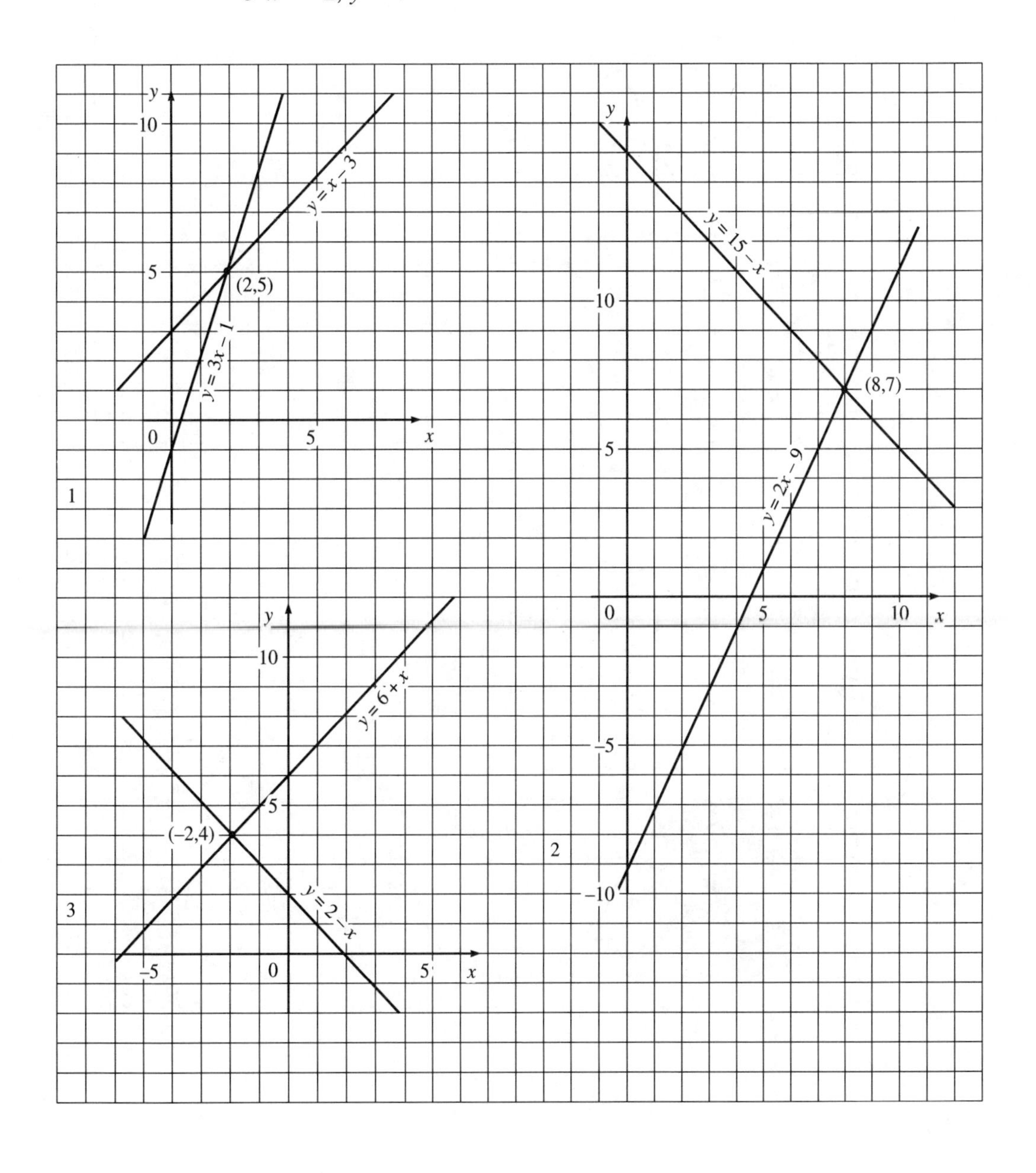

Test yourself 17.4

1 a no **b** yes **c** yes **d** no (dashed line)

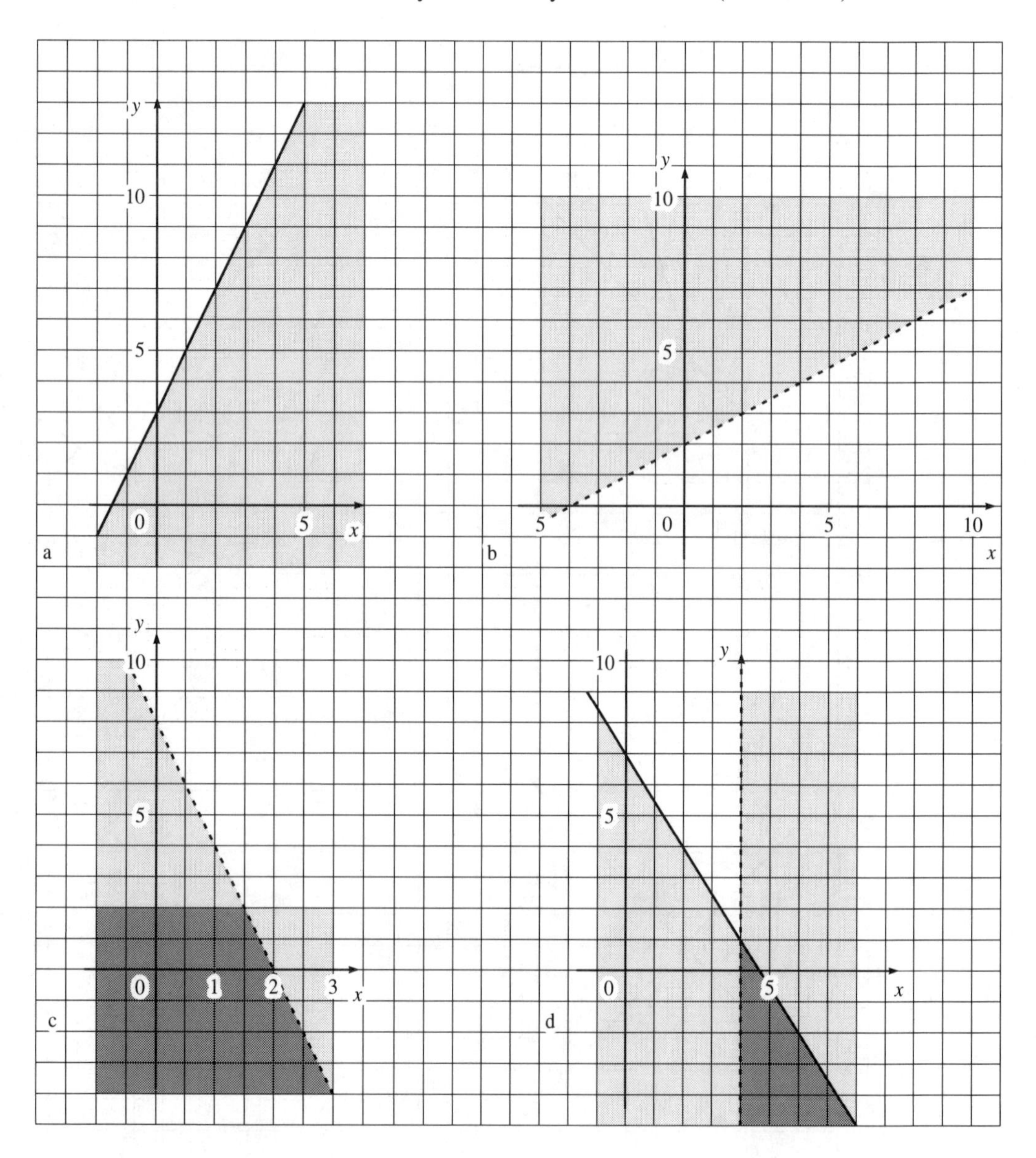

2 a $y \geqslant x$

b $y < 6 - 0.5x$

c $y \leqslant 3x - 13$

d $y < 1 + 2x$
$y \leqslant 9$

3 $a + v \leqslant 10$. Could also specify that $a \geqslant 0$ and $v \geqslant 0$ (not shown in the figure)

4 $y \geqslant 3x + 5$

$x \leqslant 10 \quad y \leqslant 50$

Could also specify that $y \geqslant 0$ and $x \geqslant 0$ (not shown in the figure)

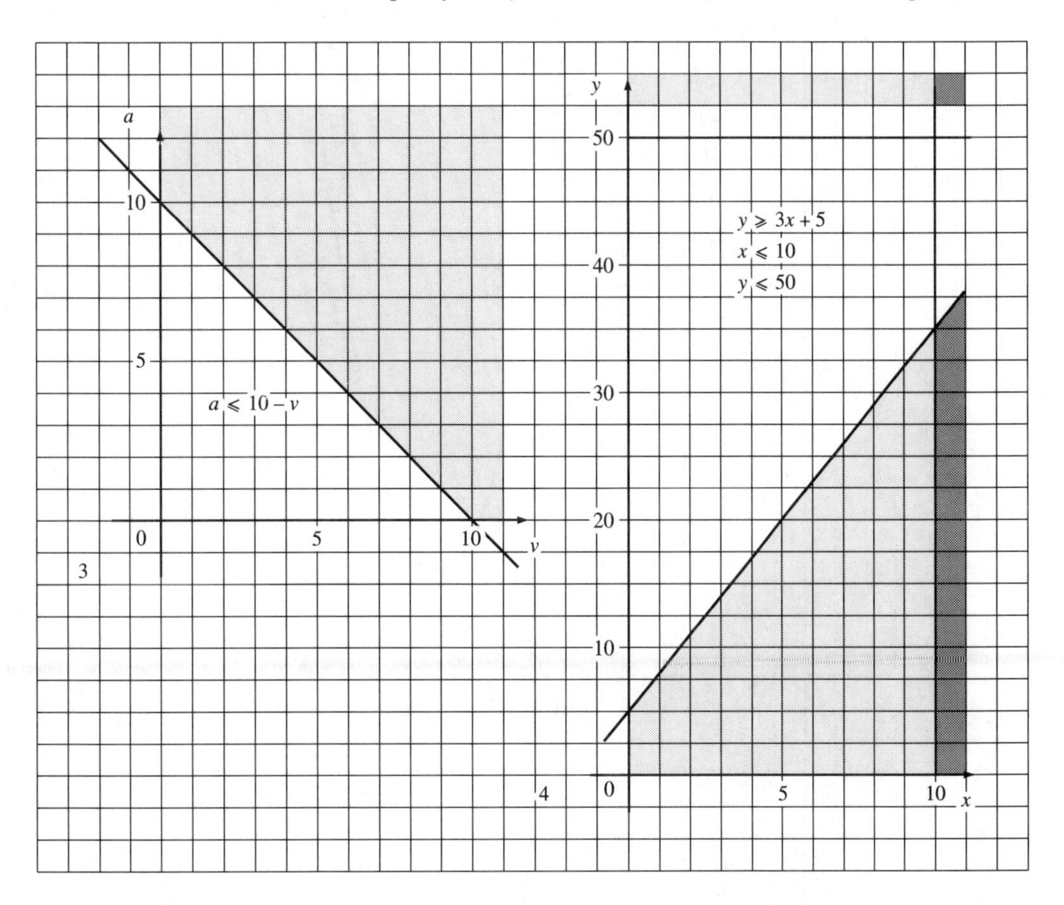

Test yourself 17.5

1

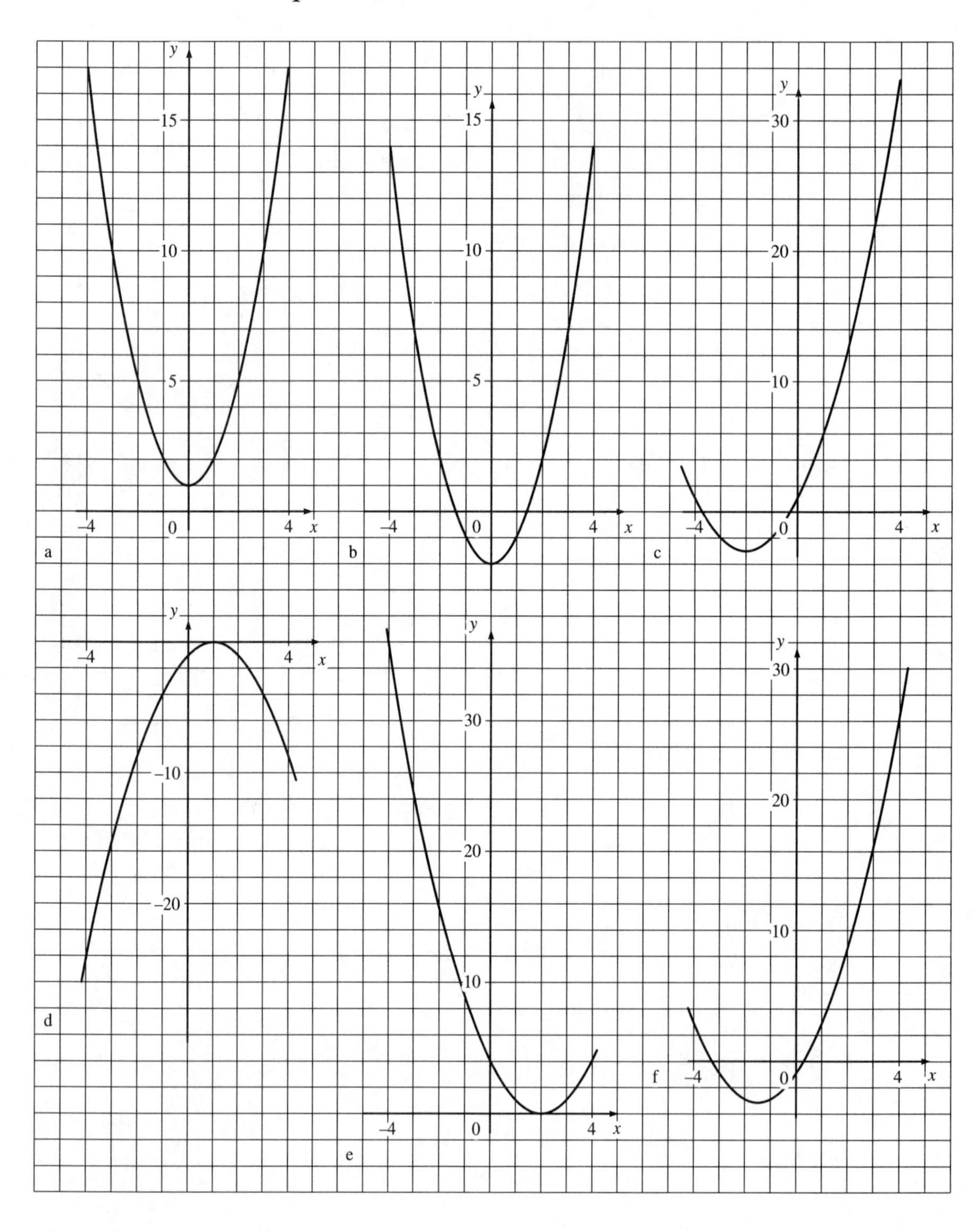

2 Figure A33a is the graph of the original data, shaded as an inequality. It appears to be part of a parabola. Figure A33b is the graph of b against v^2. This is a straight line, confirming that b is proportional to v^2. For example, doubling the speed makes the braking distance *four times* longer. From the graph, $c = 0$, $m = 0.01$. The equation is: $b = 0.01v^2$.

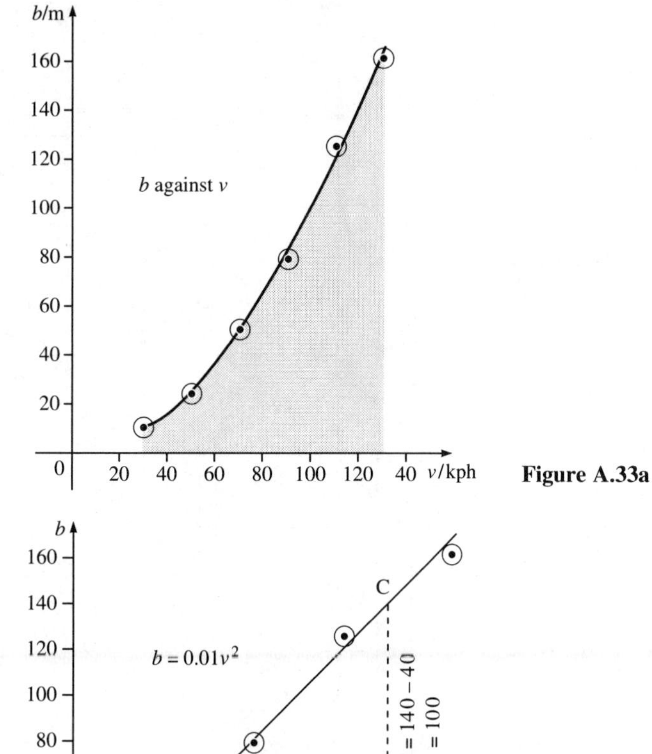

Figure A.33a

Figure A.33b

Test yourself 17.6

1 y-intercepts are:

a 0.01 **b** 1 **c** 2

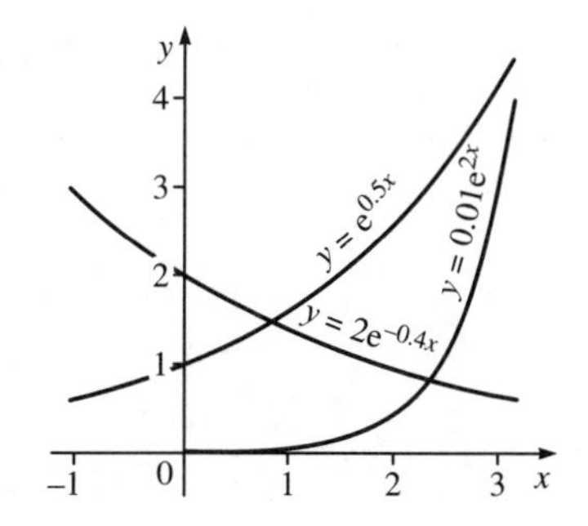

2 Plotted on semi-log paper (or by plotting natural logs on ordinary paper) we obtain a straight line. $a = 3$, $b = 0.2$, $y = 3e^{0.2x}$ (Figure A.35a)

3 $a = 2$. The line slopes down to the right, so has a negative gradient: $b = -0.5$, $y = 2e^{-0.5x}$ (Figure A.35b)

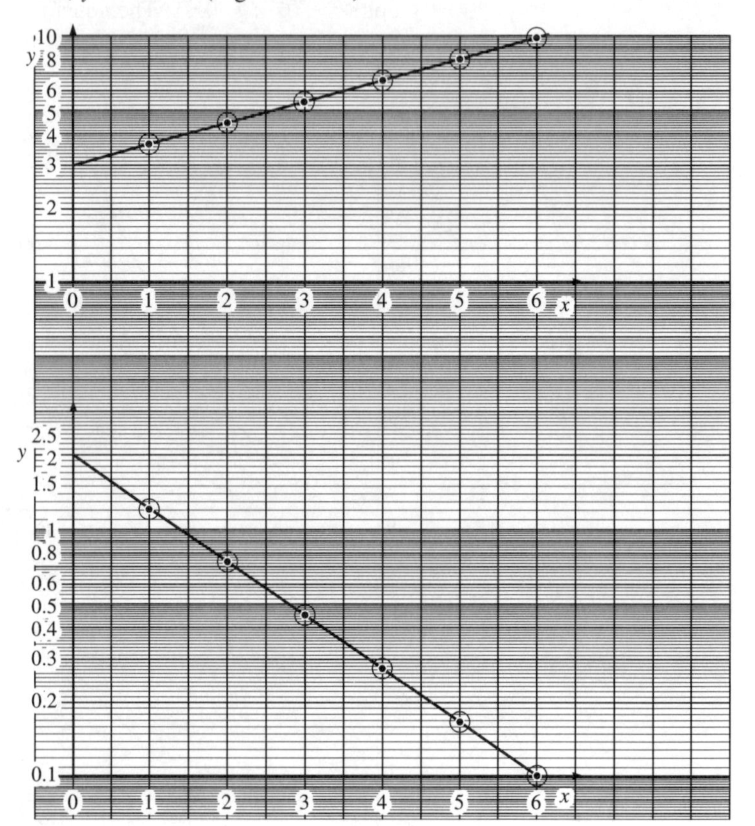

Figure A.35a

Figure A.35b

Chapter 18

Try these first (page 253)

1 4660 mm^2 **2** 8.53 mm

3 1600 mm^3 **4** 8.64 m^2

Test yourself 18.1

1 a 9.10×10 **b** 2.83×10^3

c 2.60×10^2 **d** 1.07×10^2

e 4.08×10^2 **f** 1.20×10^3

g 1.33

2 a 2852 **b** 1361

c 1.625×10^{11} km^3 **d** 48 860

3 **a** 628.3 **b** 904.8 **c** 527.8
d 1900.7 **e** 329.9 **f** 179.1

Test yourself 18.2

1 **a** 17.7 **b** 73.5 **c** 76.3

2 85.5 mm

3 90.5 mm

4 side = 49.6 mm, height = 99.2 mm

5 $30\,m^3$, 1.06 m

6 37.5 mm, $4.33 \times 10^3\,mm^2$, $1.35 \times 10^4\,m^3$

7 8.29×10^5

8 area = $9.55 \times 10^5\,mm^3$

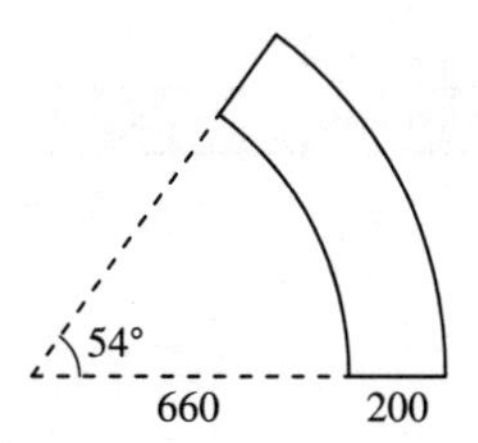

9 $6.60 \times 10^3\,mm^3$, 11.8%

10 mirror = $1.13 \times 10^4\,mm^2$
yellow panel = $2.64 \times 10^3\,mm^2$
long blue panel = $5.40 \times 10^3\,mm^2$
square blue panel = $900\,mm^2$
ratio is 1:1.86:2.23

11 $376\,mm^3$

12 $13.8\,m^3$

13 $x = 6$, when area = $216\,mm^2$ and volume = $216\,mm^3$

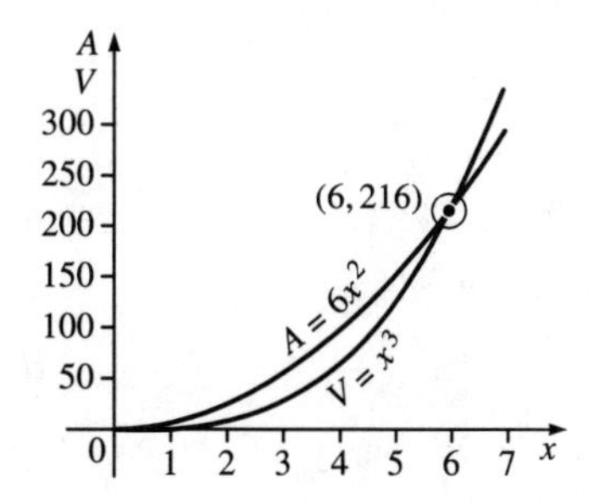

14 $w + h \leqslant 3$

$w \geqslant 0.5$

$h \geqslant 0.3$

$w \leqslant h$

Maximum width is 2.7 m, with volume 0.972 m^3

Maximum height is 1.5 m, with volume = 2.70 m^3

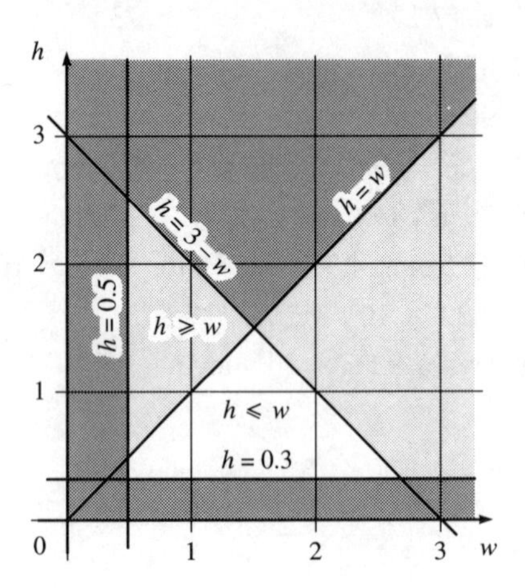

Chapter 19

Try these first (page 275)

1 a 0.9848 **b** 0.9986
c −0.7002 **d** −0.4161

2 a 68.20° or 248.20° **b** 336.42° or 203.58°

3 a $\pi/2$ rad **b** 4.08 rad

4 a 85.9° **b** 270°

5 3.15

6 43.50°

7 YZ = 3.04, area = 12.5

Test yourself 19.1

1 a 0.4226 **b** −0.5299
c −0.7431 **d** 0.2250

2 a 36.87° or 143.13° **b** 228.59° or 311.41°
c 195.66° or 344.34° **d** 0.29° or 179.71°

3 a $-1/\sqrt{2}$ **b** $\sqrt{3/2}$
c −1 **d** $\sqrt{3/2}$

Test yourself 19.2

1 a 0.0872 **b** −0.9998
c 0.9781 **d** −0.6691

2 a 45.57° or 314.43° **b** 72.54° or 287.46°
c 151.64° or 208.36° **d** 96.89° or 263.11°

3 a $-1/\sqrt{2}$ **b** $\sqrt{3/2}$
c $-\sqrt{3/2}$ **d** −0.5

Test yourself 19.3

1 a 0.35 **b** 4.10 **c** 4.71 **d** 6.00

2 a 57.30° **b** 143.24° **c** 60.00° **d** 108.00°

3 a −0.9900 **b** 0.0000 **c** −2.1850 **d** −1.0000

Test yourself 19.4

1 $a = 10.72$ **2** $b = 4.35$
3 $p = 6.45$ **4** $y = 2.71$
5 $\hat{B} = 37.33°$ **6** $\hat{L} = 21.17°$
7 $\hat{K} = 60.16°$ or $119.84°$ **8** $\hat{T} = 15.25°$

Test yourself 19.5

1 a 5.65, area = 9.38 **b** 14.61, area = 28.46
c 2.78, area = 6.86

2 a $\hat{A} = 44.42°$, $\hat{B} = 34.05°$, $\hat{C} = 101.53°$, area = 9.80
b $\hat{P} = 133.43°$, $\hat{Q} = 28.96°$, $\hat{R} = 17.61°$, area = 14.52
c $\hat{X} = 58.97°$, $\hat{Y} = 56.52°$, $\hat{Z} = 64.51°$, area = 24.71

Chapter 20

Try these first (page 293)

1 $x = 3, y = 4$ **2** $a = 4, b = -3$
3 $s = -2, t = 5$

Test yourself 20.1

1 $x = 2, y = 4$ **2** $x = 8, y = 2$
3 $a = 2, b = -5$ **4** $m = -3, n = 4$
5 $x = 4, y = 1$

Test yourself 20.2

1 $a = 2, b = 3$
2 $p = 9, q = -2$
3 $x = -4, y = 7$
4 $j = 1, k = -1$
5 $c = -8, d = 12$
6 $x = 2.5, y = 3.5$
7 $a = 11, b = 2$
8 79
9 size A = 10 mm, size B = 15 mm
10 (5,6)

Chapter 21

Try these first (page 299)

1 2, –7
2 –3, 12
3 0.569, 0.681

Test yourself 21.1

1 7, –3
2 –2, 4
3 0, 5
4 1, 2
5 3, –2
6 4, 11
7 –2, 3
8 0, 7
9 10, –9
10 5, 3
11 $x = 4$ or $x = -6$, but –6 is not allowable. Width = 4 cm, length = 2 cm.
12 $x = 3$ or $x = -4$, but –4 is not allowable. Sides are 3 m long.

Test yourself 21.2

1 1.66, –0.91
2 0.27, –1.47
3 0.25, 0.25
4 –0.31, –3.19
5 $t = 6.77$ or –14.17, but –14.17 is not allowable. The time is 6.77 s.
6 The equation is $2\pi r^2 + 20\pi r - 300 = 0$. $r = 3.53$ or $r = -13.53$, but –13.53 is not allowable. The radius of the can is 3.53 cm.

Chapter 22

Try these first (page 306)

1 a 6, 55° **b** 4, 250°

2

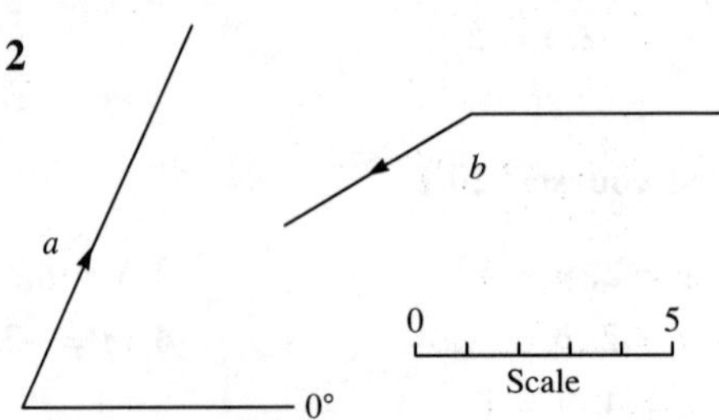

3 **e** 5.23, 114° **f** 7.91, 266°

g 2.31, 247° **h** 4.02, 138°

4 3.009, 3.993

5 (7.62, −23.20°)

6 (−3.21, 3.83)

Test yourself 22.1

1 **a** $|x| = 7.5$, $\theta_x = 35°$

b $|y| = 6.0$, $\theta_y = 135°$

c $|z| = 5.0$, $\theta_z = 290°$

2

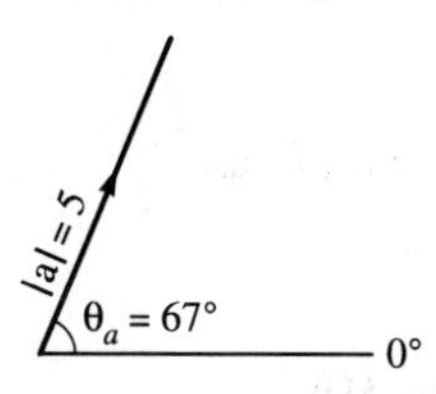

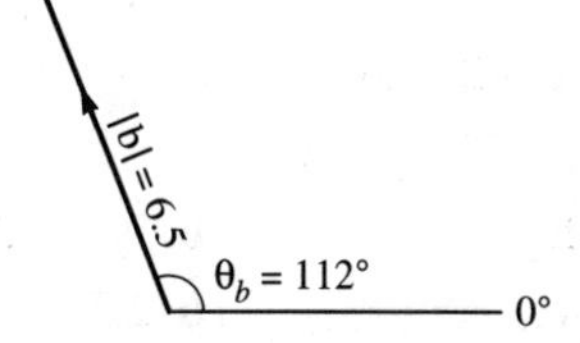

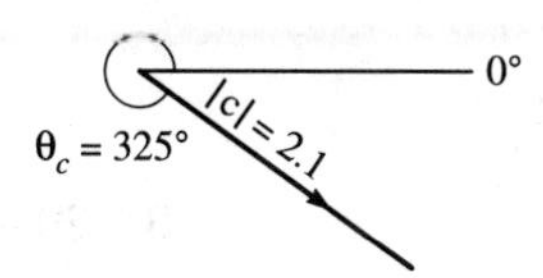

Test yourself 22.2

e 4.74, 216° **f** 5.40, 3°

g 4.17, 298° **h** 1.96, 194°

i 8.72, 247° **j** 9.98, 113°

k 2.29, 228° **l** 10.00, 140°

Test yourself 22.3

1 **a** 3.35, 0° and 2.18, 90° **b** 0.78, 0° and 8.97, 90°

c −3.86, 0° and 4.60, 90° **d** −3.21, 0° and −3.83, 90°

2 **e** 11.89, 69.62° **f** 7.11, 173.82°

3 **a** 30.29 m **b** 15.44 m

4 1.39 ms^{-1}, 21.04°

5 2.18 kg

Test yourself 22.4

1 A(8.83, 59.37°)
B(47.45, 74.85°)
C(7.70, 178.51°)
D(11.01, 289.09°)
E(94.34, 248.88°)
F(10.47, 149.39°)

2 G(12.86, 19.07)
H(1.55, 7.29)
I(–4.06, 5.20)
J(–11.28, –4.10)
K(40.80, –65.30)
L(0.91, 2.33)

3 8.60 km, 54.46° north of east

4 6.50 up, 3.75 left

Chapter 23

Try these first (page 319)

1 a 10 + j9 **b** –2 – j6

2 0.75 + j2.33, 0.75 – j2.33

3 a $7.62 \underline{/23.20°}$ **b** $7.21 \underline{/236.31°}$

4 a 8.7 + j13.4 **b** 2.55 – j0.22

5 $4.55 \underline{/44.67°}$

Test yourself 23.1

1 A 20 + j10 **B** –20 + j20
C 5 + j35 **D** 30 – j15
E 10 – j30 **F** –20 – j10
G –15 – j20 **H** –j15

2 See Figure A.41

Test yourself 23.2

1 a 9 + j9 **b** 8 + j5
c 6 + j5 **d** –2 + j2
e j3 **f** 11
g 1 + j **h** 10 + j5

2 a 2 + j5 **b** 3 + j
c –2 + j3 **d** 5 – j2
e –j6 **f** 2 – j8
g 1 + j3 **h** –6 – j

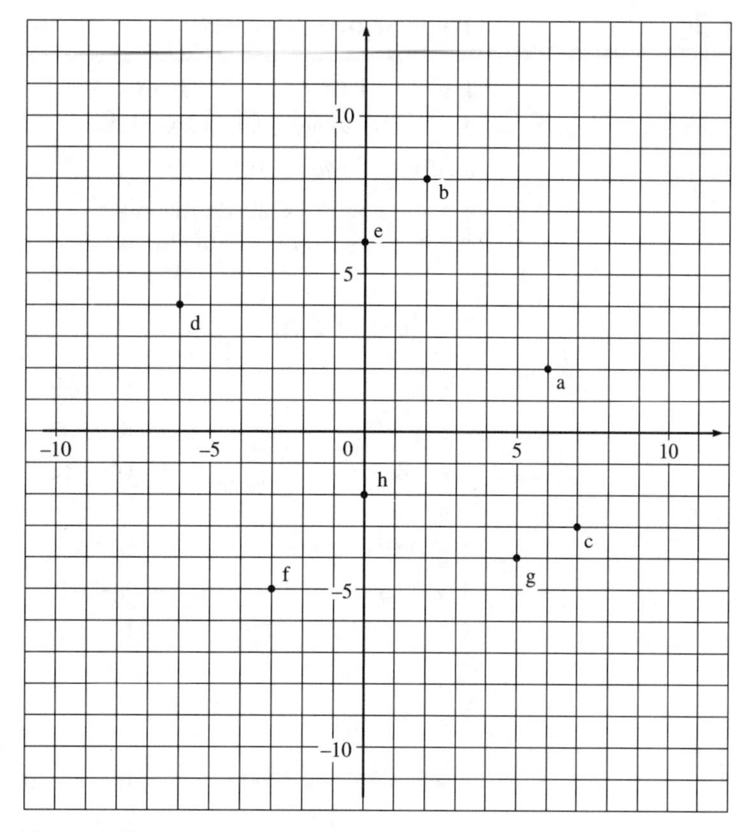

Figure A.41

Test yourself 23.3

1 **a** $7.21\,\angle 56.30^\circ$ **b** $8.82\,\angle 35.31^\circ$

c $28.60\,\angle 36.45^\circ$ **d** $9.21\,\angle 334.27^\circ$

e $10.08\,\angle 143.47^\circ$ **f** $15.62\,\angle 219.81^\circ$

2 **a** $38.43 + j54.88$ **b** $1.22 + j0.22$

c 3.5 **d** $-3.90 + j0.90$

e $48.48 + j87.46$ **f** $-0.40 + j2.07$

3 $7.85\,\angle 37.77^\circ$

Test yourself 23.4

1 $-0.200 + j1.17$, $-0.200 - j1.17$

2 $0.625 + j0.331$, $0.652 - j0.331$

Chapter 24

Try these first (page 329)

1 a $6e^{3x} + 6e^{-2x}$ **b** $6x - 5$ **c** $8 \cos 2x$

2 $v = 4.75$ V, $dv/dt = 0.498$ Vs^{-1}

3 35 ms^{-1}, $dv/dt = 10t$

4 $x = -1.333$, $y = 20.19$, minimum
$x = 6$, $y = -177$, maximum

Test yourself 24.1

1 a $6e^{3t}$ **b** $14e^{2x}$
c $14e^{4x}$ **d** $2516e^{1258n}$
e $-8e^{-4t}$ **f** $-1.5e^{-0.25x}$
g $14e^{-2t}$ **h** $e^{0.125t}$

2 a 0.593 Vs^{-1} **b** 2.35 s

3 a 3.83°C h^{-1} **b** 4.62 h

4 a -45.5 Vs^{-1} **b** -14.6 Vs^{-1}

Test yourself 24.2

1 a $\cos 36° = 0.8090$ **b** $2 \cos 18° = 1.902$
c $4.5 \cos 1.5x$ **d** $4/x$
e $8 \sin 2\theta - 8 \cos 2\theta$ **f** $1/x - 0.5e^{0.5x}$
g $4/x$ **h** $3/x + \cos x$

2 18.4 μA

3 $dy/d\theta = 15 \sin \theta + 1.875 \sin 2\theta$, $\theta = 0.5$ rad, $y = 21.37$ cm,
$dy/d\theta = 12.09$ cm rad^{-1}, $dy/dt = 6.05$ cm s^{-1}

4 $u = 2.805$ V, $A = du/dv = -22.5$

Test yourself 24.3

1 a 4 **b** $10x$ **c** 0
d $12x$ **e** 2 **f** $3x^2 - 2x$
g $8x + 3$ **h** $4x^3$ **i** $15x^2 + 1$

2 $dy/dx = 6x - 5 = 67$

3 $dy/dx = 12x^2 + 6x - 2 = 1798$

4 452 (3 sf)

5 $ds/dt = 36 - 4t$, 9 s, 162 m, -4 ms^{-2}

6 $x = 1.0954$ m, $dx/dt = 0.273\,85/t = 0.137$ mh^{-1} (3 sf)

7 $v = ds/dt = 20 - 10t$, maximum height = 2 m, $d^2s/dt^2 = -2$ ms^{-2}

Test yourself 24.4

1 **a** $x^{-0.5} = 1/\sqrt{x}$ **b** $-10x^{-3} = -10/x^3$
c $-1/6x^2$ **d** $-15x^{-4} - 3x^{-2}$
e $-3.5x^{-1.5} + 1.5x^{-0.5}$ **f** $-x^{-7}$

2 **a** (–0.214, –8.036), minimum
b (–0.417, 1.042), maximum
c (0,12), maximum, (0.25, 11.9375), minimum
d (–1.291, –8.607), minimum, (1.291, 8.607), maximum

3 8 and 8

4 200

5 maximum area = $1250\,m^2$, width = 25 m, length = 50 m

6 $dv/dx = 35 - 48x + 12x^2$; $x = 0.9592$ or 3.0408, only the first is possible, $v = 15.02\,cm^3$

7 $a = 5.77$ cm, $b = 2.89$ cm, $v = 96.35\,cm^3$

Chapter 25

Try these first (page 348)

1 $\int y = x^3 - x^2 - 6x + 11$ **2** 4
3 3.15

Test yourself 25.1

1 **a** $3x^2 + c$ **b** $5x + c$ **c** $3e^{2x}/2 + c$
d $-\cos 2\theta + c$ **e** $7x^2/2 + c$ **f** $2x^3/3 + c$
g $-\frac{1}{5} \sin 5\theta + c$ **h** $x^6 - x^5 + c$ **i** $4 \ln x + c$
j $x^2/2 + x + c$ **k** $5x^3/3 - 3x^2/2 - 8x + c$ **l** $-1/x + c$

2 **a** $2x^2 + x + 3$ **b** $5x - 1$
c $4x + e^{2x}/2 - 204$ **d** $3x^3 - 2x^2 + x + 2$
e $-\cos\theta + \sin 2\theta + 0.3407$

Test yourself 25.2

1 13.5 **2** 24
3 54 **4** 126
5 990.1 **6** 2.5481

Test yourself 25.3

1 $13.67\,ms^{-1}$ **2** 24.375 m
3 $36\,m^2$

Test yourself 25.4

1 2.67 **2 a** $2/\pi$ **b** 0

3 4.24 V **4** $11.8\,\text{ms}^{-1}$

Test yourself 25.5

1 $58.3\,\text{cm}^3$

2 $9.948\,\text{mm}^3$

3 $v = \pi \int_0^{20} \left(\sqrt{5x}\right)^2 \, dx = 5\pi \int_0^{20} x \, dx = 3142\,\text{mm}^3$

4 Consider as a cylinder, radius 2, height 0.5, plus the tapered part. For the tapered part:

$$v = \pi \int_2^3 (4/x)^2 \, dx = 16\pi \int_2^3 (1/x)^2 \, dx = 8.3776$$

Volume = 6.2832 + 8.3776 = 14.66 (4 sf)

5 $v = \pi \int_0^1 (1 - x^2) dx = 2.094$ (4 sf)

$v = \pi \int_{0.15}^1 (1 - x^2) dx = 1.627$ (4 sf)

Chapter 26

Try these first (page 363)

1 0.9596

2 2.39

3 0.8821

Test yourself 26.1

The answers to questions 1 and 2 are calculated by Simpson's Rule, using 8 strips. Your answers may vary slightly from these if you used a different rule, or a different number of strips.

1 a 0.811 **b** –0.340 **c** 0.300

2 area = $\int_0^6 \sqrt{36 - x^2} \, dx = 28$

By formula, area = $\pi r^2/4 = 28.3$

3 The mid-ordinate rule is not suitable, as it is difficult to measure mid-ordinates accurately on this diagram. Simpson's Rule does not apply as the number of strips is odd. Using the trapezium rule, the area of an arch is

28 m², but this is a slight under-estimate, as the curve is concave. Volume of material = 463.75 m³

4 There is an even number of strips (odd number of readings in the table), so use Simpson's Rule. Distance = 96 km

Chapter 27

Try these first (page 370)

1 The faster sequence is J_2, J_1, completed in 16 h.

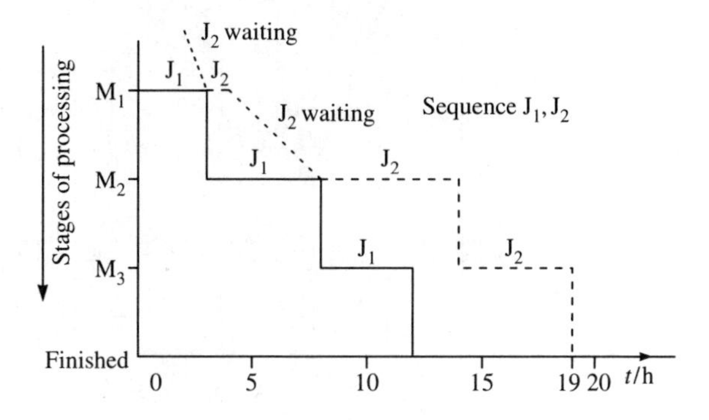

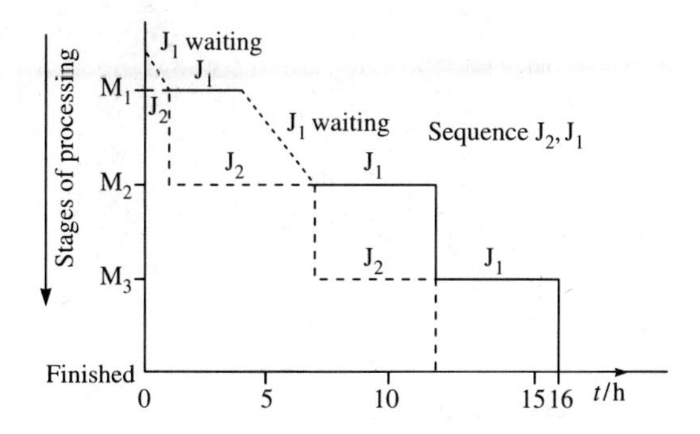

2 The renovation is completed in 21 h. The critical path is 1 → 2 → 3 → 5 → 6 (see Figure A.43), or activities B, C, G, D. Slack times are A = 12 h, E = 20 h and F = 5 h.

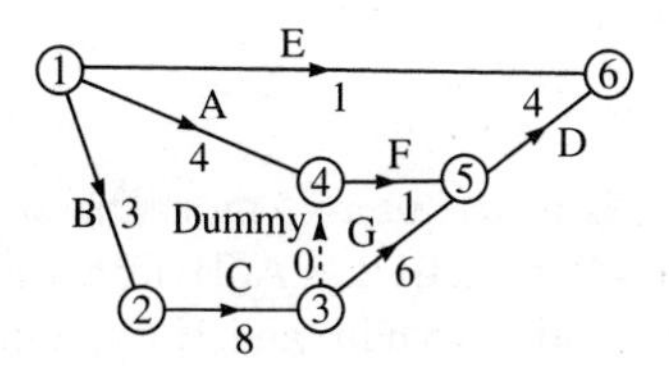

Figure A.43

3 The maximum number of heaters is 60, 20 large and 40 small. The maximum use of power is 95 kW, with 26 large heaters and 30 small.

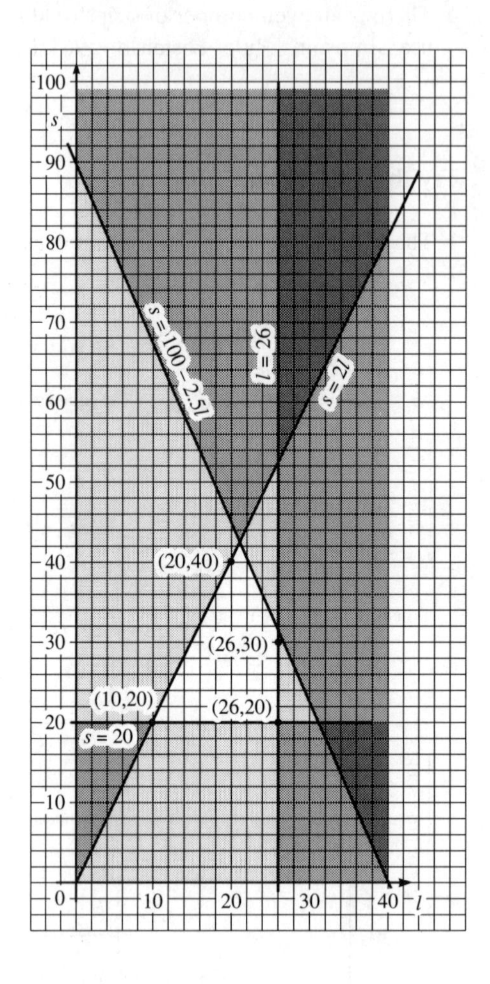

Test yourself 27.1

1 The faster sequence is J_2, J_1, taking 13 min (see Figure A.45).

2 The fastest sequence is J_2, J_1, J_3, taking 17 min (see Figure A.46).

3 The fastest sequence is 3, 5, 2, 4, 6, 1.

Test yourself 27.2

1 The meal is ready after 75 min. The critical path is $1 \rightarrow 2 \rightarrow 6 \rightarrow 7$ (see Figure A.47), or activities A, B, C. Slack times are D = 30 min, E = 20 min, F = 30 min, G = 25 min, and H = 65 min.

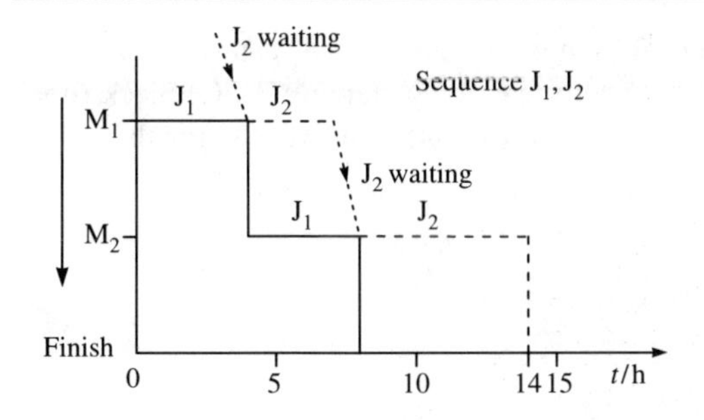

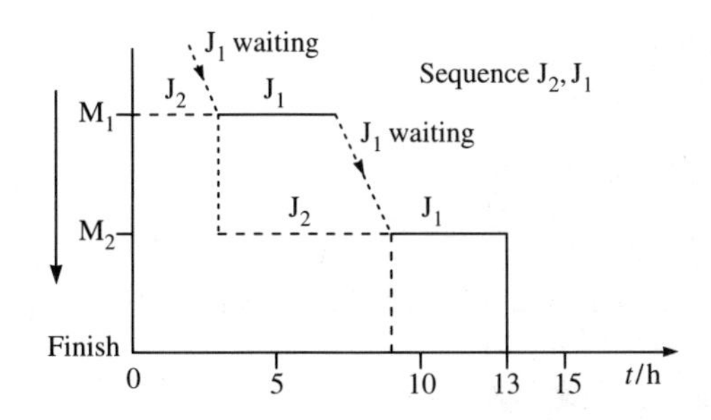

Figure A.45

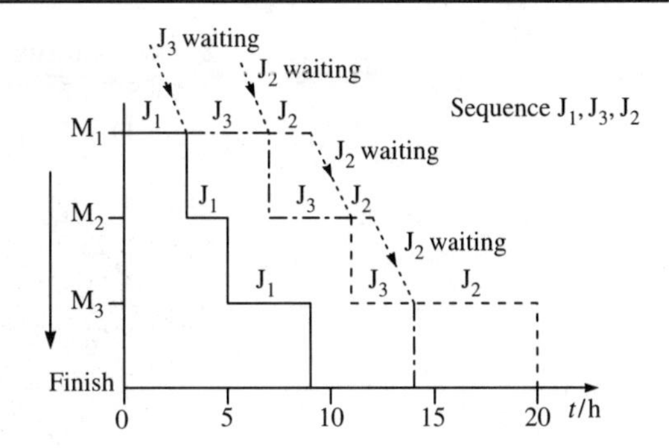

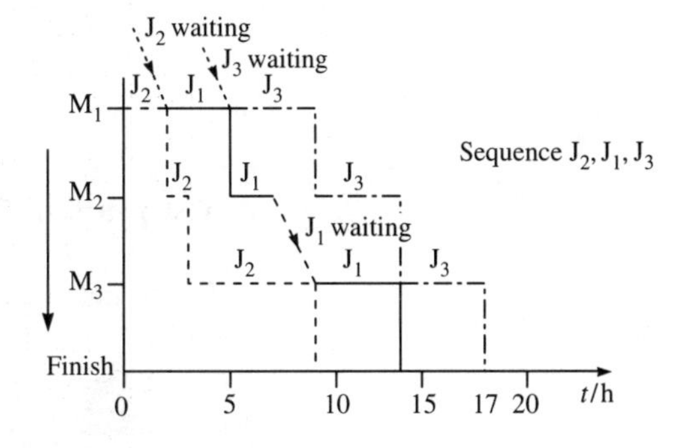

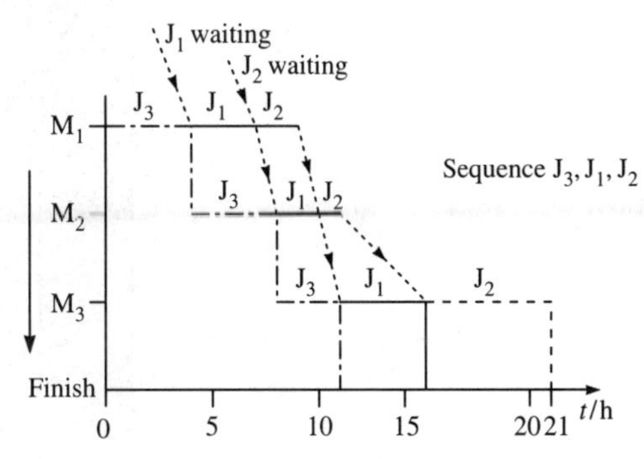

Figure A.46a

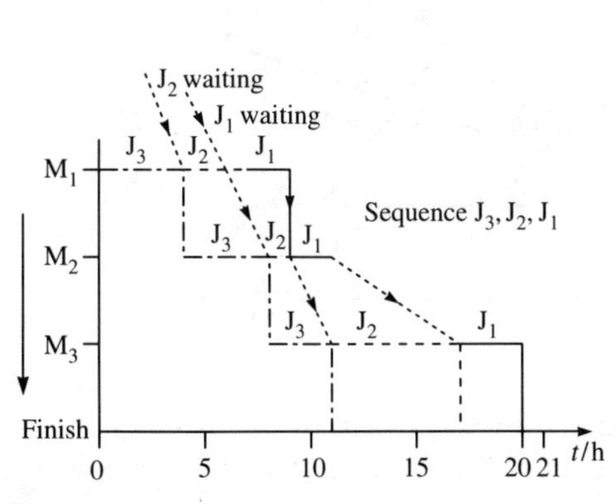

Figure A.46b

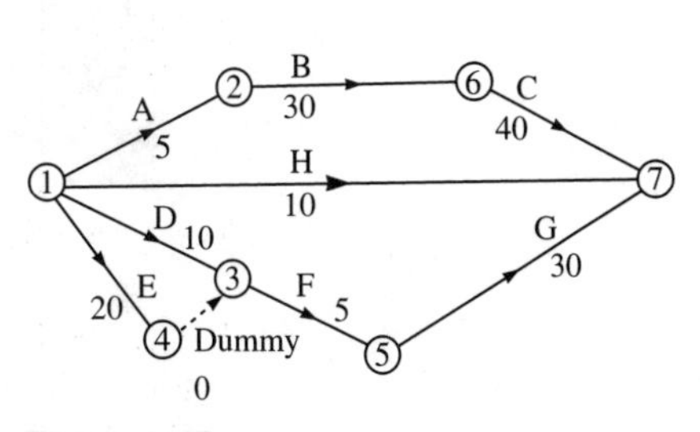

Figure A.47

2 The motorway is completed in 30 months. The critical path is 1 → 3 → 4 → 5 → 6 → 7 (see Figure), or activities A, B, dummy, C, G. Slack times are D = 7 months, E = 9 months, F = 4 months and H = 1 month.

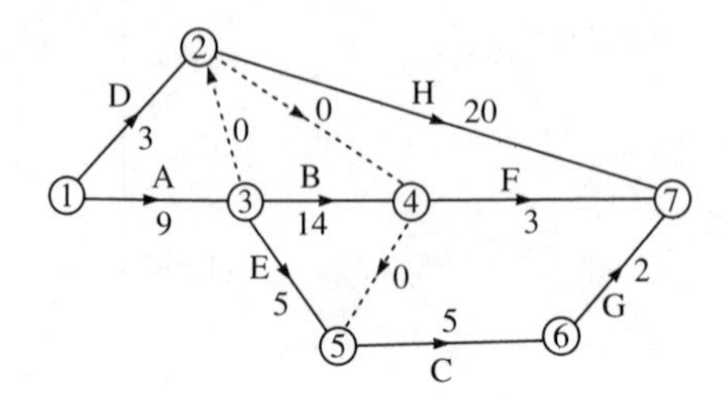

Test yourself 27.3

1 The maximum yield is 795 kg, for 36 apple trees and 5 pear trees.

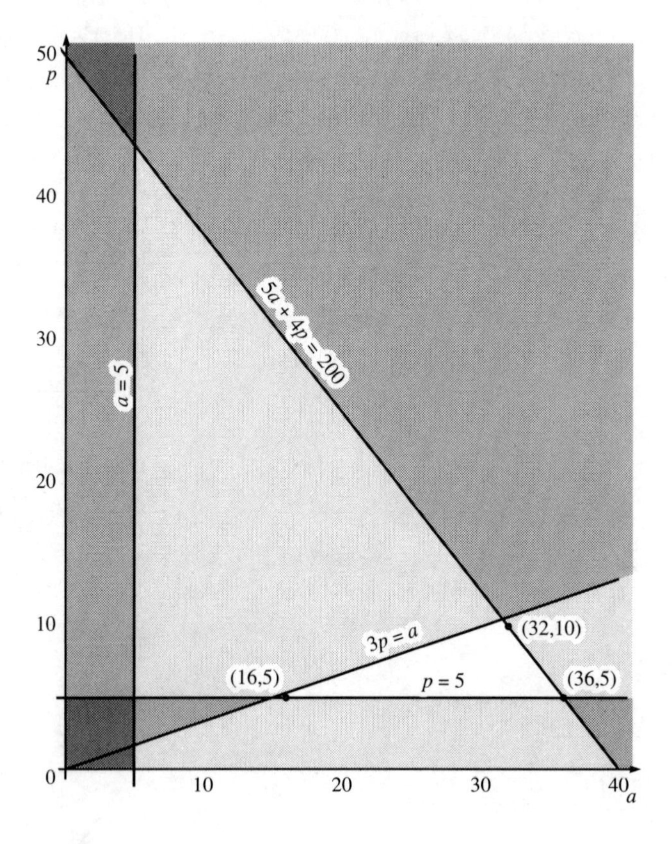

2 The minimum time is 110 s, with 8 medium rollers, no large rollers. The maximum time is 300 s, with 18 medium rollers, 6 large rollers.

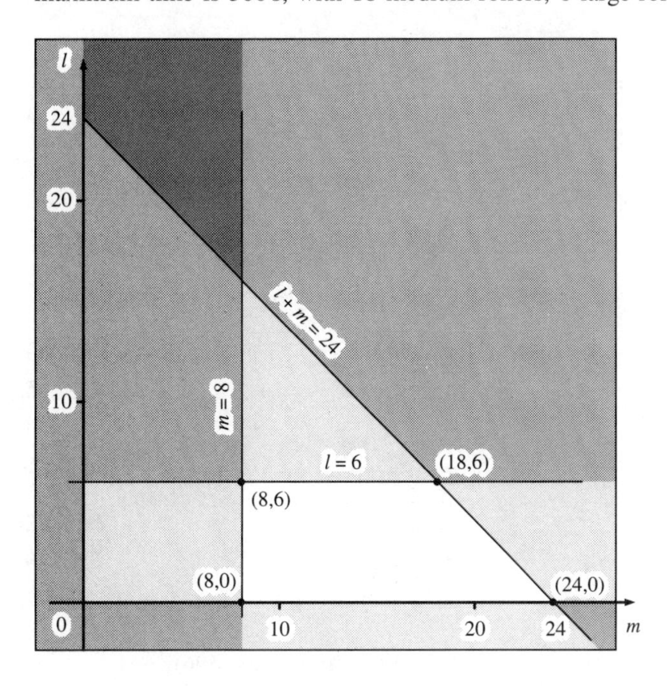

3 The maximum number of books that can be ordered is 30, all fiction, total cost £300. The order of maximum value is for 19 fiction and 8 technical books, total cost £390.

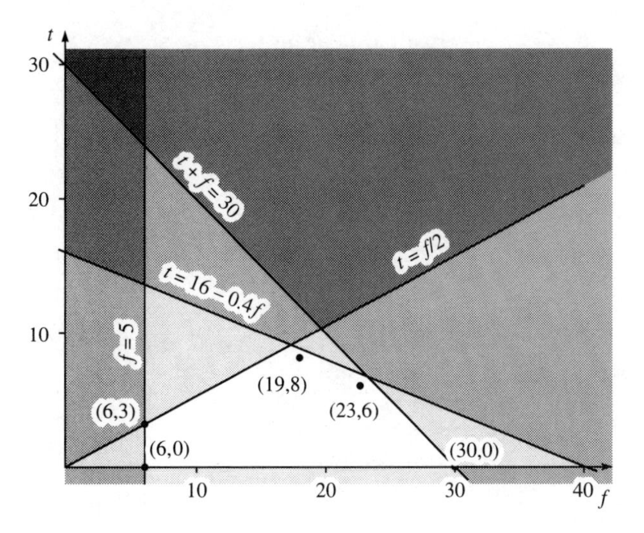

Chapter 29

Try these first (page 393)

1 1.56 **2** 1.0845
3 0.8767

Test yourself 29.1

1 $-0.7 < x < -0.8$, -0.74 **2** $-0.5 < x < -0.6$, -0.57
3 $0.4 < x < 0.5$, 0.48

Test yourself 29.2

1 2.689 10 **2** 2.354 25
3 1.369 84 **4** 2.645 75

Test yourself 29.3

1 2.6180, 0.3820 **2** 0.8526
3 0.4636 **4** 1.3797

Chapter 30

Try these first (page 406)

1 11100 **2** 38
3 1010 0010 0101 1011 **4** 1100010
5 100111

Test yourself 30.1

1 a 19 **b** 72
c 162 **d** 110
e 1315 **f** 380
g 22256 **h** 44488
2 a 23 **b** 3F
c 40 **d** 2B
e 1F4 **f** 80
g 3158 **h** FFF

Test yourself 30.2

1 a 13 **b** 55
c 73 **d** 42
2 a 10010 **b** 111111
c 1000001 **d** 110100101

3 a 4CA2 **b** 6CF8D

4 a 101110000111010 **b** 1000111100000110

Test yourself 30.3

1 a 110110 **b** 1010001 **c** 1101111

2 a 1110 **b** 1010 **c** 111111

Chapter 31

Try these first (page 419)

1 $\overline{B}\overline{C}$

2 $\overline{A}$

3 $\overline{A}\overline{C} + \overline{A}B + A\overline{B}C$

Test yourself 31.1

1 If *A* OR *B*, THEN *Z*. On Saturday morning, $Z = 0$.

2 A = the door is open
B = the food is cooking
Z = the lamp is on
If *A* OR *B* THEN *Z*

A	*B*	*Z*
0	0	0
0	1	1
1	0	1
1	1	1

Line 4 of the table never occurs, because opening the door automatically turns off the cooking circuit.

3 A = the grass is long

B = the mower has run for more than 10 min

Z = it is overheated

Same statement and truth table as in question 2. *B* changes from 0 to 1.

Test yourself 31.2

1

A	*B*	*C*	*AC*	*BC*	*Z*
0	0	0	0	0	0
0	0	1	0	0	0
0	1	0	0	0	0
0	1	1	0	1	1
1	0	0	0	0	0
1	0	1	1	0	1
1	1	0	0	0	0
1	1	1	1	1	1

The column for *Z* is identical with that of the truth table on page 425.

2

A	B	C	BC	Z	
0	0	0	0	0	
0	0	1	0	0	
0	1	0	0	0	
0	1	1	1	1	
1	0	0	0	1	Unsafe
1	0	1	0	1	
1	1	0	0	1	Unsafe
1	1	1	1	1	

3 A = person moving
B = it is night
C = test button pressed
Z = alarm sounds
$Z = AB + C$

A	B	C	AB	Z
0	0	0	0	0
0	0	1	0	1
0	1	0	0	0
0	1	1	0	1
1	0	0	0	0
1	0	1	0	1
1	1	0	1	1
1	1	1	1	1

4 A = switch A is on
B = switch B is on
Z = lamp is on

A	B	Z
0	0	0
0	1	0
1	0	1
1	1	0

5 a 1 **b** A **c** A **d** A

Test yourself 31.3

2 a $A + B$ **b** $A + BC$
c $B(A + C)$ **d** B
3 a $\overline{A}$ **b** $A + B$

Test yourself 31.4

1 A **2** $\overline{A}$
3 $A + \overline{B}$ **4** $\overline{B}\overline{C}$
5 A **6** $C + \overline{A}B\overline{D}$

7 Segment *e*: $Z = \overline{A}\,\overline{B}\,\overline{C}\,\overline{D} + \overline{A}\,\overline{B}C\overline{D} + \overline{A}BC\overline{D} + A\overline{B}\,\overline{C}\,\overline{D}$

$Z = \overline{A}C\overline{D} + \overline{B}\,\overline{C}\,\overline{D}$ (see Figure A.52a)

Segment *d*: $Z = \overline{A}\,\overline{B}\,\overline{C}\,\overline{D} + \overline{A}\,\overline{B}C\overline{D} + \overline{A}\,\overline{B}CD + \overline{A}B\overline{C}D = \overline{A}BC\overline{D} + A\overline{B}\,\overline{C}\,\overline{D}$

$Z = \overline{A}C\overline{D} + \overline{B}\,\overline{C}\,\overline{D} + \overline{A}\,\overline{B}C + \overline{A}B\overline{C}D$ (see Figure A.52b)

	$\overline{A}\overline{B}$	$\overline{A}B$	AB	$A\overline{B}$
$\overline{C}\overline{D}$	1	0	0	1
$\overline{C}D$	0	0	0	0
CD	0	0	0	0
$C\overline{D}$	1	1	0	0

a

	$\overline{A}\overline{B}$	$\overline{A}B$	AB	$A\overline{B}$
$\overline{C}\overline{D}$	1			1
$\overline{C}D$		1		
CD	1			
$C\overline{D}$	1	1		

b

Figure A.52

Chapter 32

Try these first (page 437)

1

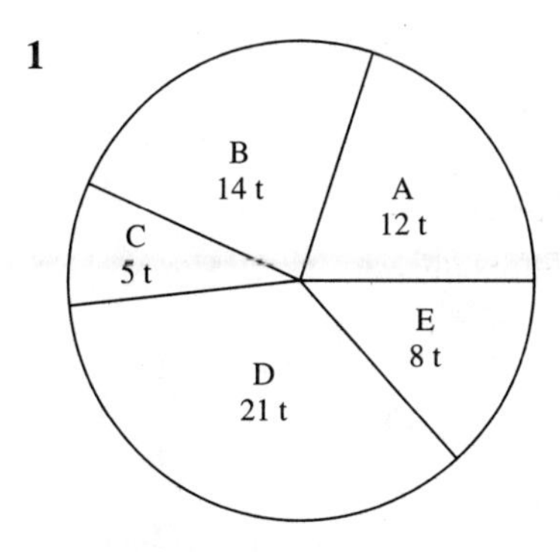

Glue production (tonnes)

2

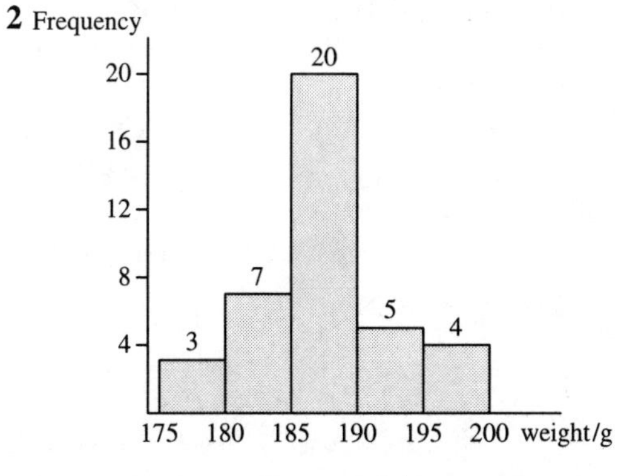

Tea cup weights

Test yourself 32.2

1 About one quarter

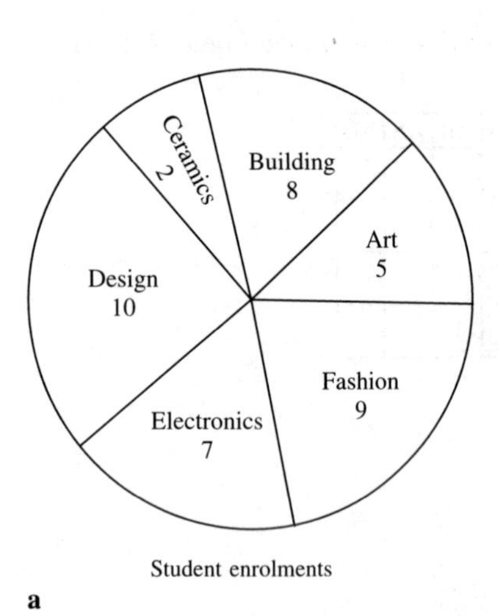

Student enrolments

a

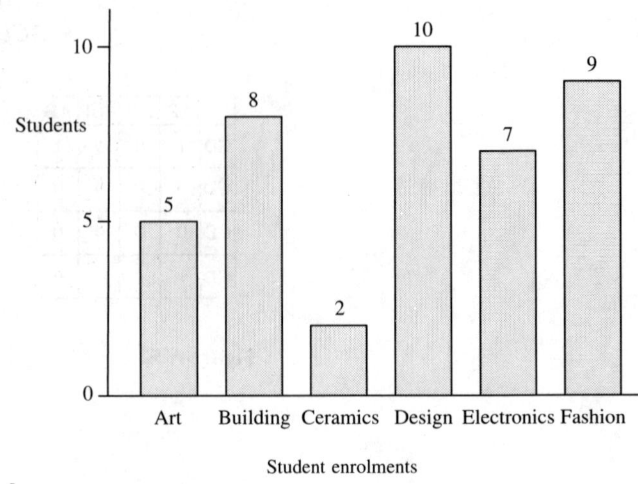

Student enrolments

b

2 About three-quarters

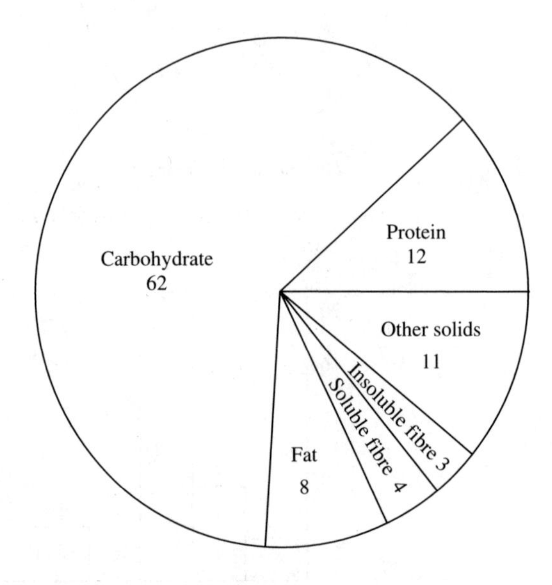

% composition of breakfast cereal
(by weight)

3 a D, C

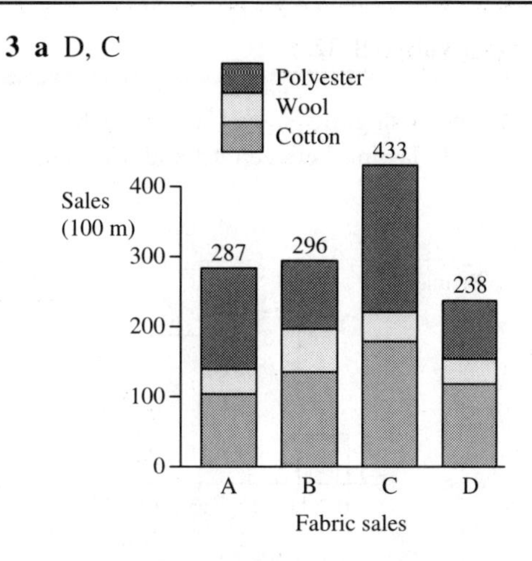

b B, A, A (polyester), C (polyester), D (cotton)

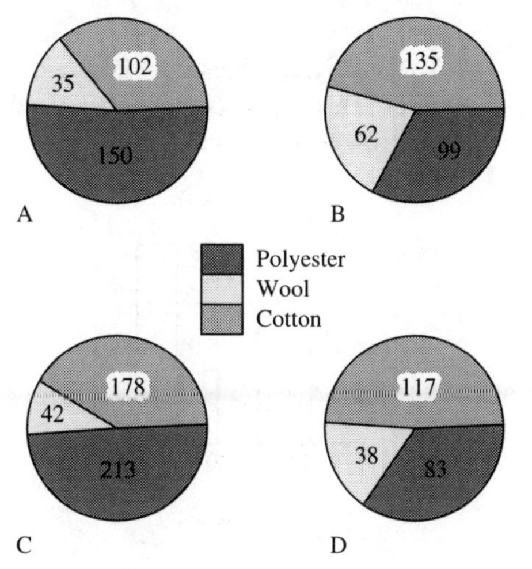

Fabric sales (100 m)

4 The survey would be biased toward the views of people who have a telephone.

5 a C, 260

b D, 86

c Proportionately larger amount of orange juice

d 553

6 a C, 132

b D, 26

c 48

d 7:2 or 3.5:1

Factory D has least total sales (8600 m)
Factory C has greatest total sales (26 000 m)

Test yourself 32.3

1 The histogram shows a typical humped distribution, peaking with most cells lasting between 12 and 13 hours.

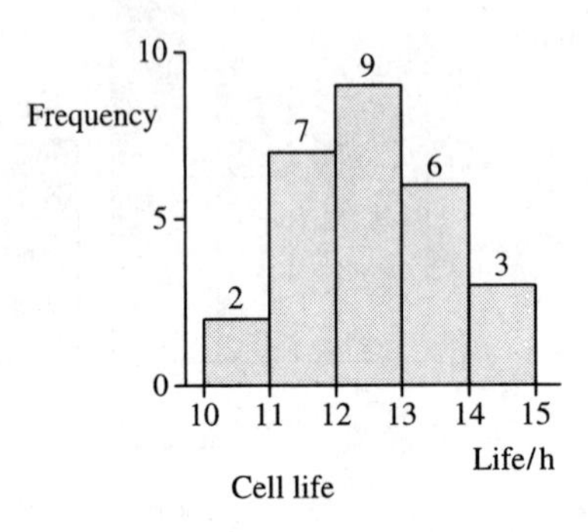

2 The histogram does not show a clear peak, and is cut off steeply on both sides. This suggests that under sized and over sized biscuits were discarded and re-cut before baking.

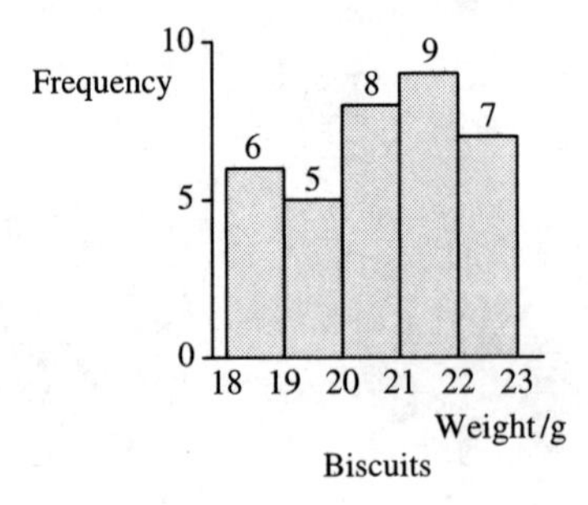

Test yourself 32.4

1 Positive correlation, 2.2 cm

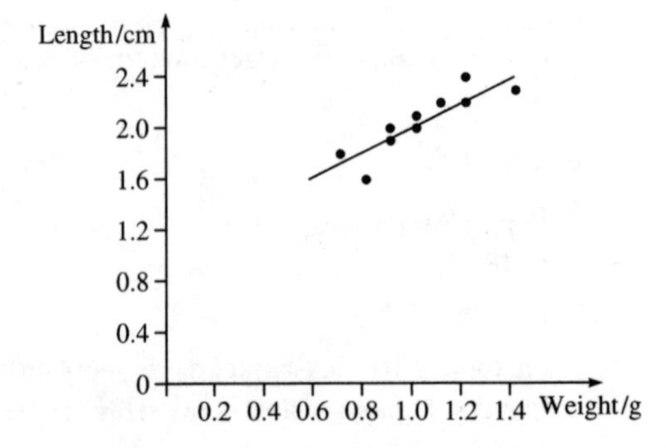

2 Negative correlation, 0.6 g

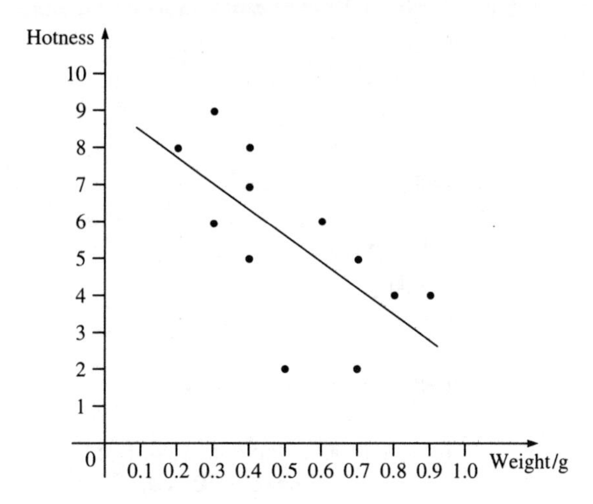

Chapter 33

Try these first (page 452)

1 10 t

2 **a** 187 g **b** 187 g
c 185 g–190 g **d** 22 g
e 184 g, 189 g **f** 5 g

Test yourself 33.1

1 Modal class = 12 h–13 h, median = 12.2 h, mean = 12.5 h

2 Modal class = 21 g–22 g, median = 20.8 g, mean = 20.7 g

3 29.8 MNm^{-2}

Test yourself 33.2

1 Range = 4.1 h, quartiles are 11.6 h and 13.5 h, interquartile range = 1.9 h

2 Range = 4.5 g, quartiles are 19.2 g and 21.9 g, interquartile range = 2.7 g

3 Quartiles are 14.5 and 22.5, interquartile range = 6

Chapter 34

Try these first (page 457)

1 5.48 t

2 s.d. of sample = 5.20 g, s.d. of population = 5.27 g. These values are more than 3 s.d.'s above the mean. Inspect production methods.

Test yourself 34.1

1 1.08 h

2 1.39 g

Test yourself 34.2

1 1.10 h

2 1.41 g

Test yourself 34.3

1 Sample A: mean = 61, s.d. = 6.06
Sample B: mean = 82, s.d. = 7.02

Means differ by nearly 2 s.d.'s. Suggests a difference between samples, that sample B has higher gain.

2 Sample A: mean = 1.82 mm, s.d. = 0.154 mm
Sample B: mean = 1.85 mm, s.d. = 0.102 mm

Means differ by less than 1 s.d. suggesting that there is no significant difference in sample means. But the s.d. of sample B is much less than that of sample A, indicating that the method used to produce sample B is given a more uniform product.

3 Both switches are more than 3 s.d.'s above the mean. A switch can be more than 3 s.d.'s above the mean in only 0.05% of cases, not in 2% of cases as here. Check for faulty production methods.

Index